CW01346460

MARITIME ECONOMICS

MARITIME ECONOMICS
A MACROECONOMIC APPROACH

ELIAS KARAKITSOS
Global Economic Research and CEPP, University of Cambridge

LAMBROS VARNAVIDES
MD & Global Head of Shipping, Royal Bank of Scotland

palgrave
macmillan

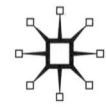

© Elias Karakitsos and Lambros Varnavides 2014

All rights reserved. No reproduction, copy or transmission of this publication may be made without written permission.

No portion of this publication may be reproduced, copied or transmitted save with written permission or in accordance with the provisions of the Copyright, Designs and Patents Act 1988, or under the terms of any licence permitting limited copying issued by the Copyright Licensing Agency, Saffron House, 6–10 Kirby Street, London EC1N 8TS.

Any person who does any unauthorized act in relation to this publication may be liable to criminal prosecution and civil claims for damages.

The authors have asserted their rights to be identified as the authors of this work in accordance with the Copyright, Designs and Patents Act 1988.

First published 2014 by
PALGRAVE MACMILLAN

Palgrave Macmillan in the UK is an imprint of Macmillan Publishers Limited, registered in England, company number 785998, of Houndmills, Basingstoke, Hampshire RG21 6XS.

Palgrave Macmillan in the US is a division of St Martin's Press LLC,
175 Fifth Avenue, New York, NY 10010.

Palgrave Macmillan is the global academic imprint of the above companies and has companies and representatives throughout the world.

Palgrave® and Macmillan® are registered trademarks in the United States, the United Kingdom, Europe and other countries.

ISBN 978–1–137–44117–1

This book is printed on paper suitable for recycling and made from fully managed and sustained forest sources. Logging, pulping and manufacturing processes are expected to conform to the environmental regulations of the country of origin.

A catalogue record for this book is available from the British Library.

A catalog record for this book is available from the Library of Congress.

Typeset by MPS Limited, Chennai, India.

CONTENTS

List of Figures ix
List of Tables xii
List of Boxes xiii
List of Abbreviations xiv

1 Introduction 1
  The scope of the book 1
    1 The benefits of a macroeconomic approach 2
    2 The structure of the book 4
    3 How this book should be read and its potential readership 7
    4 Model and data series 8
  Acknowledgements 8

Part I The microfoundations of maritime economics 9

2 The theoretical foundations of the freight market 11
  Executive summary 11
    1 A framework for maritime economics 12
    2 The traditional model 16
    3 A critique of the traditional model 18
    4 A game-theoretic approach to freight rates 21
    5 A formal statement of the bargaining problem –
      the contract or bargaining curve 22
    6 The bargaining utility function 23
    7 The solution of the bargaining game 25
    8 Economic conditions and bargaining power 27
    9 Comparison with the conventional model 29
    10 The time charter (or period) market 32
    11 Summary and conclusions 37
  Appendix: how expectations of freight rates are generated 40

3 The shipyard, scrap and secondhand markets 41
  Executive summary 41
    1 The supply of shipping services in the short run 42
      1.1 Introduction 42
      1.2 Optimal capacity utilisation and optimal speed in
           the short run 44
    2 Optimal fleet and optimal speed in the long run 47
    3 The aggregate fleet 53
    4 Investment under uncertainty 53

5	Case study	56
6	The model in perspective	58
7	The supply of vessels – the impact of shipyard capacity on NB prices	61
	7.1 Introduction	61
	7.2 A framework for analysing shipyard capacity	62
	7.3 An empirical estimate of the impact of shipyard capacity on NB prices	64
	7.4 Explaining the stylised facts	67
8	The scrap market and the net fleet	68
9	The relationship of NB and SH prices	70
	Appendix: optimal fleet capacity expansion	76
	Optimal fleet and optimal speed – long-run analysis	76
	The demand for vessels	80
	The role of expectations	81
	A comparison with conventional models	84
	Long-run equilibrium and adjustment	87

Part II The macroeconomics of shipping markets 93

4	The efficiency of shipping markets	95
	Executive summary	95
	1 Introduction	99
	2 The efficient market hypothesis	100
	3 The martingale model	104
	4 The random walk model	104
	5 The efficiency of freight markets	105
	5.1 Statement of the problem	105
	5.2 Expectations-generating mechanisms	107
	5.3 A VAR example	109
	5.4 A generalisation of the restrictions	110
	5.5 Remarks	110
	5.6 Alternative tests of market efficiency	112
	5.7 Weak form of market efficiency – cointegration tests	117
	5.8 Remarks	119
	5.9 Time-varying risk premia – an explanation of the weak form of market efficiency	120
	5.10 The empirical evidence	125
	6 The efficiency of ship prices	126
	6.1 Statement of the problem	126
	6.2 Tests of market efficiency	128
	6.3 Weak form of market efficiency – cointegrating tests	129
	6.4 Time-varying risk premia	130
	6.5 The empirical evidence	131
	7 Conclusions	132
	Statistical appendix	132
	Stationary and non-stationary univariate time series	132
	Stationarity (unit root) tests	136
	Cointegration in single equations	139

		The VAR methodology	145
		The Johansen approach to cointegration	147
5	Business cycles		153
	Executive summary		153
	1	A statistical explanation of business cycles	155
	2	An economic interpretation of business cycles	160
	3	The new consensus macroeconomics or neo-Wicksellian model	166
	4	Monetary policy in the NCM model	172
	5	The interaction of shocks and monetary policy in generating business cycles	175
	6	A reformulation of the NCM model	179
	7	Fiscal policy in business cycles	181
		7.1 Conclusions on fiscal policy	187
	8	Fiscal policy in practice: the US President's budget for FY 2013	188
		8.1 The macroeconomic effects of the President's budget	188
		8.1.1 Short-to-medium-term economic effects	188
		8.1.2 Long-term economic effects	190
	9	Potential output	191
		9.1 Statistical estimates	192
		9.2 Supply-side determinants	192
		9.3 Conclusions on potential output	197
	10	Conclusions	198
	Appendix: nominal rigidities		199
		Implicit contracts	199
		Unions or insider–outsider models	200
		Efficiency wages	202
		Real price rigidity models	204
		Nominal price and wage rigidity models	204
6	The theory of shipping cycles		209
	Executive summary		209
	1	Supply-led shipping cycles: the Tinbergen–Koopmans model	211
	2	Expectations-led shipping cycles: the Beenstock–Vergottis (BV) model	216
		2.1 A summary of the BV model	218
		2.2 The steady-state properties of the BV model	224
		2.3 The dynamic adjustment in the BV model	227
	3	An integrated model of business and shipping cycles	232
	4	The properties of the integrated model	237
	5	The interaction of business and shipping cycles	241
	6	Conclusions	249

Part III	From theory to practice		251
7	The market structure of shipping and ship finance		253
	Executive summary		253
	1	The market structure	253
	2	The capital market structure for shipping	255
	3	Banks, shipping cycles and the supply and demand equation	266
	4	An evaluation of ship finance constraints	267

		5 Optimal strategies	270
		6 Conspiracy theories in shipping	271
8	The financialisation of shipping markets		272
	Executive summary		272
		1 Asset-led business cycles	273
		2 Hedging and speculation	278
		3 The financialisation of the oil tanker market	281
		4 A structural change in the oil tanker market	284
		5 Solving the puzzle of the structural change	285
		6 The financialisation of the dry bulk market	286
		7 Conclusions	292
9	The interaction of business and shipping cycles in practice		293
	Executive summary		293
		1 Freight rates as a leading indicator of business cycles	295
		2 Shipping cycles: the stylised facts	296
		3 US business cycles	299
		4 US monetary policy in the course of the business cycle	305
		5 Why do US business cycles differ?	308
		6 The business cycles of Japan	315
		7 The business cycles of Germany	319
		8 An explanation of the stylised facts of shipping cycles	322
		9 Uncertainty-led shipping cycles	323
		10 Finance-led shipping cycles	325
		11 Summary and conclusions	327
10	Investment strategy		328
	Executive summary		328
		1 The major decisions in shipping	329
		2 Case study: when to invest in the dry market	332
		3 A non-technical assessment of methodological issues	334
	Appendix 1: the demand for dry		335
	Appendix 2: NB prices		339

Notes 342
Bibliography 356
Index 365

LIST OF FIGURES

2.1	Demand and supply of shipping service	17
2.2	The impact of higher bunker costs	18
2.3	Market demand and supply and individual owner in the short run	19
2.4	Market demand and supply and individual owner in the long run	20
2.5	Contract or bargaining curves	22
2.6	Iso-utility curves	24
2.7	Iso-utility curves in strong and weak bargaining power	26
2.8	Freight rates in 'good' and 'bad' times	26
3.1	Isoquants	43
3.2	Optimum speed and capacity utilisation	45
3.3	Dynamic adjustment of NB vessel prices	52
3.4	Dynamic adjustment of NB vessel prices	63
3.5	World orderbook	64
3.6	Order cancellations	65
3.7	Demand and supply in the shipyard industry	66
3.8	Excess supply in the shipyard market and excess demand in the freight market	66
3.9	The impact on SH prices and traded fleet when the supply is upward sloping	73
3.10	The impact on SH prices and traded fleet when the supply is downward sloping	73
3.A1	Shipyard long-run equilibrium	88
3.A2	Adjustment of fleet	88
3.A3	Adjustment of NB vessel prices	90
3.A4	Unique saddlepoint solution	90
3.A5	Dynamic adjustment of NB vessel prices	91
3.A6	Dynamic adjustment of NB vessel prices	91
3.A7	Dynamic adjustment of investment	92
3.A8	Dynamic adjustment of fleet	92
5.1	Nominal and real oil price (WTI)	156
5.2	Actual and potential output	157

5.3	Output gap	158
5.4	US real GDP %YOY	159
5.5	Generating business cycles using US GDP ARMA (12, 12) permanent shock =+1	160
5.6	Long-run equilibrium in the NCM model	177
5.7	Output gap response to a negative demand shock under optimal policy	179
5.8	Inflation and interest rate response to a negative demand shock under optimal monetary policy	179
6.1	Fleet and freight rates adjustment when $-\lambda\rho = 1$	215
6.2	Fleet and freight rates adjustment when $-\lambda\rho = 0.7$	215
6.3	Fleet and freight rates adjustment when $-\lambda\rho = 0.25$	216
6.4	Long-run equilibrium in the BV model	229
6.5	Adjustment of capacity utilisation	230
6.6	Adjustment of SH prices	230
6.7	Adjustment to long-run equilibrium	231
6.8	Adjustment for a permanent increase in demand	232
6.9	Output gap and demand for shipping	245
6.10	Interest rate and inflation	245
6.11	Freight rate and fleet capacity utilisation	246
6.12	NB and SH price adjustment	246
6.13	Adjustment with small response of freight rates to fleet capacity utilisation	248
6.14	Adjustment with large response of freight rates to fleet capacity utilisation	248
8.1	Growth of assets of four sectors in the US	274
8.2	Liabilities of US shadow and traditional banking	274
8.3	Asset leverage of US investment banks	275
8.4	Oil premium and spot oil price	282
8.5	Demand – long-run equilibrium	284
8.6	Demand for seaborne trade and demand for oil	285
8.7	Demand, supply growth in dry market	287
8.8	Fleet capacity utilisation and BDI	288
8.9	The deviations of demand from a linear trend	289
8.10	Deviations of China steel production from trend	290
8.11	Chinese cargo: deviation from trend	290
9.1	BDI and S&P 500	296
9.2	TC rates, SH prices and NB prices	298
9.3	US real GDP in the last business cycle	302
9.4	Business cycle	303

9.5	US business cycles	310
9.6	Japan business cycles	318
9.7	Germany business cycles	319
9.8	Shipping cycles	322
10.1	Average earnings in the dry market	329
10.2	Average earnings in the oil tanker market	330
10.3	The errors of a model of average earnings in the oil tanker market	331
10.A1	HOPE and SAIL demand	336
10.A2	HOPE rule of thumb demand	337
10.A3	HOPE log-linear demand model on world GDP	338
10.A4	Demand for dry (% YoY) – SAIL model	338
10.A5	HOPE NB rule of thumb	340
10.A6	Log-linear NB prices model	340
10.A7	New vessel prices – SAIL model	341

LIST OF TABLES

5.1	Accounting for growth in US potential output	196
5.2	Accounting for growth in US potential output – CBO method	197
6.1	Dynamic fleet adjustment	214
7.1	Yearly TC averages	254
7.2	The capital structure of shipping – Marsoft data	256
7.3	*Marine Money* list of public shopping companies	259
8.1	The oil tanker market in the last eight years	286
8.2	Decomposition of change in demand for seaborne trade between 2003 and 2011	286
9.1	Shipping cycles	297
9.2	The US business cycles	300
9.3	The US business cycles	308
9.4	Japan business cycles	315
9.5	Japan cycle phases	317
9.6	Germany business cycles	320

LIST OF BOXES

2.1	A macroeconomic (hierarchical) approach to shipping	14
2.2	The structure of the dry market	15
3.1	Owners and central banks	46
3.2	Optimal fleet	46
3.3	NB prices	49
3.4	Investment under uncertainty	55
3.5	Case study	56
8.1	Freight rates as asset prices	291
9.1	Freight rates as a leading indicator of business cycles	295

LIST OF ABBREVIATIONS

Lower-case letters indicate natural logs.
A = average age of fleet
CU = fleet capacity utilisation or demand–supply balance in the freight market
DL = deliveries of new vessels by shipyards, measured in tonnage
DM = demolition of ships, measured in tonnage
E_t = is the expectations operator with information available up to period t
EIA = US Energy Information Agency
FR = (spot) freight rate, expressed in US dollars per day
H^τ = a time charter contract of duration, τ, paying a fixed rate per period, expressed in \$/day
I = gross investment expenditure by owners for newbuilding vessels
K = net fleet, that is, operational fleet plus deliveries less scrapping less lay-ups and losses, expressed in dwt
KG = gross fleet, measured in tonnage
KTR = traded fleet in the secondhand market, measured in tonnage
NB = newbuilding
OC = operating cost
p = inflation rate
P = Price of newbuilding vessels
PB = price of bunkers
PE = price of equipment
PSC = price of scrap
PS = price of secondhand vessels
PT = price of steel
Q = demand for shipping services, expressed in dead weight (dwt)
Q^s = supply of shipping services, expressed in dwt
r = interest rate
rp = risk premium
S = vessel or average fleet speed
SCU = shipyard capacity utilisation or shipyard demand-supply balance
SH = secondhand
UC = user cost of capital in the newbuilding market
UCS = user cost of capital in the secondhand market
W = wages of ship crew
x = output gap in the economy
y = rate of growth of real GDP
z = a vector of variables affecting supply, such as bunker costs and port congestion
π = profit per ship per time period

1 INTRODUCTION

THE SCOPE OF THE BOOK

Following the seminal work of Tinbergen (1931, 1934) and Koopmans (1939) research in maritime economics has focused on integrating the various markets into a dynamic system. This macroeconomic or systems approach to maritime economics reached its heyday with the Beenstock–Vergottis (BV) model (1993). The BV model is the first systematic approach to explain the interaction of the freight, time charter, secondhand, newbuilding and scrap markets under the twin assumptions of rational expectations and market efficiency. The model is a landmark because it treats ships as assets and applies portfolio theory to assess their values. As asset prices depend on expectations, Beenstock and Vergottis introduce rational expectations to account for the impact of expected and unexpected changes in key exogenous variables, such as the demand for shipping services, interest rates and bunker costs.

But since the publication of the BV model, research in maritime economics has shifted from the macro approach to micro aspects. The research has been mainly empirical in nature and has concentrated, for example on the efficiency of individual shipping markets. In one of his many excellent surveys of the maritime economics literature, Glen (2006) concludes as follows:

> [t]he models developed and presented in Beenstock and Vergottis (1993) are a high water mark in the application of traditional econometric methods. They remain the most recent published work that develops a complete model of freight rate relations and an integrated model of the ship markets. It is a high water mark because the tide of empirical work has turned and shifted in a new direction. This change has occurred for three reasons: first, the development of new econometric approaches, which have focused on the statistical properties of data; second, the use of different modelling techniques; and third, improvements in data availability have meant a shift away from the use of annual data to that of higher frequency, i.e. quarterly or monthly. (Glen 2006, p. 433)

This book aims to fill this gap and make a return to a macroeconomic or systems approach to maritime economics. The brilliant book by Stopford (2009) is in the same spirit and covers all markets and their interaction at an introductory level. This book aims to be a companion to Stopford's textbook at a more advanced level.

1 THE BENEFITS OF A MACROECONOMIC APPROACH

A macroeconomic approach to maritime economics offers a number of advantages. The first relates to the microfoundations of the freight market and the second to the microfoundations of the shipyard and secondhand markets. In the traditional approach, which goes back to the Tinbergen–Koopmans (TK) model, freight rates are viewed as determined in perfectly competitive markets, where the stock of fleet is predetermined at any point in time. This implies that freight rates adjust instantly to clear the demand–supply balance, where supply is fixed (subject only to a variable fleet speed). Accordingly, freight rates respond almost exclusively to fluctuations in demand, rising when demand increases and falling when demand falls. The assumption that supply is fixed even in the short run is not appropriate, as supply is constantly changing. Charterers and owners observe the evolution of supply and must surely form expectations of future demand and supply in bargaining over the current freight rate. This calls for a dynamic rather then a static analysis in which the fleet is fixed. In Chapter 2 we suggest an alternative theorising of the freight market, which captures this dynamic analysis of freight rates. This new framework consists of a bargaining game over freight rates in which charterers and owners form expectations of demand and supply over a horizon relevant to their decisions. In this approach, freight rates are viewed as asset prices, which are determined by discounting future economic fundamentals.

This macroeconomic approach to freight rates is appropriate to recent macroeconomic developments. The business cycles of the major industrialised countries have shifted from demand-led in the 1950s and 1960s to supply-led in the 1970s and the 1980s and, finally, to asset-led cycles over the past twenty years, driven by excess liquidity. The liquidity that has financed a series of bubbles, including the Internet, housing, commodities and shipping bubbles, during this period was created as a result of a gradual process. Financial deregulation and liberalisation laid the foundations for financial engineering, while central banks have pumped more liquidity into the financial system every time a bubble has burst, thereby perpetuating the bubble era. The expanding liquidity has resulted in the financialisation of shipping markets, a topic that is analysed in Chapter 8. In the first phase of financialisation the liquidity affected commodity prices, including those of oil, iron ore and coal. The advent of investors in the commodity markets increased volatility in freight rates and vessel prices and distorted the price mechanism. Prices convey a signal of market conditions. The advent of investors in commodity markets pushed prices higher than was justified by economic fundamentals in the upswing of the cycle and lower in the downswing, thereby increasing the amplitude of the last super shipping cycle. In the second phase of financialisation, which is taking place now, the financialisation of shipping markets is affecting vessel prices turning ships into commodities. The financialisation of shipping markets makes the bargaining approach to freight rates a more plausible framework.

The second advantage of a macroeconomic approach to maritime economics relates to the microfoundations of the shipyard market, where the Beenstock–Vergottis approach remains a valid model within a macroeconomic framework. The BV model suffers from a major drawback, which has gone mainly unnoticed and unchallenged in the maritime economics literature in the last twenty years. This is that the microfoundations of the BV model involve decisions intended to maximise short-term profits (that is, profits in every single period of time) instead of maximising long-term profits (that is, profits over the entire life of the vessel). Short-term profit maximisation is imposed either explicitly as in the freight market or implicitly by invoking market efficiency, as in the secondhand and newbuilding markets. The combination of short-term profit maximisation and market efficiency destroys the simultaneity of the BV model. Decisions in the four shipping markets (freight, secondhand, newbuilding and scrap) are not jointly determined. The decisions can be arranged in such a way so that one follows from the other. This has serious implications for fleet expansion strategies, as it makes them oversimplistic. The fleet capacity expansion problem is analysed in Chapter 3 and the interaction of business and shipping cycles, is investigated in Chapters 6 and 9.

To illustrate the oversimplistic nature of the BV framework, consider an owner that maximises profits in each period of time, say a month, by choosing both the average fleet speed and the size of the fleet, so as to equate the return on shipping, adjusted for a variable risk premium, with the return on other competing assets, such as the short- or long-term interest rate. The fleet is adjusted monthly to reach the optimum via the secondhand and scrap markets. An owner adjusts his actual to the optimal fleet on a monthly basis by buying or selling vessels in the secondhand market or scrapping existing vessels according to the principle of monthly profit maximisation. Thus, an owner in the BV framework may expand the fleet one month and contract it the next month. In general, unanticipated random fluctuations in any exogenous variables, such as the demand for shipping services, interest rates and bunker costs, would trigger oscillations in the owners' fleet. Therefore, the owner in the BV model is myopic in that s/he ignores the consequences of her/his actions today for the lifetime of the vessel despite forming rational expectations. These two unsatisfactory features of the BV model of market efficiency and short-term profit maximisation are corrected in this book. In Chapter 3 it is shown that the appropriate framework for fleet expansion strategies is long-term profit maximisation. In Chapter 4 it is shown that the empirical evidence on the whole suggests that shipping markets are inefficient for practical purposes in decision making, although shipping markets may be asymptotically efficient (that is, as the investment horizon tends to infinity). The integrated model, which is laid out step by step in the preceding chapters, is analysed in Chapter 6 in an attempt to explain shipping cycles.

A macroeconomic approach also has the advantage of integrating the supply and expectations approaches to shipping cycles. The TK and BV model have shaped the theory of shipping cycles. The contribution of the TK model is that the basic cause of shipping cycles is the shipyard delivery lag. Thus, in the TK model shipping cycles arise naturally, even if demand is stable, because of the lag between placing orders and the ability of shipyards to deliver. The BV model emphasises the adjustment of expectations to exogenous shocks as the primary cause of shipping cycles. The macroeconomic approach developed in this book integrates the TK model of supply-led shipping cycles with the BV model of expectations-driven shipping cycles. As the empirical evidence of shipping markets shows that they are inefficient in the short run, the integrated model breaks away from the BV model of assuming that shipping markets are efficient. The implication is that the arbitrage conditions between newbuilding and secondhand prices and between the return of shipping and alternative assets are removed. Instead, demand and supply factors in newbuilding and secondhand markets are allowed to interact in determining prices. This has the implication that all shipping markets interact with each other. As a result, a fleet capacity expansion strategy involves expectations of future freight rates, newbuilding, secondhand and scrap prices and the net fleet, which are jointly determined. This makes fleet expansion strategies a complicated task.

The integrated model also includes the business cycle model developed in Chapter 5. In the BV model expectations are rational and drive the dynamics of the shipping model, along with the fleet accumulation dynamics, but the demand for dry is exogenous to the model. In the integrated model the demand for dry is endogenous. The implication of extending the model to cover the interaction of business and shipping cycles does not simply provide a more realistic explanation. It is shown in Chapter 6 that expectations about future freight rates, vessel prices and the demand supply balance (fleet capacity utilisation) are shaped by expectations of the future path of real interest rates and consequently on monetary policy; and, in particular, on how central banks react to economic conditions. As central banks choose their policies with the view of achieving their statutory targets, by observing current inflation and the output gap and knowing the central bank's targets, one can deduce the future path of nominal interest rates. This provides a consistent explanation of how expectations in shipping are formed integrating macroeconomics with maritime economics. This interaction is analysed in Chapter 6.

2 THE STRUCTURE OF THE BOOK

Part I deals with the microfoundations of maritime economics, which attempt to derive the general form of the underlying demand and supply functions in all four markets (freight, spot and period, newbuilding, secondhand and scrap) based on the principles of rationality, which is the basis of a scientific approach to maritime economics. Economic agents in shipping markets are assumed to be utility or profit maximisers, subject to well-defined economic and technological constraints. Chapter 2 analyses the microfoundations of the freight market (spot and period), while Chapter 3 the microfoundations of the other three markets.

Part II analyses the macro aspects of maritime economics. Chapter 4 reviews the empirical evidence of whether shipping markets are efficient and concludes that they are inefficient in a horizon relevant to decision making, although markets may be asymptotically efficient. Chapters 5 and 6 provide a macro-economic approach to maritime economics by examining all four shipping markets as a system of simultaneous equations. In Chapter 5 the model is extended to include the macroeconomy and explain how business cycles are generated using the New Consensus Macroeconomics model, as modified by Arestis and Karakitsos (2013).

Chapter 6 integrates the work from all the previous chapters. It studies the interactions of all four shipping markets and how they respond to anticipated and unanticipated shocks giving rise to shipping cycles. It investigates the causal relationship between business and shipping cycles. The analysis shows that shipping cycles are caused by business cycles. The TK model is instructive of the implications of the delivery lag. Depending on parameter values, shipping cycles can appear out of phase with business cycles, thereby giving the impression that shipping cycles move counter-cyclically to business cycles (that is, they move in opposite directions). But such behaviour does not change the direction of causality. Business cycles cause shipping cycles. Finally, Chapter 6 shows how expectations about key shipping variables can be formed consistently by expectations on economic policy (interest rates). In particular, it is shown that inflation depends on the expected path of the future output gap (the deviation of real GDP from potential output). The output gap in the economy, in turn, is a function of the expected path of future real interest rates. Central banks affect real interest rates by controlling nominal interest rates. Therefore, both inflation and the output gap are functions of the expected path of future real interest rates, which are influenced by monetary policy. Expectations of future freight rates and vessel prices depend on the expected future path of fleet capacity utilisation, the demand–supply balance in the freight market. Given a shipyard delivery lag of two years, the supply of shipping services is largely predetermined by past expectations of current demand for shipping services. Demand depends on expectations of the future output gap, and, consequently, on monetary policy. Therefore, by observing current inflation and the output gap and knowing the central bank's targets one can deduce the expected future path of key shipping variables. This illustrates the interaction of the macro-economy with shipping markets providing an integrated model.

Part III takes the big step of moving from theory to practice. Chapter 7 explains the market structure and the role and impact of ship finance. Chapter 8 explains the financialisation of shipping. The nature of the shipping markets has changed from a fundamental transport industry into an assets (or securities) market. Freight rates and vessel prices are determined as if the shipping market was a stock exchange one. This structural change occurred simultaneously in the dry and the wet market in 2003, as a result of the attraction of investors into commodities. Speculative flows into commodities distorted freight rates and vessel prices by creating a premium/discount over the price consistent with economic fundamentals (demand and supply). There was a premium in 2003–11, turning a boom into a bubble, but a discount since 2011 adding to the gloom during the depressed markets of 2011–13. The implication of this structural change is that the outlook for the dry and wet markets depends not only on economic fundamentals, but also on the risk appetite of investors.

Chapter 9 analyses the interrelationship of business and shipping cycles in practice. It describes the stylised facts of shipping cycles. It analyses the official classification of business cycles, using the US as an example because of its possible impact on world cycles, and puts forward an alternative approach that enables the distinction between 'signal' and 'noise' in identifying trends and discerning the reversal of trends. This approach helps to compare the actual and optimal conduct of US monetary policy in business cycles and shows how relatively accurate expectations of interest rates can be formulated in shipping. Chapter 9 analyses the business cycles of Japan and Germany and their interrelationship with

US business cycles and shows how these business cycles account for the stylised facts of shipping cycles in the 1980s and the 1990s. The cycles since then are explained by the behaviour of China, which has supplanted Japan in pre-eminence in world trade. Although China is now the factory of the world economy, thereby explaining the long-term growth rate of demand for shipping services, the trigger for the fluctuations in demand has been the US because of its importance in shaping world business cycles. This was true in the 1980s and the 1990s, when Japan was explaining shipping cycles and it has remained true ever since because China is an export-led economy. It will take time for China to reform its economy from an export-led to a domestic-led one, a reform that has been endorsed at the Third Plenum of the party in November 2013.

Although shipping cycles are generated by business cycles, in the real world they are also caused by large swings in expectations of demand for shipping services. Volatile expectations can be rational, as a result of cyclical developments in macroeconomic variables, or irrational, what Keynes (1936) called 'animal spirits', a situation where economic fundamentals remain unchanged and yet expectations swing from optimism to pessimism. This swing of expectations is related to uncertainty about macroeconomic developments. Therefore, a full explanation of the stylised facts of shipping cycles requires an extension of the business cycle analysis to conditions of uncertainty and the role of the availability of credit (ship finance). The theory of the fleet capacity expansion under uncertainty, analysed in Chapter 3, provides the basis for this extension. Chapter 9 explains how uncertainty about demand can lead to overcapacity, as owners may decide to wait until the recovery is sustainable before investing.

The availability of credit makes shipping cycles even more pronounced than otherwise. Banks and other credit providers are highly pro-cyclical; the loan portfolio increases in the upswing of the cycle and decreases in the downswing. This is due to the myopic attitude of credit institutions in granting credit according to the collateral value of the loan, which is highly pro-cyclical. Therefore, ship finance increases the amplitude of shipping cycles.

In every severe recession the pessimist view that shipping would never recover gains ground. This is based on the well-known conspiracy theory that it is in the interest of the country that depends most on world trade, such as Japan in the last twenty years of the twentieth century or China in the present day, to increase the fleet to keep freight rates low. Chapter 9 deals with this issue.

Instead of summarising the macroeconomic approach to maritime economics, the final chapter provides two practical examples of the approach advanced in this book. First, it shows why the majority of owners with experience can be relatively accurate in buying ships and expanding their fleet, but also that with a few exceptions they are bad decision makers in selling ships and taking profits. This is a corollary of the asymmetric and highly skewed distribution of earnings. The practical implication of the methodology developed in this book is to transform this distribution into a normal one, where outliers are covered in the tails of the normal distribution, which account for 5 per cent probability. The significance of this transformation is highlighted in the second practical example. A practical real case is studied, using pseudonyms for the actual parties involved, which compares the real world practice with the methodology of the book.

3 HOW THIS BOOK SHOULD BE READ AND ITS POTENTIAL READERSHIP

The book is particularly relevant to final year undergraduates or graduate students in maritime economics or shipping degrees as an advanced textbook. A prerequisite is some basic understanding of shipping markets and maritime economics covered in the excellent undergraduate textbook written by Stopford (2009). In addition, some economic background (of microeconomics, macroeconomics, and mathematics for economists and econometrics) might be beneficial. Nonetheless, the book is written with the aim of being self-contained. Supplementary material, which would help to fill the gaps, is provided in the form of Appendices at the end of each relevant chapter. For example, stochastic processes, stationary time series, co-integration of two or higher-order systems and the VAR methodology are explained to the extent that is needed in the Statistical Appendix of Chapter 4.

The book is also relevant to shipping professionals (owners, charterers, brokers, bankers specialising in ship finance) as the book deals extensively with the interaction of business and shipping cycles and how expectations in shipping should be formed to be of relevance to real life decision making. It is true that some shipping professionals may find the book difficult at times, but we have structured it in such a way that they can benefit without reading the full text. For the sake of the professional, each chapter has been structured as follows. At the beginning of each chapter there is an Executive Summary, which outlines the main points as plainly as possible. Wherever possible, the analysis is presented verbally with any mathematical treatment relegated to an Appendix. Wherever mathematical succinctness cannot be replaced by verbal arguments, equations are presented but without any mathematical manipulation beyond basic substitution of equations.

The book can serve as an advanced textbook in shipping courses or maritime economics in the whole or in parts. If the book is used as an advanced textbook in maritime economics it should be read as it is written starting from Part I to Part II and finally to Part III. Each part of the book can serve as a module in many relevant courses. Chapters 5, 6 and 9 can form a separate module on shipping cycles. Part III: From Theory to Practice may serve as a relevant module, and it is primarily written for the benefit of the professional. It explains how the theory advanced in the first two parts can be of practical use in decision making. Chapter 4 is probably the most difficult chapter, as it requires statistical knowledge to appreciate the empirical literature on market efficiency. Nonetheless, the professional may be satisfied with the verdict that shipping markets are inefficient for practical purposes.

For the professional we recommend that s/he goes directly to Chapter 9. This chapter requires no prior knowledge and yet it can serve as a guide to practical aspects of decision making. If the professional wants to deepen her/his theoretical understanding of the interaction of business and shipping cycles s/he can then read Chapter 6. This requires some knowledge of mathematics, although the essence can be captured in the Executive Summary. S/he can then complete her/his study by reading Chapter 5. The professional can also read independently Chapter 8 to appreciate the implications of the financialisation of shipping markets. Finally, s/he can read Chapter 10 to appreciate the contribution of the methodology advanced in the book to real life decision making.

4 MODEL AND DATA SERIES

Unless otherwise stated, all graphs and tables in the book are based on the K-model, which integrates shipping markets with the macro-economy and financial markets of the US, the UK, the euro area, Japan and China (see Arestis and Karakitsos, 2004 and 2010). The macro and financial data used and transformed in the K-model are official figures as made available by Thomson–Reuters–EcoWin Pro, a live databank (see www.thomsonreuters.com). The shipping data are available from Clarkson's Shipping Intelligence Network (see www.clarksons.net).

ACKNOWLEDGEMENTS

We are grateful to many people who have contributed to the accomplishment of this book; without their help this book would not have been written, although any errors and omissions remain our sole responsibility. Our special thanks go to Taiba Batool, the Economics Editor of Palgrave Macmillan, for her encouragement and suggestions on the structure of the book; as well as to two anonymous referees of Palgrave Macmillan who provided helpful comments on an earlier manuscript; and, finally, to Ania Wronski for being extremely helpful in the production of this book.

Many people have worked tirelessly to make comments on various draft chapters. Our particular thanks go to Costas Apodiacos and Yanni Chalkias of Victoria Steamship Ltd. Costas Apodiacos, who, in addition to commenting on various chapters, made special efforts to guide us on writing Executive Summaries for the professional and not the academic. EK would also like to thank the lifetime close friend and colleague Professor Philip Arestis for encouragement and detailed comments.

We also received valuable data and help from Matt McCleary and Campbell Houston of Marine Money, Jeremy Penn of the Baltic Exchange, Arlie Sterling of Marsoft, Martin Stopford of Clarkson's Research, and Ghikas Goumas. We are grateful to all of them.

We would also like to pay tribute to the late Professor Zenon Zannetos of MIT who was the first academic to treat shipping as a subject worthy of academic analysis.

EK is also grateful to his wife, Chloe, for forgiving him in spending almost all weekends of 2013 in writing the manuscript, his daughter and colleague Nepheli for detailed comments, and his daughter Eliza for her tolerance with a busy father.

PART 1
THE MICROFOUNDATIONS OF MARITIME ECONOMICS

2 THE THEORETICAL FOUNDATIONS OF THE FREIGHT MARKET

EXECUTIVE SUMMARY

In the traditional model of the freight market, which dates back to the 1930s, freight rates are determined in a perfectly competitive market, where the stock of fleet is predetermined at any point in time. This implies that freight rates adjust instantly to clear the demand–supply balance, where supply is fixed. Accordingly, freight rates respond exclusively to fluctuations in demand, rising when demand increases and falling when demand falls. This assumption introduces an element of irrationality on the part of both owners and charterers because the supply is constantly changing – new ships arrive in the market all the time. A cursory look at deliveries of new vessels shows that there are significant changes in supply from month to month. Hence, the assumption that supply is fixed in the short run is not appropriate. Instead, both charterers and owners form expectations of demand and supply and this requires a dynamic analysis rather then a static one in which the fleet is fixed. In this chapter we suggest an alternative theorising of the freight market, which captures this dynamic analysis of freight rates. This new framework consists of a bargaining process over freight rates in which charterers and owners form expectations of demand and supply over an investment horizon relevant to their decisions. In this framework, freight rates are viewed as asset prices, which are determined by discounting future economic fundamentals.

In the traditional model the demand for and supply of shipping services are functions of freight rates in a perfectly competitive market. But freight rates are not prices determined in auction markets, where many owners bid for the same cargo and the one with the lower bid wins the contract (Dutch auctions). The characteristics of the freight market do not accord with those of perfect competition. Intuitively, perfect competition is a market system where the actions of individual buyers and sellers have a negligible impact on the market and where both are price takers. The assumptions of perfect competition are not satisfied in the freight market. In particular, the product is not homogeneous;[1] the assumption of a very large (in theory infinite) number of buyers and sellers is not applicable, transaction costs are not zero and there is no freedom of entry and exit. Although the product is seemingly homogeneous (the capacity to transport particular categories of products or commodities), the demand for shipping services is restricted by volume, time and route – a given cargo over a particular route that meets a well-specified time schedule. Although there are many ships in the market only a few are available to satisfy the given demand specifications in time and place. These characteristics violate the homogeneous product assumption, the condition of large (infinite) number of buyers and sellers and the hypothesis of zero transaction costs. The latter should be interpreted as the penalties (legal or reputational) that a charterer would incur for waiting for a better deal (lower freight rates). Moreover, there are large barriers to entry both in capital and the operation and management of the fleet.

Rather, the freight rate over a particular cargo is the outcome of a bargaining process that happens at the same time – or approximately the same time – in different places and where information about freight rates agreed is almost instantaneously available to all other participants. Thus, the agreed freight rates do not balance demand and supply in a particular place at a particular point in time, but rather expectations of overall demand and supply in a particular segment of the market or the entire market. Accordingly, freight rates are equilibrium rates in a bargaining game where players form rational expectations about economic conditions.

The negotiations between an owner and a charterer over a contract for a particular cargo can best be viewed as a zero-sum game between the two players.[2] Both players know the freight rates that have been agreed so far. This information permeates to the rates that were agreed on the same or similar routes with ships of the same or different capacity. Players also know not just the latest rates, but their entire history. This enables them to assess and form expectations of the dynamic evolution of future freight rates. Thus, when the charterer and the owner enter the negotiations they would bargain over the deviation of the expected future freight rate from the latest or equilibrium rate. The final outcome will be influenced by the bargaining power of each party. In this context, it is better to formulate the problem as bargaining over the discounted present value of future freight rates. If the bargaining power of the charterer is stronger than that of the owner, the deviation of the agreed freight rate from the latest or equilibrium rate will be negative, implying a lower rate than the latest one. If, on the other hand, the bargaining power of the owner is stronger, the deviation of the agreed freight rate from the latest or equilibrium rate will be positive, implying a higher rate than the latest one. In some negotiations one of the parties is a big player and has the upper hand. The agreed freight rate would be to the advantage of the big player in the context of the market average. But such freight rates are outliers (they belong to the tails of the distribution). In the median negotiation the bargaining power of the charterer and the owner depends on economic conditions. In 'good' or improving economic conditions the owner has stronger bargaining power, whereas in 'bad' or worsening economic conditions the charterer's bargaining power prevails. Hence, in 'good' or improving economic conditions, freight rates would be on an uptrend; and vice versa. It is shown in this chapter that expectations about key shipping variables are formed by expectations of how policymakers (mainly central banks) would respond to current and future economic conditions.

1 A FRAMEWORK FOR MARITIME ECONOMICS

The first part of the book deals with the microfoundations of maritime economics (or shipping markets). Shipping is organised in the form of four markets: the freight market; the shipyard (or newbuilding) market; the scrap market and the secondhand market. The freight market is subdivided into the spot market and the time charter (or period) market. These markets pertain to all ship types (dry bulk, oil tankers, containers and specialised ships, like LNG) and all ship sizes. In the dry sector there are four major ship sizes: Capesize, Panamax/Kamsarmax, Handymax/Supramax and Handysize. In the oil tanker (or wet) market there are five major ship sizes: VLCC, Suezmax, Aframax, Panamax and Handysize. In this taxonomy there are four markets (freight, newbuilding, scrap and secondhand) for each type of ship and for each size.

The microfoundations developed here are common to all types and ship sizes. The microfoundations attempt to derive the general form of the underlying demand and supply functions in all four markets based on the principles of rationality, which is the basis of a scientific approach to shipping. Economic agents are assumed to be utility or profit maximisers subject to well-defined economic and technical constraints. This maximising behaviour gives rise not just to the functional form of the underlying demand and supply functions, but also to their exact determinants and in most cases to the qualitative influence (positive or negative) of each determinant on the demand or supply. These restrictions are important for drawing inferences and in empirical work on maritime economics.

This framework enables one to analyse the impact of exogenous shocks (anticipated or unanticipated) on the equilibrium of the shipping system as a whole and a single market (for example, Capes). In the first part of the book we deal with the microfoundations of each single market (freight, newbuilding, scrap and secondhand). In the second part of the book we integrate the shipping markets and examine the properties of the entire system. Moreover, we analyse the interrelationship of shipping with the macroeconomy. The economy helps to explain the demand for shipping services, which in traditional analysis is treated as exogenous to shipping markets. The integration of shipping markets with macroeconomics sheds lights on how expectations in the shipping markets are formed. In this context the assumptions of rational expectations become more palatable to swallow. For it is one thing to assume that expectations are, on average, correct for freight rates and ship prices (newbuilding and second hand) and another to assume that expectations for short-term interest rates are, on average, correct. The second assumption is more palatable to swallow, given the emphasis of major central banks on shaping and influencing interest rate expectations through announcing the targets of economic policy, the extent to which they would tolerate deviations from conflicting targets and how long they would stick with current policies (like low interest rates). In the macroeconomic approach to maritime economics expectations about policy drive expectations in shipping markets.

In this framework, developments in the major regions of the world economy shape the major forces of demand and supply in the overall shipping market (whether dry, wet or containership). These developments in the overall market infiltrate in time to the various sectors in a manner that takes into account the disequilibrium of each sector from the overall market. For example, if freight rates, the fleet size or the prices of a particular sector, say Capes, are higher than the overall market by more than justified by economic fundamentals, then Capes would adjust through time so that equilibrium is attained once more.

Hence, every market (dry, wet or containership) consists of five variables, which are determined simultaneously. These are the demand for shipping services (the cargo being transported), the stock of the net fleet, freight rates, newbuilding (NB) prices and secondhand (SH) prices. Each variable is shaped in one or more markets, but all markets are interacting with each other. In the freight market, the demand for shipping services by charterers and the supply of shipping services by owners determine the amount of cargo transported and freight rates. In the shipyard market, the demand for vessels by owners and the supply of vessels by shipyards determine NB prices and the deliveries of new vessels, which are added to the existing stock of fleet. In the secondhand market, the demand for vessels by owners and the supply of vessels by other owners, determines the SH prices and the volume of sales/purchases. In the scrap market, the demand for scrap metal by scrapyards and the supply of vessels for demolition by owners determine the price of scrap metal and the volume of ships which are demolished. The stock of the net fleet at the end of each period, say quarter, is simply the stock of old vessels at the end of the previous period, augmented by the deliveries in the current period, less the vessels for demolition in this same period.

This hierarchical structure of the dry market is presented in Box 2.1 for the dry market. A similar structure exists for the wet and containership markets. Macroeconomic developments affect the conditions in the overall dry market, which are then transmitted to each sector of the market. The four interacting markets are presented in Box 2.2. Demand and supply in each market determine the equilibrium price and quantity. With the help of these boxes it is easy to appreciate the interactions of the various markets. A booming world economy spurs world trade and the demand for shipping services. With a fixed stock of net fleet, but assuming some spare fleet capacity, the cargo being transported increases and freight rates go up. If freight rates cover the operational costs, then owners increase the supply of shipping services to meet the higher demand by increasing the speed of the vessels. A sustained increase in demand that cannot be met by higher speed induces owners to buy in the secondhand market and order new ships. This increases SH prices, lowers the fleet demolition and increases the demand for new vessels. Shipyards respond by increasing NB prices, as in the short run they cannot meet the higher demand for vessels; it takes time, approximately two years, to deliver a new vessel.

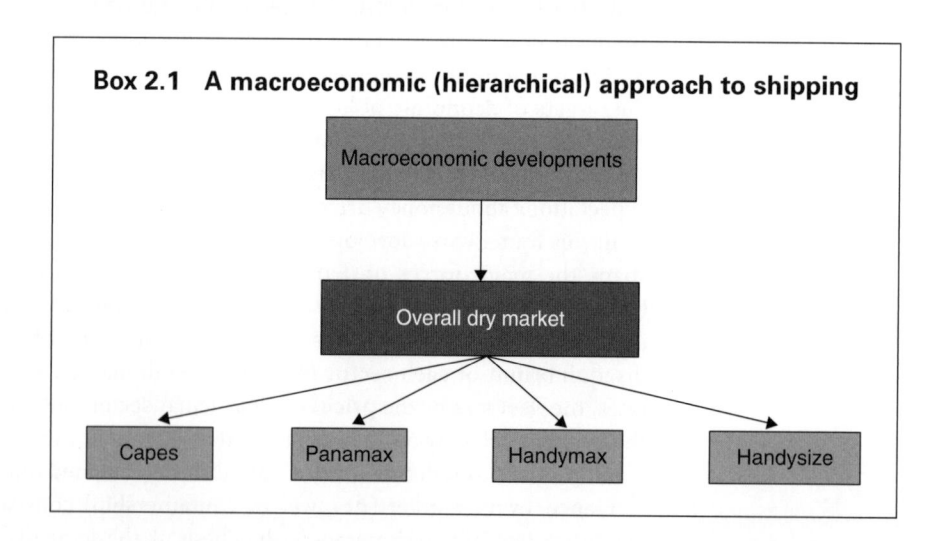

Box 2.1 A macroeconomic (hierarchical) approach to shipping

Macroeconomic developments → Overall dry market → Capes, Panamax, Handymax, Handysize

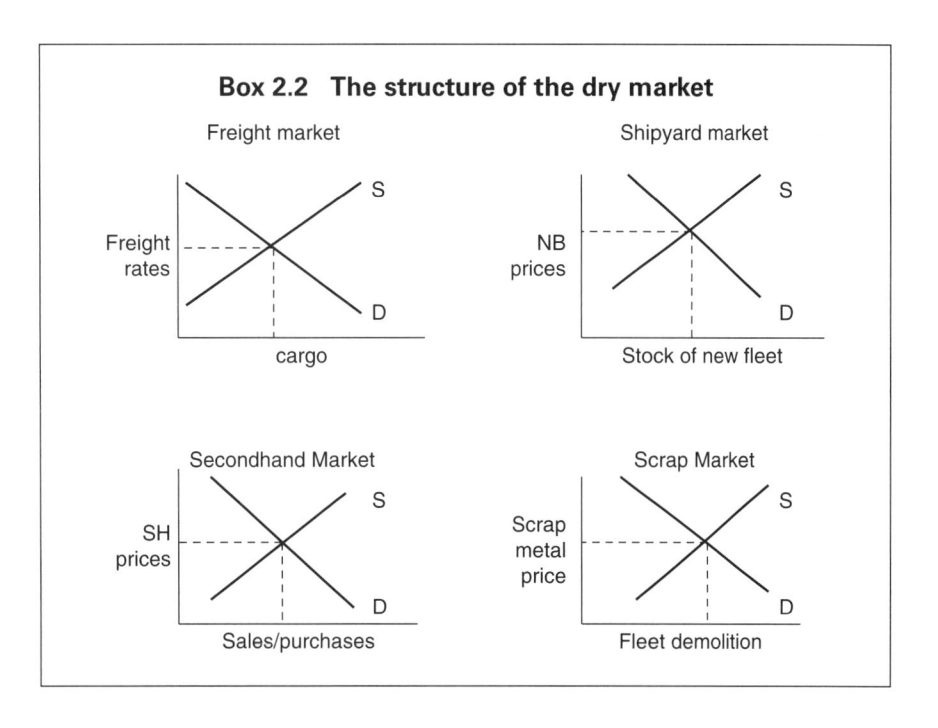

> **Box 2.2 The structure of the dry market**
>
> [Four supply-and-demand diagrams:]
>
> - Freight market: Freight rates vs. cargo, with S and D curves
> - Shipyard market: NB prices vs. Stock of new fleet, with S and D curves
> - Secondhand Market: SH prices vs. Sales/purchases, with S and D curves
> - Scrap Market: Scrap metal price vs. Fleet demolition, with S and D curves

When the world economy slows down or falls into recession the demand for shipping services falls. The cargo being transported is reduced and freight rates slide. The demand for secondhand ships diminishes, SH prices fall and demolition picks up steam. Owners cancel orders and NB prices fall. This means that shipping cycles are primarily caused by economic (or business cycles).

The macroeconomic approach to maritime economics differs from the traditional approach, where the freight market is isolated from the rest of the system. So, in the traditional approach the system is not simultaneous; it is post-recursive, namely it can be arranged in a particular order to be solved. First the freight market is solved and then the equilibrium level of freight rates and cargo enter the other three markets, which are then solved simultaneously. It should be borne in mind that the Beenstock and Vergottis (1993) model, which is regarded as a high-water mark in shipping systems, is entirely post-recursive. This destroys the simultaneity of the shipping system and gives the impression that shipping decisions are easy to take and depend entirely on developments in the shipping market, which are governed exclusively by exogenous developments in the demand for shipping services. This approach has given rise to the relative isolation of maritime economics from other branches of economics and a tendency for treating ship sizes as segmented markets. Instead, the macroeconomic approach gives priority to the overall market and postulates a hierarchical approach to shipping. Thus, the market of each ship size co-moves (or, in the jargon of econometrics, is co-integrated) with the overall market. Therefore, shocks to the overall market are transmitted to each ship market. In time, each ship-size market moves to equilibrium with the overall market.

Part I of the book consists of two chapters. Chapter 2 analyses the theoretical foundations of the freight market, which is split into two markets: the spot market and the time charter market. In this chapter we offer a new framework for analysing spot freight rates. This framework has some implications for the nature of the risk premium in the time charter market.

Chapter 3 examines the theoretical foundations of the shipyard, scrap and secondhand markets. It starts with a single owner's decision problem of the optimal fleet and explains how the demand for newbuilding vessels is derived at the individual and aggregate level. It then goes on to consider the shipyard market and examines the influence of supply in determining the price of new vessels and vessel deliveries. It then analyses the secondhand market and shows how the demand and supply functions are obtained. Finally, it considers the scrap market and explains how the net fleet is determined.

2 THE TRADITIONAL MODEL

The traditional model of freight rates goes back to Tinbergen (1931, 1934), Koopmans (1939), Hawdon (1978), Strandenes (1984, 1986) and Beenstock and Vergottis (1993). The demand for and supply of shipping services are functions of freight rates in a perfectly competitive market. The cargo is measured in tonne-miles in recognition of the fact that both the volume of the cargo to be transported and the distance covered matter. The demand for shipping services is assumed to be a negative function of freight rates, while the supply of shipping services is a positive function. The demand for shipping services is assumed to be very inelastic with respect to freight rates, as charterers have a lot to lose if the entire cargo that is earmarked for transport is not shipped and does not arrive on time at the destination port. The supply of shipping services, on the other hand, is supposed to be a non-linear function of freight rates. At low freight rates the supply of shipping services is very elastic, as there is a glut of vessels. A small increase in freight rates attracts many shipowners willing to take the existing cargo. Alternatively, an increase in the demand for shipping services is met largely by an increase in the volume to be transported at unchanged or slightly higher freight rates. But as the demand for shipping services keeps on increasing a smaller proportion of extra volume is transported, while freight rates increase at a bigger proportion. As the demand for cargo rises to the point where all ships are fully utilised, it becomes impossible to meet the extra demand. Charterers are bidding for higher freight rates to see that their cargo is transported. In the limit the same cargo is transported in aggregate but at much higher freight rates.

The non-linear supply function is thus the result of a fixed supply in the short run, as it takes approximately two years for shipyards to respond to a higher demand for vessels by the owners. At some low level of freight rates the supply curve becomes perfectly elastic, as below that level some owners do not cover the average variable cost and go bust. But as long as they cover the average variable cost, it is worthwhile remaining in business in the short run. In the long run, though, owners must cover the average total cost, which includes fixed costs and the cost of debt service, in order to remain in business. As bankruptcies rise, the total fleet in the market diminishes and the minimum freight rate goes up.

In a perfectly competitive freight market the volume to be transported and the freight rates are those that equilibrate the demand for and supply of shipping services. In this framework, the equilibrium condition that demand must equal supply provides an equation for determining freight rates, while either the demand for shipping services or the supply of shipping services is used to determine the equilibrium cargo transported. In empirical work, the demand and supply are transformed into functions of capacity utilisation measured in millions of dead weight (dwt). In this framework, the equilibrium freight rate is a positive non-linear function of the fleet capacity utilisation and a negative linear function of bunker costs, while the demand for shipping services is used to determine the equilibrium cargo being transported.

The mechanics of this theory are illustrated in Figure 2.1. The supply of shipping services, labelled S, is plotted as a curve, which is relatively flat at low levels of demand and becomes very steep at high levels. The demand for shipping services, labelled D, is negatively sloped but very steep, implying that it is highly inelastic. At low levels of fleet utilisation equilibrium is attained at A, whereas at high levels of utilisation at C. Because of the curvature of the supply curve, an increase in demand, reflected as a parallel shift from D1 to D2, results in a new equilibrium at B (low level of fleet utilisation) or at D (high level of fleet utilisation). The increase in freight rates from A to B is small compared to that from C to D. Similarly, the equilibrium cargo transported is larger from A to B compared with that from C to D. In the limiting case of perfectly inelastic supply (a vertical segment), no extra cargo is transported; the entire increase in demand is met with higher freight rates.

The supply of shipping services is also a negative function of bunker costs. As the price of oil rises, owners are willing to reduce the supply of shipping services at the same freight rate. Higher bunker costs shift the supply curve to the left, as to transport any given cargo owners would demand higher freight rates to cover the dearer bunker costs. As a result, the equilibrium cargo is lower and the equilibrium freight rate is higher. This is illustrated in Figure 2.2. The supply curve shifts to the left from S_1 to S_2 in response to higher bunker costs. Initial equilibrium is at A and final equilibrium is at B, which implies higher freight rates and slightly smaller cargo, because the demand curve is very inelastic.

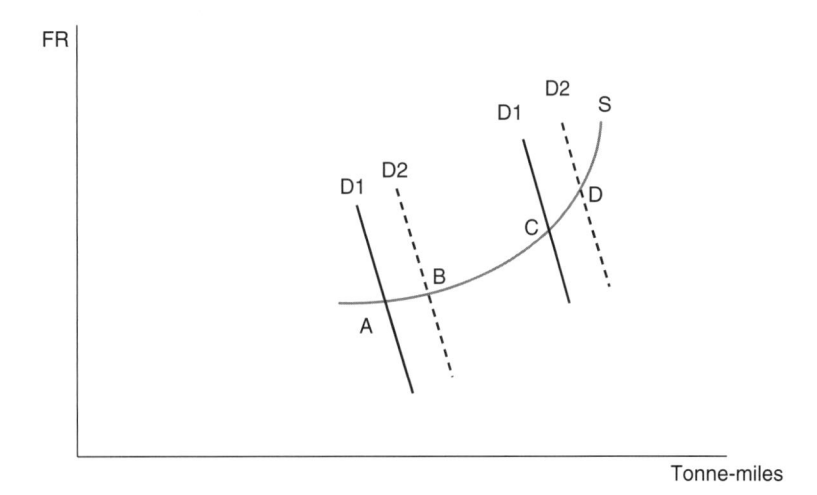

Figure 2.1 Demand and supply of shipping services

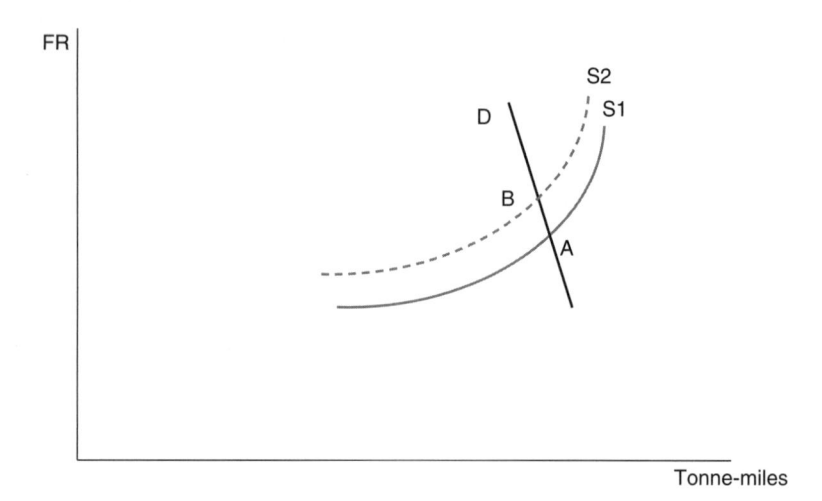

Figure 2.2 The impact of higher bunker costs

3 A CRITIQUE OF THE TRADITIONAL MODEL

The basic assumption of the traditional model (1) is that freight rates, at any point in time, clear a perfectly competitive market for shipping services. A market is said to be perfectly competitive if it satisfies the following conditions. First, the product is homogeneous. Second, there are a large number of charterers and owners. Third, both charterers and owners possess perfect information about the prevailing price and current bids, and they are profit maximisers in that they take advantage of every opportunity to increase profits. Fourth, there are no transaction costs. Fifth, there are no barriers to entry or exit in the shipping industry.

The condition of homogeneous product ensures the uniformity of charterers and owners. With respect to the owner the condition implies that the shipping services of one owner are indistinguishable from the shipping services of others. Charterers have no reason to prefer the shipping services of one owner to those of another. Hence, trademarks, patents, special brand labels and so on do not exist. The uniformity of charterers ensures that an owner will sell to the highest bidder. Customer relationships and rules of thumb, such as 'first-come-first-served', do not exist.

The condition of a large number of charterers and owners ensures that both are price takers. An owner can sell as much shipping services as s/he likes without affecting the prevailing market freight rates. Owners observe the market freight rates and adjust the shipping services sold so that these services maximise their profits without affecting the prevailing freight rates. A charterer can buy as much shipping services as s/he likes without affecting the prevailing market freight rates. The charterers are price takers in that they adjust the quantities purchased so that they maximise their profits without affecting the prevailing market freight rates. Hence, the large number of charterers and owners ensures that the impact of individual actions on the market freight rate is negligible. An owner can increase his shipping services considerably, but this would have an imperceptible movement along the market demand curve.

The perfect information condition ensures that there are no uninformed charterers or owners. Hence, owners cannot succeed in charging more than the prevailing market freight rates because charterers would resort to other owners. The owner that charges even slightly above the prevailing market freight rates would end up selling nothing. Similarly, charterers cannot get away by paying less than the prevailing market freight rates. A charterer that is offering a marginally lower freight rate than the market rate would see his entire cargo not being transported.

The condition of no transaction costs ensures that the freight rate agreed between a charterer and an owner cannot deviate from the prevailing market freight rate. The condition of no barriers to entry or exit ensures that in the long run the shipping industry does not earn supernormal profits, namely higher profits than other industries. Excessive profitability in the shipping market over a short period of time would attract new owners (or the existing ones would expand capacity) until normal profits are restored. Normal profit is the minimum profit that the owner must earn to remain in business. Normal profits include payment for risk bearing, for providing organisation and for managerial services.

Figure 2.3 portrays the mechanics of the entire market and a typical owner. The left-hand panel plots the market demand and supply functions in the short run. The right-hand panel portrays the supply side of an individual owner in the short run. In this framework, the market demand function is obtained as the sum of the demand functions of individual charterers.[3] Similarly, the market supply function is the sum of the supply functions of individual owners. The intersection of the market demand and supply functions provides the equilibrium point E with freight rate, P, and aggregate cargo, Q. Each owner in the market confronts a perfectly elastic demand curve at the market freight rate, P, which is given to each owner. Thus, the individual demand curve is D_i, which is horizontal and cuts the vertical axis at P, the market freight rate. A perfectly elastic demand curve means that the owner cannot charge a freight rate higher than the market rate, as charterers would not buy any services from her/him. The average and marginal revenue of each owner is equal

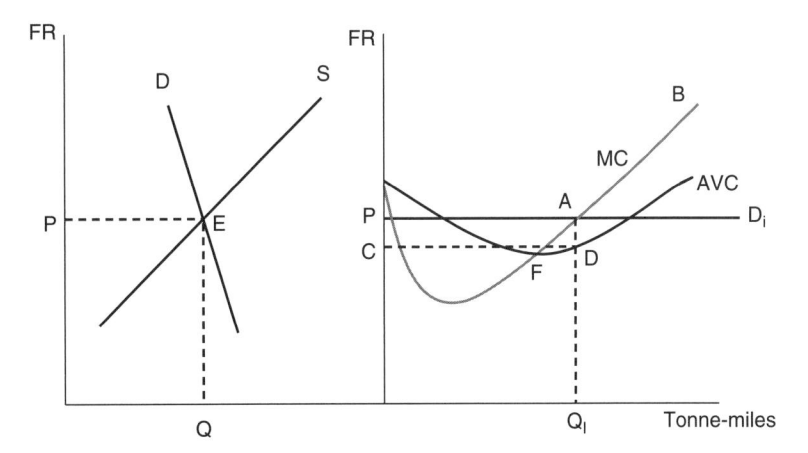

Figure 2.3 Market demand and supply and individual owner in the short run

to the price P.[4] The marginal cost curve, labelled MC in Figure 2.3, cuts the average variable cost (AVC) curve from below at the minimum.[5]

Each owner maximises profits and the condition for maximum is that s/he equates marginal revenue with marginal cost. As the price is equal to marginal and average revenue, the condition for maximum is usually stated that the price equals marginal cost. In Figure 2.3, equilibrium for the owner is achieved at A, the intersection of the individual demand curve, D_i, with the marginal cost curve. The owner supplies Q_i shipping services to the market at the market freight rate P. The sum of Q_i for all *i* owners is equal to Q in terms of the first panel in Figure 2.3. The supply function of the individual owner is the segment of the MC curve from the intersection of the marginal cost curve with the average variable cost upwards. Thus, the individual short-run supply function is the segment of the MC curve labelled FB. For freight rates below point F, the supply is zero, as the owner does not cover the average variable cost. In the short run, the owner has to cover just the average variable cost and not the average total cost, which includes fixed costs. In the case portrayed in Figure 2.3 the owner earns supernormal profits, which are measured by the rectangle CPAD. This is equal to the product of the cargo transported times the per unit profit. The latter is equal to the difference between the price and the average variable cost AD. These supernormal profits can only be earned in the short run. If these supernormal profits are earned by the typical owner, then more owners will enter the market in the long run until the marginal cost curve cuts the average cost curve at the given market freight rate. This is pictured in Figure 2.4. The MC curve cuts the AC curve at point A. In the long run, the owner has to cover not just the AVC, but also the average total cost, which includes fixed costs. Hence, the long-run supply curve of the owner is equal to the portion of the marginal cost above the market freight rate. This is now the line AB. The owner's supply below the market freight rate P is zero. At freight rates below P the owner goes bust.

But the characteristics of the freight market do not accord with those of perfect competition. In particular, the product is not homogeneous, the assumption of a very large number of buyers and sellers is not applicable, transaction costs are

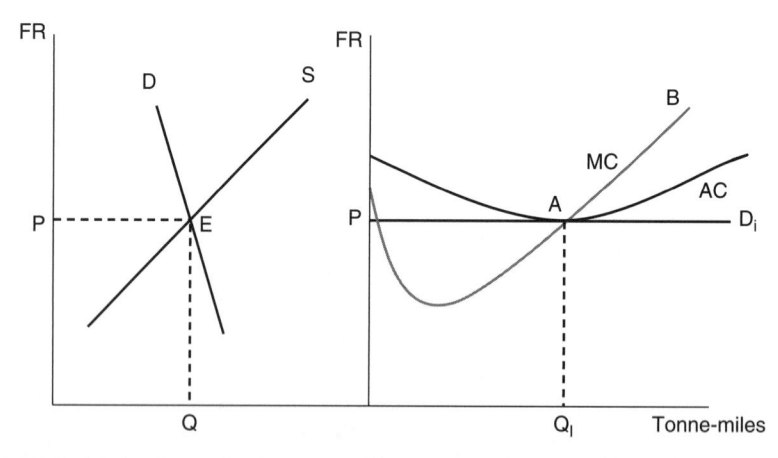

Figure 2.4 Market demand and supply and individual owner in the long run

not zero and there is no freedom of entry and exit. Although the product is seemingly homogeneous (the capacity to transport particular categories of products or commodities), the demand for shipping services is restricted by volume, time and route – a given cargo over a particular route that meets a well-specified time schedule. Although there are many ships in the market only a few are available to satisfy the given demand specifications. These characteristics violate the homogeneous product assumption, the condition of large number of buyers and sellers and the hypothesis of zero transaction costs. The latter should be interpreted as the penalties (legal or reputational) that a charterer would incur for waiting for a better deal (lower freight rates). Moreover, there are large barriers to entry both in capital and the expertise required to operate a fleet.

4 A GAME-THEORETIC APPROACH TO FREIGHT RATES

The basic assumption of the traditional model is that freight rates, at any point in time, clear a perfectly competitive market for shipping services. But freight rates are not prices determined in auction markets, where many owners, as assumed in a perfectly competitive environment, bid for the same cargo and the one with the lower bid wins the contract (Dutch auctions). Rather, the freight rate over a particular cargo is the outcome of a bargaining process that happens at the same time or approximately the same time in different places and where information about freight rates agreed is almost instantaneously available to all other participants. Thus, the agreed freight rates do not balance demand and supply in a particular place at a particular point in time, but expectations of overall demand and supply in a particular segment of the market or the entire market. Accordingly, freight rates are equilibrium rates in a bargaining game. The bargaining power is not uniform in the shipping industry, as some charterers may have stronger power than others; and similarly some owners have more power than others. The varying bargaining power among charterers can be described by a distribution (whether it is normal or not). There is a similar distribution for owners. Both distributions reflect the structural characteristics of the industry. The shape of each distribution may be invariant (or slowly changing) to external (industry or macro) factors, but the relative bargaining power of the median charterer and the median owner would be responsive to these factors. Some of these industry (or macro) factors may be random or simply unpredictable, such as wars, strikes and weather conditions, but economic conditions may be more predictable. It is the latter that is of interest here, as both players form rational expectations about economic conditions.

These expectations determine the relative bargaining strength of the two players in the bargaining game. In 'good' economic conditions the median owner has stronger bargaining power than the median charterer; in 'bad' economic conditions the situation is reversed. In good economic conditions the outcome of the bargaining game for each cargo trade pushes the freight rate marginally up, most of the time. This creates an upward trend of freight rates and leads reacting players to form expectations of rising future freight rates. Forward-looking agents do not need this evidence to form expectations of rising future freight rates. They can discount the implications of changes in current economic fundamentals on the future course of freight rates. This theory of freight rates is as useful to the two parties in the bargaining process (that is the owner and the charterer) as ornithology is to birds, to paraphrase a popular quote attributed to Richard Feynman.

This theory implies that both charterers and owners form expectations of future freight rates on the basis of the latest information. The current asset price reflects the implications of future economic fundamentals and discounted back to today of new information in current economic conditions. For example, as new information comes in (for example, China would not reflate its economy), both agents infer the implications for future demand for and supply of shipping services and compute the present value of the gains or losses and hence the equilibrium freight rates that would result from the new information. This is the pricing principle for all risky assets, such as equities, bonds and commodities. Thus freight rates and equity prices are determined by the same principles. Both are risky assets. A risky asset is different from a risky business. Shipping has always been a risky business. However, in the past ten years freight rates have also become a risky asset.

5 A FORMAL STATEMENT OF THE BARGAINING PROBLEM – THE CONTRACT OR BARGAINING CURVE

These ideas can be organised more formally with the help of Figure 2.5. Let X_2 be the payoff (benefit) of the owner and X_1 that of the charterer resulting from an agreement on the equilibrium freight rate of, say, $10,000. The payoff of each player can be measured in *cardinal* or *ordinal* utility. Cardinal utility is measured in monetary terms, whereas ordinal utility relies on the postulate of rationality, according to which a decision maker is able to rank outcomes in order of preference.[6] The decision maker possesses an ordinal utility measure if s/he does not need to assign numbers that represent (in arbitrary units, for example,

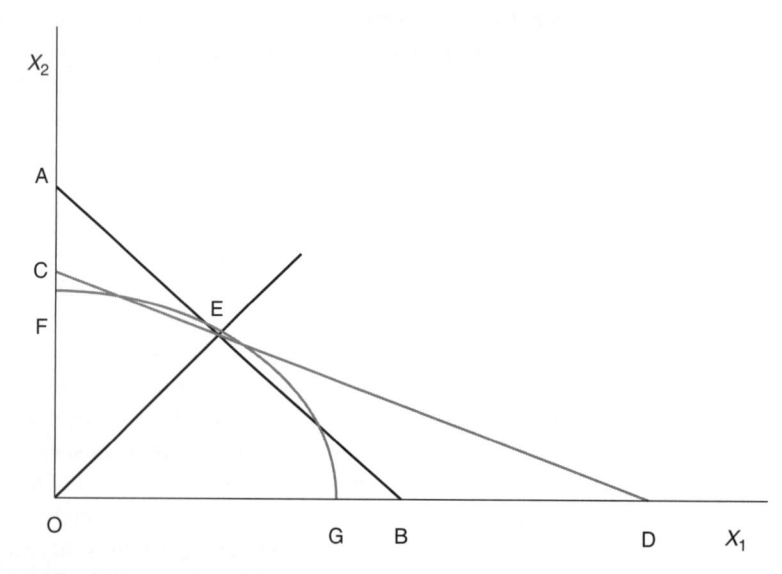

Figure 2.5 Contract or bargaining curves

monetary terms) the degree or amount of utility that s/he derives from alternative outcomes.[7]

The payoff to the owner consists of the (cardinal or ordinal) value that accrues to the business of the owner by coming to an agreement with the charterer over the contract. This is the value not simply of the freight rate but rather of all additional benefits to the business. For example, the contract might be to a port from which another contract is guaranteed; gaining one more contract might help in assuring banks that loans would continue to be performing; a successful bargaining might influence slightly future freight rates. In measuring the total benefits to the owner we also add the value of the equilibrium freight rate, $10,000. Similarly, the benefits to the charterer accrue from the sale of the goods at the destined port, which, most of the time, are of considerably higher value than the freight rate; avoiding the legal penalties of shipping the goods late or simply the reputation costs involved in arriving late. Such benefits to the charterer are net of the equilibrium freight rate, $10,000.

The owner and the charterer bargain over the known equilibrium freight rate. The feasible set of alternatives defines the *contract* or *bargaining curve*. Such curves may be straight lines, such AB or CD or the convex-curve FG in Figure 2.5. If the bargaining curve is AB, the gains of one player are exactly offset by the losses of the other player. In the bargaining curve CD, a particular loss of the owner is offset by a more than proportionate increase in the gain of the charterer. The bargaining curve is assumed to be continuous, increasing in value from O to A on the owner's utility scale, while decreasing in value on the charterer's utility scale from B to O.[8] If no agreement is reached, the alternative is disagreement, denoted as Point O in Figure 2.5, in which the payoffs of the two players are zero. This means that both players have an incentive to reach an agreement. If neither party dominates the negotiations, equilibrium is achieved at point E, which lies at the intersection of the 45^0 line with the bargaining curve, be it a straight-line or convex curve. This implies that the agreed freight rate is equal to the known equilibrium rate of $10,000. As the bargaining solution moves from E to A, the owner wins at the expense of the charterer. The agreed freight rate is equal to the equilibrium rate plus the deviation from the equilibrium rate.[9] On the other hand, as the bargaining solution moves from E to B the agreed freight rate is smaller than the equilibrium rate of $10,000.

6 THE BARGAINING UTILITY FUNCTION

The value of the feasible set to be chosen depends on the utility that each player attaches to the various payoffs and on the bargaining power s/he commands. Although the owner and the charterer bargain over a known equilibrium rate, the solution of the bargaining problem must be satisfactory to both; otherwise there is no agreement and point O is the outcome; both players are worse off. This implies that there exists a utility function that is an increasing function of the payoffs of the two players. Let U denote such a utility function as in equation (2.1) immediately below:[10]

$$U = g(X_1, X_2) \tag{2.1}$$

A larger payoff for either the owner or the charterer leads to higher utility (or satisfaction). The rationale is that for a given level of utility for one of the players, the other's utility becomes higher as her/his payoff increases. This is consistent with the concept of equilibrium: a player cannot become better off without the other becoming worse off.

A particular level of utility or satisfaction can be derived from different combinations of X_1 and X_2. For a given level of utility U^0 equation (2.1) becomes

$$U^0 = g(X_1, X_2) \qquad (2.2)$$

where U^0 is a constant. Since the utility function is continuous, (2.2) is satisfied by an infinite number of combinations of X_1 and X_2. Two such curves are plotted in Figure 2.6 for given U^0 and U^1, where $U^1 > U^0$. Such curves are called iso-utility curves, as all points on a curve give the same level of utility. Iso-utility curves correspond to higher and higher levels of satisfaction, as one moves in a north-easterly direction in Figure 2.6. For each level of U, equation (2.2) defines a different iso-utility curve. The family of all possible iso-utility curves, each one corresponding to a different utility level, is called the *utility map*.

Iso-utility curves are negatively sloped in the entire admissible set of combinations of X_1 and X_2. The slope of the tangent to a point on an iso-utility curve, such as point C in Figure 2.6, measures the rate at which X_1 must be substituted for X_2 (or X_2 for X_1) in order to maintain the same level of utility U^1. The negative of the slope is defined as the *rate of payoff substitution*[11] and is equal to the ratio of the two first-order partial derivatives. The requirement that iso-utility curves are negatively sloped throughout implies that each player must be compensated more as his payoff tends to zero in order for the utility of the bargainers to remain unchanged and for an agreement to be reached. Thus, as one moves from A to B along the iso-utility curve U^0 the owner is willing to give up a smaller payoff of his own for every unit increase in the payoff of the charterer. If the iso-utility is steep,

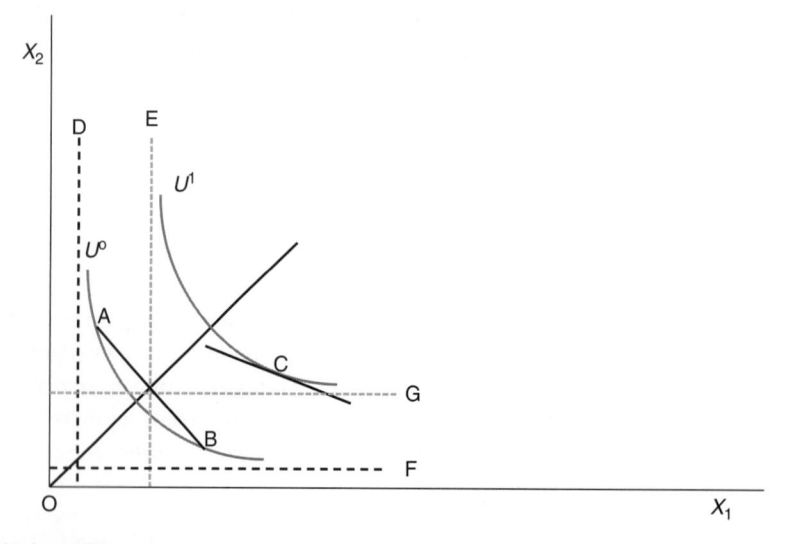

Figure 2.6 Iso-utility curves

the rate of payoff substitution is high. In this case the owner has a large payoff and the charterer a low payoff. The owner is willing to give up a large payoff of his own to let the charterer increase his payoff by one unit and complete the agreement. If the iso-utility curve is flat, the rate of payoff substitution is low. The owner is willing to give up only a small amount of his own payoff to satisfy the charterer by one extra unit.[12]

7 THE SOLUTION OF THE BARGAINING GAME

Although the owner and the charterer bargain over a known equilibrium rate, the solution of the bargaining problem must be acceptable to both. Acceptance means that neither player becomes worse off. Accordingly, the solution of the bargaining problem implies a move *along* the same iso-utility curve. As iso-utility curves that lie north-easterly imply higher utility, the solution of the bargaining game is to reach the highest iso-utility curve that is permissible by the contract or bargaining curve. This implies a tangency point between the contract or bargaining curve and the highest iso-utility curve.[13] But there is a second element that must be satisfied to achieve an optimum solution of the bargaining game. This is to take account of the relative bargaining power of each player in the negotiations and means that an extra restriction must be imposed on the admissible family of the utility function (2.1). A utility function that satisfies this restriction and enables the formulation of such a bargaining game has the following form

$$U = A \, X_1^{1-\beta} \, X_2^{\beta} \quad 0 < \beta < 1, \quad A > 0 \tag{2.3}$$

The bargaining power of the owner is measured by $1/\beta$. The higher the bargaining power of the owner, the steeper the iso-utility curves become with respect to the X_2 axis for a given value of X_1.[14]

Figure 2.7 plots two iso-utility curves for the same level of utility U^0 but for different levels of bargaining power for the owner. The one labelled W represents weak bargaining power for the owner and consequently strong power for the charterer. The other, labelled S, represents strong bargaining power for the owner. When $\beta = 0.5$ the two players have equal bargaining power. Although both iso-utility curves are drawn for the same level of utility, for each value of β the utility function (2.3) defines a different utility map. The steeper curve represents stronger bargaining power for the owner, as for each level of X_2 the payoff of the owner is larger than for weak bargaining power.[15]

Figure 2.8 illustrates the solution of the bargaining game for three different levels of bargaining power of the owner. The contract or bargaining curve is the straight line CD representing a larger payoff (cardinal or ordinal) for the charterer than the owner. Point G represents the solution of the bargaining game when the owner has strong bargaining power, whereas point B the solution with weak bargaining power. The solution at G implies a higher payoff for the owner than at B. Both points are 'acceptable' solutions as they imply a move *along* the highest iso-utility curve, which at the same time satisfies the contract or bargaining curve (that is, the feasible set). The bargaining implies that one of the players gives something

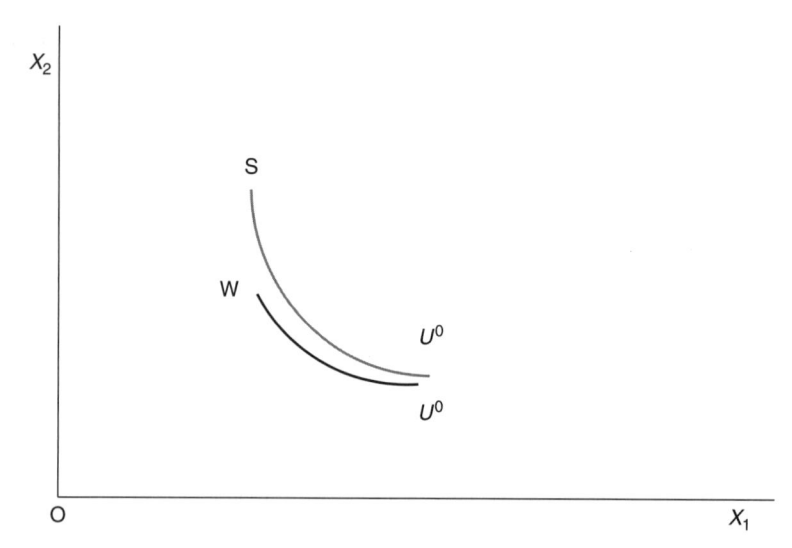

Figure 2.7 Iso-utility curves in strong and weak bargaining power

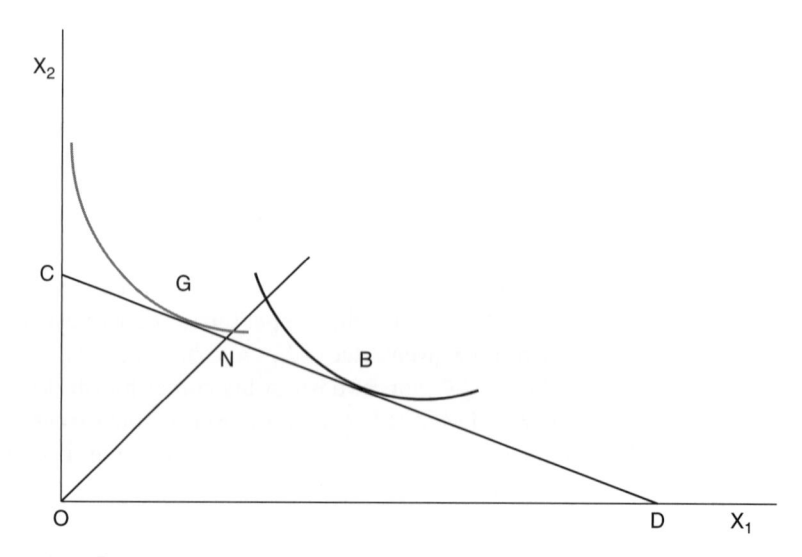

Figure 2.8 Freight rates in 'good' and 'bad' times

up for the benefit of the other, but the joint utility level of the bargaining game remains the same. Both solutions represent a tangency point between the contract curve and the highest iso-utility curve. If the charterer pushed for a solution that implies a larger payoff for her/him (a move in the south-easterly direction), the resulting solution of the game will lie on a lower iso-utility curve. Mutatis mutandis, if the owner pushed for a higher payoff for herself/himself (a move in the north-westerly direction from G), the resulting solution of the game will lie on a lower iso-utility curve. The solutions at B and G are equilibrium points because a move away from them implies that a player cannot become better off without the other becoming worse off in a way that makes the joint welfare worse.

Point N represents the solution when both players have equal bargaining power. This is simply the point of intersection between the 45° line and the contract curve. For $\beta = 0.5$ the utility function (2.3) defines another utility map (not shown in the graph). Point N, in addition to being the intersection point between the 45° line and the contract or bargaining curve, is also a tangency point between the highest iso-utility curve for $\beta = 0.5$ (not shown in the graph) and the contract curve.

The analysis so far is static, in that it is a snapshot of an individual transaction in the dry market; in other words, what happens at a point in time. But it is easy to see how this snapshot would develop through time and how it would be aggregated to explain market freight rates. For the latter there are particular routes that are monitored and rates reported are weighted averages. For the overall market, there are well-known indices (the Baltic Indices) that comprise several routes for each segment of the market. Hence, the real problem is how the freight rate of a particular cargo is determined and how it evolves through time. The analysis so far explains a snapshot, which can easily be extended to explain its dynamics. For example, assume that economic conditions are 'good' and the implied solution of the game is a point that lies north-westerly of point N, like G in Figure 2.8. Other bargaining games between a different owner and a different charterer would be subjected to the same conditions, giving rise not to exactly the same but to a similar solution such as G. Each bargaining game would be a stepping stone towards raising the equilibrium freight rate from $10,000 to, say, $11,000, thus shifting the contract or bargaining line CD in Figure 2.8 to the right for the next bargaining game between the same or another pair of owner and charterer. If the new players' perception about economic conditions continues to be 'good', the new agreed freight rate would be established north-westerly of point N in Figure 2.8, at a point such as G, but on a higher iso-utility curve, as the contract curve has shifted to the right. The higher iso-utility curve associated with the $11,000 contract curve implies that the payoff of owner i and charterer j are larger than the equilibrium at $10,000, thus giving rise to an upward trend in freight rates.

8 ECONOMIC CONDITIONS AND BARGAINING POWER

So far, the nature of the equilibrium (that is, the solution of the game) has been analysed by varying the bargaining power, but without explaining the causes of strong or weak bargaining. In this section we dwell on this issue. The bargaining power is not uniform in the shipping market. Some charterers may have more bargaining power than others. Similarly, some owners may have more power than others. For example, charterers are reluctant to place high-value cargoes on vessels owned by parties perceived to be weak and where the vessel may be arrested. Similarly, end-use charterers do not like chartering vessels where there is a chain of subcharterers. Owners who meet these criteria would have stronger bargaining power than those who do not. Owners with weak bargaining power will accept lower time charter rates from strong parties, such as the two large Japanese charterers, BHP and Cargill; or, in the tanker market, the large oil companies. This may result in lower income for the weak owners but greater certainty of income.

As a result, the bargaining power of each group (charterers and owners) should be viewed as a separate distribution. Each distribution would shift as macro factors change. Some of these macro factors may be random or simply unpredictable, such as wars, strikes or weather conditions. However, other factors may be predictable economic conditions. In 'good' or improving economic conditions it is likely that the median owner would have the upper hand in the negotiations and impose a positive deviation from the equilibrium or latest freight rate in the bargaining game. Such a solution is portrayed as G in Figure 2.8. In 'bad' or worsening economic conditions the median owner is likely to be in weaker bargaining power than the charterer. Such a solution is portrayed as B in Figure 2.8, as the deviation of the agreed freight rate from the latest or equilibrium rate is likely to be negative.

Economic conditions are shaped by expectations of the evolution of the determinants of future freight rates. Profit-maximising behaviour implies that the determinants of future freight rates depend on fleet capacity utilisation and a set of variables that affect the supply of shipping services, such as bunker costs and port congestion. When economic conditions are 'good' or improving, future freight rates are expected to rise. In 'bad' or worsening economic conditions, future freight rates are expected to fall.[16] The bargaining is taking place over the present value of the future freight rate.

We can thus summarise the determinants of the current (log) value of freight rate, fr, expressed in deviation from long run equilibrium, as

$$fr_t = b_1 E_t(fr_{t+1}) + b_2 E_t(cu_{t+1}) + b_3 E_t(z_{t+1}) \tag{2.7}$$

$$0 < b_1 = \frac{1}{1+r} < 1, \quad b_2, b_3 > 0$$

$$cu_t = q_t - k_t$$

E_t is the expectations operator with information available up to period t
cu = (log) fleet capacity utilisation or demand–supply balance in the freight market
q = (log) demand for shipping services
k = (log) net fleet, that is, operational fleet defined as past deliveries less scrapping less lay-ups and losses
z = a vector of variables affecting supply, such as bunker costs and port congestion.

According to Equation (2.7) the current freight rate depends positively on expectations, held today with information available up to now, of next period's freight rate, discounted back to today at the discount rate r, the fleet capacity utilisation and bunker costs.

Equation (2.7) can be solved forward to yield

$$fr_t = E_t b_1^j fr_{t+j} + E_t \left\{ b_2 \sum_{j=0}^{\infty} b_1^j cu_{t+j} + b_3 \sum_{j=0}^{\infty} b_1^j z_{t+j} \right\} \tag{2.8}$$

The first term on the right-hand side approaches zero as *t* tends to infinity. Thus (2.8) simplifies to

$$fr_t = E_t \left\{ b_2 \sum_{j=0}^{\infty} b_1^j cu_{t+j} + b_3 \sum_{j=0}^{\infty} b_1^j z_{t+j} \right\} \quad (2.9)$$

The forward solution suggests that current freight rates depend on expectations of the present value of the evolution of fleet capacity utilisation and bunker costs. If such expectations are positive then economic conditions are 'good' and the bargaining power is on the owner's side. In this case the bargaining solution would lead to equilibrium G in Figure 2.8. If these expectations are negative then economic conditions are 'bad' and the bargaining power is on the side of the charterer with equilibrium attained at B in Figure 2.8.

9 COMPARISON WITH THE CONVENTIONAL MODEL

It is instructive to compare the freight equation of the bargaining game with the traditional freight model of Tinbergen (1931, 1934) and Koopmans (1939) so that the similarities and differences can be evaluated. In the traditional model the supply of shipping services, Q^s, is a function positively related to the fleet, K, measured in tonnage, and the freight rate, F, but negatively related to the price of bunkers, PB. Thus,

$$Q^s = K^a PB^{-b} FR^{\gamma} \quad (2.10)$$

The demand for shipping services, Q, is perfectly inelastic with respect to the freight rate and it is exogenously given, determined outside the shipping model. The freight market is assumed to be perfectly competitive and equilibrium implies that

$$Q = Q^s \quad (2.11)$$

Denoting by lower case letters the logs of these variables and substituting (2.10) into (2.11) and solving for the freight rate gives

$$fr = \frac{1}{\gamma} q - \frac{a}{\gamma} k + \frac{\beta}{\gamma} pb \quad (2.12)$$

If the supply of shipping services is proportional to the fleet size (that is if $a = 1$), and Tinbergen (op. cit.) obtains a value of 0.94 in his empirical estimates, then making use of the fleet capacity utilisation definition, equation (2.12) can be rewritten as

$$fr = \frac{1}{\gamma} CU + \frac{\beta}{\gamma} pb \quad (2.13)$$

The similarities and differences between the traditional freight model and the game bargaining model are now apparent through a comparison of equation (2.13) and (2.9) and the assumption that the vector *z* consists of just one variable, the price of bunkers. Both models suggest that freight rates depend on the same variables, fleet capacity utilisation and bunker costs. But whereas the traditional model implies that current freight rates depend on the current values of these variables, in the game bargaining model freight rates depend on expectations of the future values of these variables. Hence, the fundamental difference between the two models stems from the perfectly competitive assumption in the traditional model. Freight rates move instantly to clear an exogenous demand for shipping services with supply that is fixed by the fleet and the price of bunkers. In the game bargaining model, the owner and the charterer bargain on the present value of the expected future freight rate with their bargaining power dependent on expectations of the present value of the future path of the fleet capacity utilisation and bunker costs.

Practitioners' long-term expectations are determined by strategy; medium-term expectations by economic policy; and short-term expectations by liquidity. Strategic decisions usually portray a steady state of the world without explaining the dynamic adjustment path to that state and the associated risks which may make it impossible to reach that steady state. For example, statements such as 'shipping would thrive, if every Chinese had the income to buy a car' describe such a steady state. The forecast is based on the logic that if China grows faster than the western world, it would eventually catch up with it, reaching the consumption levels and patterns that we now witness in the West. But such a statement does not explain how we can get there and therefore ascertain the risks. The impact of economic policy is felt after a year, but becomes very small after two or three years and dissipates to zero in five years. Financial conditions, such as liquidity and the availability of credit, have a big impact in the short run. All three factors (strategy, economic policy and liquidity) have an impact on expectations, but the question is how they can be combined in a consistent way.

Structural models of the macro-economy and of the dry market allow for a consistent interaction of all three forces of expectations. Structural models are dynamic and, for given initial conditions, they can be solved for the implicit steady state. But as at the same time such models are also dynamic, they inevitably describe the adjustment path to that steady state. Most structural dynamic macro models also capture the restraint to growth imposed by the lack of liquidity. Such models can be successfully employed to assess the impact on expectations. Thus, expectations about future freight rates are formed by projections of the demand for dry for the particular segment of the market (for example, Capes, Panamax, Supramax, Handysize), the supply of shipping services in that segment of the market and the resulting fleet capacity utilisation. Expectations of the demand for dry, in turn, are formed on the basis of the macro-environment of the major economies and the resultant pattern of world trade. Such expectations can be generated as the projections of two interacting models – a macro model of the world economy and a structural model of the dry market (see the Appendix for how such expectations are formed). Once the two interacting models have been solved simultaneously, expectations about future freight rates depend only on the truly exogenous variables of the two models (that is, the variables that cannot be explained by the two models) and the projections about economic policy. Of the two sets of variables economic policy is the more important, as there is likely to be consensus, at least in the direction of policy, between the expectations of the owner and the charterer. For example, both players can easily see the implications that 'if China were to reflate its economy, the demand for dry would increase and freight rates would recover'. Structural macro models enable players to form more complicated hypotheses about future freight rates. For example, 'as the commodities bubble has burst, China would manage to bring inflation under control, which would enable the policymakers to stimulate the economy'.

Therefore, expectations about the conduct of economic policy and, in particular, monetary policy in the major economies are the most important single factor that influences the bargaining power of the two players. This can be judged by observing current economic conditions, as players can deduce the optimal economic policy (that is, the optimisation of the policymakers' objective function subject to the way the instruments of economic policy affect the economy, namely subject to the macro model, see for details Chapter 5); and then deduce the impact of this policy change on the main macro variables and hence on the demand for dry; and finally through the dry cargo model the impact on future freight rates.

Expectations about the conduct of economic policy also help to resolve the puzzle of how to forecast the most widely accepted leading indicator of world economic activity: freight rates.[17] If freight rates are leading even stock prices in forecasting turning points in world economic activity, then it looks impossible to forecast freight rates. The current framework provides a solution to this puzzle by postulating that expectations of future freight rates can be formulated by forecasting the future economic policy of the major economies.

The presence of exogenous variables in the structural macro model is usually by necessity. Variables are treated as exogenous simply because they cannot be accurately predicted by other economic variables. For example, the price of oil is a difficult variable to forecast with any reasonable degree of accuracy. Accordingly, it is wise to treat it as an exogenous variable. This is usually not because economists cannot hypothesise the determinants of the oil price. Rather, it is the case that the impact of the explanatory variables on oil varies through time because other factors, such as conflicts or collusion among the oil cartel members, are unpredictable. Therefore, it is better to treat expectations of truly exogenous variables as conditioning the main scenario around the systematic forecast of economic policy. In this context, expectations of exogenous variables pose a risk around the main scenario of economic policy. Risk scenarios can be computed by simulating the model under a different oil path.

Although this issue is dealt with more thoroughly in Chapter 6, it might be useful to apply this methodology to equation (2.9) for the sake of completeness. It is assumed that demand for shipping services, q, depends on aggregate demand in the world economy, x. Accordingly, expectations of demand for shipping services is a function of the aggregate demand in the world economy

$$E_t q_{t+1} = E_t x_{t+1} \qquad (2.14)$$

For reasons that will become clear in Chapter 3, expectations of the evolution of the fleet for a short horizon (for example, one year) relevant to the bargaining of the owner and the charterer reflect past expectations of demand for shipping services. Assuming for simplicity that all vessel deliveries took place in the last year we have:

$$E_t k_{t+1} = E_{t-1} q_t \tag{2.15}$$

As it will be shown in Chapter 5, expectations of aggregate demand in the economy depend on expectations of real interest rates, r. Thus,

$$E_t x_{t+1} = E_t (r_{t+1}) \tag{2.16}$$

The fleet capacity utilisation rate is an endogenous variable, whereas the price of bunkers is exogenous. Hence, expectations for the fleet capacity utilisation can be derived by substituting (2.14)–(2.16) into the definition of fleet capacity utilisation

$$E_t c u_{t+1} = E_t q_{t+1} - E_t k_{t+1} = E_t r_{t+1} - E_{t-1} r_t \tag{2.17}$$

By using (2.17) the freight rate is determined by

$$fr_t = b_2 \sum_{j=0}^{\infty} b_1^j [E_t r_{t+j} - E_{t-1} r_{t+j-1}] + b_3 E_t \sum_{j=0}^{\infty} b_1^j z_{t+j} \tag{2.18}$$

Therefore, the freight rate in the bargaining game is shaped by the present value of expectations of future real interest rates, past expectations of real interest rates and current expectations of future bunker costs and possibly other exogenous variables, such as port congestion.

10 THE TIME CHARTER (OR PERIOD) MARKET

The time charter market is peripheral to the rest of the shipping model, as a time charter contract reduces the demand and the supply in the spot market by the same amount, leaving the equilibrium spot rate unchanged. Although the time charter market does not help to understand the interactions of the various shipping markets, it plays a crucial role in the real world where the management of cash flows is a key decision in shipping. We deal with this issue in Part III of the book. In this section we simply examine the theoretical underpinnings of the relation between spot and time charter rates. It is shown here that the time charter rate is equal to a weighted average of expected future spot rates and risk premiums with declining weights so that the near future carries more importance than the distant one. The horizon of expectations is dictated by the duration of the time charter contract. Thus, the one-year time charter rate is equal to the expected spot rate for next year and a risk premium. Spot rates, on the other hand, are determined through the interaction of the four main markets, namely freight, newbuilding, secondhand and scrap markets. In our framework both spot and time charter rates are assets as their prices today are determined by discounting future economic fundamentals. This is an innovation, as in conventional models spot rates are not asset prices. They are prices clearing a perfectly competitive market.

In a time charter hire contract, control of the ship is passed from the shipowner to the charterer for a fixed rent for the duration of the contract. Variable costs, such as bunker costs, port charges, canal dues and so on, are borne by the charterer, while fixed costs, such as wages, insurance and debt service, remain with the owner.[18] The rent is payable per period of time (for example monthly) and is fixed for the duration, τ, of the contract. The length of time charter contracts varies from one year to ten years or more. The fixed rent payable to the owner is known as the time charter (or period) rate for the duration of the contract and is denoted by H^τ. The time charter will normally vary with the duration of the contract. The relation between spot (or voyage) freight rates and time charter (or period) rates, which differ only in duration, is called the *term structure* of freight rates. Zannetos (1966) was the first to recognise the similarity between the term structure of freight rates and the *term structure* of interest rates, which has made it possible to borrow the well-developed theories of the term structure of interest rates and apply them to freight rates. The borrowed methodology means that scholars in maritime economics have a common approach to the term structure of freight rates and we see no reason to challenge it. But the borrowed methodology has brought with it the same controversies that pervade to the term structure of interest rates, mainly whether or not markets are efficient.

In the term structure of interest rates riskless profit arbitrage is forcing equality of 'returns' of different maturities.[19] Accordingly, in equilibrium long rates are a weighted average of expected short rates with or without a risk premium. In the freight market riskless profit arbitrage ensures the equality of the discounted present value of two alternative strategies. The first involves a direct time charter contract of a fixed duration, while the second rolling short-term contracts in the spot market that cover the duration of the time charter plus a risk premium, rp. The risk premium is frequently called the *term premium* as the risk pertains to investment alternatives that differ only in respect of the term (that is, the period) to maturity. This arbitrage relation equates the discounted present value of the time charter rates with the discounted present value of the rolling spot rates and is expressed as

$$\sum_{i=0}^{p-1} \frac{H^\tau}{(1+r)^i} = \sum_{i=0}^{p-1} \frac{E_t FR_{t+i}}{(1+r)^i} + rp_t \qquad (2.22a)$$

The left-hand side of equation (2.22) is the discounted present value of a time charter contract of duration, τ, paying a fixed rate, H^τ, per period (for example, monthly, expressed as \$/day) at the discount rate, r. To compare like with like, the time charter contract of duration τ should be expressed in the duration units (for example, months) of the spot contract, s. Thus, if a single voyage takes two months (that is, $s=2$), then a time charter contract of 36 months involves 18 rolling spot contracts. The number of rolling spot contracts, p, in a time charter contract of duration τ is equal to $p=\tau/s$. The right-hand side of equation (2.22a) is the discounted present value of the expected rolling spot rates for the p periods at the same discount rate, r.

It can easily be seen how the equilibrium condition in equation (2.22a) would be satisfied. Consider the case where the discounted present value of the time charter contract is higher than the discounted present value of the rolling spot contracts. This implies that the time charter rate is higher than spot rates and the demand for time charter contracts would increase, while the demand for spot contracts would decline. As a result, the price of time charter contracts would increase and their return would decline, while the price of spot contracts would decrease and their return would rise. The freight market consists of contrarian owners and herd (or noise) owners. Contrarian owners are risk-neutral (or less risk-averse than herd owners) and act as speculators in the freight market, hiring ships from more risk-averse owners at the time charter rate and profiting from spot rates in a strong market. In a weak market contrarian owners reverse positions.

Equation (2.22a) can be solved for H^{τ}. The solution is straightforward but tedious (see the Appendix in Kavussanos and Alizadeh, 2002a).

$$H_t^{\tau} = k\sum_{i=0}^{p-1}\delta^i (E_t FR_{t+i} + rp_t), \quad k = \frac{1-\delta}{1-\delta^p}, \quad \delta = \frac{1}{1+r} < 1 \quad (2.23)$$

Therefore, the arbitrage relation implies that the time charter rate is equal to a weighted average of expected spot rates and time-varying risk premiums with declining weights into the future as $\delta < 1$. This means that more importance is attached to nearby spot rates and risk premiums than in the distant future.

A number of competing theories of the term structure of interest rates give rise to the formulation that the long rate in equilibrium is equal to a weighted average of short rates with or without a risk premium and these are discussed next.

In the *pure expectations hypothesis* (PEH) owners are risk neutral and therefore they are concerned only with the expected return and not the risk of the return. According to the PEH, freight rates are expected to move so as to equalise the expected holding period yield of all chartering alternatives that differ with respect to maturity. In this case the risk (or term) premium is zero in (2.23). Therefore, the time charter rate is simply a weighted average of expected future spot rates.

In the *expectations hypothesis* (EH) the risk premium is simply a constant irrespective of the term to maturity.

In the *preferred habitat hypothesis* (PHH) each owner is trying to match his assets with liabilities and therefore for some owners the risk premium is positive, while for others it is negative. For the market as a whole the risk premium can be positive or negative.

In the *liquidity preference hypothesis* (LPH) the risk premium is a constant, but varies with the term to maturity.

In the *time-varying risk hypothesis* (TVRH) the risk premium varies over time and varies also for time charter contracts of different maturities.

In the Capital Asset Pricing Model (CAPM) the spot contracts can be viewed as a portfolio. Thus, an owner can choose one-, two- or five-year time charter contracts. Accordingly, his portfolio consists of either 12, 24 or 60 rolling spot contracts. His own portfolio, H^{τ} is related to the market portfolio H^m by the CAPM equation

$$E_t H_{t+1}^{\tau} = FR_t + \beta_t^{\tau}[E_t H_{t+1}^m - FR_t], \quad \beta_t^{\tau} = \text{cov}[H_{t+1}^{\tau}, H_{t+1}^m]/\sigma_{m,t+1}^2 \quad (2.24)$$

The second term in the above equation captures the risk premium. This consists of the excess return of the market portfolio over the spot freight rate and the covariance of the owner's time charter contract portfolio with the market portfolio. None of these factors are expected to be constant through time and therefore the CAPM provides a theory for a time varying risk premium in equation (2.23).

Whereas the risk premium in the term structure of interest rates is expected to be positive, Zannetos (1966) was the first to recognise that the risk premium in the freight market is expected to be negative. Zannetos (op. cit.) argues that an owner operating in the spot market takes various risks and therefore should be compensated for risk taking. Therefore, the spread between time charter and spot rates should be negative implying a negative risk premium (or a discount) in equation (2.23). According to Zannetos (op. cit.), the risks of operating in the spot market include the unemployment risk, namely the risk that a vessel may not be fully employed, including ballast (relocation) risk to secure another contract in another port, high administrative costs and brokerage fees to ensure chartering in the spot market and high capital costs, as banks would charge a higher interest rate on loans for a ship operating in the spot than the time charter market. Therefore, risk-averse owners would prefer to operate in the time charter market and would demand a compensation for operating in the spot market.

Although most risk factors point to a negative risk premium the liquidity risk, in general, suggests a positive risk premium. Time charter contracts should be viewed as illiquid financial instruments and therefore they should demand compensation. Veenstra (1999a) takes this view and argues that in a time charter contract there is a loss of liquidity for the owner as it is difficult and costly to terminate an existing contract. However, this is not correct as both the owner and the charterer are exposed to liquidity risk and they take opposite positions. Therefore, without reference to their respective risk appetite it is impossible to give a verdict of whether the risk premium in the market should be either positive or negative. From a purely theoretical point of view, Adland and Cullinane (2005) assess all factors that determine the risk premium and conclude that on balance it should be negative.

Empirical evidence also supports the hypothesis of a negative risk premium (discount). The unconditional variance in the bulk market decreases with maturity. Thus the spot market exhibits the highest standard deviation of earnings in the dry and wet markets. The standard deviation decreases as the maturity moves from one to five years. Glen, Owen and van der Meer (1981) and Kavussanos and Alizadeh (2002a) find empirical support for a negative risk premium.

The game-theoretic framework of this chapter goes one step further by arguing that spot rates depend on economic conditions, which affect the relative bargaining power between the owner and the charterer. According to this framework, the risk premium also depends on market conditions. In good markets the discount increases and in bad markets the discount decreases. Adland and Cullinane (2005) heuristically argue along the same lines, although they do not offer a framework to justify their claims. They do add, though, a very useful point that the negative risk premium depends also on the duration of the time charter contract. The finding that the negative risk premium depends on market conditions is also consistent with empirical studies of the owners' risk preferences.

Lorange and Norman (1973) find that the risk appetite of Norwegian owners depends on market conditions. In good markets owners are risk lovers or risk neutral, but in bad market conditions owners become risk averse. Eckbo (1977) confirms the findings of Lorange and Norman (1973), concluding that owners have decreasing absolute risk aversion with respect to market conditions.

The arbitrage relation (2.22a) can be expressed in an alternative way that relates the time charter rate to the spot freight market determinants. The profit function[20] of operating in the spot market is

$$\pi_t = k(FR_t^{1+\gamma}/PB_t^{\gamma}) - OC_t \qquad (2.25)$$

where FR is the freight rate, PB is the price of bunkers and OC is operating cost. The expected profit function of operating in the one-year time charter market is

$$E_t\pi_{t+1}^H = H_t - E_tOC_{t+1} \qquad (2.26)$$

where π^H is the profit of operating in the time charter market and H is the one-year time charter rate. By leading equation (2.25) by one period and taking expectations we have

$$E_t\pi_{t+1} = k \cdot E_t(FR_{t+1}^{1+\gamma}/PB_{t+1}^{\gamma}) - E_tOC_{t+1} \qquad (2.27)$$

The arbitrage relation implies that in equilibrium the expected profit from operating in the one-year time charter market must be equal to the profit accruing by operating for one year in the spot market. Therefore, equating the right-hand side of equations (2.25) and (2.26) and solving for H, the alternative arbitrage condition is obtained.

$$H_t = k \cdot E_t(FR_{t+1}^{1+\gamma}/PB_{t+1}^{\gamma}) \qquad (2.28)$$

Taking logs on both sides of equation (2.28) we obtain the following equation

$$\ln H_t = (1+\gamma)E_tFR_{t+1} - \gamma \cdot E_tPB_{t+1} + \lambda \qquad (2.22b)$$

Therefore, this alternative arbitrage relation implies that the one-year time charter this year is equal to the expected spot rate next year less the expected bunker costs next year plus a risk premium, λ.

Despite the importance of the time charter market in the real world, it is not an essential component of the interaction of shipping markets. All four markets (namely spot freight, newbuilding, secondhand and scrap) interact with each other and are simultaneously determined. The time charter market does not interact with any one of the four markets; rather, it is determined as a residual from the other four markets. In other words, the entire system of five markets (including the time charter market) is post-recursive. The time charter market is determined once the other four markets are solved simultaneously. The post-recursive nature of the time charter market follows from the simple logic that a time charter contract leads to a reduction in demand and supply in the spot market by the same amount, leaving the equilibrium spot rate unaffected.

A negative risk premium (discount) in time charters means that the yield curve (a curve that relates the spot rate with TC rates at a point in time) has a negative slope in normal conditions. This is the opposite of the yield curve of interest rates. A negatively sloped yield curve for freight rates means that it pays to be in the spot market under normal conditions. A very steep yield curve implies that spot freight rates are abnormally high and are likely to fall in the future. On the other hand, an inverted yield curve implies that spot rates are abnormally low and they are likely to increase in the future. The shape of the yield plays a pivotal role in cash flows management (see Part III).

11 SUMMARY AND CONCLUSIONS

This chapter has reviewed the traditional model of freight rates, based on Tinbergen (1931,1934) and Koopmans (1939). This model has remained unchallenged and has provided the theoretical foundations of all subsequent empirical work. It has been argued in this chapter that the model violates the basic assumptions of perfect competition. As a result, an alternative model has been suggested, based on a game-theoretic approach, to shed light on the issue of how freight rates are determined in the real world. The chapter is, by definition, theoretical and therefore may be unappealing to the layman. A big effort has been made, though, to keep it simple by splitting the analysis between the main text, which hopefully can be read without any prior knowledge, and footnotes and Appendices, which deal with more technical factors and important theoretical details that are necessary for theoretical completeness, but which sometimes require some mathematical knowledge.

The main thesis is that freight rates are determined in a bargaining game between the owner and the charterer over a given equilibrium rate, which is usually the latest one available to both players. Both players have an incentive to reach an agreement and consequently they are willing to compromise, as the alternative implies a worse state for each one of them. The feasible set of alternatives along which the two players can bargain is defined in terms of the payoff (cost/benefit) that each one derives from reaching an agreement. Such payoff may be measurable in monetary terms or may be subjective to the value that each player assigns to it. There is a bargaining because a higher payoff for one player implies a smaller payoff for the other. This is reflected in the contract or bargaining curve being negatively sloped. The gain for one player may be equal in absolute value to the loss of the other (that is, a zero-sum game); but this need not be the case, if the benefits extend to more than the freight rate and each player assigns different subjective value to the benefits.

At the same time, both players assign a particular utility to the possible payoffs. In doing so, they also pay attention to the benefits of the other player, so that an agreement can be reached. This involves choices between alternative outcomes, which we assume that rational players can make. This enables the formulation of a utility function that describes the joint utility that is achieved by reaching an agreement. Such utility is, by definition, a function of the payoff of both players. This does not follow from an altruistic approach, but from the need to reach an agreement with the other party. The preferences about alternative payoffs define a utility map that consists of a family of iso-utility curves, each one of them for a different level of joint utility. Each iso-utility curve is not simply negatively sloped throughout, which means that one loses while the other one gains. Rather, each iso-utility curve is concave, meaning that when a player starts the bargaining with a high payoff he is willing to give up a lot of his own payoff for a unit increase in the payoff of the other player so that an agreement can be reached. But as his payoff decreases he is willing to sacrifice less and less of his own benefit to satisfy the other player.

The solution of the bargaining game is obtained through the optimisation of the utility function subject to the feasible set, defined by the contract or the bargaining curve. This implies a tangency point between the highest iso-utility curve and the contract or bargaining curve. The compromise involves a movement along the same (highest) iso-utility curve, which ensures that the utility of the bargaining game remains unchanged. One of the players sacrifices his own payoff in the knowledge that this provides the same level of utility to both players viewed jointly together. Which player would be willing to compromise so that an agreement can be reached depends on the relative bargaining of each player. While many factors can affect this relative bargaining power, economic conditions are perhaps the single most important factor in most situations and relevant in explaining shipping cycles. In this context, in 'good' or improving economic conditions, the owner has stronger bargaining power than the charterer and this implies a positive deviation of the agreed freight rate from the equilibrium or latest rate. In 'bad' or worsening economic conditions the charterer has stronger bargaining power and the deviation of the agreed freight rate from the equilibrium or latest rate is negative.

Whether economic conditions are 'good' or 'bad' depends on expectations about future freight rates and purely exogenous variables, such as bunker costs and port congestion. Accordingly, in the game bargaining model current freight rates depend on expectations of future freight rates and bunker costs. But future freight rates depend on future conditions in the freight market. Profit maximising behaviour entails that conditions in the freight market depend on the fleet capacity utilisation rate (the demand–supply balance) in the freight market and exogenous variables, such as bunker costs. Accordingly, expected future freight rates are a function of expected fleet capacity utilisation and bunker costs.

This enables a comparison between the traditional model and the game bargaining model of freight rates. Both models suggest that freight rates depend on the same variables, fleet capacity utilisation and bunker costs. But whereas the traditional model implies that current freight rates depend on the current values of these variables, the game bargaining model implies that freight rates depend on expectations of the future values of these variables. Therefore, freight rates in the game bargaining model are viewed as asset prices. The characteristic of asset prices is their current value, which is equal to the discounted present value of future economic fundamentals. The fundamental difference between the two models stems from the perfectly competitive assumption in the traditional model. Freight rates move instantly to clear an exogenous demand for shipping services with supply that is fixed by the fleet and the price of bunkers. In the game bargaining model, the owner and the charterer bargain on the present value of expected future freight rates with their bargaining power dependent on expectations of the present value of the future path of the fleet capacity utilisation and bunker costs.

But such expectations are difficult to form as freight rates are a leading indicator, as stock prices are, of future economic activity. A leading indicator is, by definition, difficult to predict and many owners find it hard to know what to watch when attempting to predict the direction of future freight rates. Luckily, economic policy is the variable that should be monitored to deduce where freight rates are heading. Expectations of the fleet capacity utilisation involve separate expectations for demand and supply. The latter reflects past expectations of demand, while the former depends on macroeconomic developments. Thus the expected fleet capacity utilisation is equal to current and past expectations of macroeconomic developments that depend on economic policy, namely interest rates and government budget deficits.

Such expectations can be formulated in a consistent manner through the use of structural dynamic models for the world economy and the shipping market (dry, wet or containers), such as the K-model (see Arestis and Karakitsos, 2004 and 2010). It is shown in this chapter that expectations of future freight rates depend only on the conduct of economic policy in the major economies and on purely exogenous variables. Economic policy, though, is systematic and therefore it is predictable with a reasonable margin of error. This forms the basis of the main scenario, while assumptions about exogenous variables pose the risk around it. Structural models are of great help here because they can be used to assess quantitatively the impact of risk. Therefore economic policy is the only factor that can be of help in the guidance of where freight rates are heading and this is the variable upon which to formulate expectations of future freight rates.

For practitioners strategy affects long-term expectations, economic policy medium-term expectations and liquidity short-term expectations. Such taxonomy may be useful and may provide some insight into how to look at the impact of future developments. However, this taxonomy has the drawback that such expectations may be inconsistent with each other. Structural models have the advantage that they make expectations based on short-, medium- and long-term factors. The latter are consistent with each other as they combine liquidity effects and describe the dynamic adjustment path to a steady state that is adapting as initial conditions change.

This chapter, finally, analyses the time charter market. This market is peripheral to the rest of the shipping model, as a time charter contract reduces the supply and the demand in the spot market by the same amount, leaving the equilibrium spot rate unaffected. Spot rates, on the other hand, are determined through the interaction of the four main markets, namely freight, newbuilding, secondhand and scrap markets. The time charter rate is equal to a weighted average of expected future spot rates and risk premiums with declining weights; so that the near future carries more importance than the distant one. The horizon of expectations is dictated by the duration of the time charter contract. Thus, the one-year time charter rate is equal to the expected spot rate for next year and a risk premium.

APPENDIX: HOW EXPECTATIONS OF FREIGHT RATES ARE GENERATED

A structural model of the dry market purports to explain freight rates, the demand for dry, the supply of dry, prices of secondhand and new vessels for the entire market and its various segments. These five variables are interdependent and therefore are determined by solving simultaneously the five equations. These five variables for each segment of the market and for the entire market are called endogenous variables, because they are explained by the model of the simultaneous equations. The endogenous variables in the dry market model depend also on a set of macro-variables for the major economies and global variables, such as world trade. Let Y denote an $m \times n$ matrix of m endogenous variables, each up to n lags. This includes the five variables for each segment and the entire dry market at time t; Z is a similar $k \times n$ matrix for the k-macro and global endogenous variables; Π is a $p \times n$ matrix of the instruments of fiscal and monetary policy in each major economy and X are the truly exogenous variables, which are not determined by either model, but for which values exist up to time t and assumptions are made for the period $(t+1,\ldots,T)$; $e(1)$ and $e(2)$ are white noise vectors for the endogenous variables in the dry and macro models respectively. These represent the cumulative influence of all non-systematic factors, whose impact cancels out on average. Then the system of the two models can be represented as:

$$Y = G[Z] + e(1) \qquad (2.\text{A}1)$$

$$Z = F[\Pi, X] + e(2) \qquad (2.\text{A}2)$$

The policymakers in each country minimise a loss function J of the deviations of the targets of economic policy from their bliss or desired values subject to the way their instruments affect the targets, which is portrayed by the economic model F. The minimisation of the loss function J subject to F is carried out by choosing Π over the horizon $(t+1,\ldots,T)$. This optimisation results in the optimal policy in this time period. This is formally presented as:

$$\min_{\Pi} J(Z, \Pi) | F(Z, \Pi, X) = 0 \qquad (2.\text{A}3)$$

The optimisation yields the trajectory of the Π-vector (Π^*) over the period $(t+1,\ldots,T)$, while assumptions of X complete the information set that is needed to form expectations with information at time t, denoted by E_t, for all endogenous variables in the system Y and Z. Thus,

$$E_t(Y_j) = f(\Pi^*, X) + u_j \quad \text{for } (j = t+1,\ldots,T) \qquad (2.\text{A}4)$$

This means that future economic policy and truly exogenous variables are only required in forecasting future freight rates.

3 THE SHIPYARD, SCRAP AND SECONDHAND MARKETS

EXECUTIVE SUMMARY

The supply of shipping services by owners in the freight market is interlinked with the demand for vessels in the shipyard industry. The variable that drives the supply of shipping services and the demand for vessels is the demand for cargo by charterers. An increase in demand spurs owners to increase the supply of shipping services by increasing the average fleet speed; by putting back to the market laid-up vessels; and by placing orders for new vessels or buying ships in the secondhand market. Although a single owner has the option of buying a ship in the secondhand market, for the industry as a whole there is no such option, as transactions in the secondhand market involve a change of ownership, but no change in the aggregate stock of net fleet. At the industry level, the total supply of shipping services is equal to the stock of vessels that have been produced by shipyards, less the obsolete fleet that has been scrapped. The interrelationship of the supply of shipping services and the demand for new vessels is intertemporal. In the short run, the stock of fleet is fixed and owners can meet the extra demand for cargo by increasing the fleet capacity utilisation (fewer laid-ups or fast steaming). In the medium to long term, the stock of fleet is a choice variable. Owners choose the optimal fleet level, so that they can maximise profits in the long run. This gives rise to the demand for new vessels, which is inversely related to the price of new ships in the long run. But in the short run, the demand for vessels is positively related to the price – owners buy ships when prices increase, motivated by making capital gains. This is the typical pattern of the demand for an asset – demand increases in the short run in response to rising prices, but falls in the long run.

The supply of new vessels depends on the available shipyard capacity and the gestation lags to turn an order into a ready floating vessel. In traditional models NB prices are a mark-up on costs. But the volatility of NB prices is too high to be explained by the low volatility of costs. A mark-up approach is thus not an appropriate method to explain the formation of NB prices. Instead, this chapter puts forward an alternative framework, where shipyards adjust the profit margin to what the owners are prepared to pay. This is reflected in the secondhand prices. This approach unifies the theory of freight rates and vessel prices according to which both are asset prices.

The shipyard market determines the equilibrium level of NB prices and the gross (before scrapping) stock of fleet through the forces of a downward-sloping demand curve and an upward-sloping supply curve. In this chapter the demand for vessels is viewed as a dynamic optimisation problem of capacity expansion. The optimisation determines jointly the target or desired stock of fleet (that is, the demand for new ships) over an appropriately long investment horizon (which usually covers the length of a shipping cycle) and the desired pace of vessel deliveries (that is, the investment flows). The dynamic adjustment path to the target (optimum) stock of fleet depend on the owner's expectations of the evolution of the fleet capacity utilisation rate (that is, the demand for shipping services and the fleet) and freight rates over the investment horizon. It is proved in the Appendix that these expectations are formed by how owners expect policymakers (that is, central banks and governments) to react to current and future economic conditions. The methodology developed in this chapter for the demand for fleet is applicable to both newbuilding and secondhand ships, as it deals with a single owner's decision problem of the optimal fleet.

This chapter is organised as follows. The next section deals with the single owner's decision problem on the optimal fleet and the optimal speed. This framework enables the study of the dynamic adjustment of NB prices in response to exogenous shocks, such as the demand for vessels. Section 3 explains the aggregation of these decisions that lead to the demand for vessels at the industry level. Section 4 examines how the owner's investment decision on the fleet is affected under conditions of uncertainty. Uncertainty leads owners to slow the pace of adjustment to the optimal fleet level (wait for a while) and even reduce the optimal stock of fleet. Section 5 offers a Case Study for a single owner with conditions as those that prevailed in September 2012. In other words, we apply the methodology developed in Section 2 for an owner in possession of information available up to September 2012 and analyse his conditional decision on the optimal fleet under conditions of uncertainty. Section 6 puts the model developed in this chapter for ship prices in perspective with respect to the literature.

Section 7 derives the supply of vessels at the shipyard and industry level and investigates the influence of supply on the dynamic adjustment of prices. It is shown that shipyards increase the price volatility through their delivery lag. Section 8 analyses the determination of the aggregate (net) fleet by analysing the scrap market.

As the model requires knowledge of advanced mathematics the reader can ignore the Appendix and concentrate on the main text, where every effort is being made to explain the optimisation process and the logic of the underlying mathematical formulae. The Appendix provides in detail the mathematical treatment of the optimisation problem.

1 THE SUPPLY OF SHIPPING SERVICES IN THE SHORT RUN

1.1 INTRODUCTION

The supply of shipping services can be distinguished as *short-run* and *long-run* depending on whether the fleet is fixed or variable. The *short-run* supply is obtained for a given stock of fleet, while in the *long-run* supply the stock of fleet is allowed to vary. In this section we derive the *short-run* supply function of shipping services from profit maximisation. We begin with a general framework that can handle both the short and the long run.

An owner supplies shipping services according to the level of demand. These shipping services (the volume of cargo offered) depend on the stock of fleet and the speed at which it is operated. Let Q denote the supply of shipping services, measured in tonne-miles; K the stock of fleet, measured in million DWT or CGT; and S the average speed of the fleet, measured in miles per hour. Then, the supply of shipping services by a single owner is

$$Q_t = F(K_t, S_t) \tag{3.1}$$

Equation (3.1) can be considered as a production function in which the inputs K and S are used to produce cargo (output), Q. The function F is assumed to be continuous and to have first- and second-order partial derivatives, which are also continuous. Each partial derivative is also a function of K and S. The first-order derivatives can be interpreted as the marginal product of fleet (the contribution of one additional vessel to the volume of cargo offered) and the marginal product of the average speed (the contribution of one additional mile of speed to the volume of cargo offered by the owner). Although the marginal product of fleet is equal to the tonnage capacity of the vessel, its contribution to the supply of cargo depends on the speed at which the vessel is operated. Therefore, the marginal product of fleet is also a function of K and S.

In supplying shipping services (cargo), the owner can partly substitute some fleet with a higher average speed. Under the assumptions of equation (3.1) the owner can provide a given amount of tonne-miles through an infinite combination of K and S. Thus, for each level of Q, equation (3.1) defines a negatively sloped curve in K and S. Figure 3.1 plots two such curves corresponding to Q^0 and Q^1 where $Q^1 > Q^0$. These curves are called isoquants ('equal quantities'), as they give combinations of K and S that produce the same quantity of shipping services.

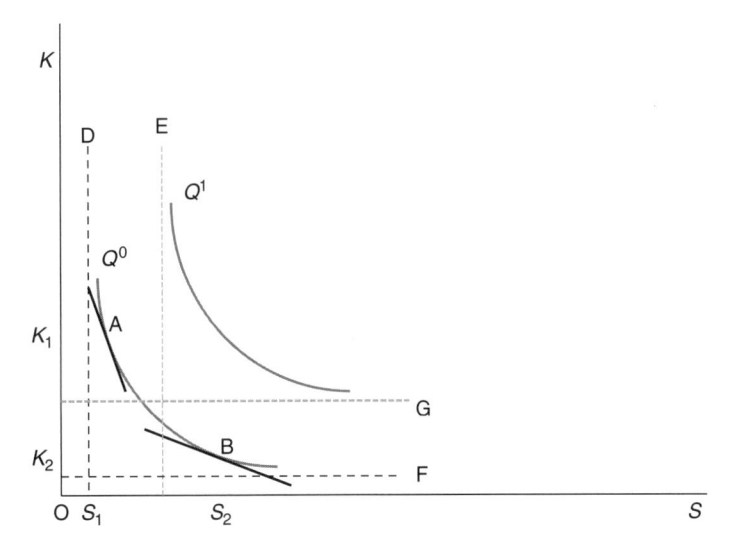

Figure 3.1 Isoquants

Because K can be substituted for S, each isoquant is throughout negatively sloped. This implies that as the fleet is reduced by one vessel, the owner can still supply the same cargo, by increasing the average speed of the fleet. Thus, at point A, slow steaming implies an average speed S_1. This speed requires a large fleet K_1 to produce Q^0 tonne-miles of cargo. At point B, a smaller fleet K_2 requires higher speed, S_2 to produce the same tonne-miles Q^0.

Moreover, each isoquant is concave. The concavity assumption implies that as the fleet is progressively reduced by one vessel at a time, the owner would need to increase the average speed of the fleet proportionately more for each vessel reduction in the fleet to supply the same cargo. The concavity assumption implies that the elasticity of substitution[1] between fleet and average speed is not uniform along an isoquant. At low levels of speed the elasticity is high and at high levels of speed the elasticity is low. Thus, the elasticity at A is high, whereas at B it is low.

Clearly, there is a limit beyond which the owner cannot offset the reduction of the fleet by one vessel by increasing the average speed of the fleet, if for no other reason, because it would be technologically infeasible. The substitution works the other way round too. If bunker costs rise steeply and the owner chooses slow steaming to reduce these costs, the fleet would have to expand to maintain the same amount of shipping services. But again there is a limit as to how much slow steaming is possible. This suggests that each isoquant converges to an asymptotic, a line parallel to each axis. The distance of each asymptotic from each axis depends on technological factors.

1.2 OPTIMAL CAPACITY UTILISATION AND OPTIMAL SPEED IN THE SHORT RUN

At any point in time, t, the fleet is a quasi-fixed factor of production and therefore the owner can choose the fleet capacity utilisation (that is, actively trading fleet less laid-up vessels) and the average speed to supply shipping services. Therefore, the supply of shipping services in the *short run* is an increasing function of both capacity utilisation and average speed. The owner can increase the supply in the *short run* by either increasing the speed or making higher utilisation of the existing fleet, until full capacity is reached.

The optimum speed, along with the optimum capacity utilisation, can be computed as follows. Let c denote the fuel consumption required to cover a particular distance, d, for example, 10 tons per 100 miles. Then, c is a function of speed, S, with positive first and second derivatives.

$$c = f(S) \quad f' > 0, \quad f'' > 0 \qquad (3.2)$$

The function f is defined by engine technology. Any technological improvement in engine design makes the engine more fuel-efficient (that is, less fuel is required to cover a particular distance at a particular vessel speed). For any given engine technology, the higher the speed, the greater the fuel consumption. This implies that the first- and second-order derivatives of f are positive.

The total fuel consumption, C, of cargo Q is equal to the distance, d, measured in miles, times c, that is,

$$C = c \cdot d \qquad (3.3)$$

The fuel bill, FB, is then defined as the product of fuel consumption, C, times the price of fuel, namely bunker costs, PB, measured in dollars per ton. Thus,

$$FB = C \cdot PB \qquad (3.4)$$

Accordingly, the owner in deciding the *short-run* supply of shipping services, Q, faces the budget constraint:

$$CT = PB \cdot C + r \cdot K + b \qquad (3.5)$$

In equation (3.5), CT is the total cost for purchasing and operating the fleet, K; r is a weighted average of the interest rate charged on loans to finance the fleet, K, and the opportunity cost of the owner's equity; and b is the operating expenses of the fleet. For convenience, b is treated as fixed, as for example the salaries of the crew required to operate a given fleet can be treated as given at any point in time.

By substituting (3.3) and (3.2) into (3.5) and solving for K the budget constraint is obtained:

$$K = \frac{1}{r}[CT - b - PB \cdot d \cdot f(S)] \qquad (3.6)$$

For each level of total cost, such as TC^1, equation (3.5) or (3.6) defines combinations of fleet and average speed along which the total cost is unchanged. For this reason, equation (3.5) or (3.6) is called the 'isocost' curve, namely equal cost. Figure 3.2 plots one isocost curve for total cost, TC^1. For a higher total cost, TC^2,

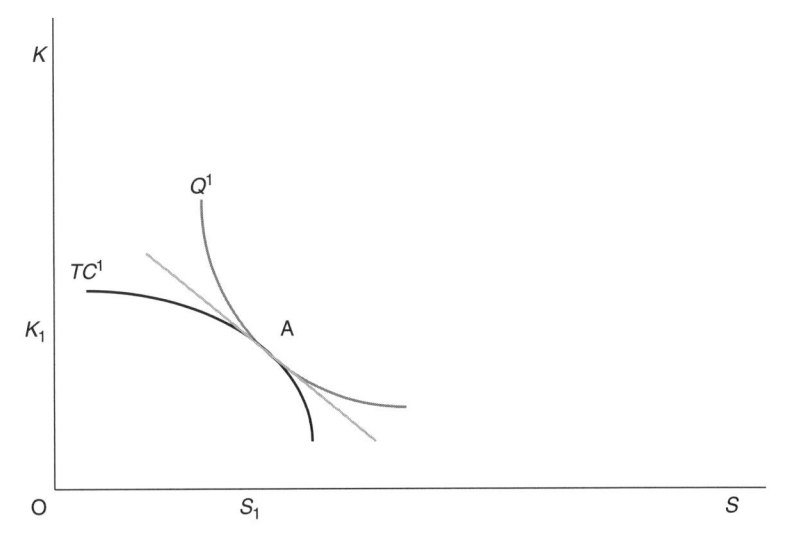

Figure 3.2 Optimum speed and capacity utilisation

equation (3.6) defines another isocost curve (not shown in the graph), which is parallel to TC^1, but lies further away from the origin.

The slope of the isocost curve is negative as all components of the first derivative are positive:

$$\frac{dK}{dS} = -\frac{1}{r} \cdot PB \cdot d \cdot f'(S) < 0 \qquad (3.7)$$

The second derivative is also negative:

$$\frac{d^2 K}{dS^2} = -\frac{1}{r} \cdot PB \cdot d \cdot f''(S) < 0 \qquad (3.8)$$

Box 3.1   Owners and central banks

The owner must take decisions in the same manner as a central bank. A central bank decides about the level of interest rates now on the basis of its forecast for inflation and the output gap two years ahead. As the two-year forecast changes through time the central bank adjusts its interest rates.

In a similar fashion the owner must decide on the fleet today on the forecast of the demand for shipping services, freight rates, vessel prices and the user cost of capital two years ahead. As the forecast changes the owner makes additions to the fleet or hedges the excess fleet.

Box 3.2   Optimal fleet

The optimal fleet capacity expansion problem can be split into two stages: where we want to go (the target or optimal fleet) and how fast to get there (how to adjust from the current actual fleet to the target, that is, the optimal path).

The target fleet is a function of expected demand and the expected freight rate per unit of the user cost of capital. The latter includes the price of the ship, the cost of capital, the depreciation rate of the ship, the economies of scale resulting from the additional vessel to the fleet and the capital gains when the ship is sold in the secondhand market.

The demand for cargo is more important than relative prices (freight rates relative to the user cost of capital) when the elasticity of substitution between fleet and average speed is less than one.

On each point of the optimal path the cost of one extra vessel (that is, the vessel price plus the cost of adjustment) is equal to the present value of the discounted future stream of marginal profits throughout the lifetime of the vessel.

This implies that the isocost curve is convex.

The owner's decision on the optimal fleet capacity utilisation and optimal average speed can be viewed as either a maximisation or minimisation problem. The owner maximises the supply of shipping services (output) subject to a given cost outlay, TC^1. Alternatively, the optimisation problem is defined as the minimisation of cost subject to the constraint of producing a given supply of shipping services Q^1. The optimisation yields the optimum average speed and the optimum fleet capacity utilisation. Both the maximisation and the minimisation process give the same result. In the maximisation problem, the optimum is obtained at the point of tangency between the highest isoquant for a given isocost curve.[2] In the minimisation problem, the optimum is obtained at the point of tangency between the lowest isocost curve, for a given isoquant. In terms of Figure 3.2, the optimum is defined at point A with optimum average speed S_1 and the optimum fleet capacity utilisation, K_1. In the maximisation problem point A represents the combination of K and S such that output reaches its maximum for a given cost, TC^1. In the minimisation problem point A represents the combination of K and S such that the cost is at a minimum for given output, Q^1.

At the optimum, the ratio of the marginal product of fleet to the marginal product of speed is equal to the ratio of their prices. The optimality condition can alternatively be expressed as follows: the owner would expand output (shipping services) up to the point where the ratio of the marginal product of fleet to its reward, r, is equal to the ratio of the marginal product of speed to its price. Each ratio is equal to λ, which measures the contribution to output of the last dollar spent upon each input.

If bunker costs increase, the isocost curve would be tilted crossing the K-axis at a higher point. The new optimum would be at the tangency of the new isocost curve and the highest isoquant. At the new optimum the owner would choose a lower average speed at a slightly higher fleet capacity utilisation.

2 OPTIMAL FLEET AND OPTIMAL SPEED IN THE LONG RUN

In the short run, the stock of fleet is fixed and the supply of shipping services is a function of the fleet capacity utilisation (fleet less laid-up vessels) and average speed. But in the long run, the stock of fleet can increase to meet the demand for cargo. The *short-run* supply of shipping services is analysed in the previous section. In the Appendix we offer a rigorous analysis of the dynamic optimisation problem for a single owner who chooses the optimal stock of fleet and the optimal average speed at which to operate the fleet to maximise the present value of the entire future stream of net cash flows in the lifetime of the vessel.

The stock of fleet is the total tonnage capacity of ships, expressed in dwt or cgt. At any point in time, this tonnage is available to an owner for offering shipping services. Gross investment is the extra tonnage capacity that is added to the stock of fleet in a period of time, say a year. Gross investment consists of the net addition to the fleet and replacement investment, new ships that are intended to replace obsolete ones that have been sold for scrap. Thus, at any point in time the stock of fleet is the accumulation of net investment (gross investment less depreciation).

Some of the complications with the theory of (gross) investment arise from the fact that while the demand for a product is a *flow* demand, that is, demand per unit of time, the demand for vessels is usually expressed as a *stock* demand with the implicit assumption that there is a fixed relation between the stock of fleet and the flow of services derived from it. However, as was pointed out by Haavelmo (1960) the demand for investment cannot simply be derived from the demand for fleet without any further assumptions. The demand for the stock of fleet can lead to any rate of investment from almost zero to infinity depending on the extra assumptions on the costs the owner faces in adjusting the stock of fleet and the delivery lags of the shipyard industry in supplying new vessels. From this viewpoint the demand for investment can be split into two stages: First, what determines the target or desired (the *optimal*, or *steady state*) level of the fleet stock, K^*. Second, how does the owner adjust from its *actual*, K, to its *optimal* capital stock, K^*? In other words, the investment decision can be thought of as where we want to go and how fast we want to get there. The optimisation process yields the optimal fleet size and the optimal dynamic path (that is, the desired pace of deliveries by owners).

In deriving an optimal fleet strategy we take into account that the supply of shipping services (volume of cargo offered) depends partly on the elasticity of substitution between fleet capacity and average fleet speed. An owner can meet a higher demand for shipping services by increasing, within limits, the average speed of a given fleet.

In the long run, the owner can choose the fleet accumulation process and the speed to maximise profits over a planning horizon $(1, \dots, T)$. In this optimisation process there are only two choice variables (the so-called decision variables or controls), namely the adjustment path to an optimal stock of fleet and the average speed. The optimisation determines jointly the optimal stock of fleet, the optimal adjustment path to this level of fleet and the optimal speed.

The optimal average speed is obtained at the point where the *current* marginal product of speed (the contribution of an additional mile of speed to the volume of cargo offered) is equal to the *current* value of the real rate of bunker costs, adjusted for technology factors (see equation (A.7) in the Appendix). The optimality condition for the speed involves a comparison of *current* values of the marginal product of speed with real bunker costs and therefore does not require expectations of these variables in the future. In every period, the owner observes the real bunker costs (the bunker price and the freight rate) and decides on the optimal speed. In the next period he can choose a different speed if either the freight rate or the price of bunkers have changed. Therefore, the choice of the optimal speed involves a static (a snapshot) optimisation problem rather than a dynamic one (evolving through time).

It is shown in the Appendix (see equation (A.20)) that the desired or target level of the fleet depends on economic fundamentals, namely the demand for cargo, Q; relative prices, which are defined as the freight rate, F, per unit of the user cost of capital, UC; the ship's technological improvement, A, and the elasticity of substitution between fleet and average speed, σ, in providing shipping services. The elasticity of substitution reflects the ability of the ship for slow and fast steaming. The user cost of capital includes the price of the vessel adjusted for the cost of capital and depreciation, the economies of scale resulting from one additional vessel in the fleet and the expected capital gains by selling the ship in the secondhand market.

> **Box 3.3 NB prices**
>
> In line with freight rates, NB prices are determined as if shipping was a stock exchange. NB prices are asset prices.
>
> In the short run, the owners' demand for vessels increases as vessel prices rise. Investors buy an asset when they expect prices to increase for capital gains. But at some point in the long term they sell the asset because it is very expensive (they expect capital losses). This contrasting behaviour of the demand for vessels in response to vessel prices in the long and the short term is typical in asset demand functions.
>
> We find empirical support for the positive and negative impact of prices on the demand for vessels, which provides justification to the claim that vessel prices are asset prices.

The desired or target fleet rises as the demand for cargo increases, relative prices (namely, freight rates relative to the user cost of capital) rise and the ship's technology improves. Higher freight rates for a given user cost of capital increase the demand for vessels. Similarly, an increase in the user cost of capital, for given freight rates, decreases the demand for vessels. As the user cost of capital is a composite variable, movements in the constituent components may offset each other, leaving the user cost of capital unchanged. For example, higher interest rates or faster depreciation (when a secondhand ship is considered) raises the price of the vessel and hence the user cost of capital. But such increases may be offset by expected gains in the secondhand market or larger economies of scale leaving the user cost of capital unchanged.

In the demand function for vessels, the demand for cargo has an elasticity of 1, whereas relative prices have an elasticity of σ. Relative prices (freight rates per unit of the user cost of capital) are more important than the demand for cargo, only when $\sigma > 1$. When $\sigma = 1$ relative prices are equally important as demand; for $\sigma < 1$ the demand for cargo is more important than relative prices; in the limiting case in which $\sigma = 0$, only the demand for cargo matters.[3]

It is shown in the Appendix that the demand for vessels by owners is a decreasing function of vessel prices in the long run. Thus, higher vessel prices lead to a drop in the demand for vessels, other things being equal. But in the short run, the impact of vessel prices is the other way around. The owners' demand for vessels increases as vessel prices rise. This contrasting behaviour of the demand for vessels in response to vessel prices in the long and the short term is typical in asset demand functions. Investors buy an asset when they expect prices to increase for capital gains. But at some point in the long term they sell the asset because it is very expensive (they expect capital losses).

The condition for the optimal fleet implies that the owner must take decisions in the same manner as a central bank. A central bank decides about the level of interest rates now on the basis of its forecast for inflation and the output gap two years ahead. As the two-year forecast changes through time the central bank adjusts its interest rates. In a similar fashion the owner must decide on the fleet today on the forecast of the demand for shipping services, freight rates, vessel prices and the user cost of capital two years ahead. As the forecast changes the owner makes additions to the fleet or hedges the excess fleet.

At any point on the optimal investment path, the cost of one extra vessel (that is, the cost of capital) is equal to the ship price and the costs of adjusting the fleet to the optimal level, mainly the cost of a fast or slow delivery schedule and the cost of operating the fleet at a loss for a period of time. Adjustment costs incur as the owner tries to expand the fleet rapidly. If the owner tries to reach an optimum fleet size of ten ships in a short period of time, then he would be facing increasing costs (a higher price per ship). Usually, for a big order an owner would get a discount, but only if the desired schedule of deliveries meets the shipyard capacity. If the owner demands a faster delivery than the optimal response of the yard, then the cost per ship would increase. Adjustment costs should also include the cost of operating the fleet at a loss for a short period of time. This would be typical if the owner is a contrarian investor, buying ships early in the cycle in anticipation of an increase in the demand for shipping services in the future. In this framework, the costs of adjusting the fleet depend positively on the amount of investment, I. Costs are increasing with investment at an accelerating pace, the higher the investment (that is, the faster the fleet adjustment), the bigger the costs. But the costs of adjustment would also depend on the stock of fleet, K. A big fleet implies economies of scale (lower management fees and lower labour costs) and higher bargaining power with shipyards. The bigger the owner's fleet is, the larger the shipyard discount for a given order and hence the lower the costs of adjustment. Thus, the costs of adjustment are a decreasing function of K.

The optimal fleet is obtained at the point where the cost of one extra vessel (that is, the vessel price plus the cost of adjustment, for example, the cost of operating the vessel at a loss for a short period of time) is equal to the present value of the discounted future stream of marginal profits throughout the lifetime of the vessel. The marginal profit consists of the marginal revenue of fleet (which is equal to the marginal product of fleet times the freight rate) and the economies of scale resulting from one additional vessel in the stock of fleet. The discount factor in the calculation of the present value consists not only of the interest rate, r, to finance the capital expenditure, but also of the depreciation rate, δ, since the stock of fleet diminishes in value at that rate (see equation (A.12) or (A.32) in the Appendix).

The optimality condition for the fleet implies that freight rates and the fleet capacity utilisation rate (that is, demand as a proportion of fleet capacity) are the two key economic variables that affect the decision on the size of the optimal fleet. The elasticity of freight rates is equal to one, whereas the elasticity of the fleet capacity utilisation rate is $1/\sigma$ (see Equation (A.32) in the Appendix). Hence for $\sigma < 1$, the fleet capacity utilisation is more important than freight rates and the opposite is true for $\sigma > 1$. When $\sigma = 1$, both are equally important.

In contrast to the optimal average speed, which does not require the formation of expectations, the optimal fleet depends on a comparison of observed variables (the price of ship, the cost of fleet adjustment, economies of scale and the elasticity of substitution between fleet and average speed) with expectations of variables in the future. The owner must form expectations of the demand of shipping services per unit of fleet capacity (that is, the fleet capacity utilisation rate) and freight rates throughout the lifetime of the vessel. It can be shown (see equation (A.31) in the Appendix) that expectations of these two key shipping variables can be deduced from expectations about future economic policy. These forward-looking expectations imply that the owner calculates the impact on shipping variables from the policy reaction of central banks and governments to current and future economic conditions. The interaction of a structural model for shipping with a structural macro model provides a practical and consistent way for forming such forward-looking expectations. For example, if growth in the world economy falters, then the owner anticipates loosening of fiscal and monetary policies in the major economic regions of China, the US, Europe and Japan. This would stimulate world trade and hence the demand for shipping services in time. But the impact on freight rates and vessel prices would be felt the moment the policy reaction is announced by tilting the balance of bargaining power towards the owners. This is a fundamental principle of asset prices. They adjust instantly in theory, very fast in the real world, to news on economic fundamentals. This adjustment entails a short run 'overshooting' of the new long run equilibrium of freight rates and vessel prices.

The dynamic adjustment of NB prices, which involves overshooting, can be studied with the help of Figure A5 in the Appendix, reproduced here for convenience as Figure 3.3. In the long run, the demand and supply functions in the shipyard industry have the normal slopes; the demand curve is negatively sloped and the supply positively sloped (see equations (A.42) and (A.43) in the Appendix). The intersection of the demand curve labelled D_1 and the supply curve, S, defines the initial long-run equilibrium, namely when the adjustment is complete. At point A the long-run equilibrium stock of fleet for an owner is K_1 and equilibrium vessel price is P_1. Now assume that demand in the world economy is boosted by expansionary economic policy in some or all of the major economic regions and that owners form expectations rationally using a structural model, such as the K-model. The owners can compute through the K-model the impact of the expansionary economic policy on the demand for ships and hence the new long-run equilibrium demand curve. This new demand curve is labelled as D_2 in Figure 3.3. The new long-run equilibrium would be established at point C, after a long period of time. But the dynamic adjustment to this new long-run equilibrium is subject to restrictions (see the Appendix for a rigorous analysis of these restrictions on the dynamic adjustment path to long run equilibrium).

Shipyards tend to stabilise the market because they increase the supply of vessels when the price is higher than equilibrium and reduce the supply when the price is lower than equilibrium. Owners, on the other hand, tend to destabilise the market because in the short run they buy more ships when vessel prices increase and fewer ships when vessel prices decline. As result of this contrasting behaviour of shipyards and owners, the shipyard market has only one unique stable path to long-run equilibrium C. This is indicated by the negatively sloped line labelled EE, in Figure 3.3. Along this path the market is stable because the demand for vessels by owners decreases as prices are falling, whereas shipyards increase the supply because the prices along EE are higher than equilibrium. The stability of the system requires that vessel prices increase sharply in the short run so that they can decline along the stable dynamic adjustment path EE. In terms of Figure 3.3 the vessel price would jump from P_1 to P^* before the fleet had any time to adjust, that is, when it is still at K_1. But from P^* vessel prices would decline, gradually reaching the new long-run equilibrium at point C. Therefore, the optimal adjustment path of NB vessel prices involves an overshooting of the new long-run equilibrium ($P^* > P_2$). The overshooting is the result of the perverse effect of vessel prices on the demand for vessels in the short run under rational expectations: rising vessel prices induce owners to buy more ships (the demand for vessels is a positive function of prices).

In the real world, the overshooting is not instantaneous but involves a short period of time in which vessel prices increase rapidly. This usually occurs when there is a 'squeeze' between the demand for and supply of vessels.[4] After a long period of stagnating freight rates and asset prices, demand for shipping services rises significantly and abruptly, whereas the previous low demand for ships has led shipyards to scrap capacity. The higher demand for shipping services and the resulting excess demand for ships results in significantly higher freight rates and vessel prices. This is the period that involves overshooting of freight rates and vessel prices. Once shipyards expand capacity to meet the excess demand for ships, P^* is reached and subsequently there are capital losses for the owners, as vessel prices decline to the new long-run equilibrium at C. Figure 3.A6 in the Appendix depicts the adjustment path of prices following a sharp increase in the demand for vessels associated with easy fiscal policy in China and the US.

The time path of gross investment follows the adjustment path of vessel prices. Investment rises on impact to its maximum pace and then declines towards its higher new long-run equilibrium (see Figure 3.A7). The investment path also involves an instantaneous overshooting of the new long-run equilibrium. Despite the overshooting of gross investment, the fleet adjusts gradually and monotonically from the initial equilibrium to the new equilibrium with higher fleet (see Figure 3.A8).

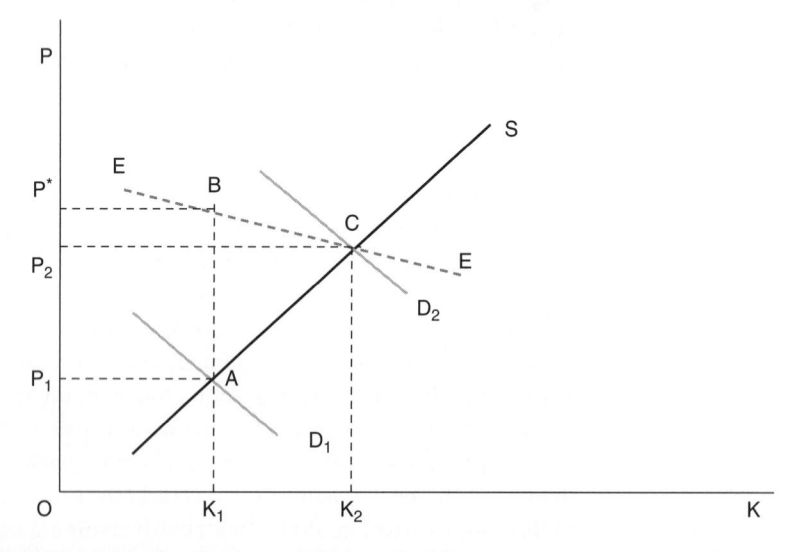

Figure 3.3 Dynamic adjustment of NB vessel prices

The model can describe accurately what happened in the boom of 2003–08. The surge in prices in 2006–08 reflects overshooting of the long-run equilibrium. The level of NB prices in 2008 corresponds to P* in Figure 3.3. NB prices would have declined to their new long-run equilibrium, even if there were no global recession. The recession shifted the demand curve back, thus accentuating the fall of NB prices. The post-2008 decline of NB prices implies losses for the late-coming owners. The demand for shipping services was persistently high in 2003–06, surprising both owners and shipyards. Contrarian owners bought ships early in the cycle and benefitted from the boom in freight rates and vessel prices. Owners and shipyards that were backward looking responded with a lag. The more risk averse the owners and shipyards, the greater the likelihood that they entered too late in the cycle. The stylised facts of the dry market show that the majority of owners and shipyards entered very late (from 2008 onwards) after demand waned.

3 THE AGGREGATE FLEET

The analysis so far derived the individual owner's demand for ships. The derivation of the demand for vessels at the industry level involves a simple aggregation over the individual demand curves. Thus, if K_i is the individual owner's demand for ships and there are n owners in the market then the aggregate demand for ships, denoted by K, is simply the sum the individual owner's demand functions.

$$K_t = \sum_{i=1}^{n} K_i \qquad (3.11)$$

Similarly, the aggregate supply of shipping services is equal to the sum of the individual owner's supply functions. Thus, if Q_i^s is the individual owner's supply function and Q is the aggregate supply function, then

$$Q_t^s = \sum_{i=1}^{n} Q_i^s \qquad (3.12)$$

These aggregation rules enable us to move quickly from the individual level to the aggregate level. We suppress the extra cumbersome notation since it is obvious when the analysis applies to an individual owner and when to the aggregate level.

4 INVESTMENT UNDER UNCERTAINTY

Under conditions of certainty and perfect foresight, owners would expand capacity until the benefits of one extra vessel are equal to its cost. The cost of capital includes the vessel price along with the cost of adjusting the fleet. The marginal benefits are equal to the present value of the sum of the expected marginal revenue of one extra vessel derived from the entire lifetime of the vessel along with the economies of scale resulting from the additional vessel in the fleet (for example, lower management fees, bigger bargaining power when negotiating with shipyards). The model implies that the owner would expand capacity gradually from K_1 to K_2 in terms of Figure 3.3. But this is optimal because the owners' expectations, being rational, are correct on average, but not on every single occasion. In other words, this process is optimal in repeated sampling, as if the experiment were to be repeated an infinite number of times under the same conditions. How much should the owner expand the fleet under conditions of uncertainty? Does uncertainty bias the results towards a smaller fleet expansion than the optimal?

The extension of the theory of investment under uncertainty is not straightforward – see Driver and Moreton (1991) for a comprehensive analysis and the problems arising thereof. Part of the problem stems from the role of non-linearities in computing the expected value of the utility or profit. This expectation, in general, is not equal to the value obtained when all stochastic variables are set equal to their mean values (that is, certainty equivalence does not hold) because of non-linearities in either the profit function or the utility function (for example, risk aversion). In general, investment depends on current and expected demand (or output) and cost conditions, but an explicit solution can be obtained only under restrictive assumptions. One such case arises when the profit function is linear in capital, as it is for a firm that operates under constant returns to scale in competitive markets (Abel, 1983). Another case is when the profit and adjustment cost functions are quadratic under risk neutrality, with constant interest rates and the price of capital (Blanchard and Fischer, 1990).

The irreversibility of investment becomes important under conditions of uncertainty (see, for example, McDonald and Siegel, 1986; Nickel, 1978; Pindyck, 1988 and for a comprehensive survey of the literature Pindyck, 1991). Irreversibility arises because capital is industry- or firm-specific and therefore cannot be used elsewhere, so that investment expenditure is considered as sunk cost. Although in shipping the existence of the scrap market does not imply that the entire capital expenditure should be considered as sunk cost, it is still true that the optimal investment rule must take into account the opportunity cost of delaying investment. This can be thought of as the price of 'exercising' an option. Thus, for an owner considering a capacity expansion problem the optimal investment rule is to expand capacity up to the point where the expected cash flow from an additional ship is just equal to the sum of the purchase price, the installation cost and the price of exercising the option to invest (see, Pindyck, 1988; Dixit, 1992; Dixit and Pindyck, 1994).

An important issue is the relationship between investment and uncertainty. Under risk aversion and incomplete markets an increase in uncertainty is likely to lead to lower investment (that is, a negative relationship) (see, for example, Craine, 1989). However, under risk neutrality the relationship between uncertainty and investment is ambiguous – see Driver and Moreton (1991) for taxonomy of the factors that affect this relationship. Hartman (1972) and Abel (1983 and 1984) found a positive relationship under symmetric and convex adjustment costs, perfect competition and a convex profit function. On the other hand, the literature on irreversible investment with asymmetric costs of adjustment and imperfect competition (for example, Pindyck, 1988) suggests a negative relationship between investment and uncertainty.

Cabellero (1991) has provided an explanation of these conflicting results and Driver et al. (1991) have extended them. Asymmetric adjustment costs are not sufficient for a negative relationship between uncertainty and investment. However, when combined with imperfect competition they produce a negative relationship. These results are intuitively easy to understand. Under perfect competition today's investment decision depends exclusively on the expected path of the price of capital and the expected profitability of capital. However, the latter does not depend on the level of the capital stock. Hence an increase in price uncertainty raises the level of investment since the marginal profitability of capital is convex with respect to price. On the other hand, under imperfect competition the marginal profitability of capital depends on the level of capital stock – an increase in the stock of capital reduces profitability. When this is combined with asymmetric costs of adjustment (that is, when it is more expensive to reduce capital than to increase it, a weak version of irreversibility) it is better to have a shortage of capital than too much of it. Thus, an increase in uncertainty reduces investment. However, the relationship between investment and uncertainty is not robust to the various factors upon which it depends – see Paraskevopoulos et al (1991) for an extension to a loose oligopoly and Driver and Moreton (1991) for an evaluation of the literature.

In the real world the net present value rule does not work. The return on capital by far exceeds the user cost of capital (around three or four times; see Summers, 1987 and Dertouzas et al., 1990). The difference represents the *required rate of return* or *the hurdle rate*, which owners demand as a compensation for undertaking the risk of the investment project under conditions of uncertainty. The hurdle rate can be thought of as the price of 'exercising' the option to invest (see Pindyck, 1988 and Dixit and Pindyck, 1994). Hence, under conditions of uncertainty the net present value rule has to be adjusted to include this hurdle rate. The hurdle rate depends on the risk of the investment project. If firms are uncertain about the level of expected demand or its volatility then the hurdle rate rises. On the other hand, an increase in the profitability of investment reduces the hurdle rate.

Box 3.4 Investment under uncertainty

Under conditions of uncertainty the optimal fleet should be a percent of the target or desired capacity. This percentage is a function of the probability of the main scenario for the evolution of the demand for dry and the cost components over the investment horizon.

The optimal rules derived in the Appendix give rise to a specific strategy that entails that the current optimal fleet should be a percentage of the target or desired capacity under conditions of uncertainty. This percentage is a function of the probability of the main scenario for the evolution of the demand for dry and the cost components over the investment horizon. The optimal rule involves the computation of the optimal path of NB prices and freight rates under two alternative policy assumptions: main scenario and risk scenario. In the main scenario, the most likely response of the policymakers in the main economic regions (the

US, China, Europe and Japan) to the current and future economic conditions forms the basis upon which the macro environment will develop in the next two years. This macro environment is then used as an input in the shipping model to evaluate the impact on the key shipping variables. In the risk scenario, an alternative path of the key shipping variables is obtained by assuming a different policy response to the current and future economic conditions. The results of the main and risk scenarios are then compared in terms of the gap they create for NB and freight rates so that a strategy can be formulated. The yardstick for deciding whether to expand the fleet capacity under conditions of uncertainty is given by the risk–reward ratio, which is equal to the ratio of NB prices two years ahead in the main scenario and the risk scenario. If this risk–reward ratio is higher than the subjective risk-reward, then the owner should proceed with the investment up to a percent of the target fleet equal to the probability of the main scenario.

5 CASE STUDY

In order to put the optimal rules into perspective, it is assumed that the owner stands in September 2012 with all information available until that time. Most owners feel that four years on from the 2008 crisis this is the time for putting in place the desired or target fleet to take advantage of the new business cycle that started in mid-2009. However, there are considerable concerns as to the exact timing of the fleet expansion. The owner faces the problem of deciding on the target fleet until the end of the shipping cycle, some time in 2016–18. The owner makes a joint decision on how quickly to reach that level; namely to decide on the desired pace of deliveries (the investment path).

Box 3.5   Case study

'Smart' owners form forward-looking expectations of how central banks and governments would respond to current and future economic conditions. This enables them to be 'contrarian' investors – buying ships near the bottom and selling near the peak.

Most owners and shipyards form backward-looking expectations by extrapolating past patterns of demand. Such owners are 'herd' investors – following the trend. The more risk-averse they are, the more likely they will lose money by entering too late. Together with shipyards, they are responsible for prolonged periods of overcapacity and excess price volatility in the shipyard industry.

The macro environment in September 2012 can be summarised as follows. China's growth peaked in the first quarter of 2010 and the slowdown has gathered steam since then, as the fiscal stimulus of 2009 faded and the world economy slowed because of austerity measures adopted around the world and primarily in Europe. The process for the once in a decade change of political leadership in China started in September 2012, thus posing a risk of policy inaction amidst the slowdown of the economy. The new leadership is not expected to be in place before the end of March 2013 and new policies are unlikely to be implemented before the second half of 2013. The US faces political uncertainty in view of a potential fiscal cliff (automatic spending cuts and tax increases, as the tax cuts of the Bush era are expiring at the end of 2012) resulting from a political gridlock following elections in November 2012. In Europe, the risk of a euro break-up has subsided following some structural reforms of the monetary union in July 2012 and the promise of Draghi to do everything in his power to deter a euro break-up. But the euro area is falling deeper into recession as a result of austerity measures adopted in the periphery. In the main scenario, China is assumed to adopt easy fiscal policy and tight monetary policy. The fiscal stimulus is expected to be 2 per cent of GDP, but implemented in the second half of 2013 with its major impact being felt in the economy in 2014. In the US, the total fiscal cliff of $600 billion per annum might be avoided, but some tightening of $235 billion (or 1.5 per cent of GDP) is likely to take place. In Europe, the recession is likely to be deeper than assumed by the ECB, the EC and the consensus, around −0.5 per cent in 2013. The consequences for the world economy of this policy mix are slower growth in 2013, but with a rebound in 2014 and beyond as the fiscal tightening dissipates in Europe while a positive trade multiplier is set between the US and China with Japan and Germany taking immediate advantage. This would resuscitate the demand for shipping mainly in 2014 and beyond. The probability of the main scenario in September 2012 was 60 per cent, which means that there is a risk of 40 per cent that things would not develop as planned.

Assume that the economic environment of depressed freight rates and vessel prices that prevailed in September 2012 represents a long-run equilibrium with vessel prices $P_1 = \$375$ per dwt expected to be hit in 2013 compared with $467 per dwt in September 2012, investment I_1 and fleet K_1. Assume that the main scenario develops as planned. The owner can compute through a structural model, such as the K-model, the impact of the macro environment on the demand for ships and hence the new long-run equilibrium demand curve. The new long-run equilibrium entails higher vessel prices, $P_2 = \$550$ per dwt, larger gross investment, I_2, and a bigger fleet, K_2. The adjustment of vessel prices involves an overshooting in the short run from $P_1 = \$375$ to $P^* = \$700$ and a subsequent decline in vessel prices from $P^* = \$700$ to $P_2 = \$550$ (see Figure 3.A6 in the Appendix). Because of the price overshooting, the owner cannot afford to wait until the expectations materialise to expand the fleet. If he delays, he may miss the big capital gains that occur from the low of P_1 to P^*. If the owner is very risk averse and enters after P^*, he will suffer capital losses on the additional vessels as prices would fall from P^* to P_2, the new long-run equilibrium. The adjustment path of investment resembles the pattern of vessel prices (see Figure 3.A7 in the Appendix); whereas the fleet converges gradually and monotonically to K_2 (see Figure 3.A8 in the Appendix).

The practical conclusion of this optimal fleet capacity expansion is that the owner must adjust its fleet before the market had a chance to price in the effects of easy fiscal policy on the equilibrium demand for ships. This is consistent with the popular wisdom among owners that once the market starts moving, it is very difficult to catch up with. The real profits from the asset game can be captured only if the owner is prepared to take the risk and expand the fleet on the expectation that the demand for ships would increase in 2014 and beyond. The owner should expand the fleet now (September 2012) to the probability of the main scenario 60 per cent, as any further expansion does not justify the extra risk. Hence, the owner should have in place 60 per cent of the optimal level of fleet by 2014. If the optimal fleet is ten ships the owner should have in place six ships by 2014.

6 THE MODEL IN PERSPECTIVE

As a ship is an asset that provides a flow of services throughout its lifetime it is common in the maritime literature to consider ship prices as asset prices. Asset pricing usually involves a net present value rule and in this respect all approaches of modelling ship prices look the same. The debate in modelling ship prices is sometimes presented as a net present value rule vs a demand–supply framework.

But the two are interlinked when considering the entire market. For the market as a whole, when the price exceeds the benefits the demand for the asset would drop and hence its price. When the price is lower than the benefits, the demand for the asset would increase and hence its price. Thus in equilibrium the price of an asset should equal its benefits. Therefore, a demand–supply framework is always implicit in the net present value rule, as it is the mechanism that ensures the equality of cost and benefits. Considered from this perspective, it is inappropriate to view the debate on ship prices as stemming from a net present value rule vs a demand–supply framework.

Nonetheless, the two approaches are different when the net present value rule is assumed rather than derived explicitly from an optimisation problem (an atheoretical adoption of the net present value rule). This can be shown easily by considering that a rational investor would not pay more for an asset than the benefits derived from it throughout the lifetime of the asset.

Consider the definition of the one-period return on a shipping investment, r^s, for a ship bought in period t at price P_t and where π_{t+1} is the profit in period $t+1$ and E is the mathematical expectations operator as with information at time t. Then the one-period return on shipping is defined as

$$r^s = \frac{E_t \pi_{t+1}}{P_t} + \frac{E_t P_{t+1} - P_t}{P_t} \qquad (3.13)$$

The first term on the right-hand side represents the expected profit in the next period obtained from operating the vessel and expressed as a percentage of of the current ship price. The second term represents the expected capital gains (or losses) from selling the ship in the next period in the secondhand market.

This equation becomes a theory if it is assumed that investors would be attracted in the shipping market as long as the return on shipping exceeds the return on alternative assets, r (perhaps adjusted by a time-varying risk premium, RP), and quit the market when the return on shipping is lower than the risk-adjusted r. Hence, in equilibrium $r^s = r + RP$. Using this equilibrium condition the above equation can be solved for P.

$$P_t = \frac{E_t \pi_{t+1} + E_t P_{t+1}}{1 + r_t + RP_t} \qquad (3.14)$$

This is a first-order difference equation and its solution is

$$P_t = C(1+r+RP)^t P_{t+n} + \sum_{s=t}^{\infty} \frac{1}{(1+r+RP)^{s-t+1}} E_t \pi_{s-t+1} \qquad (3.15)$$

where C is an arbitrary constant. If $C \neq 0$, then with time the first term on the right-hand side tends to plus or minus infinity as it is raised to ever-increasing powers of t. The case when $C > 0$ has been interpreted in the literature of finance as corresponding to 'rational bubbles' – the price of the asset rises ad infinitum in time. At some point in time $C < 0$ and as a result the bubble bursts. Rational bubbles can be excluded on the basis that investors are rational. This means that $C = 0$, and therefore the above equation simplifies to

$$P_t = \sum_{s=t}^{\infty} \frac{1}{(1+r_s+RP_s)^{s-t+1}} E_t \pi_{s-t+1} \qquad (3.16)$$

This equation states that the price of an asset (ship) is equal to the present value of expected future profits. This is the net present value rule. It is not entirely atheoretical as it assumes that investors would invest in ships until the return on shipping is equal to the return of alternative investments or simply to the cost of capital, being either the short- or the long-term interest rate (with or without a time-varying risk premium). For investors to enforce the equality of returns ($r^s = r + RP$), they must be rational, have full knowledge of all investment opportunities and exploit such opportunities whenever they arise. Investors would have full knowledge of all investment opportunities, if the asset prices 'fully reflect' all available information. This is the definition of the Efficient Market Hypothesis (EMH). A market is said to be 'efficient' with respect to an information set, if the price 'fully reflects' that information set. This requires that the price would be unaffected by revealing this information set to all market participants, as there is no incentive to trade upon this information – it is already included in the price. A necessary condition for market efficiency is that all related markets are perfectly competitive.

When investors act according to the EMH, they instantaneously enforce the arbitrage condition of the equality of returns ($r^s = r + RP$). In our framework this arbitrage condition is not imposed. In the calculation of the net present value rule in our model a subjective discount rate is used, which may include the cost of money, r, and a risk premium, but this does not need to be equal to the return on shipping. The imposition of the arbitrage condition has serious implications for the shape of the underlying demand for vessels. When the arbitrage condition is imposed, and therefore when the shipping market is efficient, the underlying demand for the asset is perfectly elastic (horizontal in the price–quantity space) at the level of price at which the arbitrage condition ($r^s = r + RP$) holds true. When the arbitrage condition is not imposed, as in our framework, the demand for vessels is a normal downward-sloping curve. Therefore, the real debate on shipping prices is not about the net present value rule vs. a demand–supply framework, but about the underlying elasticity of the demand for vessels.

There is one more fundamental difference between our model of vessel prices and those based on the EMH. The present value rule, summarised in equation (3.16) lacks a theory on how profits in shipping are to be determined. An easy way out of this problem (for example, Strandenes, 1984, 1986) is to make use of the definition of profits using the time charter equivalent rate of duration τ, H^τ. Profits are equal to the time charter equivalent rate less fixed operating costs (such as labour costs), denoted by OC

$$\pi_t = H_t^\tau - OC_t \qquad (3.17)$$

Using data on OC one obtains data for profits and can test the validity of the net present value rule in equation (3.16). But this formulation has the drawback that it does not identify the determinants of profits, which are essential in the calculation of expectations that would enable the owner to decide whether the investment is worthwhile. The use of time charter equivalent freight rates includes *market* expectations of freight rates and profitability. But market expectations are not a good guide for an owner, as he would only make more than normal profits if he does not follow the 'herd' – the market.

The advantage of our model is that it derives the net present value rule without invoking the EMH. The ship price, whether NB or SH, is an asset price. But this is because in our framework the demand for vessels is positively related to the price in the short run, but inversely in the long run. Therefore, owners buy a ship on the expectation of making capital gains and this implies that demand for vessels rises when the price increases – the characteristic feature of an asset demand. But at some point in time, the price is very expensive and the demand for vessels becomes a normal downward sloping curve.

Moreover, our approach has the advantage of deriving the determinants of profits. These are the freight rate and the fleet capacity utilisation rate. Finally, our theory enables the quantification of the importance of these economic fundamentals. Whether freight rates or the fleet capacity utilisation rate are more important depends on the elasticity of substitution between fleet and average speed in producing shipping services. In addition to economic fundamentals, the demand for vessels depends on the ship's technological improvements and the economies of scale that result from one additional vessel to the fleet.

Our model provides a much better framework to analyse shipping prices than models based on the EMH. If the EMH is rejected by empirical tests one can claim that this is due to the omission of a risk premium that should be added to r in equation (3.16) or to an inappropriate definition of the risk premium. Time-varying risk premia are commonly invoked in the literature of vessel prices, whether NB or SH, to justify the failure of the EMH. In our opinion the debate on whether ship prices are efficient or not is futile. The requirement that the price 'fully reflects' a given information set is an exacting requirement, suggesting that no market can be efficient in the real world. The EMH should better be viewed as being asymptotically true. Viewed from this angle the EMH is one of the strongest hypotheses in economics.

The conditions of the EMH are not satisfied in the shipping industry. There are barriers to entry in the shipping industry stemming from the large size of the investment. Investors must have considerable expertise in shipping. They must know the shipping technology; they should be able to find crews with expertise who would be loyal and take risks; they should have management skills; they should be connected with brokers to find the cargo; they should have competent technical staff to take care of damages. These requirements usually exclude outside investors. The experience shows that whenever charterers or speculators are attracted to the ownership side of the shipping market, they have suffered huge losses. A good recent example of this was the decision by Vale to enter shipowning by ordering a large number of very large ore carriers (VLOCs), with a view to keeping prices between Brazil and China for iron ore movements nearer to the price of Australian ore shipments to China. Not only were these vessels ordered at high prices; they also suffered from a ban on entering ports in China. Although the final reckoning has not taken place it already seems to have been a very suboptimal investment.

7 THE SUPPLY OF VESSELS – THE IMPACT OF SHIPYARD CAPACITY ON NB PRICES

7.1 INTRODUCTION

The supply of vessels depends on the shipyard capacity. In the short run, shipyard capacity is fixed and the supply is a positive function of the price of vessels because the marginal cost is increasing. In the long run, though, shipyards can expand capacity and lower the cost per vessel. The amount of capacity that shipyards should install depends on current and expected demand. If shipyards form rational expectations, the capacity should be, on average, correct to meet the demand for vessels. If shipyards are backward looking and form expectations by extrapolating past demand, however, then they will face prolonged periods of excess demand and excess capacity. An excess demand for vessels would accentuate vessel price increases and an excess supply would accentuate price falls. Therefore, backward-looking expectations by shipyards increase the volatility of vessel prices.

On the other spectrum, some owners are forward looking, taking decisions today by forming expectations of how policymakers would react to economic conditions two years ahead (see Appendix 1 for more details). These owners are 'contrarian' investors, meaning that in every cycle they buy ships near the bottom of the market and sell at or around the peak. But not all owners are contrarian investors. The majority are backward looking; as a result, they are 'herd' investors – following the trend. Herd investors can still make profits, but they have to be quick and therefore risk takers. The only category of owners that would lose money is backward looking and risk averse. The more risk averse an owner, the more time he needs to be convinced that the current trend is sustainable and therefore the more likely that he would enter near the peak of the cycle. Contrarian investors are responsible for trend reversals, whereas herd-investors for the continuation of a trend.

The stylised facts of the shipyard industry provide support to the hypothesis that shipyards form backward-looking expectations by extrapolating past levels of demand, whereas some owners are forward looking, forming expectations by how policymakers would react to economic conditions two years ahead. In the boom years of the 2003–08 bull market, namely 2005–08, there was more demand for vessels than shipyard capacity could meet. As a result, newbuilding prices increased more rapidly than would have been justified if shipyards operated with spare capacity. In time, shipyards put in place the capacity that was needed to meet the huge demand of the boom years. Unfortunately, this extra capacity was installed when the demand for vessels started to decline. Backward-looking owners who missed the boom of 2005–08 queued to place orders in the aftermath of the boom. These orders created some euphoria in the shipyard industry and have kept it going in the period since 2008. However, the orderbook has been declining constantly, showing that the forward-looking owners have not participated in this rally of orders. Consequently, a huge spare shipyard capacity has been created, at the very time when the demand for vessels continues to abate. This spare shipyard capacity exerts more downward pressure on newbuilding prices than is justified by the fall in the demand for vessels. In this section we set out a model to evaluate the impact of the shipyard capacity on newbuilding prices. We find that so far (September 2012) the current shipyard overcapacity has exerted a 20 per cent drop in newbuilding prices from the peak in 2008 out of a total fall of 50 per cent. This represents 40 per cent contribution of shipyard excess capacity on newbuilding prices, thereby suggesting that the shipyard excess capacity would act as a drag on NB prices when the demand for dry and freight rates recovers. The implication of this analysis is that if the demand for vessels is resuscitated with a revival of the freight market in 2014–17, then newbuilding prices will recover with a lag, say, in 2015–17.

7.2 A FRAMEWORK FOR ANALYSING SHIPYARD CAPACITY

In Appendix 1 we derive the long-run demand for vessels through the optimising behaviour of owners.[5] In deriving the optimal dynamic adjustment path of newbuilding prices we consider that the supply is an increasing function of prices, but for simplicity we treat shipyard capacity as fixed (that is, that the supply curve does not shift).[6] In the Appendix we examine the conditions and the conclusions that can be drawn from allowing shipyards to expand or contract capacity according to demand. This new framework enables us to decompose the total change in prices to demand and supply factors. In other words, we can isolate the impact of excess shipyard capacity on prices.

The conceptual problem that we want to analyse can be stated more succinctly with the help of Figure 3.4. Assume that the initial long-run equilibrium is at A with demand D_1 and supply S_1. In this long-run equilibrium vessel prices are P_1 and the fleet produced by all shipyards is K_1. Consider next a fiscal stimulus by China and the US that shifts the demand for vessels to D_2. In the analysis of Appendix 1, which assumes no change in shipyard capacity, the new equilibrium is attained at point C with higher vessel prices at P_2 and a bigger fleet at K_2. It is worth remembering that because of the forward-looking behaviour of owners and their short-run destabilising behaviour, according to which they buy more vessels when prices are increasing, the dynamic adjustment path of prices would jump on impact to B, where prices overshoot their long-run equilibrium by P* − P$_2$. Because of this short-term destabilising behaviour of owners, vessel prices would decline subsequently following the unique saddlepoint path EE to C. This analysis is consistent with the well-founded empirical result that vessel prices increase in the short run more than justified by economic fundamentals and decline subsequently as they approach their new long-run equilibrium. Nonetheless, at the new long-run equilibrium vessel prices are higher than the initial equilibrium. As a result of the short-run overshooting of prices owners that do not succeed in placing their orders before the market captures the impact of easy fiscal policy in China and the US would likely suffer capital losses on their new vessels.

But this is not the end of the story. In time, shipyards would gradually increase capacity as a result of higher vessel demand. In Figure 3.4 this increase in shipyard capacity is portrayed as a shift to the right of the supply curve from S$_1$ to S$_2$. The new long-run equilibrium would be established at F, instead of C. At this new long-run equilibrium vessel prices at P$_3$ are even lower than at C and the fleet produced at K$_3$ is even bigger than at C. Prices would still jump to P* initially and then would decline to P$_3$ following the new saddlepoint path E$_1$E$_1$. The saddlepoint path rotates clockwise around point B as the supply curve shifts to the right. Therefore, with fixed shipyard capacity prices would increase from P$_1$ to P$_2$ via P*, as a result of the fiscal stimulus in China and the US. By allowing shipyards to expand capacity, vessel prices would increase from P$_1$ to P$_3$ via P*. Thus, the impact of shipyard capacity is a drop in prices equal to P$_3$ − P$_2$. Moreover, the drop in vessel prices is steeper when shipyard capacity expands than when it remains fixed. With fixed capacity, prices fall by P$_2$ − P*, whereas with variable capacity they fall by P$_3$ − P*. In the next section we provide an empirical estimate of P$_3$ − P* and decompose it to the factors that cause this change.

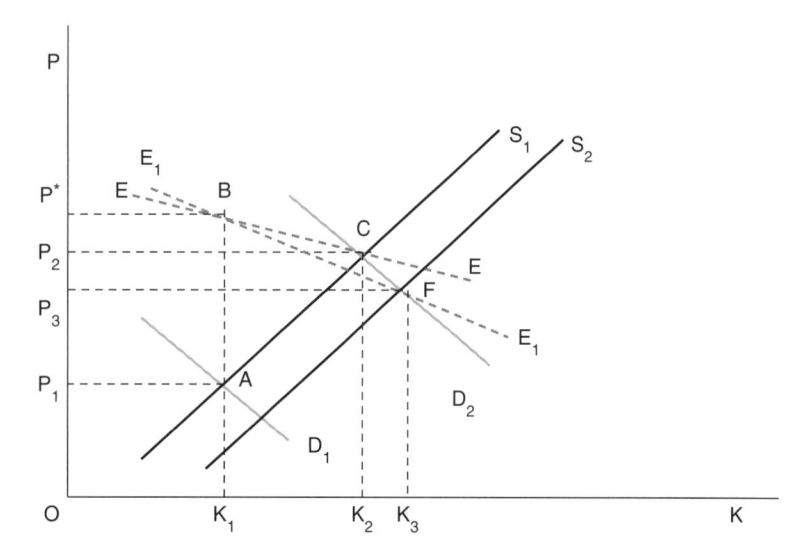

Figure 3.4 Dynamic adjustment of NB vessel prices

7.3 AN EMPIRICAL ESTIMATE OF THE IMPACT OF SHIPYARD CAPACITY ON NB PRICES

The demand for and supply of vessels in the entire shipyard industry can be calculated from data on orders, contracts and deliveries. Vessel deliveries are an accurate measure of supply per period of time (S), whereas the change in the orderbook less cancellations is a measure of demand per period of time. Supply is the minimum of the demand for vessels (D) and the shipyard capacity (SCU). When the demand for vessels exceeds the shipyard capacity, the supply of vessels by all shipyards is constrained by the limited capacity. When there is spare capacity in the shipyard industry the supply of vessels is constrained by the level of demand. Hence,

$$S = SCU \ \ if \ D \geq SCU \ \ or \ \ S = D \ \ if \ SCU \geq D \quad (3.18)$$

Figure 3.5 plots the orderbook of all shipyards for all type of vessels in the dry, wet and specialised markets, along with a projection from the K-model. This is an appropriate measure upon which to build demand in the shipyard industry, as some shipyards may have the technology to shift production from one type of vessel to another according to demand. The orderbook (dotted line) was increasing on a linear trend until 2003, but it accelerated in the period 2003–08. The acceleration of the orderbook in this period implies that the rate of growth (solid line) increased on a linear trend until 2008, but declined thereafter.

However, the orderbook is not an accurate measure of the cumulative demand for vessels because of order cancellations. Figure 3.6 plots the order cancellations for each month from 1996 onwards. It can be seen that there were no order

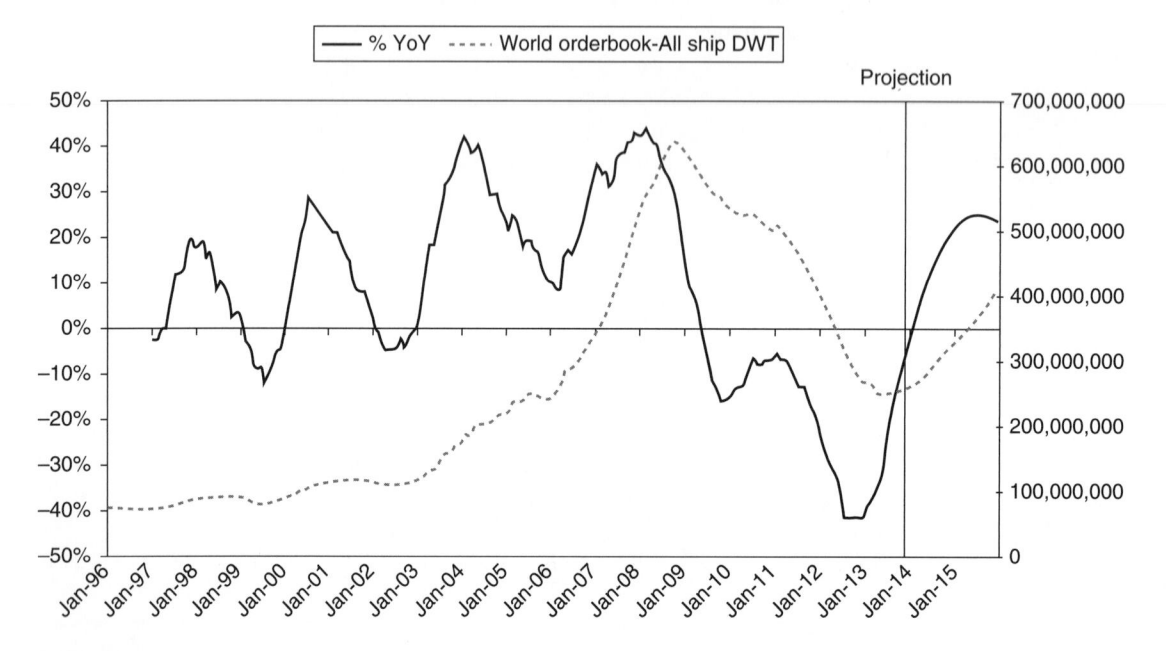

Figure 3.5 World orderbook
Source: K-model.

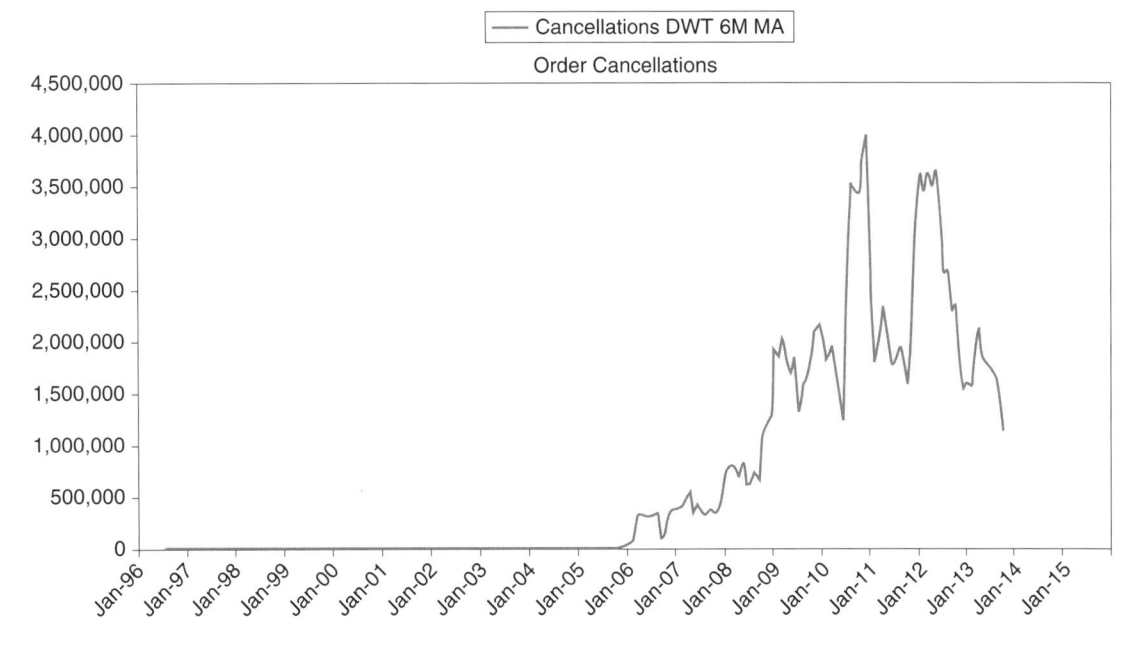

Figure 3.6 Order cancellations
Source: K-model.

cancellations until 2006, but they soared afterwards as many smart owners felt that the boom of 2005–08 was ephemeral.

Figure 3.7 shows the evolution of the demand for and the supply of vessels since 1996. With the exception of 12 months (from mid-2003 to mid-2004) there was spare capacity in the shipyard industry until early 2006. There was an excess demand for vessels in the period April 2006 to October 2008, which peaked in October 2007. Whereas many smart owners deemed the boom in demand as ephemeral and accordingly cancelled orders at an increasing pace from 2008 onward (see Figure 3.7), the shipyard industry considered it as permanent, thereby providing empirical support to the claim that some owners are forward looking, whereas the entire shipyard industry is backward looking. This led shipyards to expand capacity to meet the demand for vessels. In reality, the demand boom was only transient. Demand declined, while supply soared. Hence, the two-and-a-half year boom in the shipping industry triggered an excess demand for shipyard capacity for a while, but overcapacity in the last four years.

Figure 3.8 shows the demand–supply balance in the shipyard industry, namely the deviation of the demand for vessels from the shipyard supply in Figure 3.7. Shipyard capacity was sufficient to meet the demand for vessels until the spring of 2006. The demand boom in 2005–08 created an average excess demand for shipyard capacity of 11 million dwt per period of time. But the capacity expansion by all shipyards created an average excess supply of 35 million dwt per period by mid-2012. This represents an average swing of 46.5 million dwt per period from peak to bottom (see Figure 3.8).

As a result, in mid-2012 there were fears among many owners that even when demand for vessels recovers the huge shipyard overcapacity would continue to

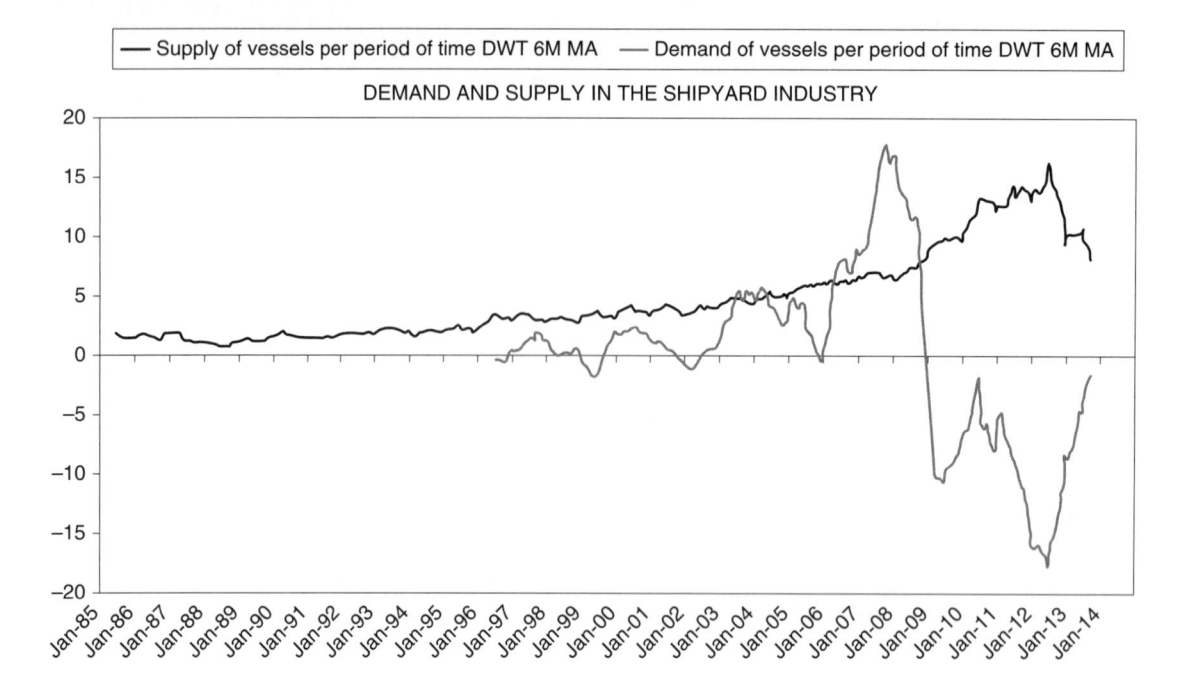

Figure 3.7 Demand and supply in the shipyard industry
Source: K-model.

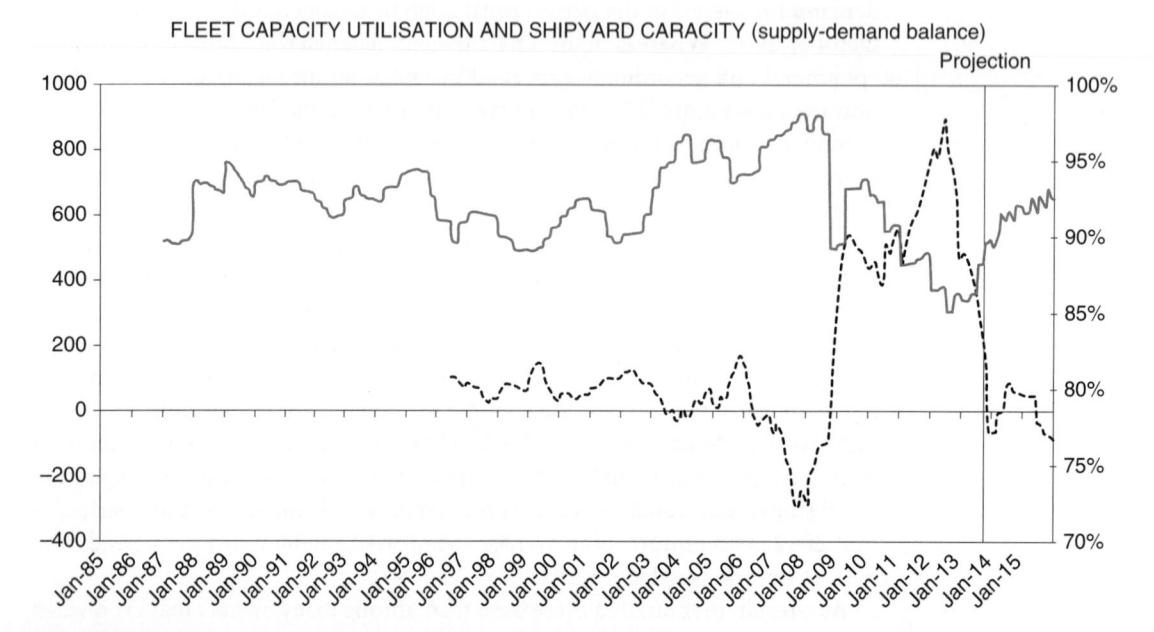

Figure 3.8 Excess supply in the shipyard market and excess demand in the freight market
Source: K-model.

exert downward pressure on prices for years to come. Figure 3.8 also plots the dry fleet capacity utilisation rate of the dry market. This is a measure of the demand–supply balance in the freight market. The two excess demand functions are interlinked.[7] An excess demand in the market for dry shipping services, where freight rates are determined, would lead to an excess demand for vessels in the shipyard market. With the latest data, as of September 2012, the estimated elasticity of the shipyard capacity is 6.4 million dwt.[8] This means that an increase in the fleet capacity utilisation in the dry market by 1 per cent would increase the shipyard demand–supply balance by 6.4 million dwt. The implication of this elasticity is that the dry fleet capacity utilisation rate should increase by 5.5 per cent to absorb the current shipyard overcapacity, other things being equal.

To enable us to calculate the impact of shipyard capacity on NB prices we combine the optimal rules for capacity expansion by owners (equation A.12 or A.13 in the Appendix) with the model that relates the demand–supply balance in the shipyard industry with the fleet capacity utilisation (equation 19) and, finally, the basic law in economic theory that vessel prices increase proportionately to the excess demand for vessels and fall when there is excess supply (equation 20).[9] This three-equation system provides the long-run impact on vessel prices. The dynamic adjustment path to the new long-run equilibrium depends on its own momentum, negatively on the degree of the previous disequilibrium of price vessel inflation and positively on the acceleration of all explanatory variables.[10]

Equation (3.20) provides the long-run impact on vessel prices, whereas equation (3.21) shows the dynamic adjustment path. In terms of Figure 3.4 the first equation provides an estimate of $P_3 - P_1$ and $P_3 - P^*$, whereas the second equation the dynamic adjustment from P_1 to P^* to P_3, namely from A to B to F. Vessel prices in the dry market peaked at $925 per dwt in August 2008 and fell to $467 per dwt in September 2012; this is exactly a 50 per cent drop in vessel prices in the four-year period from mid-2008 to mid-2012 (see Figure 3.A6). Thus, $P_3 - P^*$ in terms of Figure 3.4 is equal to −50 per cent. This differential in vessel prices can be decomposed to the four factors (shipyard capacity, secondhand prices, the prices of the three major bulks and bunker costs) that explain the total. The K-model estimate of the impact of the shipyard capacity on vessel prices is 0.005. As shipyard capacity fell by 46.5 million dwt from mid-2008 to mid-2012 this implies that the impact of shipyard capacity on newbuilding prices in the dry market is 19.5 per cent. Therefore, out of the 50 per cent fall in vessel prices in the last four years nearly 20 per cent is due to the deterioration of the shipyard capacity and 30 per cent is due, firstly, to the drop of secondhand and the three major bulk prices and, secondly, and to a lesser extent to the increase in bunker costs.

7.4 EXPLAINING THE STYLISED FACTS

The shipyard industry is characterised by high price volatility and prolonged periods of overcapacity. The price volatility is not consistent with mark-up pricing and confirms the new nature of vessel prices as asset prices. The overcapacity is the result of a contrasting decision-making process between owners and shipyards. 'Smart' owners take decisions today on expectations of how the policymakers in major economies would react to current and future economic conditions. On the other hand, shipyards and 'herd' owners take decisions by extrapolating the historical level of the demand for ships. Hence, 'smart' owners are forward looking, whereas shipyards and 'herd' owners are backward looking. This is the consequence of the opposing forces of NB prices on the demand for vessels and the upward-sloping supply curve. Whereas the demand for vessels is a negative function of the price of vessels in the long run, in the short run owners buy ships when prices rise and sell when prices fall. This short-term destabilising behaviour of owners, coupled with their forward-looking expectations and the backward-looking behaviour of shipyards, resulted in an average excess demand for shipyard capacity of 11 million dwt per quarter in the 2005–08 demand-boom, but spare capacity of 35 million dwt per quarter in the last four years. This represents a swing in capacity of 46.5 million dwt by mid-2012.

In this chapter we set up a model that relates the excess demand in the shipyard industry to the fleet capacity utilisation rate, a measure of the demand–supply balance in the dry freight market. The model captures the principle that an excess demand in the market for shipping services, where freight rates are determined, would lead to an excess demand for vessels in the shipyard market. In September 2012, the estimated elasticity of the shipyard capacity was 6.4 million dwt. This means that an increase in the fleet capacity utilisation in the dry market by 1 per cent would increase the shipyard demand–supply balance by 6.4 million dwt. The implication of this elasticity is that the fleet capacity utilisation should increase by 5.5 per cent to absorb the shipyard overcapacity that was created in 2008–12, other things being equal.

Vessel prices in the dry market peaked at $925 per dwt in August 2008 and fell to $467 per dwt in September 2012; this is exactly a 50 per cent drop in vessel prices in the four-year period from mid-2008 to mid-2012. The model of equations (3.9) and (3.10) explains the factors that have contributed to this fall in NB prices. The model suggests that of this 50 per cent fall in vessel prices nearly 20 per cent is due to the deterioration in the demand–supply balance in the shipyard industry and 30 per cent is due to, firstly, to the combined drop of second-hand prices and the prices of the three major bulk prices; and secondly, but to a lesser extent, to the increase in bunker costs. This contribution of shipyard excess capacity on newbuilding prices accounts for 40 per cent of the fall in NB prices. The implication of this analysis is that if the demand for vessels resuscitated with a revival of the freight market in 2014–17, then the existing shipyard capacity would act as a drag to the rebound of NB prices.

8 THE SCRAP MARKET AND THE NET FLEET

In deriving the *net* fleet in this chapter, we have assumed that a constant proportion of the fleet is replaced every year, as it becomes technologically obsolete. But in the real world legislation about ship safety and strict environmental laws to reduce sea pollution as well as economic factors would make the hypothesis of proportionality invalid. Legislature changes are purely exogenous to the shipping market and should be treated as an exogenous shock that affects the net fleet. Thus, in this section we concentrate on the economic factors, which are purely endogenous to the shipping market and affect the net fleet. There is almost unanimity amongst economists on the scrapping decision of an individual owner and through the aggregation of these decisions on the industry level of scrapping. A ship should remain in the active fleet if it is economically viable. The economic viability of a ship is ensured if profits are positive and the firm is solvent. The condition of solvency is important, as otherwise there is no economic decision – the firm (often each ship) is bankrupt and the ship is foreclosed. A simple rule that guarantees positive profitability is the identity that relates profits to the time-equivalent charter rate, H, operating costs, OC, and debt service, DS. The debt service includes any margin calls (injections of new capital by the owner) or a higher margin over the interest rate (usually Libor) that a bank may require. Thus if

$$\pi_t = H_t^\tau - OC_t - DS_t > 0 \qquad (3.22)$$

the ship is economically viable. The validity of the above equation is ensured, as in a time charter contract the charterer is responsible for variable costs, such as fuel, port charges and canal dues, while fixed costs, such as wages and debt service (that is, interest payments) are borne by the owner.

However, the above equation does not provide a framework for comparing the scrap value of a ship with its trading value, as the decision to scrap a ship depends on whether it is worth more 'dead' than 'alive'. This implies comparing the ship's secondhand value with its scrap value. The scrap value of a ship is determined by the amount of metal in the ship times the scrap price per tonne. The trading value of a ship depends on the price it can fetch in the secondhand market and it is a function of its age and embodied technology. Therefore, an alternative rule for economic viability involves a comparison of the secondhand price, PS, with the scrap price (PSC). Accordingly, the supply of ships for scrap (DM) in period t as a proportion of the net fleet put in place until the end of period $t-1$, K_{t-1}, is an increasing function of the ratio of the scrap price to the secondhand price and an increasing function of the age of the ship, A, for an individual owner or the age profile of the fleet at the industry level.

$$\frac{DM_t}{K_{t-1}} = f\left(\frac{PSC_t}{PS_t}, A_t\right) \quad f_1 > 0, f_2 > 0 \qquad (3.23)$$

As the price of scrap rises relative to secondhand prices a higher proportion of the fleet is scrapped for a given age profile. In contrast, higher secondhand prices for a given scrap price and age profile reduce the proportion of the fleet for demolition. For a given ratio of prices, as the age of the fleet increases, a greater proportion of the fleet is scrapped.

The above equation is equivalent to equation (3.22), as the secondhand price reflects the discounted value of the stream of profits of a ship. It is worth remembering that in the dynamic optimisation problem of fleet capacity expansion, the optimal price applies both to newbuilding and secondhand prices. But equation (A.32) has to be adapted, as the decision is whether to keep the vessel in the fleet or scrap it rather than acquire it. Accordingly, the cost of fleet adjustment is zero in (A.32). The economies of scale term in (A.32), that is, G_K, is still relevant, as scrapping a ship might create diseconomies of scale. For example, the demolition of one ship might make the expenses of the management office too high for the remaining fleet.

$$PS_t = E_t \sum_{s=t}^{T} \frac{1}{(1+r_s+\delta)^{s-t+1}} \cdot \left(\frac{a}{A^\rho} \cdot FR_{s-t+1} \cdot \left(\frac{Q}{K} \right)_{s-t+1}^{1/\sigma} - G_K(I,K) \right) \quad (3.24)$$

Equation (3.24) states that in equilibrium the secondhand price should be equal to the discounted expected future profits, where the discount factor includes the high rate of depreciation of the ship, as it is getting old, and the interest rate for servicing the debt. Future profitability is computed by forming expectation of future freight rates and fleet capacity utilisation. As the ship is getting old the elasticity of substitution between fleet and speed, σ, should be smaller than that for new technology ships and may even be smaller than one. This implies that an owner has to put more importance to the expected fleet capacity utilisation rather than freight rates in computing the equilibrium secondhand ship price. If the actual secondhand price in the market is smaller than the equilibrium price obtained from (3.24), then the discrepancy is due to market inefficiencies and the owner should use the equilibrium secondhand price in equation (3.24) in the scrapping decision of equation (3.23), as inefficiencies are likely to be corrected in the future.

Gross investment, I, is defined as the expenditure that an owner incurs per time period for net fleet expansion and replacement investment to buy ships that have been scrapped. The gross fleet is equal to the accumulation of deliveries, DL, whether for net fleet capacity expansion or for the replacement of the scrapped vessels. Accordingly, the *net* fleet in period t is defined as the fleet in the previous period plus deliveries during the period less scrapping in that period

$$K_t = K_{t-1} + DL_t - DM_t \quad (3.25)$$

Thus, the deliveries and demolition in period t would be incorporated into the new fleet at the end of period t. These definitions hold true for the individual owner and the industry as a whole through aggregation. By dividing both sides of (3.14) by K_{t-1} and rearranging terms we get

$$g_t \equiv \frac{K_t - K_{t-1}}{K_{t-1}} = \frac{DL_t}{K_{t-1}} - \frac{DM_t}{K_{t-1}} \quad (3.26)$$

Equation (3.26) states that the rate of growth of the net fleet, g, is equal to the deliveries in period t as a percentage of the fleet put in place until the end of period $t-1$, K_{t-1}, less the demolition as a percentage of the fleet. Equation (3.26) shows that the correct way of expressing equation (23) is as a percentage of the net fleet at the end of period $t-1$.

9 THE RELATIONSHIP OF NB AND SH PRICES

The dynamic optimisation problem of fleet capacity expansion gives rise to the demand for ships whether they are newbuilding or secondhand and derives their corresponding demand prices. Thus, the demand for newbuilding ships is given by equation (A.20) reproduced here for convenience as equation (3.27), while the demand for secondhand ships is given by equation (3.28)

$$K^d = [a/A^p]^\sigma \cdot Q \cdot [FR/UC]^\sigma \quad UC_t = (r_t + \delta_1) \cdot P_t + G_K(I_t, K_t) - \dot{PS}_t \quad (3.27)$$

$$KTR^d = [a/A^p]^\sigma \cdot Q \cdot [FR/UCS]^\sigma \quad UCS_t = (r_t + \delta_2) \cdot PS_t + G_K(I_t, K_t) - \dot{PSC}_t \quad (3.28)$$

A comparison of the two demand functions shows that each one depends on the same economic fundamentals, namely the demand for shipping services and relative prices (that is, the freight rate per unit of the user cost of capital, UC). But in each case the user cost of capital is different, as the price is different, the depreciation rate is different and the expected capital gains in the newbuilding and secondhand markets are also different. Thus, in the newbuilding market the user cost is defined in terms of the newbuilding price, whereas in the secondhand market in terms of the secondhand price. The depreciation factor is δ_1 for new ships, whereas it is δ_2 for old ships; clearly, $\delta_1 < \delta_2$. The capital gains in the newbuilding market are a function of the five-year or ten-year secondhand price. But for an old ship the capital gains are most probably determined by the price of scrap. Similarly, in each demand function the technology, A, embedded in each ship category is different and the elasticity of substitution between fleet and speed, σ, is also different.

The supply of new ships by shipyards (that is, deliveries in a period of time) is a positive function of the price of new ships and a negative function of the variable factors of production, such as steel, labour and equipment.[11] Thus, if PT is the steel price, W is the wage rate and PE is the price of equipment, then the shipyard supply of new ships is

$$K^s = f\left(\frac{P}{PT}, \frac{P}{W}, \frac{P}{PE}\right) \quad (3.29)$$

Although a multitude of reasons may trigger owners to sell their ships in the secondhand market we can discern four major ones. The first stems the policy by some owners of keeping their fleet within a particular age profile. Thus, when ships reach a particular age they are sold in the secondhand market with the intention of being replaced by new ones. This suggests that the supply of secondhand vessels is a function of the age profile of the fleet. The second category is owners who have pessimistic expectations on economic fundamentals and would put their ships for sale in the secondhand market. This decision is motivated by the wish to avoid capital losses or negative cash flows in the future. The third category consists of ships which are in need of repair and owners believe that it is not worthwhile to bear these expenses or simply because they lack the financial means. The fourth category consists of non-profitable vessels. The owners of these vessels must have made bad decisions in the past. For example, they may have bought them at high prices on a large proportion of debt relative to equity (high leverage). Lower prices now would have resulted in margin calls by banks and requests for injection of new capital that the owners are unable to meet. Alternatively, freight rates might have fallen or interest rates may have risen and owners are unable to service their debt. All these bad decisions reflect expectation errors in forecasting economic fundamentals. As the demand for new vessels depends on expectations of demand for shipping services and relative prices (freight rates per unit of the user cost of capital), these unfortunate owners must have made expectation errors in forecasting these key economic fundamentals. As freight rates depend on expectations of fleet capacity utilisation (demand as proportion of the fleet) and the user cost of capital involves newbuilding as well as secondhand prices, the supply of secondhand vessels depends on unanticipated developments in fleet capacity utilisation, newbuilding prices and secondhand prices. Thus the supply of second-hand prices is

$$KTR^s = f(A, S_t - E_{t-1}S_t, E_t Q_{t+1}, E_t(FR/UC)_{t+1}, PS) \qquad (3.30)$$

$$f_1 > 0, f_2 < 0, f_3 < 0, f_4 < 0, f_5 > 0$$

In the above equation A stands for the age profile of the fleet. As the average age of the existing fleet increases, the supply of secondhand ships increases. The expectation errors for capacity utilisation (CU), newbuilding prices (P) and secondhand prices (PS) are captured by the vector S in (3.30). For example, if the actual fleet capacity utilisation in period t turns out to be lower than what was expected with information at $t-1$, then the second term in (3.30) is negative and this increases the supply of secondhand ships. Similarly, lower new and secondhand prices than anticipated also increase the supply of secondhand ships. However, in the long-run equilibrium these expectation errors are zero and the supply depends on the same economic fundamentals that determine the demand for secondhand ships (equation (3.28)), but with signs reversed, as owners on the supply side have opposite expectations to owners on the demand side. Finally, the supply of secondhand ships depends on their price. Normally, the supply curve is upward sloping; an increase in the secondhand price induces owners to increase the supply of ships in the secondhand market, in particular if they are motivated by replacing their old ships with new ones or have made bad decisions in the past and their ships have to be sold. But for those owners that can afford to stay in the market increasing prices may be a signal of improving economic fundamentals in the future. Accordingly, they would reduce their supply in an environment of rising secondhand prices. In this case the supply curve is downward sloping. Therefore, the outcome of an increase in secondhand prices on the supply is ambiguous – it can be positive or negative. The supply curve can be upward or downward sloping.

Equilibrium in the newbuilding and secondhand markets is achieved when demand is equal to supply in each market. The equilibrium in the shipyard market determines the stock of new fleet and newbuilding prices, while equilibrium in the secondhand market determines the volume of purchase and sale of existing fleet ships and the secondhand price. The interrelationship of newbuilding and secondhand prices is analysed with the help of Figures 3.4, 3.9 and 3.10. Initial equilibrium is attained at point A in each graph. An increase in the demand for shipping services induces owners to demand new ships from the shipyards. This shifts the demand curve from D_1 to D_2 in Figure 3.4. In the long run, shipyards increase capacity and this shifts the supply curve to the right. The new long-run equilibrium is attained at F with higher newbuilding prices and a bigger fleet. The dynamic adjustment path involves an overshooting of the long-run equilibrium; the price rises from P_1 to P^* but then falls to P_3. The fleet expands from K_1 to K_3.

In the secondhand market the increase in demand shifts to the right the demand curve from D_1 to D_2 in Figure 3.9. But for reasons already explained the supply of secondhand ships is reduced. This shifts the supply curve to the left from S_1 to S_2. If the supply curve has the normal upward slope then the volume of

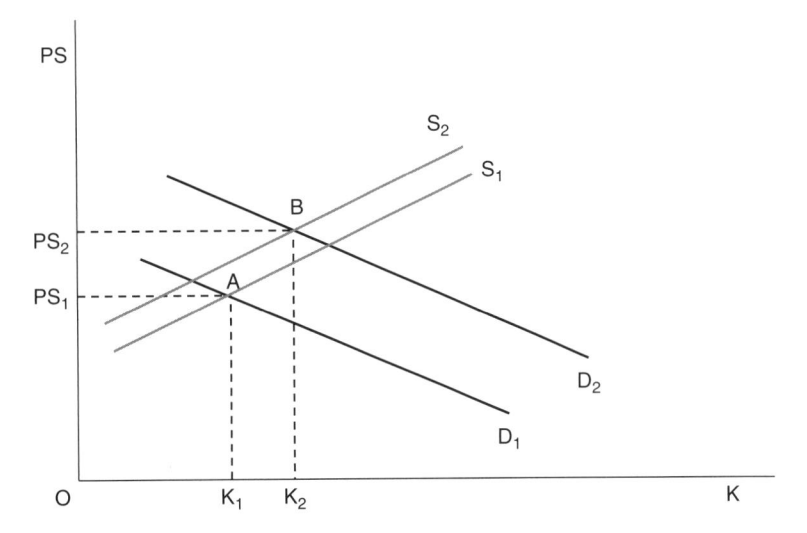

Figure 3.9 The impact on SH prices and traded fleet when the supply is upward sloping

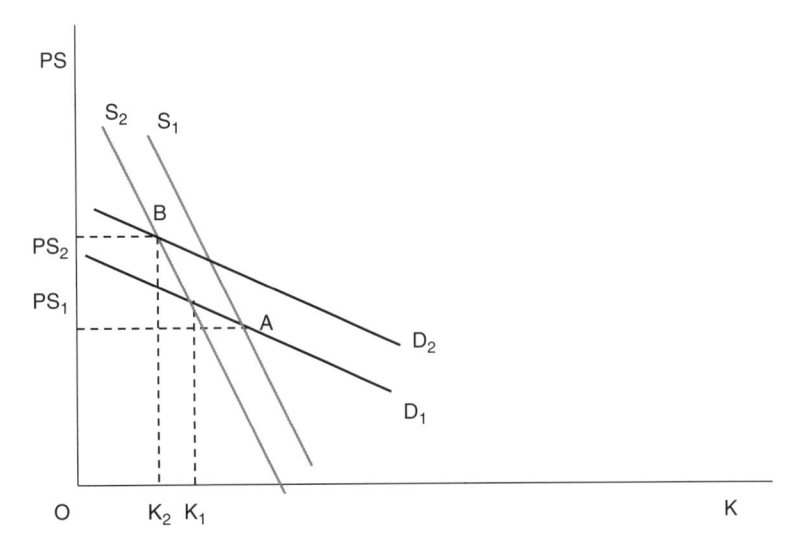

Figure 3.10 The impact on SH prices and traded fleet when the supply is downward sloping

purchase and sale is likely to increase, as the shift in demand outweighs the shift in supply. The new equilibrium is attained at B in Figure 3.9. The secondhand price increases from PS_1 to PS_2 and the traded fleet increases from K_1 to K_2. The secondhand price increases under the twin influence of an increase in demand and a reduction in supply. But it is unclear whether the increase in secondhand prices exceeds that of newbuilding ones because the latter overshoots the new long-run equilibrium.

If the supply curve of secondhand ships is downward sloping, then the new long-run equilibrium is attained at B in Figure 3.10. The volume of traded fleet is reduced from K_1 to K_2 and the price increases from PS_1 to PS_2. The increase in the secondhand price with a downward-sloping supply curve is even more pronounced than in the case where the supply is upward sloping (compare Figures 3.9 and 3.10). But again it is unclear whether secondhand prices increase more than newbuilding ones.

In the above framework, an owner has to solve the dynamic optimisation problem twice and compare the outcomes in order to derive the proportion of the fleet to be devoted to new and secondhand ships. It is thus useful to have a unifying framework that would enable the owner to decide the proportion of each ship category in the total fleet in one step, thus avoiding the cumbersome task of solving the optimisation problem twice. This can easily be achieved by comparing the ratio of the two prices.[12] Thus the proportion of secondhand to new fleet is a function of their relative prices.

$$\frac{K^{sh}}{K^n} = f\left(\frac{PS}{P}\right) \quad f_1 < 0 \qquad (3.31)$$

As the price of secondhand ships rises relative to newbuilding ones, the proportion of secondhand ships to new ones declines. This simple relationship compares the expected profitability of each ship category. This can easily be verified by comparing the determinants of each price.

$$PS_t = E_t \sum_{s=t}^{T} \frac{1}{(1+r_s+\delta)^{s-t+1}} \cdot \left(\frac{a}{A^\rho} \cdot FR_{s-t+1} \cdot \left(\frac{Q}{K}\right)_{s-t+1}^{1/\sigma} - G_K(I,K)\right) \qquad (3.32)$$

$$P_t + G_I(I,K) = E_t \sum_{s=t}^{T} \frac{1}{(1+r_s+\delta)^{s-t+1}} \cdot \left(\frac{a}{A^\rho} \cdot FR_{s-t+1} \cdot \left(\frac{Q}{K}\right)_{s-t+1}^{1/\sigma} - G_K(I,K)\right) \qquad (3.33)$$

For new ships the cost includes the price and the cost of the fleet adjustment. For secondhand ships the cost of fleet adjustment is much smaller, if not zero. The technological differences are also embedded in each formula. The ratio of (3.32) to (3.33) that determines the proportion of secondhand to new fleet can be simplified if we assume, in line with Beenstock and Vergottis (1993), that new and secondhand ships are perfect substitutes. In this simplified case it is as if there was a futures market that relates new and secondhand prices. This implies that the expectation of the secondhand price for period t as with information at the time the order for the new ship was placed is an unbiased predictor of the newbuilding price today.

$$P_t = E_{t-m}PS_t + u_t \qquad (3.34)$$

If m is the delivery lag, then the order for the new ship was placed at $t-m$. According to (3.34) the price of a new ship is equal to the expectation of the secondhand price plus an error term, u. The unbiasedness property of the prediction requires that the error terms for each time period (in hypothetical repeated sampling) follow a normal distribution with zero mean. In reality we observe just one observation out of the entire distribution of error terms for each time period. When we compare the distribution of error terms between two successive time period the unbiasedness property requires in addition that they have the same variance and the errors are independent of each other. The independence is ensured if they are not correlated. The unbiasedness property is usually referred to that the error term u is identically and independently distributed and is denoted as iid.

In the real world new and secondhand ships are not perfect substitutes and as it is obvious from (3.32) and (3.33); they differ by age, reflected in the depreciation rate, δ; technological factors captured in the coefficients A and σ, which for convenience are denoted by ρ; the forecast error u, and a risk premium a. Therefore, the price of a new ship should exceed the secondhand price by all these factors.

$$P_t = (1+\delta+\rho+u+a)\cdot PS_t \quad \text{or} \quad \frac{PS_t}{P_t} = \frac{1}{1+\delta+\rho+u+a} \qquad (3.35)$$

With the exception of the risk premium all other factors can be considered as fixed coefficients in a shipping cycle. The risk premium, on the other hand, captures the economic fundamentals that differentiate the demand for new ships relative to secondhand ones. The analysis in this chapter suggests that the risk premium is a function of the following economic fundamentals.

$$a = f(\sigma_Q, \Delta Q^s - \Delta Q^L, \Delta CU_t - \Delta SCU_t) \quad f_1 > 0, f_2 > 0, f_3 > 0 \qquad (3.36)$$

The first factor is the degree of uncertainty regarding the strength of the demand for shipping services. This is captured by the standard deviation of the forecast error of demand, σ_Q. For example, assume that an owner expects demand to grow by 5 per cent, with a margin of error (one standard deviation) of 1 per cent. Thus with 95 per cent probability the demand would grow in the range 5 per cent plus or minus two standard deviations, namely in the range of 3 per cent to 7 per cent. If uncertainty regarding the accuracy of the demand forecast, σ_Q, increases to 2 per cent, then the range would widen to 1–9 per cent. This would induce the owner to prefer a secondhand ship than a new one, as it involves a smaller capital expenditure and therefore a smaller risk on the investment. Hence, an increase in σ_Q raises the risk premium a and therefore reduces the secondhand price relative to the new one. This according to equation (3.31) increases the proportion of secondhand to new fleet.

The second factor that affects the risk premium is whether the owner expects the increase in demand to be permanent or transitory. If the increase in demand is perceived as transient, lasting for one or two years, then the owner would again prefer a secondhand ship to a new one, as it is immediately available to take advantage of the boost in demand with a smaller capital expenditure, which increases profitability. If we denote by ΔQ^s the expected increase in demand in the short term (one or two years) and by ΔQ^L the long-term increase in demand (more than the delivery lag), then if $\Delta Q^s > \Delta Q^L$ the increase in demand is perceived as transitory. This raises the risk premium and diminishes the secondhand price relative to the new one, thereby inducing the owner to increase the proportion of secondhand to new fleet. If $\Delta Q^s < \Delta Q^L$, the risk premium falls and lowers the new price relative to the secondhand one, which, in turn, raises the proportion of new relative to secondhand fleet. If $\Delta Q^s = \Delta Q^L$ the result is ambiguous. The risk premium remains unchanged and hence so do relative prices. The outcome depends on the owner's degree of risk aversion, his investment horizon and the availability of own funds. A risk-neutral owner may have a long investment horizon and prefer a high debt to equity ratio (that is, high leverage) and therefore buy a new ship. A risk-averse owner may have small appetite for risk (that is, low leverage) and a shorter investment horizon and buy a secondhand ship. Therefore, the outcome is ambiguous when $\Delta Q^s = \Delta Q^L$.

The third factor that affects the risk premium is the expectation of the spillover effect from the demand for shipping services to the demand for vessels. An anticipated increase in the demand for shipping services would raise the expected fleet capacity utilisation rate, ΔCU. This excess demand for shipping services would spill over to an excess demand for vessels (from the freight market to the shipyard market). But if there is shipyard excess capacity, ΔSCU, then $\Delta CU - \Delta SCU > 0$, and the spillover would fail to lift new prices, as shipyards would compete for orders before raising their prices. This would raise the risk premium and reduce the secondhand price relative to the new one and induce owners to expand the proportion of the secondhand to new fleet.

The overall conclusion is that the relationship between secondhand prices and newbuilding ones is ambiguous. Sometimes, secondhand prices may be a leading indicator of newbuilding ones and on other occasions it may be a lagging indicator.

APPENDIX: OPTIMAL FLEET CAPACITY EXPANSION

OPTIMAL FLEET AND OPTIMAL SPEED – LONG-RUN ANALYSIS

As with the short-run analysis of shipping services (analysed in section 1) of the main text, the long-run supply of cargo can be viewed as a production function in which the stock of fleet and the average fleet speed are combined to produce the supplied cargo, Q, measured in tonne-miles. The production function of the short-run analysis of section 1 is reproduced here for convenience:

$$Q_t = F(K_t, S_t); \quad F_K > 0, F_S > 0, F_{KK} < 0, F_{SS} < 0 \qquad (A.1)$$

The marginal product of fleet and marginal product of speed (first partial derivatives) are positive, while the second-order derivatives are negative, reflecting diminishing marginal productivity.

Let FR_t stand for the freight rate per tonne-mile in period t, p_t for the NB price of a vessel, I_t for gross investment, namely the addition by one ship in the fleet with all other symbols as defined in the previous chapter. Then, in any given period, t, the owner's profit (net cash flow) is defined as:

$$\pi_t = FR_t \cdot Q_t - PB_t \cdot C_t - p_t \cdot I_t - G(I_t, K_t) \quad (A.2)$$

The product $FR \times Q$ is the total revenue to the owner by selling Q tonne-miles of services at the freight rate FR. The second term on the right-hand-side (RHS), namely the product $PB \times C$, is the fuel bill; the third term on the RHS (that is, the product $p \times I$) is the outlay in each period towards buying one extra ship; and G is the cost of adjusting the fleet rapidly or slowly.

Gross investment is equal to net investment plus replacement investment. The latter is assumed to be proportional to the stock of fleet. If δ represents the depreciation of the stock of fleet per period of time, say one year, then replacement investment is δK_{t-1}. Hence, net investment, ΔK, in discrete time is equal to:

$$\Delta K_t = K_t - K_{t-1} = I_t - \delta K_{t-1} \quad (A.3a)$$

In continuous time, net investment is:

$$\dot{K} = I - \delta K \quad (A.3b)$$

A dot over a variable denotes the first-order time-derivative. This approximates the change in the stock of fleet per unit of time, say one year.

By substituting (A.3b) into the G function in (A.2), the cost of adjusting the fleet is

$$G(I - \delta \cdot K) \quad G' > 0, G'' > 0 \quad (A.4)$$

Adjustment costs incur as the owner tries to expand the fleet rapidly. If the owner tries to reach an optimum fleet size of ten ships in a short period of time, then he would be facing increasing costs (a higher price per ship). Usually, for a big order an owner would get a discount, but only if the deliveries are spread over time to meet the shipyard capacity. If the owner demands a faster delivery than the capacity of the yard, then the cost per ship would increase. Adjustment costs should also include the cost of operating the fleet at a loss for a short period of time. This would be typical if the owner is contrarian, buying ships early in the cycle in anticipation of an increase in the demand for shipping services in the future. In this framework, the costs of adjusting the fleet depend positively on the amount of gross investment, I; the first derivative is positive. Moreover, the higher the investment (that is, the faster the fleet adjustment), the bigger the costs; the second derivative is also positive. This assumption implies cost convexity: costs are increasing with investment at an accelerating pace. But the costs of adjustment would also depend on the stock of fleet, K, as this determines the owner's bargaining power with shipyards. The bigger the owner's fleet is, the larger the shipyard discount for a given order and hence the lower the costs of adjustment. Thus, G is a decreasing function of K. The function G and the first and second order derivatives are non-negative so that the cost is zero for zero investment.

The owner is assumed to operate in a perfectly competitive freight market so that he can sell as much cargo as he can at the equilibrium freight rate, FR. The owner aims at maximising the present value of its profits (net cash flow) over an infinite horizon:

$$V_t = \max \int_{s=t}^{\infty} \pi_s \cdot \exp\{-r_t \cdot (s-t)\} \cdot ds \qquad (A.5)$$

where V_t is the value of the firm at time t and r_t is the interest rate at which the firm discounts future net cash flows in date s back to the present date t. The discount rate, r, is a weighted average of the interest rate charged on loans to finance the fleet, K, and the opportunity cost of the owner's equity. The owner chooses the time path of the average speed and gross investment over the planning horizon (that is, for $s \geq t$) subject to the production function (A.1), the dynamic constraint, which describes the evolution of the stock of fleet described by (A.3b) and the condition that K_t is given (or predetermined) at the beginning of each period in which investment and production of vessels takes place. The level of the fleet stock at time t, K_t, is treated as a predetermined variable. In other words, the stock of fleet at time t is given and therefore represents an initial condition. The evolution of the fleet stock is obtained on the assumption that the capital stock depreciates at a constant proportional rate, δ.

To solve the firm's maximisation problem the average speed and gross investment must be chosen to maximise the current value Hamiltonian, H_t

$$H_t = \pi_t + P_t \, \dot{K}_t$$

where P_t is the shadow price of fleet. Substituting (A.2) and (A.3b) into the above relationship the current value Hamiltonian becomes

$$H_t = FR_t \cdot F(S_t, K_t) - PB_t \cdot d \cdot f(S) - p_t \cdot I_t - G(I_t, K_t) + P_t \cdot (I_t - \delta \cdot K_t) \qquad (A.6)$$

The first-order conditions for maximum imply setting the partial derivatives of (A.6) with respect to S_t and I_t equal to zero

$$F_S(S_t, K_t) = PB_t \cdot d \cdot f'(S)/FR_t \qquad (A.7)$$

$$p_t + G_I(I_t, K_t) = P_t \qquad (A.8)$$

Equation (A.7) states that the owner would adopt an optimal average speed in each period t at the level at which the marginal product of speed is equal to real cost of bunker costs, adjusted for technology, in the same period. Therefore, the optimality of the dynamic problem is the same as that in the static problem of the previous chapter. The owner does not need to form expectations of the state-variables, for example, freight rates and NB prices. The optimality condition for the adjustment path of gross investment, on the other hand, involves an intertemporal trade-off between future benefits and costs.

In addition to choosing I_t and S_t the firm's intertemporal maximisation problem requires that the shadow price of fleet, P, obeys the condition:

$$\dot{P} - r_t \cdot p_t = -\frac{\partial H_t}{\partial K_t} \tag{A.9}$$

which implies that

$$\dot{P} = (r_t + \delta) \cdot p_t - FR_t \cdot F_K(S_t, K_t) + G_K(I_t, K_t) \tag{A.10}$$

The last condition is a differential equation in P whose solution is:

$$P_t = \int_t^\infty \{FR_t \cdot F_K(S_t, K_t) - G_K(I_t, K_t)\} \cdot \exp[-(r_t + \delta)(s - t)]ds \tag{A.11}$$

This equation states that the shadow price of fleet, P, is equal to the present value of the future stream of marginal profits from one extra ship added to the fleet at time t, thus providing justification to the present rule used in section 1. The marginal profit consists of the *value* of the marginal product (or the marginal revenue) of fleet and the reduction in the adjustment cost due to the additional vessel. The discount factor consists not only of the interest rate, r, but also of the depreciation rate, δ, since the stock of fleet diminishes in value at that rate.

Substituting (A.11) into (A.8) the optimality condition implies that the firm will expand fleet capacity until the marginal cost of investment (that is, the price of a vessel and the cost of adjustment) is equal to the present value of marginal profits of the fleet stock:

$$p_t + G_I(I_t, K_t) = \int_t^\infty \{FR_t \cdot F_K(S_t, K_t) - G_K(I_t, K_t)\} \cdot \exp[-(r_t + \delta)(s - t)]ds \tag{A.12}$$

If the shadow price of fleet, P_t, is interpreted as the price a marginal vessel can be bought or sold in the second-hand market, PS, then the optimality condition (A.10) can alternatively be written as

$$\{FR_t \cdot F_K(S_t, K_t) - G_K(I_t, K_t) - \delta \cdot p_t + \dot{PS}\} / p_t = r_t \tag{A.13}$$

This equation states that the firm will expand fleet capacity up to the point where the real profit from one extra vessel is equal to the cost of capital, r_t, which is a weighted average of the interest rate charged on loans to finance the fleet and a risk premium (for example, the opportunity cost of the owner's equity). The real profit of one extra ship consists of four components. First, the increase in revenue resulting from one extra vessel, $FR_t \times F_K(S_t, K_t)$; second, economies of scale, that is, the reduction in the cost of adjustment due to an additional ship, $-G_K(I_t, K_t)$; third, the extra loss due to the depreciation of the vessel, $\delta \dot{p}_t$; fourth, the capital gains (or losses), $\dot{PS}$, from selling the vessel in the secondhand market.

On the interpretation of the shadow price of fleet as the price at which the vessel can be sold or bought in the secondhand market the optimality condition (A.10) can be written in a fourth alternative way as

$$F_K = \{G_K(I_t, K_t) + (r_t + \delta) \cdot p_t - \dot{PS}_t\} / FR_t \equiv UC_t / FR_t \quad (A.14)$$

Equation (A.14) states that the firm would expand capacity up to the point where the marginal product of fleet is equal to the real user cost of capital, UC_t / FR_t, where the user cost of capital is defined as

$$UC_t = (r_t + \delta) \cdot p_t + G_K(I_t, K_t) - \dot{PS}_t \quad (A.15)$$

The user cost of capital increases with the price of newbuilding ships, the cost of capital and the depreciation rate, but decreases with the capital gains (when the ship is sold in the secondhand market) and with economies of scale (recall that $G_K < 0$).

By multiplying both sides of (A.14) by the freight rate the optimality condition is restated as the marginal revenue of fleet equals the user cost of capital, UC.

$$FR_t \cdot F_K = UC_t \quad (A.16)$$

THE DEMAND FOR VESSELS

To relate the optimal (target or desired) stock of fleet to its determinants consider the Constant Elasticity of Substitution (CES) production function

$$Q = A[a\ K^{-\rho} + (1-a) S^{-\rho}]^{-1/\rho} \quad (A.17)$$

where $A > 0$ is a constant reflecting the ship's technological improvement; $0 < a < 1$ is a constant reflecting the contribution of the fleet to the supply of shipping services, while $(1 - a)$ reflects the contribution of speed; $\rho > -1$ is the elasticity of substitution between fleet and average speed. The parameter ρ is related to σ through the formula

$$\sigma = 1/(1+\rho) \quad \text{and} \quad \rho = (1-\sigma)/\sigma \quad (A.18)$$

Hence, the inequality $\rho > -1$ implies that $\sigma > 0$. The marginal product of fleet is

$$F_k(K_t, S_t) = [a/A^\rho][Q_t/K_t]^{(1/\sigma)} \quad (A.19)$$

Substituting (A.19) into (A.14) and solving for K gives the desired level of the stock of fleet (or the demand for vessels)

$$K^* = [a / A^\rho]^\sigma \cdot Q \cdot [FR / UC]^\sigma \qquad (A.20)$$

Thus, equation (A.20) states that the demand for vessels depends on economic fundamentals, namely the demand for freight, Q; the user cost of capital, UC, which includes the price of newbuilding ships and capital gains (or losses) in the secondhand market; the freight rate, FR; the ship's technological improvement, A; and the elasticity of substitution between fleet and average speed, σ. The demand for vessels rises as the demand for freight increases, relative prices (namely, freight rates relative to the user cost of capital) rise and economies of scale increase. The demand for vessels varies inversely with the price of ships through the user cost of capital. It is also negatively related to the cost of capital and the depreciation rate.

The demand for freight has an elasticity of 1, whereas relative prices have an elasticity of σ. Only when $\sigma = 1$ the demand for freight and relative prices are equally important in determining the demand for vessels;[13] for $\sigma < 1$ the demand for cargo is more important than relative prices in the demand for vessels; in the limiting case in which $\sigma = 0$, only the demand for cargo matters. The demand for vessels is perfectly inelastic with respect to freight rates and the prices of ships. In this case ships are no longer assets.

THE ROLE OF EXPECTATIONS

From the first-order condition of the optimal speed, equation (A.7), it is clear that expectations play no role whatsoever. In arriving at the optimal average speed the owner would increase the speed up to the point where the marginal product of speed is equal to the real value of bunker costs (that is, bunker costs deflated by the freight rate) adjusted for technical factors. This condition does not require the firm to take a deep look into the future, namely to form expectations, as at any point in time the *current* marginal product of speed is equated to the *current* real bunker rate.

However, the condition for investment is more complicated because the stock of fleet is a quasi-fixed factor of production. Owing to fleet's durability the benefits of investing require calculation of the marginal product of fleet throughout the lifetime of the stock of fleet, as with information available at time t. This forces the firm to take a deep look into the future and requires the formulation of expectations of the time path of all variables involved in the calculation of the marginal products of the fleet, namely the demand for shipping services, freight rates, vessel prices and the user cost of capital.[14] The role of expectations in the optimality condition of the fleet becomes apparent if equation (A.12) is rewritten in discrete time and the owner is assumed to maximise the expected value of V in equation (A.5), with information available at time t, that is, $E_t(V_{t+1})$:

$$p_t + G_I = E_t \left\{ \sum_{s=t}^{\infty} \frac{1}{(1 + r_s + \delta)^{s-t+1}} (MRK_{s-t+1} + MCR_{s-t+1}) \right\} \qquad (A.21)$$

The marginal cost is equal to the price of vessel, p, and the marginal cost of adjustment, G_I. These represent sunk costs, not entirely because of the scrap market, which cannot be recovered and require no expectations, as they are known today. The marginal benefits are equal to the present value of the sum of the expected marginal revenues of one extra vessel derived from its entire lifetime, MRK, and the reduction in the adjustment cost due to the additional vessel in the fleet, MCR (that is, economies of scale).

The marginal revenue of one extra ship embodies the firm's production function and the demand curve it faces. For a price-taking firm the marginal revenue of one extra vessel is simply the freight rate multiplied by the marginal product of vessel. This becomes concrete if we use the marginal of vessel

$$MRK = FR_t \cdot F_K(K_t, S_t) = \frac{a}{A^\rho} \cdot FR_t \cdot \left(\frac{Q}{K}\right)^{1/\sigma} \qquad (A.22)$$

The reduction in the adjustment cost due to one additional vessel is

$$MCR = -G_K > 0 \qquad (A.23)$$

As it is clear from (A.21) expectations are important for investment decisions. The computation of the expected stream of future marginal revenues of the fleet requires assumptions about technology and about the demand curves facing the firm in the freight market as well as input markets. In general, it will involve expectations of the time path of the demand for shipping services (output), the freight rate, NB vessel prices, the interest rate and the depreciation rate.

To illustrate some of the points related to expectations denote the marginal revenue of capital, MRK, simply as λ. Assume that λ follows a stochastic process which, for expositional convenience, can be specified as a first-order univariate autoregression,

$$\lambda_t = \mu\, \lambda_{t-1} + \varepsilon_t \qquad (A.24)$$

where μ is an expectations parameter and ε is an expectations error. Under rational expectations, ε_t is orthogonal to all variables known to the firm in period t. Hence, the expectation of λ is

$$E_t(\lambda_{t+s}) = \mu^{s+1} \lambda_{t-1} \qquad (A.25)$$

This shows the ease with which expectations can be formed as they depend on only one parameter, μ, and the information set contains only one variable, λ.

The exercise is conceptually not more complicated, if it is assumed that λ is generated as a linear combination of some observable vector-Z_t which evolves according to a vector auto-regressive process. Formally, this can be written as

$$\lambda_t = a^T Z_t \qquad (A.26)$$

where a is a vector of known constants and

$$Z_t = A\, Z_{t-1} + \varepsilon_t \qquad (A.27)$$

where the vector-Z contains all the variables that enter equation (A.20), namely demand for shipping services (freight), freight rates, NB vessel prices, interest rates and the depreciation rate. This scheme is slightly more elaborate than the previous one as it makes λ a function of a few key economic variables.

However, the disadvantage with these expectations schemes is that they are mechanical, in the sense that they extrapolate the most recent information. Although they assume that expectations are rational in reality they are backward rather than forward looking. A forward-looking behaviour will take into account the impact on the vector-Z of the response of the policymakers to current and future economic conditions. In this approach firms base their expectations of the variables that enter the marginal revenue of fleet on their own estimates of what the policymakers would do in the future. Under rational expectations

$$E_t(Z_{t+s}) = Z_{t+s} + \varepsilon_{t+s} \quad \text{for all } s \qquad (A.28)$$

The expectation error ε is purely random and therefore assumes random values from a normal distribution with zero mean and a constant variance. Equation (A.26) states that expectations are, on average, correct, as the mean value of the error term ε is zero. To make sure that owners compute, even on average, accurate expectations they must form expectations of the vector-Z based on a structural model, like the K-model, which explains the data generation process of Z. In the vector-Z of K-model all shipping variables, are explained in terms of economic fundamentals, such as GDP, consumption, exports, industrial production, inventories, inflation, commodity prices, interest rates and exchange rates. Let Y denote the vector of all these economic fundamentals, then the shipping variables included in the vector-Z are explained by the model:

$$Z_t = Z(Y_t, Y_{t-i}) + \varepsilon_t \quad \text{for all } i \qquad (A.29)$$

In equation (A.29) ε stands for all other non-systematic variables that affect the vector-Z, which are not included in the vector-Y. The values of ε are randomly drawn from a normal distribution with zero mean and constant variance. The vector-Y is explained by a macro model, such as the macro K-model, where fiscal and monetary policy, denoted by U, plays an important role. Thus, the macro model can be written as

$$Y_t = Y(U_t, U_{t-i}) + e_t \quad \text{for all } i \qquad (A.30)$$

In equation (A.30) e_t stands for all other non-systematic variables, not included in vector-U that affect the vector-Y, again drawn randomly from a normal distribution with zero mean and constant variance.

According to this framework, the owner forms expectations of the vector-Z by substituting equations (A.29) and (A.30) into (A.28) and setting the expected value of all error terms equal to zero:

$$E_t(Z_{t+s}) = E_t Z[Y(U_{t+s})] \quad \text{for all } s \qquad (A.31)$$

According to equation (A.31) expectations of all shipping variables that enter vector-Z depend on expectations about future economic policy. These forward-looking expectations imply that the owner calculates the impact on shipping variables from the policy reaction to current and future economic conditions. For example, if growth in the world economy falters, then the owner anticipates loosening of fiscal and monetary policies in the major economic regions of China, the US, Europe and Japan. This would stimulate the demand for shipping services, which would increase freight rates and vessel prices at the time the news become available. This is a fundamental principle of asset prices. They adjust instantly to news on economic fundamentals. This adjustment entails a short run overshooting of the new long-run equilibrium of freight rates and vessel prices.

A COMPARISON WITH CONVENTIONAL MODELS

It is instructive to compare our model with conventional models found in the literature, such as Beenstock (1985), Beenstock and Vergottis (1989a, 1989b, 1993) and Strandenes (1984, 1986). To put our model in perspective consider that the owner is maximising the expected value of the firm over the investment horizon $(1, \ldots, T)$ with information available at time t and take the discrete version of (A.12), while substituting the marginal product of fleet from (A.19). Then the price of a vessel, according to our model, is equal to

$$p_t + G_I(I,K) = E_t \sum_{s=t}^{T} \frac{1}{(1+r_s+\delta)^{s-t+1}} \cdot \left(\frac{a}{A^\rho} \cdot FR_{s-t+1} \cdot \left(\frac{Q}{K} \right)_{s-t+1}^{1/\sigma} - G_K(I,K) \right) \qquad (A.32)$$

Strandenes hypothesises that the demand for new ships is a positive function of the value of the firm, V, relative to the price of new ships. Thus,

$$K^* = (V/p)^{k'} \qquad (A.33)$$

The value of the firm, V, is equal to the discounted stream of profits anticipated over the lifetime of the vessel. The parameter k' measures the elasticity of demand of new vessels. According to equation (A.33) the demand for new vessels increases with expected profitability and decreases with the price of ships.

The discounted stream of profits over the lifetime of the vessel is assumed to be

$$V = k \cdot \frac{1}{r+\delta} [H^\infty - OC] \qquad (A.34)$$

where k is a time-dependent constant that controls for the effect of the delivery lag on the value of the new building contract, H^∞ is the perpetual time charter equivalent and OC = fixed operating costs, such as wages. The perpetual time charter is not necessary, as a contract of any duration can instead be assumed. For example, Strandenes assumes that the economic lifetime of a vessel is around 15 years and therefore inserts H^{15} (a 15-year time charter contract) in (A.34). The accounting identity that relates profits, π, to the time charter equivalent for the duration of the contract, τ, and fixed operating costs is given by

$$\pi = H^\tau - OC \tag{A.35}$$

If capital markets are competitive, arbitrage will ensure that the newbuilding price of a ship would be equal to the present value of expected short-term profits, π^s, and long-term profits, π^L. Thus,

$$p_t = k\left(\frac{1}{r+\delta}[a \cdot \pi_t^s + b \cdot \pi_t^L]\right) \tag{A.36}$$

We would expect that the sum of $a + b = 1$, if the shipping market is efficient. Strandenes finds support of the efficient market hypothesis for some ship categories.

A comparison of equation (A.36) with (A.32) shows the similarities and differences between our approach and that of Strandenes. In both models the price of a new ship reflects the discounted value of the future stream of profits. But whereas in Strandenes, profitability is related to time charter equivalent rates of different contract duration, our analysis derives the determinants of this profitability through an explicit dynamic profit maximisation over the lifetime of the vessel. The explicit dynamic framework shows that profitability depends on expectations of freight rates and demand per unit of fleet capacity with a coefficient that depends on the substitution between fleet and average speed in producing shipping services. With low elasticity of substitution (that is, when $\sigma < 1$) the demand for shipping services per unit of fleet capacity is more important than freight rates, which has a unitary elasticity under all circumstances. As σ increases, the importance of demand per unit of fleet capacity decreases. When $\sigma = 1$, both freight rates and the demand per unit of fleet capacity are equally important. For $\sigma > 1$, freight rates are more important than demand per unit of fleet capacity. This qualifies the Strandenes result as only modern and technologically advanced vessels provide the option of slow and fast steaming and, in particular, containers. Another feature of our analysis is that the present value rule takes into account the economies of scale that result from adding one extra vessel to the fleet. Finally, it is not simply that the price of a vessel should equal the discounted stream of profits, so defined, but also the extra cost of adjusting the fleet.

A comparison of equation (A.33) with (A.20) shows the similarities and differences of the demand functions of ships. In both models the demand for ships varies inversely with the price and positively with profitability. In Strandenes the demand for ships depends on a direct comparison of profitability to the price of a ship. In our analysis the negative relation between the demand for ships and their price depends on its impact on the user cost of capital that hinges upon the elasticity of substitution between fleet and average speed in producing shipping services. A more expensive ship may be worthwhile than a cheaper one if the higher price is offset by increasing economies of scale, lower depreciation or lower interest rates and, more importantly, by expectations of capital gains when the ship is sold in the secondhand market.

The Beenstock and Vergottis model is analysed in detail in Chapter 6. Here we simply compare the equations for the price of new ships and the demand for ships. In Beenstock and Vergottis (1993) new and old ships are perfect substitutes, but for age. This enables Beenstock and Vergottis (BV) to hypothesise that there is an efficient futures market where shipyards, owners and speculators alike, buy and sell new building contracts. In this efficient futures market the price of new ships is related to the price of secondhand ships, *PS*, through the following equation

$$p_t = E_t PS_{t+m} + k_1 \qquad (A.37)$$

where k_1 is a risk premium, which nonetheless is treated as a constant and m is the shipyard delivery lag. This formulation has the unappealing feature that the demand for vessels is perfectly elastic at the expected future secondhand price minus the risk premium. This implies that owners do not have a desired or optimal stock of fleet, but would accept any fleet that satisfies the arbitrage equation (A.37). The perfect substitutability of secondhand and new ships is conflicting with empirical evidence. For example, a high degree of uncertainty as to expected level of demand for shipping services or a perception by owners of a temporary rather than a permanent increase in demand would make them to opt for secondhand ships because in the first case they commit less capital and in the second case they take immediate advantage of the increase in demand. This can result in a premium of SH prices over NB ones.

In the BV model, shipyards fix the price of new ships as a mark-up on costs. This is counterintuitive, as the demand for ships by owners is generally expected to play a significant role in the price formation. In contrast, in the BV model it is secondhand prices that absorb fluctuations in demand, which means that SH prices are demand determined. But even that is counterintuitive, as owners are supposed to diversify their wealth according to portfolio theory by comparing the return on ships with that on alternative investments, such as stocks or commodities. Thus, if r^s denotes the return on shipping and r the return on alternative investments, the proportion of global wealth allocated to shipping depends on these returns

$$p \cdot K / W = f(r^s, r) \quad f_{r^s} > 0, f_r < 0 \qquad (A.38)$$

where the numerator of the left-hand side is the proportion of wealth invested in shipping and W is global wealth.

In the distant future this may be an accurate description of investment in the shipping market, if outside investors, such as private equity funds or hedge funds, come to dominate the market. At the moment, the influence of these outsize investors is rather small, but nonetheless growing. In their framework, the return on shipping is equal to that on alternative investments plus a risk premium. The influence of outside investors would ensure through arbitrage the equality of returns, but for a risk premium

$$r^s = r + p \cdot K / W \qquad (A.39)$$

where the last term in (A.38) represents the risk premium in shipping. The risk premium increases with the vessel price and the fleet and decreases with global wealth.

Nonetheless, it is not clear that the influence of speculators will induce equality of returns in shipping and alternative investments, as hypothesised by BV in equation (A.39). The experience of the boom and bust cycle of 2003–13 shows that the influence of speculators enhanced the upside on shipping returns in the boom, but at the cost of aggravating the downside of returns in the bust. Therefore, the influence of outside investors is likely to increase vessel price volatility and therefore force divergence rather than convergence of returns.

LONG-RUN EQUILIBRIUM AND ADJUSTMENT

Equations (A.3b) and (A.10) is a pair of differential equations that describe the dynamic adjustment of the fleet and the price of vessels. Equation (A.3b) describes the actual dynamic adjustment path of the fleet provided by all shipyards. If it is assumed that the actual fleet delivered is the result of profit maximisation on behalf of shipyards, then equation (A.3b) captures the *industry* supply curve of vessels by all shipyards.[15] Optimising behaviour by shipyards implies that the industry supply curve of the fleet is a positive function of the price of vessels. Shipyards would supply more vessels, if the vessel price increases. Accordingly, equation (A.3b) can be rewritten as

$$\dot{K}_t = I[P(t)] - \delta K_t \quad \text{with} \quad \frac{\partial I}{\partial P} > 0 \qquad (A.40)$$

By setting the time derivative of the fleet (that is, vessel deliveries) equal to zero, equation (A.30) describes combinations of fleet and vessel prices along which the supply of vessels by shipyards is optimal, in the long run. Thus, in long-run equilibrium, the supply of fleet is:

$$I[P(t)] = \delta K_t \qquad (A.41)$$

Equation (A.41) can be solved for the vessel price, P, which gives the industry supply curve of the fleet as a positive function of the price of vessels:

$$P = I^{-1}(\delta \cdot K), \quad \frac{\partial P}{\partial K} > 0 \qquad (A.42)$$

Similarly, equation (A.10) describes the dynamic adjustment path of vessel prices, which is consistent with optimising behaviour on the part of the owners. In long-run equilibrium, the time derivative of vessel prices is equal to zero. Thus, setting equation (A.10) to zero and solving for p gives the demand for fleet consistent with long-run equilibrium:

$$p = \frac{FR \cdot F_K - G_K}{r + \delta}, \quad \frac{\partial P}{\partial K} < 0 \qquad (A.43)$$

This is a negative function of K on the assumption that the numerator is negative. The marginal product of the fleet is negative for a decreasing-returns-to-scale production function. But since G_K is also negative, the first term in the numerator of (A.43) must exceed in absolute value the second term. Thus, on these assumptions a negatively sloped long-run demand curve for vessels is derived.

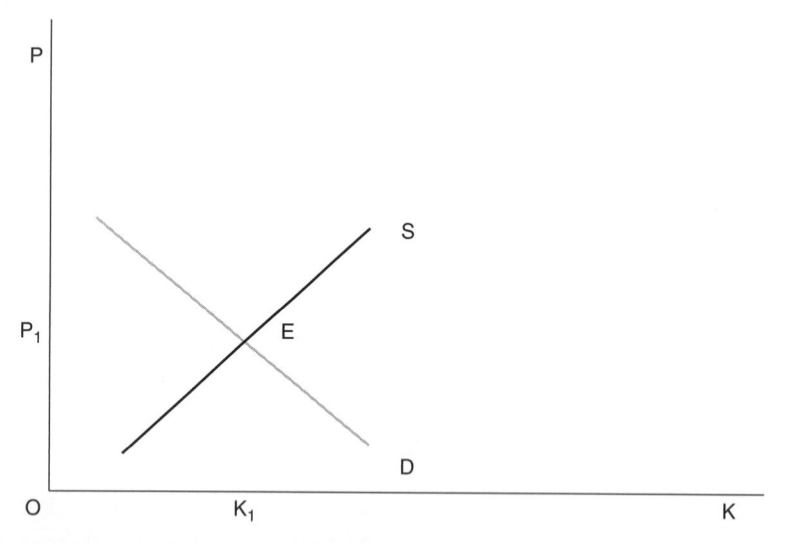

Figure 3.A1 Shipyard long-run equilibrium

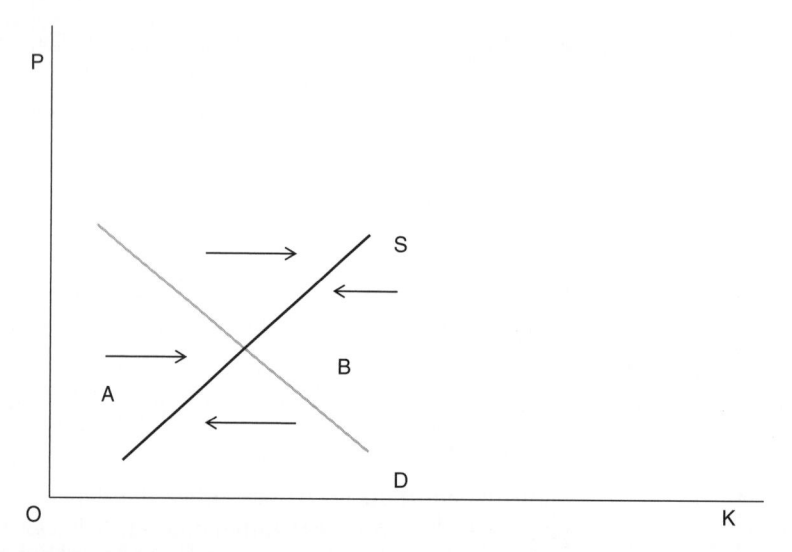

Figure 3.A2 Adjustment of fleet

Figure 3.A1 plots the long-run demand curve for fleet and the long-run supply curve of the fleet (equations A.42 and A.43). The intersection of the demand and supply curves defines the equilibrium stock of fleet and equilibrium vessel price in the long-run equilibrium, namely when the adjustment is complete. Thus, long-run equilibrium is defined at E with optimal fleet K_1 and optimal vessel price P_1.

A major issue is whether this long-run equilibrium is stable. Stability means that if the system starts from any point other than E in Figure 3.A1, would it end up at E or whether the system would converge to a zero or an infinity combination of (P, K).

Figure 3.A2 describes the adjustment of the fleet by considering the implications of (A.40) when the system is in disequilibrium. Suppose that the system starts at point A in Figure 3.A2, which lies above the supply curve. At point A, the vessel price is higher than that consistent with long-run equilibrium and therefore $I[P(t)] - \delta K > 0$. This implies that the time derivative of K (equation A.40) is positive and hence the fleet is expanding. The horizontal arrows for points above the supply curve indicate fleet expansion. Hence, if the system started at any point above the supply curve, the fleet will expand; the fleet adjustment is stable – it moves back to the supply curve. Similarly, if the starting point is below the supply curve, such as point B, then the vessel price for a given fleet is lower than that consistent with optimising behaviour. In this case $I[P(t)] - \delta K < 0$ and therefore the time derivative of K in equation (A.40) is negative. Hence, the system is stable, as point-B implies a fleet contraction – a move back to the supply curve.

Now consider the dynamic adjustment of NB vessel prices, which is described by equation (A.10), and plotted in Figure 3.A3. Assume that the starting point is any point above the demand for fleet curve, such as point F, in Figure 3.A3. At point F the price of vessels is higher than that consistent with long-run equilibrium. Accordingly, the time derivative of NB prices, given by equation (A.10), is positive, as $(r + \delta) P - F F_K - G_K > 0$, which implies that the time derivative of prices is positive. This, in turn, means that the vessel price would increase, thereby deviating further from the demand curve. Thus, when vessel prices are higher than consistent with equilibrium, owners expect prices to rise further and hence to make capital gains by buying more ships, thus pushing prices further up. Consequently, owners are destabilising the market – higher vessel prices lead to even higher prices. The vertical arrows point to instability – a move away from the demand curve. This means 'noise' investors rather than contrarian investors dominate the market. Noise investors extrapolate current trends in forming expectations about future prices, whereas contrarian investors buy ships in the neighbourhood of the bottom and sell near the peak.

Figure 3.A4 combines the adjustment of vessel prices and fleet from any initial disequilibrium. The demand and supply curves divide the (P, K) space into four quadrants. In quadrant (I) the direction of the two arrows points towards slightly larger fleet, but to an ever-increasing vessel price. If the system started in disequilibrium in quadrant (I), then it would diverge to an infinite price; the system is not stable. In quadrant (III) the system is again not stable; vessel prices and fleet would shrink towards zero. But the system is not necessarily unstable in quadrants (II) and (IV). If we start with a price that is lower than the long-run equilibrium, such as point A in quadrant (IV), then the system is unstable. But if we started

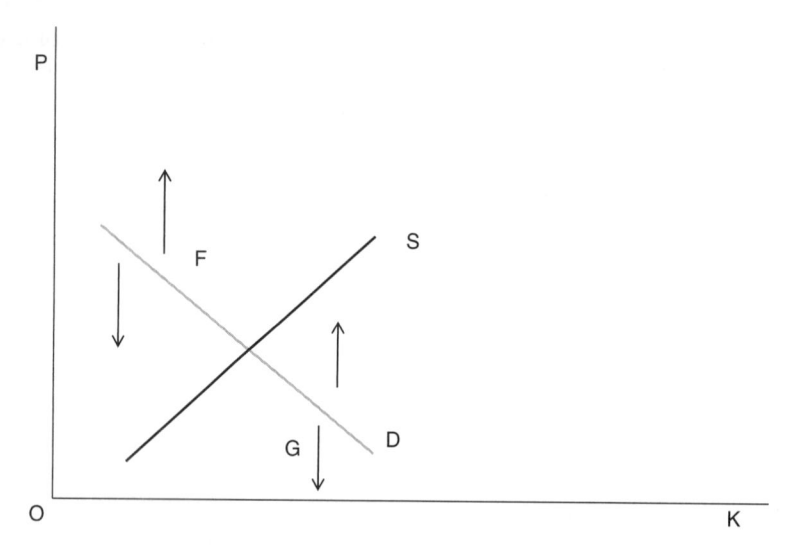

Figure 3.A3 Adjustment of NB vessel prices

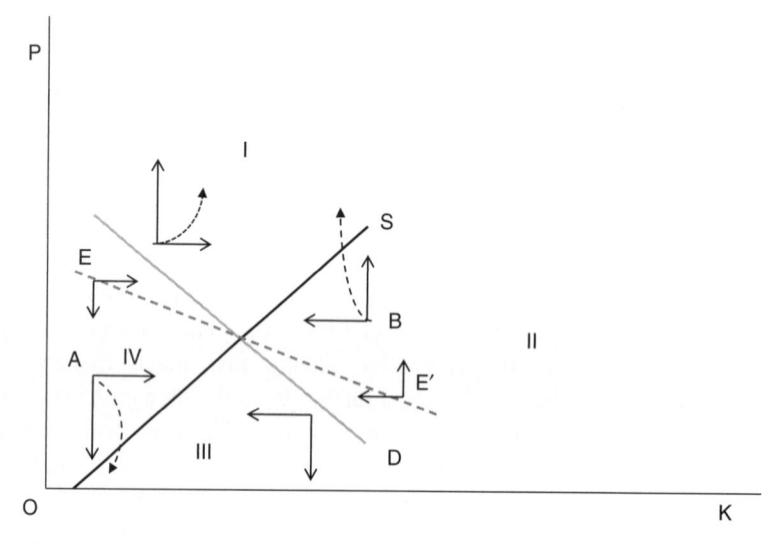

Figure 3.A4 Unique saddlepoint solution

with a higher price than the long-run equilibrium, such as point E, then the system is stable; it would converge to the long-run equilibrium. Similarly, if we started with a very high vessel price, such as point B in quadrant (II), then the system is unstable. But if we started with a low vessel price, such as point E′, the system is stable; the price is lower than the long-run equilibrium and owners expect to make capital gains by buying ships. The system is stable for any negatively sloped line that passes through the long-run equilibrium, such as the line EE′. This is a unique stable path, which must satisfy the saddlepoint property.[16] For this reason, the line EE' is called the 'saddlepoint' equilibrium path.

The saddlepoint path implies the following adjustment path for vessel prices, investment and the stock of fleet. The vessel price would jump from P_1 to P^*

before the fleet had any time to adjust, that is, when it is still at K_1 in Figure 3.A5. But from P* the vessel price would decline gradually, reaching the new long-run equilibrium at point C. Therefore, the optimal adjustment path of NB vessel prices involves an overshooting of the new long-run equilibrium (P* > P_2). The time path of gross investment follows the adjustment path of vessel prices (see Figure 3.A6). Investment rises on impact to its maximum pace and then declines towards its higher new long-run equilibrium (see Figure 3.A7). The investment path also involves an instantaneous overshooting of the new long-run equilibrium. Despite the overshooting of gross investment, the fleet (capital) adjusts gradually and monotonically from the initial equilibrium to the new equilibrium with higher fleet (see Figure 3.A8).

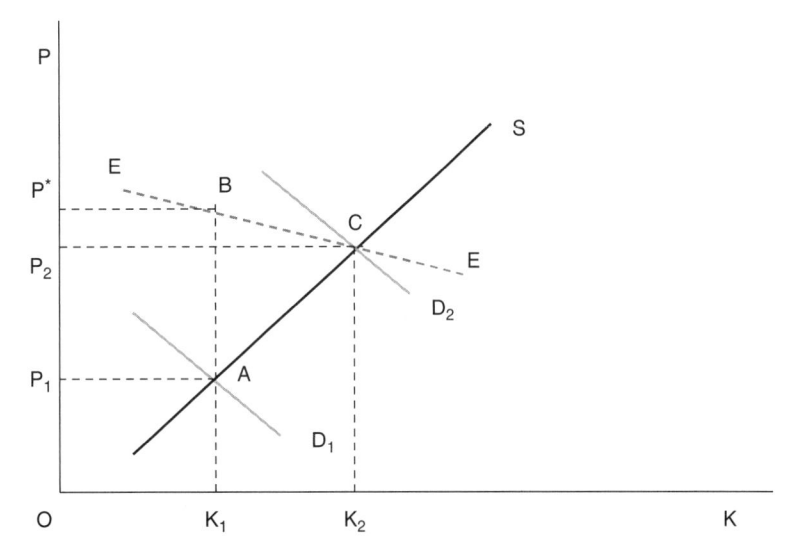

Figure 3.A5 Dynamic adjustment of NB vessel prices

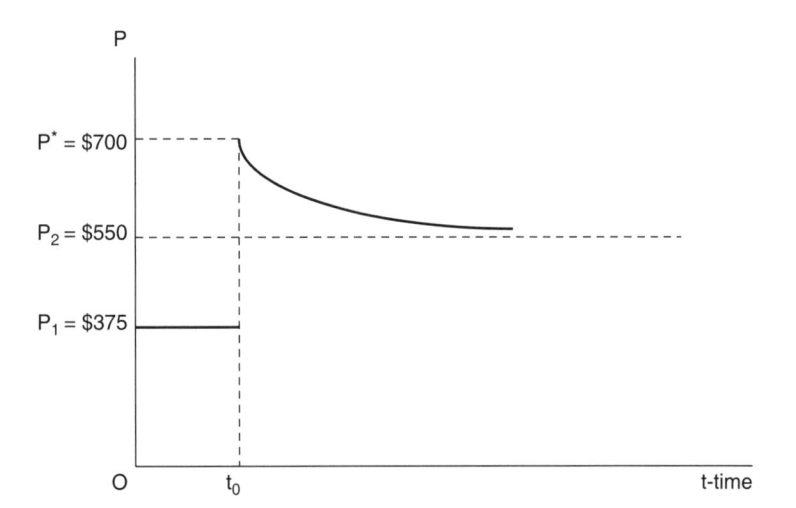

Figure 3.A6 Dynamic adjustment of NB vessel prices

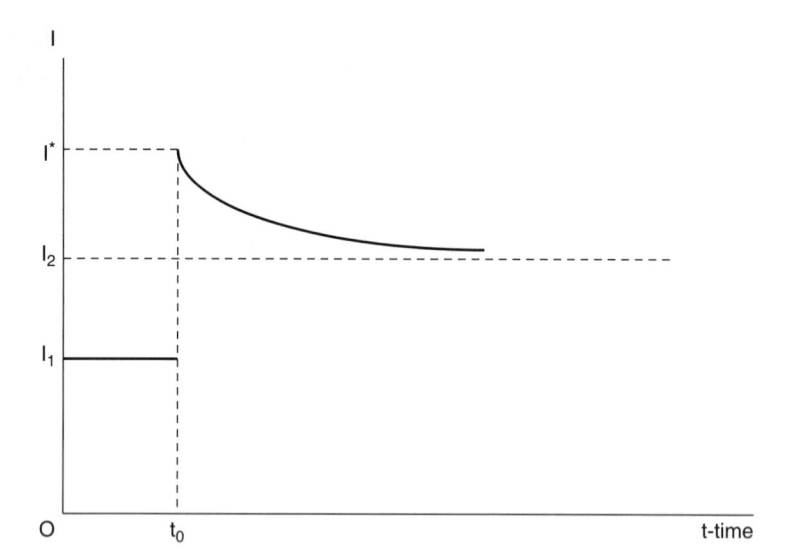

Figure 3.A7 Dynamic adjustment of investment

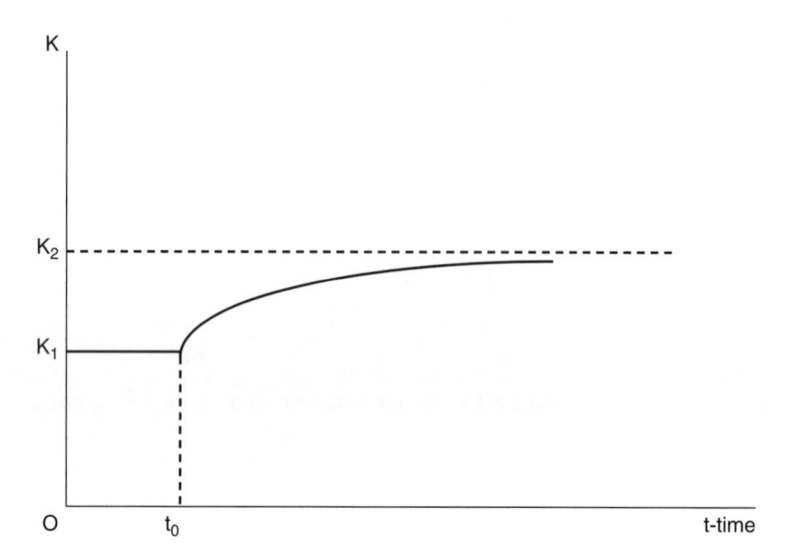

Figure 3.8A Dynamic adjustment of fleet

PART II
THE MACROECONOMICS OF SHIPPING MARKETS

4 THE EFFICIENCY OF SHIPPING MARKETS

EXECUTIVE SUMMARY

In this chapter we explore the issue of whether freight rates and ship prices (newbuilding and secondhand) are 'efficient'. This is an issue to which academic economists in the field of maritime economics, following similar lines of research in the field of finance, have devoted a great deal of time and effort. The issue is interesting from not only a theoretical point of view, but also a practical one. For the practitioner owner a key question is whether to employ the vessels in the time charter (period market) or the spot market. If freight markets are efficient, the decision that is taken will make no difference to profitability. If markets are inefficient, profit opportunities arise. The question is under what circumstances and when to switch from one market to the other to maximise profits. Similarly, if ship prices are inefficient, an owner can devise strategies to maximise profits. When ship prices are lower than their fundamental value, defined under the hypothesis of market efficiency, excess profits can be made by buying and operating these ships. When ship prices are higher than their fundamental value, it might be profitable to charter vessels rather than buying them.

These micro issues also have macro implications. Market inefficiency gives rise to misallocation of resources. For example, if expectations of rising freight rates do not reflect economic fundamentals, buying a ship on such expectations will lead to a misallocation of resources.

With the term 'market efficiency' economists are asking whether the difference in profit between two alternative strategies (the so-called excess profit or excess return) is zero. If the excess profit is zero, the underlying market where the strategies are developed, is said to be efficient. If the excess profit is non-zero (positive or negative), the market is inefficient. This sounds very simple, but the Efficient Market Hypothesis (EMH) is posed *ex ante* (the profit opportunity at the time the decision had to be made) and not *ex post* (with hindsight). A simple example, analysed in the main text, highlights the difference. Consider the recent event of the Greek bailout by the troika in May 2010. The event marked the reversal of economic policy in the major economies from restoring growth and eliminating unemployment to the pre-crisis levels, to reining in public finances. It is the major cause of the distressed shipping markets from the spring of 2010 to the first half of 2013, as demand fell behind supply, predetermined from projections of rosy demand conditions before the crisis and in the aftermath following the swift recovery of demand until the Greek crisis. If markets were efficient in the short run, then time charter rates would have adjusted instantly to the low level of demand from this change of policy in such a way that in terms of expected profit or loss it would have made no difference in choosing to stay in the spot or the time charter market. So, why markets may be inefficient?

Many economic agents simply do not appreciate the implications of the new information. Continuing with our example, one would not expect many owners to have appreciated the full implications of the Greek bailout for the locking in of time charter rates. Only the very well-informed owners (with good consultants) would have been able to grasp the significance of this event for shipping. Even from those owners that did, only a few might have been prepared to act upon it. The reason may well be that owners are not always prepared to take the risk. What was the *ex post* (with hindsight) profit? The average three-year time charter rate in the first half of 2010 for a 52,000 tonne Supramax was $17,300, but the average in the spot market over the same period was $18,800. An owner may have viewed the cost of moving from the spot to the time charter market as giving up $1,500 per day for an event that might have come to nothing at the end. *Ex post*, this would have been a very good decision as an owner operating in the spot market in the three-year period from the second half of 2010 to the end of the first half of 2013 would have made for the same ship only $9,600 per day (assuming no risk of default on the charterer). Therefore, *ex post* the strategy of moving to a three-year time charter contract would have resulted in a net gain of more than $7,600 per day. But the issue of market efficiency is posed not *ex post*, but *ex ante*. In other words, the issue of market efficiency is whether this profit opportunity could have been recognised at the time with information available then, so that freight rates would have adjusted to eliminate the profit opportunity. For market efficiency to hold economic agents must learn instantly all new information, absorb its implications for profit opportunity or loss and react instantly to take advantage of it. These conditions are unlikely to be valid in the real world. This would mean that markets might be inefficient in the short run (that is, in every period of time), but not necessarily in the long run.

In the long run, *ex ante* excessive profits are zero, but in the short run some economic agents may be able to exploit them. But as more and more agents learn about these profit opportunities the *ex ante* excess profitability is eliminated. In the context of our earlier example the adjustment of the time charter rates would have been gradual rather than instantaneous to the new long-run equilibrium of lower demand for shipping services. Accordingly, owners that would have acted swiftly and locked in three-year contracts in the second half of 2010 would have been better off than staying in the spot market. The condition of market efficiency in the long run implies that these profits would, in time, decline and ultimately be eliminated. Hence, in the real world there may be evidence of a weak form of market efficiency. The example also illustrates that the failure of market efficiency in the short run may be due to how risk affects shipping decisions.

Although these conclusions may seem uncontroversial and common sense, at least to the layman, they are much more difficult to prove in lab conditions. This chapter explains the different tests that have been devised to examine the validity of the Efficient Market Hypothesis in the context of freight rates and ship prices.

In describing these tests the aim is not simply to show whether shipping markets are efficient or not, but how some of these elaborate models can help to improve decision making in shipping.

This chapter begins by explaining the Efficient Market Hypothesis and rational expectations, which is an indispensible constituent component of market efficiency. Then the chapter explains the two models that support the Efficient Market Hypothesis, namely the random walk model and the martingale model. Two tests of the validity of the Efficient Market Hypothesis emerge naturally out of this literature: tests of unpredictability of excess profitability (or excess returns) and tests of informational efficiency or orthogonality of past information with future prices. In an efficient market, where economic agents are rational, actual prices should reflect the fundamental value of the asset. In freight markets this amounts to the time charter rate being equal to a weighted average of rolling spot rates for the duration of the time charter contract. For ship prices, the fundamental value is the present value of expected future profitability. In an efficient market excess returns over those implied by the fundamental value should be independent (or uncorrelated) of historical information available at time t or earlier, which implies the unpredictability of excess returns and the orthogonality of past information with future prices.

A second set of battery tests can be devised by comparing the actual price of the asset with that implied by its fundamental value. Regression tests can be conducted to test whether the two prices (actual and fundamental) are equal. If they are, this is evidence of market efficiency. If they are not, it does not necessarily follow that markets are inefficient. The reason is that expected profits require the specification of expectations-generating mechanisms. Two broad categories can be distinguished: forward-looking and backward-looking expectations. Rational expectations fall squarely in the first category, but backward-looking ones are not inconsistent with the Efficient Market Hypothesis, as the information set available at time t includes all past values of all relevant variables. From this angle, a test of market efficiency is a test of the joint hypothesis of market efficiency and the specific expectations scheme. Thus, if the joint hypothesis is rejected by the data, it is not clear which part of the joint hypothesis is responsible for it: is the market inefficient or the assumed expectations scheme is wrong?

The development of the VAR methodology (see the main text and the Statistical Appendix) has helped to shed light on this issue because this methodology ensures that all past information is used efficiently in forming expectations. From this point of view, a third battery of tests emerges. The validity of the Efficient Market Hypothesis implies some (non-linear) restrictions on the VAR. This means that a test of market efficiency is to compare the statistical fit of the restricted with the unrestricted model. Market efficiency requires that the two models should give the same fit. If the unrestricted model performs better than the restricted one, then the test indicates the rejection of market efficiency because an unjustified imposition of restrictions reduces the explanatory power of the model. Such a test may be computationally cumbersome to perform because it requires the estimation of two models: the restricted and the unrestricted one. An alternative test is to work with the unrestricted model and test whether the restrictions are met.

What are the implications of the restrictions on the coefficients of the VAR for the freight market? First, there are no excess profits to be made by owners in choosing a time charter contract over a series of rolling spot contracts that span the duration of the time charter contract. The difference is zero and hence either scheme gives the same profit. Second, no other information is needed to forecast the spread between the time charter and the spot rates. In other words, the forecast error in predicting future changes in spot rates is independent of information available at time *t* or earlier. Therefore, the orthogonality principle of market efficiency is valid. Third, the validity of the non-linear restrictions is a test of the joint hypothesis of the expectations theory of the term structure of freight rates and the VAR system of generating expectations. Therefore, the interpretation of the non-linear restrictions taken all together is that freight markets are efficient.

The VAR methodology gives rise to a third battery set of testing for market efficiency. This method is based on volatility tests in the form of variance inequality or variance bounds tests. The basic idea of volatility tests is that spot freight rates are too volatile to reflect changes in economic fundamentals, namely that the time charter rate is equal to a weighted average of expected changes in spot rates. Therefore, evidence of excess volatility is not consistent with economic fundamentals, thereby rejecting the Efficient Market Hypothesis.

The variance inequality test of market efficiency is an alternative to the regression tests explained earlier. Both are derived on the principle of informational efficiency or orthogonality of rational expectations. This means that the regression test and the variance inequality test (in theory, though not in practice) are equivalent and follow from each other.

On balance, these tests do not support the hypothesis of market efficiency for freight rates and ship prices. But, as has been mentioned, this does not mean that that these markets are not efficient in the long run. In the context of our earlier example a key question is: how long does it take for these profit opportunities to be eliminated? Would owners that locked in after the first six months following the Greek crisis have failed to capitalise on relatively robust time charter rates? In other words, how long does it take for disequilibrium to be corrected before the new long-run equilibrium (no profit opportunity) is attained? These questions can be tackled using the framework of cointegration. In the Statistical Appendix to this chapter some effort is made to explain this framework intuitively and mathematically. Broadly speaking, two variables are cointegrated if they are driven together not because of a common trend but because they are moving in such a way as to restore a new long-run equilibrium. In the freight markets this means that the time charter and spot rates are moving in such a way that their spread is eliminated in the long run. This would be true if the spread between the time charter and the spot rates is equal, in the long run, to a weighted average of expected changes in spot rates over the lifetime of the time charter contract. Therefore, the efficiency of freight rates is viewed as the long-term condition of zero excess profitability of staying in the time charter market over a strategy of opting for rolling spot contracts for the duration of the time charter contract. The existence of these long-term relations (cointegrating vectors) is only a necessary condition for market efficiency. The validity of the restrictions on the coefficients of the cointegrating would provide a sufficient condition for long-term market efficiency. But in some cases, such as in ship prices, the theory does not impose numerical restrictions on the coefficients of the cointegrating vectors, other than that these coefficients should be either positive or negative. Although this drawback makes the tests based on cointegration inconclusive on the issue of market efficiency, they are extremely useful in developing models that help to improve decision making in shipping.

The evidence of excess profitability in the short run is not necessarily against market efficiency if risk is allowed to affect decisions in shipping. If the excess profitability is viewed as compensation for risk taking, the hypothesis of market efficiency cannot be rejected both in theory and in empirical tests. This is the literature on time-varying risk premia in the theory of the term structure of freight rates and vessel prices. The contribution of this literature, therefore, is not in testing for market efficiency, but in defining and modelling risk and explaining how shipping decisions are affected by risk. The starting point is that risk can be approximated with the conditional variance of past forecast errors. This concept is intuitively appealing. Consider an owner that takes seriously the advice of an economic consultant and assume that in the recent past the *ex post* variance of the forecast errors is small (the consultant's forecasts are relatively accurate). In this case the owner's perception of risk is low (or is reduced), as the owner can rely more on the consultant's advice to formulate a strategy. Moreover, risk depends on the degree of the owner's risk aversion. The second point of this methodology is that high risk is rewarded with high return. The excess profitability is a function of risk, where the risk is measured by the degree of the owner's risk aversion and the variance of the forecast errors. The more risk averse is an owner and the less accurate the previous forecasts are, the greater is the required profit (the risk premium) needed to compensate the owner for moving from the time charter to the spot market. This interpretation follows from viewing the equation for excess profitability as the demand for spot contracts (a risky asset) against the demand for time charter contracts (the safe asset) in a 'mean-variance' model of asset demands with two assets. Similarly, the excess return of investing in ships over the return on money is a function of risk, where risk is measured again by the conditional variance of the forecast errors. According to the Capital Asset Pricing Model (CAPM) of asset pricing, the higher the risk is, the higher the excess return.

The principles of cointegration and time-varying risk premia help greatly in specifying models of freight rates and ship prices that enable owners to predict them. Therefore, these models help to improve shipping decision making.

1 INTRODUCTION

As we have seen in Chapter 2, freight rates have become asset prices and vessel prices (newbuilding and secondhand) are undisputedly asset prices. Thus, it is plausible to ask whether freight rates and ship prices can be predicted, as improved shipping decision making requires that both are predictable. By nature, asset prices are widely thought to be unpredictable because they discount the implications of 'news' on economic fundamentals, which are extremely volatile. But at a deeper level, the extent to which asset prices are unpredictable is due to the belief that asset markets are 'efficient'. Hence, it is important to review the Efficient

Market Hypothesis and the two widely used models that support it: the martingale and the random walk models. We then examine models of freight rates and vessel prices that are consistent with market efficiency and discuss the tests that can be conducted to test for efficiency in shipping markets. Finally, we present the empirical evidence on the efficiency of shipping markets. This chapter is organised as follows. After this short introduction, we review the Efficient Market Hypothesis and in sections 3 and 4 the martingale model and the random walk models that support it. In section 5 we analyse the tests for the efficiency of freight markets and present the empirical evidence. Section 6 deals with the same issues of ship prices, while the last section concludes.

2 THE EFFICIENT MARKET HYPOTHESIS

The *Efficient Market Hypothesis* (EMH) is the simple statement that asset prices fully reflect all available information (Fama, 1970). More precisely, a market is said to be efficient with respect to given information set, if its prices remain unaffected by revealing that information to all market participants (Malkiel, 1992). A market that would obey this property must be a perfectly competitive one in which market participants must be fully rational. These rational traders rapidly assimilate all available information and prices are accordingly adjusted. If all available information is immediately reflected into current asset prices, then only new information or 'news' on economic fundamentals, such as dividends, can cause changes in prices. This implies that prices are unforecastable because news, by definition, cannot be predicted and hence it is impossible to make economic profits by trading on the basis of that information set.

These ideas of market efficiency can be more formally presented by using the concepts of mathematical expectations and rational expectations.[1] As a first approximation, investors form rational expectations when they forecast an asset price as the mathematical expectation and they apply the principles of mathematical expectations to make such a forecast. What are these principles? Mathematical expectations have three properties: unbiasedness, orthogonality and the law of iterated expectations. Let us examine these properties separately. Under rational expectations the mathematical conditional expectation of an asset price is equal to its actual price, P_t, plus a random error, ε_t

$$E(P_{t+1}|I_t) = P_{t+1} + \varepsilon_{t+1} \quad \varepsilon_t \sim IID(0,\sigma^2) \qquad (4.1)$$

where $IID(0,\sigma^2)$ denotes that the random variable ε_t is independently and identically distributed through time with mean 0 and variance σ^2 and E is the mathematical conditional expectation of P_{t+1}. The mathematical expectation of the price next period is conditional on all information available in period t, contained in the information set I_t. An alternative simpler notation of $E(P_{t+1}|I_t)$ is $E_t(P_{t+1})$. Equation (4.1) states that forecasts are on average correct. Any particular forecast may be wrong by the value of ε_t and this may be large or small, but in repeated exercises forecast errors cancel out. From a statistical point of view equation (4.1) implies that the conditional expectation is an **unbiased** forecast,[2] which is the first principle of mathematical expectations.

The forecast error (or innovation) is uncorrelated with all information available at time *t*. This is stated mathematically as

$$E(\varepsilon'_{t+1} I_t) = 0 \tag{4.2}$$

and is known as the *informational efficiency or orthogonality* property of conditional expectations. The justification of this property is simply that if they were correlated then that information could be used to improve the forecast. Since the information set I_t contains also realisations of the error term ε (that is, past forecast errors, also called innovations), then it is obvious that the orthogonality property implies also that the error term is not serially correlated. If the error term is serially correlated, then past innovations can be used to improve the forecast of the asset price. This can easily be seen if it is assumed that the error term follows a first-order autoregressive scheme

$$\varepsilon_{t+1} = \rho \varepsilon_t + u_{t+1} \tag{4.3}$$

where *u* is a white noise process. Rewrite equation (4.1) as

$$P_{t+1} = E_t(P_{t+1}) + \varepsilon_{t+1} \tag{4.4}$$

Lagging this equation and multiplying by ρ and then subtracting it from the original equation using also (4.3) gives:

$$P_{t+1} = \rho P_t + [E_t(P_{t+1}) - E_{t-1}(P_t)] + u_{t+1} \tag{4.5}$$

The first term on the right-hand side shows that prices are forecastable. Tomorrow's price depends on the price today. The other terms contain no information which can be used to improve the price forecast. Hence, the orthogonality property rules out serial correlation in the residuals.

Consider now using equation (4.4) to make a forecast into the distant future, say two periods ahead. This is written mathematically as

$$E_t[E_{t+1}(P_{t+2})] = E_t(P_{t+2}) \tag{4.6}$$

In deriving this result the **rule of iterated expectations** has been used which states that

$$E_t E_{t+1} E_{t+2} \ldots = E_t \tag{4.7}$$

The meaning of the rule of iterated expectations is that with the information available today the investor does not know how the forecast will be revised next period when more information becomes available. Hence, the forecast two periods ahead is based only on the information available today.

It is obvious that investors are rational when they treat their forecast as the mathematical expectation and they apply the principles of mathematical expectations.

However, this is not sufficient to claim that they form rational expectations in the sense of Muth (1961) because their subjective expectations may differ. Expectations are formed rationally, in the sense of Muth, when investors, in the aggregate, do not make systematic errors. This requires that their *subjective* expectations are equal to each other (homogeneous expectations) and, in turn, are equal to the conditional mathematical expectations which are based on the **true** economic model. Thus investors are forming rational expectations when they use all available information to form their 'best' forecast of asset prices and in processing this information they are making use not only of the **true** economic model but also of the three properties of mathematical conditional expectations, namely, unbiasedness, informational efficiency or orthogonality and the law of iterated expectations.

The extreme version of the EMH can now simply be stated as the requirement that all investors know the true information set I_t and that this is available at no cost to all of them. The extreme version of EMH, namely that asset prices fully reflect all available information, is obviously false since a precondition is that information and transactions costs are always zero. Even in theory, if there are costs in gathering and processing information some profits can still be earned (see Grossman and Stiglitz, 1980).

However, these profits are obviously a reward for the information and processing costs and thus cannot be abnormally high. From this perspective a market is efficient if prices reflect information to the point where the marginal benefits simply match the marginal cost and hence a normal profit is earned. In the real world, though, it is difficult to define what would constitute normal profits since costs cannot be measured precisely. Hence, the issue is not so much whether a market is efficient in absolute terms, because the answer then is that it is not efficient. The issue is whether a market is efficient in relative terms. The advantage of the EMH is that it provides a benchmark against which the *relative* efficiency of a market can be assessed. For example, one can compare the efficiency of the futures vs spot markets, the efficiency of an emerging market relative to the New York stock exchange, the efficiency of the freight forward agreements (FFAs) vs spot freight rates.

All empirical tests of the EMH attempt to measure whether profits can be made on trading or whether prices are forecastable for a given information set. Some tests concentrate on whether fund managers in the real world can earn excess or abnormal returns or whether such returns can be earned on a hypothetical trading rule. The advantage of the former is that it concentrates on real trading, but suffers from the drawback that the information set used by portfolio managers is not observable. A hypothetical trading rule is therefore superior but it requires an explicit definition of the information set, the normal and excess return, and the trading costs. Following Roberts (1967), Fama (1970) defined three different information tests and hence three forms of efficiency.

Weak-form Efficiency: The information set includes only the history of prices or returns.
Semi-strong Efficiency: The information set includes all publicly available information.
Strong-form Efficiency: The information set includes in addition private information available to just few participants (insiders' information).

Fama (1991) introduced a slightly different taxonomy of efficiency with the hindsight of twenty years of research conducted between his first and second review articles on market efficiency. Instead of weak-form tests he proposed *tests for return predictability*. These include not only the history of prices or returns but also other variables like dividend yields, price–earning ratios and interest rates. These tests include not only time series tests but also cross-sectional predictability of returns, that is, tests of asset-pricing models and the anomalies (like the size effect) as well as seasonal patterns (like the January effect). In this category are also included tests of excess volatility. Instead of semi-strong efficiency he proposed *event studies* and instead of strong efficiency *tests for private information*.

The second element that needs to be specified for an empirical test of the EMH is a model of 'normal' returns. For a long time the standard assumption was that the normal return is constant over time. However, in recent years the emphasis has been on equilibrium models with time-varying expected returns. Finally, abnormal or excess returns must be defined. These are easily defined as the difference between the actual return on a security and its normal return.

From the above framework it can easily be seen that the issue of whether a market is efficient in an absolute sense is impossible to settle. Laboratory conditions for testing whether a hypothetical trading rule can lead to excess profits are difficult to set. If a trading rule fails to generate excess returns this does not mean that there is no trading rule which can generate excess return. Even worse, assume that someone discovers such a rule and makes excess profits. Would she still be able to make excess returns with the same rule after a while? The information set would soon include this rule and as more people are using it any excess profits will be eliminated. Thus a better test may involve the predictability of asset prices. But even this route is not promising. Assume that prices are forecastable, but the explained variance is small (less than 10 per cent), as is indeed the case with short horizon returns. The question immediately arises as to whether this is significant and that in turn depends on whether profits can be made. Hence, we are back to square one!

But the worst problem in any form of test of market efficiency is the *joint hypothesis* problem. Any test of market efficiency is conditional on the equilibrium model of normal returns. If efficiency is rejected, this could be due to the market being truly inefficient or to the incorrect model being used. This *joint hypothesis* problem means that the market efficiency as such can never be rejected. However, this may not be important, if one accepts that absolute efficiency is an ideal world and that the advantage of the EMH lies in specifying a benchmark against which relative efficiency can be assessed. We now turn to models, which can explain the time series properties of asset prices from the point of view of the EMH. Two such models are explored, the martingale and random walk models. We then examine the issue of whether the EMH implies models that reflect economic fundamentals. We show that the EMH is compatible with models that reflect economic fundamentals. Then we proceed to analyse models that contradict the EMH based on some form of irrationality, like fads. Finally, we examine the empirical evidence of which side may be right.

3 THE MARTINGALE MODEL

The appeal and popularity of the martingale model is that it captures the notion of a *fair* game and hence the property of the EMH that no profits can be made. In broad terms, a fair game is one which is neither to your advantage nor to your opponent's and hence it implies a zero expected profit. A fair game does not mean that one cannot win or lose. An investor may win or lose large sums in a particular period, but over a long horizon these cancel out and the average return is zero. More formally, a stochastic process which satisfies the following condition

$$E_t[P_{t+1}|P_t, P_{t-1},\ldots] = P_t \qquad (4.8)$$

or equivalently,

$$E_t[P_{t+1} - P_t|P_t, P_{t-1},\ldots] = 0 \qquad (4.9)$$

is called a martingale. If P is the price of an asset then the martingale model implies that the expected price tomorrow, conditional on the information of the entire history of the asset price, is equal to today's price. The martingale model implies that the market is (weakly) efficient, the current price fully reflects all available information. The current price includes all relevant information since in the absence of news the expected price tomorrow is equal to that of today. No other present or past information helps to improve the forecast. Hence, from a forecasting point of view the martingale model implies that the 'best' forecast of tomorrow's price is simply today's price, where 'best' means minimal mean square error. The martingale model also implies that asset prices are unpredictable since the actual price tomorrow will differ from today's price by a random variable, news on economic fundamentals, which sometimes would be positive and sometimes would be negative, but on average would be zero.

The fair game assumption is captured in the alternative definition of a martingale, equation (4.9). If the asset pays no dividends, then its return is simply the capital gains or losses, $P_t - P_{t-1}$, and hence a martingale defines a fair game in that the expected profit is always zero. Hence, the martingale model captures the other property of market efficiency, namely, that no profits can be made.

4 THE RANDOM WALK MODEL

The simplest version of the random walk model is that the error terms are independently and identically distributed (IID) and that the dynamics of the asset price are governed by the law

$$P_t = \mu + P_{t-1} + \varepsilon_t \quad \varepsilon_t \sim IID(0, \sigma^2) \qquad (4.10)$$

where μ is the expected price change or drift. A random walk without drift is obtained when $\mu = 0$. Clearly P_t is a martingale and ΔP_t is a fair game. The independence assumption of ε_t implies not only that ε_t is uncorrelated but that any non-linear function is also uncorrelated. Clearly the higher moments, including the variance, are non-linear functions. Hence, the random walk model imposes further restrictions than the martingale model by requiring that also higher moments are also uncorrelated. Thus, the variance of ε_t is also uncorrelated in the random walk model, whereas no such restriction is imposed in the martingale model. The variance is important because the risk premium can be approximated by it. Thus the random walk model implies that both the price and its variance are unpredictable, whereas the martingale model postulates that the price is unpredictable, but its variance can be predicted from past variances, as for example in the simple case of

$$\sigma_t^2 = \beta \sigma_{t-1}^2 \qquad (4.11)$$

This restrictive assumption of the random walk model can be relaxed by assuming that ε_t is not identically distributed through time, although it is still independent (*INID*). The assumption of identically distributed ε_t through time is clearly not valid because it implies that the probability law of, for example, daily returns remained invariant in the many changes in the economic, social and institutional environment that occurred, say, in the last hundred years. However, the independence assumption still implies that prices are unpredictable.

5 THE EFFICIENCY OF FREIGHT MARKETS

5.1 STATEMENT OF THE PROBLEM

In Chapter 2 we show that a time charter rate is equal to a weighted average of expected spot rates and risk premia with rolling spot contracts that span the duration τ of the time charter contract (cf. equation (2.23) in Chapter 2). This equation is derived on the hypothesis that freight markets are efficient in the meaning of this chapter and that enables us to invoke the riskless profit arbitrage relation upon which equation (2.23) is derived. In this section we examine the validity of the hypothesis that freight markets are efficient.

All theories of the term structure of freight rates assume that freight markets are efficient and express the time charter rate as a weighted average of expected spot rates and risk premia. Although there are six competing hypotheses in explaining the term structure of freight rates (borrowed from the term structure of interest rates), all these theories differ only in respect of the treatment of the risk premium. In the pure expectations theory the risk premium is zero. In the expectations theory the risk premium is a constant. In the preferred habitat theory the constant risk premium can be positive or negative. In the liquidity preference theory the risk premium varies with the term to maturity. In the time-varying risk theory the risk premium is varying with time and the term to maturity. In the CAPM the risk premium varies with the excess return of the market time charter portfolio over the spot rate and the covariance of the individual owner's time charter contract portfolio with the market portfolio.

The empirical tests of the efficiency of freight markets are also borrowed from the field of finance (see Mankiw and Miron, 1986, and Campbell and Shiller 1987, 1988 and 1991). Hale and Vanags (1989), Veenstra (1999a) and Kavussanos and Alizadeh (2002a) test the validity of the EMH with the last two papers applying the Campbell and Shiller methodology to the freight market. This methodology implies a transformation according to which the spot freight rate is subtracted from both sides of equation (2.23) in Chapter 2. The transformation enables the spread between the time charter rate and the spot rate to be expressed as a weighted average of expected future changes in spot freight rates (ignoring for the time being the risk premium)[3]

$$S_t^* = H_t^\tau - FR_t = k \sum_{i=1}^{\tau-1} (\delta^i - \delta^\tau) E_t(\Delta FR_{t+i}) \qquad (4.12)$$

where S is the spread (or difference) between time charter and spot rates expressed in units of the spot contract; and Δ is the difference operator (i.e. $\Delta FR_t = FR_t - FR_{t-1}$). Thus, if time is measured in months, the first difference of spot rates is the month-to-month change in the spot rate. The spread calculated in (4.12), denoted by S^*, assumes that freight markets are efficient (as it is simply a transformation of equation (2.23) in Chapter 2) and for this reason it is called the perfect foresight or theoretical spread in accordance with the assumed expectations scheme (rational expectations or a backward-looking mechanism such as AR or VAR), see the Statistical Appendix for an explanation). The empirical test of the validity of the Efficient Market Hypothesis in the freight market consists in comparing the perfect foresight or theoretical spread with the actual one. If the freight markets are efficient then the two spreads should be equal (in a stochastic sense, namely up to a small error term); otherwise markets are inefficient. This can be tested empirically by running a regression of S^* on S and any other relevant information, which is included in Λ, and imposing restrictions on the coefficients of the regression. Recall that according to the EMH any information contained in the information set Λ is not relevant in explaining the difference between the perfect foresight or theoretical spread and the actual spread. Thus, in the regression

$$S_t^* = \alpha + \beta \cdot S_t + \gamma \cdot \Lambda_t + u_t \qquad (4.13)$$

the following restrictions should be obeyed for the validity of the EMH in the freight market

$$\alpha = 0, \quad \beta = 1 \quad \text{and} \quad \gamma = 0 \qquad (4.14)$$

Unfortunately, equation (4.13) cannot be tested as it stands because it involves unobserved variables on the right-hand side of the equation, namely expected future changes in spot freight rates. Solutions to this problem can be found, but the drawback is that a test of the EMH becomes a test of a joint hypothesis of market efficiency and the scheme assumed in generating expectations of changes in spot rates. In this case, if the restrictions in (4.14) are satisfied empirically, then the freight markets are efficient. But if the restrictions are rejected, it is not clear which part of the joint hypothesis is responsible for it. It may be that markets are still efficient, but the expectations generating mechanism is wrong or it could be that the latter is correct and markets are inefficient. Even with this drawback it is worthwhile investigating whether markets are efficient.

5.2 EXPECTATIONS-GENERATING MECHANISMS

Three assumptions can be made about the expectations-generating mechanism: autoregressive models (AR), rational expectations, and Vector Auto Regression (VAR) models.[4] The first method assumes that the first difference of the spot freight rate follows an autoregressive form of order n, $AR(n)$. But the essence of the approach can be captured by an $AR(1)$ process. Thus, the first difference of the spot freight rate is generated by

$$\Delta FR_t = \rho \cdot \Delta FR_{t-1} + \varepsilon_t \quad \varepsilon_t \approx IID(0,\sigma^2) \quad (4.15)$$

The error terms are assumed to be independently and identically distributed around a zero mean, exhibit zero autocorrelation between any pairs of the residuals and have a constant variance through time. Expectations of changes in spot freight rates are then computed by taking expectations of both sides of (4.15) and noting that $E_t(\varepsilon_{t+1}) = 0$, (i.e. the mean of the error term is zero). This gives rise to the so-called *chain rule of forecasting*:

$$E_t \Delta(FR_{t+i}) = \rho^i \cdot \Delta FR_{t+i} \quad (4.16)$$

By substituting (4.16) into (4.12) we get:

$$S_t^* = k \sum_{i=1}^{\tau-1} (\delta^i - \delta^\tau) \rho^i (\Delta FR_{t+i}) \quad (4.12a)$$

Using equation (4.12a), the theoretical spread has now an empirical counterpart. Accordingly, equation (4.13) can be estimated and the restrictions (4.14) can be tested empirically as all variables are now observable.

Such an approach has been used, for example, by Hale and Vanags (1989). Most authors though do not prefer this approach as such expectations are backward rather than forward looking. The assumption of rational expectations seems more suitable, as it is an indispensible constituent component of the EMH (see section 2 in this chapter for more details) and it has been used widely in the literature (for example, Glen et al., 1981; Beenstock and Vergottis, 1993; Veenstra, 1999a, 1999b; Kavussanos and Alizadeh, 2002a).

The rational expectations hypothesis implies that expected future changes in spot rates are equal to actual ones, but for an error term, ε, that is a white noise process. Thus,

$$E_t(\Delta FR_{t+i}) = \Delta FR_{t+i} + \varepsilon_{t+i} \quad (4.17)$$

Substitution of (4.17) into (4.12) gives:

$$S_t^* = k \sum_{i=1}^{\tau-1} (\delta^i - \delta^\tau)[(\Delta FR_{t+i}) + \varepsilon_{t+i}] \quad (4.12b)$$

Equation (4.12b) enables the empirical testing of the joint hypothesis of market efficiency and rational expectations through equation (4.13). When rational expectations are used to compute the theoretical spread, S^*, (i.e. (4.12b)) the spread is called the *perfect foresight spread* (see Campbell and Shiller, 1987).

Kavussanos and Alizadeh (2002a) estimate equation (4.13)[5] with rational expectations (equation 4.17) for three different ship sizes in the dry market (Handy, Panamax and Capes). They use two lags of ΔFR and S as the information contained in the supplementary information set Λ. Their empirical results suggest that the restrictions in (4.14) either each one on its own or jointly together are not satisfied by the data. The null hypothesis that the restrictions are satisfied is rejected by the data. But, as has already pointed out, rejection of the restrictions (4.14) does not necessarily imply that the EMH is invalid, as this is a test of the joint hypothesis of market efficiency with rational expectations. It is worth noticing that by ignoring the risk premium this test of Kavussanos and Alizadeh (op. cit.) amounts to a test of the EMH through the pure expectations theory of the term structure of freight rates.

The third expectations-generation mechanism is through VAR models advocated by Campbell and Shiller (1987). According to this method, expectations for future changes in spot rates, ΔFR, can be obtained as forecasts from a system of two equations – one for the spread and one for the changes in freight rates. The VAR system can be described by the following two equations:

$$S_t = \sum_{i=1}^{n} \lambda_i \cdot S_{t-i} + \sum_{i=1}^{n} \mu_i \cdot \Delta FR_{t-i} + \eta_{1t} \quad (4.18a)$$

$$\Delta FR_t = \sum_{i=1}^{n} \xi_i \cdot S_{t-i} + \sum_{i=1}^{n} \zeta_i \cdot \Delta FR_{t-i} + \eta_{2t} \quad (4.18b)$$

In the system of equations (4.18) the state variables (S and ΔFR) depend on n past values of themselves and is called vector autoregression as it extends an autoregressive process for one variable (univariate) to an arbitrary number of variables (multivariate). Although the VAR is of order n, as it involves n-lags, it can be reduced (see the Statistical Appendix) to a first order by stacking together all lagged values of the two variables other than one in the so-called *companion matrix*, $\mathbf{A}$:

$$\mathbf{Z_t} = \mathbf{A}\mathbf{Z_{t-1}} + \boldsymbol{\eta_t} \quad (4.19)$$

Where $Z_t^T = [S_t, \Delta FR_t, ..., S_{t-n}, \Delta FR_{t-n}]$ is $(2n \times 1)$ vector of current and lagged values of the state-variables (S and ΔFR); $\mathbf{A}$ is an appropriately partitioned $(2n \times 2n)$ matrix of coefficients; and $\eta_t^T = [\eta_{1t}, \eta_{2t}, 0,...,0]$ is a $(2n \times 1)$ vector of residuals and zero elements. Equation (4.19) can be used to generate forecasts of S and ΔFR in the same way as an AR process (equation (4.16)), which can be used to compute the theoretical value of S. Then an equivalent form to (4.13) can be obtained in matrix notation, which can be tested empirically.

5.3 A VAR EXAMPLE

A simple example illustrates the VAR methodology and the nature of the restrictions. Consider for simplicity that the time charter rate is a weighted average of three rolling spot rates that span the duration of the time charter contract:

$$H_t^3 = \frac{1}{3}[FR_t + E_t(FR_{t+1}) + E_t(FR_{t+2})] \qquad (4.20)$$

As a result, the spread defined by equation (4.12) becomes:

$$S_t = \frac{2}{3}E_t(\Delta FR_{t+1}) + \frac{1}{3}E_t(\Delta FR_{t+2}) \qquad (4.21)$$

Now assume that both the spread and expected changes in spot freight rates are generated by a two-equation (bivariate) vector autoregression model of order 1:

$$S_{t+1} = a_{11}\Delta FR_t + a_{12}S_t + \eta_{1t+1} \qquad (4.22)$$

$$\Delta FR_{t+1} = a_{21}\Delta FR_t + a_{22}S_t + \eta_{2t} \qquad (4.23)$$

The above system can be written in matrix notation:

$$\mathbf{Z_{t+1}} = \mathbf{A}\mathbf{Z_t} + \mathbf{\eta_{t+1}} \quad \mathbf{Z}_t^T = [S_t, \Delta FR_t] \quad \mathbf{A} = \begin{bmatrix} a_{11} & a_{12} \\ a_{21} & a_{22} \end{bmatrix} \quad \mathbf{\eta}_{t+1}^T = [\eta_{1t+1}, \eta_{2t+1}] \qquad (4.24)$$

Using (4.22) and (4.23) we can generate the expectations needed in (4.21). The first term on the right-hand side of (4.21) is given by the right-hand side of (4.23) with the expectation of the error term set to its mean value of zero. The second term on the right-hand side of (4.21) can be computed from (4.23) by taking expectations of both sides:

$$E_t(\Delta FR_{t+2}) = a_{21}E_t(\Delta FR_{t+1}) + a_{22}E_t(S_{t+1}) \qquad (4.25)$$

The expectation of the spread in period $t+1$ (the second term in (4.25)) can be computed from (4.22), whereas the expectation of the change in spot freight rates (the first term in (4.25)) from (4.23). Substituting these values into (4.25) we get:

$$E_t(\Delta FR_{t+2}) = (a_{21}^2 + a_{22}a_{11})\Delta FR_t + a_{22}(a_{21} + a_{12})S_t \qquad (4.26)$$

Substituting (4.26) and the expectation of (4.23) into the equation of the expectations theory of the term structure of freight rate (4.21) we get:

$$S_t = \left[\frac{2}{3}a_{21} + \frac{1}{3}(a_{21}^2 + a_{22}a_{11})\right]\Delta FR_t + \left[\frac{2}{3}a_{22} + \frac{1}{3}a_{22}(a_{21} + a_{12})\right]S_t \qquad (4.27)$$

If the joint hypothesis of the expectations theory of the term structure of freight rates and the VAR system (4.22) and (4.23) for generating expectations, is valid, then the coefficients of both sides of (4.27) must be equal. This implies that the coefficient of ΔFR_t should equal zero, whereas the coefficient of S_t on the left hand side of (4.27) should equal the coefficient of S_t on the right-hand side:

$$\left[\frac{2}{3}a_{21} + \frac{1}{3}(a_{21}^2 + a_{22}a_{11})\right] = f_1(\mathbf{a}) = 0, \quad \left[\frac{2}{3}a_{22} + \frac{1}{3}a_{22}(a_{21} + a_{12})\right] = f_2(\mathbf{a}) = 1 \quad (4.28)$$

5.4 A GENERALISATION OF THE RESTRICTIONS

The above example illustrates that to define the set of restrictions even for a very simple model can be a rather tedious process. Hence, the manipulation of matrices is recommended as it is more convenient. Define $\mathbf{e1} = [1, 0]$ and $\mathbf{e2} = [0, 1]$ as two selection vectors. Then the forecasts using the bivariate VAR model (4.22) and (4.23) can be written in matrix notation as follows:

$$E_t Z_{t+1} = A Z_t \quad E_{t+1} Z_{t+2} = A^2 Z_t \quad (4.29)$$

Using the selection vectors, the definition of the spread and the forecasts generated by (4.29) can be written as:

$$S_t = \mathbf{e1}^T Z_t \quad E_t(\Delta FR_{t+1}) = \mathbf{e2}^T A Z_t \quad E_{t+1}(\Delta FR_{t+2}) = \mathbf{e2}^T A^2 Z_t \quad (4.30)$$

Substituting all three relationships of (4.30) into the expectations theory of the term structure of freight rates (equation (4.21)) we get:

$$\mathbf{e1}^T Z_t = \left[\frac{2}{3} \mathbf{e2}^T A + \frac{1}{3} \mathbf{e2}^T A^2\right] Z_t \quad (4.31)$$

If the expectations theory of the term structure of freight rates and the VAR model for generating expectations taken together are valid, then by equating again the coefficients of both sides of (4.31) implies the following non-linear restrictions, denoted by $f(\mathbf{a})$:

$$f(\mathbf{a}) \equiv \mathbf{e1}^T - \mathbf{e2}^T \left[\frac{2}{3} A + \frac{1}{3} A^2\right] = 0 \quad (4.32)$$

Equation (4.32) generalises the restrictions on the VAR for the validity of market efficiency. These are non-linear restrictions. The non-linearity arises because of A^2. It is clear that if the expectations of the changes in freight rates were generated by a simple AR(1), there would still be restrictions on the coefficients, but this time they would be linear.

5.5 REMARKS

At this stage some remarks may clarify many of the ideas behind the mathematical manipulations. First, what is the meaning of the restrictions (4.32) on the coefficients of the VAR? First, there are no excess profits to be made by owners in choosing a time charter contract over a series of spot contracts that span the duration of the time charter contract. The difference is zero and hence either scheme gives the same profit. Second, no other information is needed to forecast the spread. In other words, the forecast error of ΔFR_{t+1}, ΔFR_{t+2} using the VAR is independent of information available at time t or earlier (i.e. of ΔFR_{t-j}, S_{t-j} for $j \geq 0$). Therefore, the orthogonality principle of market efficiency, explained in this chapter, is valid. Third, the validity of the non-linear restrictions is a test of the joint hypothesis of the expectations theory of the term structure of freight rates and the VAR system of generating expectations. Therefore, the interpretation of the non-linear restrictions taken all together is that freight markets are efficient.

Second, it is important to compare the rational expectations approach with the VAR method of generating expectations. Under rational expectations the expectations of changes in spot freight rates are equal to the actual values plus a white noise process:

$$E_t(\Delta FR_{t+j}) = \Delta FR_{t+j} + \varepsilon_{t+j} \tag{4.33}$$

Substituting the values of (4.33) into the expectations theory of the term structure of freight rates (equation (4.21)) we get:

$$[(2/3)\Delta FR_{t+1} + (1/3)\Delta FR_{t+2}] + [(2/3)\varepsilon_{1t+1} + (1/3)\varepsilon_{2t+1}]$$
$$\equiv S_t^* + [(2/3)\varepsilon_{1t+1} + (1/3)\varepsilon_{2t+1}] \tag{4.34}$$

The term in the parenthesis on the left-hand side of (4.34) is defined as the *perfect foresight spread* and denoted by S^*. If the joint hypothesis of the expectations theory of the term structure of freight rates and rational expectations is valid, then the perfect foresight spread should equal the actual one plus a moving average of the rational expectations errors, ε_t:

$$S_t^* = S_t + [(2/3)\varepsilon_{1t+1} + (1/3)\varepsilon_{2t+1}] \tag{4.34a}$$

Therefore a simple test of the validity of the expectations theory and rational expectations in this VAR example is to estimate the regression we applied in the general case (4.13):

$$S_t^* = a + \beta S_t + \gamma \Lambda_t + u_t \tag{4.35}$$

and examine whether the following restrictions are valid:

$$a = 0, \quad \beta = 1, \quad \gamma = 0 \tag{4.36}$$

Notice that any other information included in Λ, such as past values of the spread or past changes in spot freight rates, is not relevant in forecasting the spread.

The theoretical spread, on the other hand, is computed as forecasts of the VAR (equations (4.22) and (4.23) in this example). In other words, the perfect foresight spread is the actual spread assuming no errors in forecasting, whereas the theoretical spread is the best forecast of the spread based on past values of the spread and changes in spot freight rates. Under rational expectations the past values of these variables do not contribute to the forecast of the spread (orthogonality property). The perfect foresight spread is forward looking (that is, it discounts some future long-run equilibrium as a result of changes in current or expected economic fundamentals). The theoretical spread is backward looking (that is, the forecast relies on the past, the historical values of the state variables). Therefore, from a purely theoretical perspective the rational expectations hypothesis is more suitable than a VAR approach in testing for the efficiency of freight markets. The rational expectations hypothesis does not require an explicit forecasting scheme, whereas the VAR approach restricts the information set to the state variables (that is, the spread and changes in spot freight rates). Many more variables could have been used in forecasting changes in spot freight rates. Therefore the rational expectations approach is less stringent than the VAR approach in testing for market efficiency.

Third, whereas under rational expectations economic agents use the full information set available to forecast changes in spot freight rates, with a VAR (or an AR) system they use a limited information set, Λ, based only on the spread and changes in spot rates. For this reason, sometimes it is said that explicit expectations generating mechanisms, such as AR or VAR, are *weakly rational*.

Fourth, although both a univariate autoregressive $AR(n)$ process and a multivariate $VAR(n)$ process are backward looking and impose conditions on the coefficients, there is an important difference between them. In the AR process expectations of changes in spot freight rates depend on their own past values. But an inspection of the equation of the theory of the term structure of freight rates shows that these expectations depend on the spread (see equation (4.21)). Thus, the VAR model is more suitable than the AR one in forecasting future changes in spot freight rates even from a purely theoretical point of view. But statistical reasons also confirm the superiority of VAR over AR. A regression of changes in spot freight rates on past values may result in a loss of 'statistical efficiency' (see the Appendix for more details). A VAR system corrects for this possible loss of statistical efficiency.

Fifth, an inspection of equation (4.34) shows that the error term is a moving average of the rational expectations predictions. This may invalidate the assumption of a constant variance (homoscedasticity) in the residuals with the implication that the estimated standard errors of the coefficients are biased when equation (4.35) is estimated by Ordinary Least Squares (OLS). Hence, OLS is not an appropriate method in testing for market efficiency. The so-called problem of heteroscedasticity (i.e. lack of homoscedastic residuals) can be corrected by using the Generalised Method of Moments (GMM), which can be viewed as an extension of the linear instrumental variables regression (see Hansen, 1982).

5.6 ALTERNATIVE TESTS OF MARKET EFFICIENCY

The VAR methodology gives rise to three more alternative tests of the efficiency of freight rate markets. The first test involves a comparison of the 'statistical-fit' between the unrestricted and restricted VAR systems. The second test involves estimating the unrestricted VAR and testing whether the set of non-linear restrictions embedded in (4.32) is valid. The third test involves a variance equality test between the restricted and unrestricted VAR. We also review variance inequality (or variance bounds) tests.

In the first approach one can estimate the VAR (that is, the system of equations (4.18)) without any restrictions imposed on the coefficients (the so-called *unrestricted* VAR). Then estimate the same VAR with the restrictions imposed (the so-called *restricted* VAR) and compare the statistical fit of the two systems. Market efficiency requires that the two models should give the same fit. If the unrestricted VAR performs better than the restricted VAR, then the test indicates the rejection of market efficiency. The null hypothesis of market efficiency against the alternative of inefficiency is carried out with a log-likelihood test. The method and the meaning and computation of likelihood ratio tests are explained next.

On the assumption that the residuals η_{1t} and η_{2t} in the unrestricted VAR (the set of equations (4.18)) are white noise processes, the variance-covariance matrix of the unrestricted VAR, Σ_u, is given by:

$$\Sigma_u = \begin{bmatrix} \sigma_{11}^2 & \sigma_{12} \\ \sigma_{21} & \sigma_{22}^2 \end{bmatrix} \qquad (4.37)$$

In the above matrix the diagonal elements represent the variance of η_{1t} and η_{2t}, respectively, whereas the off diagonal elements the covariance of η_{1t} and η_{2t}. Obviously, the covariance of the two residuals is the same (i.e. $\sigma_{12} = \sigma_{21}$). The restricted VAR has a variance-covariance matrix of the same form as (4.37), denoted by Σ_r. On the basis of the two variance-covariance matrices a likelihood ratio test can be computed to compare the fit of the unrestricted and the restricted VAR. The likelihood ratio test, *LR*, is defined as:

$$LR = T \ln[\det(\Sigma_r)/\det(\Sigma_u)] \qquad (4.38)$$

In the above expression, T is the number of observations in the sample and $\det(\Sigma)$ is the determinant of the variance-covariance matrix of the restricted and unrestricted VAR. Thus, the determinant of the unrestricted variance-covariance matrix is given by:

$$\det(\Sigma_u) = \sigma_{11}^2 \sigma_{22}^2 - \sigma_{12}^2 \qquad (4.39)$$

If the restricted VAR, which satisfies the conditions of market efficiency, fits the data equally well as the unrestricted VAR, then the likelihood ratio test should be statistically equal to zero (as the ratio is a log-difference). Accordingly, the null hypothesis is that freight markets are efficient. The alternative hypothesis assumes that markets are inefficient. In the alternative hypothesis the restricted VAR should perform poorer than the unrestricted one, because one imposes non-valid restrictions. Consequently, if the null hypothesis of market efficiency is rejected, the residuals of the restricted VAR should, on average, be larger than those of the unrestricted one and therefore the likelihood ratio statistic should be statistically different from zero; and conversely, if the null hypothesis of market efficiency is valid the residuals from the unrestricted and restricted VAR should be approximately equal to each other and the likelihood ratio statistic should be zero.

It can be shown that the likelihood ratio is distributed as a chi-squared (χ^2) distribution with degrees of freedom equal to the number of the restrictions. Therefore, the null of market efficiency against the alternative of inefficiency is tested by comparing the LR statistic of the sample with the χ^2 of the tabulated distribution. The null is accepted if the LR is less than the χ^2 for a level of statistical significance, such as 5%.

The second method is much simpler than the first, as it bypasses the step of estimating both the unrestricted and the unrestricted versions of the VAR. The method requires estimating only the unrestricted VAR and testing whether the restrictions summarised in equation (4.32) are valid. To illustrate the second method, assume that the expectations theory of the term structure of freight rates is described by equation (4.20). In other words, the time charter contract is a weighted average of just three rolling spot contracts. According to the expectations theory of the term structure of freight rates, this means that the time charter rate is equal to the current change of spot rates and expectations of these changes for the next two periods. Next assume that these expectations are generated by a two equation VAR of order one, as in equations (4.22) and (4.23). This means that only the spread and the change in spot rates are required to form expectations. The assumption that the VAR is of order one means that only current values and lagged once values of the spread and changes in the spot rates are required to form expectations. If the joint hypothesis of the term structure of freight rates and the postulated mechanism for generating expectations is valid, then the restrictions summarised in equation (4.28) should hold. The null hypothesis of market efficiency is that the restrictions are valid and can be tested through a Wald statistic (see below).

The second approach to testing for market efficiency involves estimating only the VAR system of equations (4.22) and (4.23) without bothering about equation (4.20) that describes the term structure of freight rates. The Wald statistic is an appropriate test of the validity of the restrictions (4.28) as it is a parametric statistical test which is applied whenever a hypothesis can be expressed as a statistical model with estimated coefficients from a sample of data. The Wald-statistic tests the true value of the coefficients based on the sample. Under the Wald statistic the maximum likelihood estimate of a set of coefficients are compared to their theoretical value, such as (4.28), under the assumption that the difference between them is approximately normally distributed. Typically, the square of the deviation of an estimate, b^e, of a coefficient b divided by the variance of the estimate [i.e. the $s\,(b^e - b)^2/\text{var}(b^e)$] follows a χ^2 distribution. In the univariate case and when the restrictions are linear the Wald test is equivalent to a t-student statistic. Thus the Wald test is appropriate in the multivariate case and whenever the restrictions are non-linear.

The intuitive interpretation of the Wald test is that only the spread and changes in spot rates (current and lagged once) are needed for the validity of the expectations theory of the term structure.

In spite of its appeal, the Wald statistic has poor small sample properties and it is not invariant to the way the non-linear restrictions are expressed. As a result of these deficiencies, a Wald statistic may reject the null of market efficiency for a coefficient very close to unity, leaving the modeller wandering whether the hypothesis of market efficiency still holds.

An alternative to a Wald test in this second method (estimating the unrestricted VAR and testing for the validity of the restrictions) is the so-called *block exogeneity* test. This test involves the concept of 'Granger causality'. A time series X is said to Granger-cause Y, if it can be shown that X provides statistical significant information about future values of Y. The test is usually conducted through *t*-statistics on the individual coefficients of X and lagged values of X. The contribution of X as a group (current and past values) is tested through an *F*-statistic. The test is only valid for stationary time series (see the Appendix for a definition and statistical tests of stationarity). Accordingly, the variable X enters the regression of Y in level if it is stationary or in as many differences as required to render it stationary. The number of lags to be included is judged by the use of the Akaike Information Criterion or the Schwartz Information Criterion (see Statistical Appendix).

The Granger-causality is applied in the current context as follows. According to the expectations theory of the term structure, for example in the simplified case of equation (4.21), only the spread is required to form expectations of changes in spot rates. Therefore, S_t and its past values should 'Granger-cause' changes in spot rates. If in the VAR model S_t on its own or as a group (current and lagged values of S) do not Granger-cause changes in spot rates (ΔFR_t) then the hypothesis of market efficiency is rejected by the data. It may be important to note that the block exogeneity test is only a test of the weak form of market efficiency.

The third method of testing for market efficiency is based on volatility tests in the form of variance inequality or variance bounds tests. The basic idea of volatility tests is that spot freight rates are too volatile to reflect changes in economic fundamentals, namely that the time charter rate is equal to a weighted average of expected changes in spot rates. Therefore, evidence of excess volatility is not consistent with economic fundamentals, thereby rejecting the EMH.

A common view among practitioners in the shipping market is that spot freight rates are unpredictable as they are affected by a multitude of unpredictable events, whereas time charter rates are based on expectations about long-term economic fundamentals. Therefore, the theoretician's view that time charter rates are a weighted average of expected changes in spot rates is at odds with the practitioner's view. Therefore, in broad terms volatility tests of market efficiency are effectively testing the theoretician's view that markets are efficient against the practitioner's view that markets are inefficient. To be able to argue that spot freight rates are excessively volatile one must have a yardstick against which to measure volatility. The benchmark volatility arises when spot rates reflect basic economic fundamentals. In this framework 'news' on economic fundamentals gives rise to benchmark volatility. Therefore, volatility tests examine whether the variance of the spread or the variance of time charter rates is consistent with the volatility implied by economic fundamentals. One can measure this benchmark volatility within the rational expectations framework or within the framework of AR or VAR models. This gives rise to the variance bounds tests.

As we have seen in section 2, rational expectations imply that economic agents make no systematic errors in forecasting economic variables. This is summarised in equation (2.1). Applying equation (2.1) to the spread version of the expectations theory of the term structure (cf. equation (4.12)) defines the perfect foresight spread, denoted by S^*:

$$S^*_{t+1} \equiv E_t(S_{t+1}) = S_{t+1} + u_{t+1} \quad u_t \approx NIID(0, \sigma^2) \tag{4.40}$$

As a result, rational expectations are 'optimal' forecasts (that is, on average economic agents make accurate predictions). The optimality property of rational expectations depends on the assumptions of unbiasedness and informational efficiency or orthogonality (see equations (2.1) and (2.2)).

The variance of the perfect foresight spread is defined by the following equation:

$$\operatorname{var}(S^*_{t+1}) = \operatorname{var}(S_{t+1}) + \operatorname{var}(u_{t+1}) + 2\operatorname{cov}(S_{t+1}, u_{t+1}) \tag{4.41}$$

As we have seen, the informational efficiency or orthogonality property of rational expectations (cf. equation (2.2)) implies that the forecast error is independent (or uncorrelated) of all information available at time t, I_t:

$$E_t[(S^*_{t+1} - S_{t+1})|I_t] = 0 \tag{4.42}$$

Since S_t is part of the information set (that is, $S_t \subset I_t$), then S_t is independent of u_t. This means that the covariance of S_{t+1} and u_{t+1} is zero (that is, $\operatorname{cov}(S_{t+1}, u_{t+1}) = 0$). Accordingly, equation (4.41) is simplified to:

$$\operatorname{var}(S^*_{t+1}) = \operatorname{var}(S_{t+1}) + \operatorname{var}(u_{t+1}) \quad or \quad \operatorname{var}(S^*_{t+1}) = \operatorname{var}(S_{t+1}) + \operatorname{var}(S^*_{t+1} - S_{t+1}) \tag{4.43}$$

As the variance of the forecast error u_t is at least zero or positive and equal to σ^2 it follows that:

$$\operatorname{var}(S^*_t) \geq \operatorname{var}(S_t) \quad or \quad VR \equiv \operatorname{var}(S^*_t)/\operatorname{var}(S_t) \geq 1 \tag{4.44}$$

In other words, if the expectations theory of the term structure of freight rates is valid and expectations about future changes in spot freight rates are rational then the variance ratio, VR, should not be statistically greater than 1.

The variance inequality test of market efficiency is an alternative to the regression test summarised in (4.13). Both are derived on the principle of informational efficiency or orthogonality of rational expectations (that is, equation (4.42). This means that the regression test and the variance inequality test (in theory, though not in practice) are equivalent and follow from each other. This can be verified as follows. Under the orthogonality condition of rational expectations we expect the conditions in (4.14) to be valid and therefore equation (4.12) simplifies to (4.40) from which the variance inequality test (4.44) is derived.

The inequality variance tests can be expressed in terms of the time charter rate instead of the spread. In this case the starting point is not (4.12), but equation (2.23) in Chapter 2. In other words, instead of expressing the expectations theory as a weighted average of expected *changes* in spot freight rates, the theory is expressed as the time charter rate is a weighted average of the *levels* of expected spot rates. Thus, ignoring, for the time being, the risk premium, the expectations theory of the term structure of freight rates is expressed as a weighted average of expected spot rates:

$$H_t^* = k\sum_{i=0}^{p-1}\delta^i(E_t FR_{t+i} + u_t), \quad k = \frac{1-\delta}{1-\delta^p}, \quad \delta = \frac{1}{1+r} < 1 \quad (4.45)$$

Assuming that expectations of spot freight rates are formed rationally we have:

$$\varepsilon_{t+1} = FR_{t+1} - E_t(FR_{t+1}) \quad (4.46)$$

By substituting (4.46) into (4.45) we eliminate the unobserved variable and therefore we can generate historical data for the theoretical time charter rate, H^*. If the joint hypothesis of the expectations theory and rational expectations is valid then in theory the time charter rate and the actual one should move together according to the equation:

$$H_t^* = H_t + \sum \varepsilon_{t+j} \quad (4.47)$$

Since FR is independent of ε, it follows that the $\text{cov}(H, \varepsilon) = 0$. As the variance of ε in (4.47) is zero or positive, the variance of the theoretical time charter should be at least equal to the actual one. Therefore, the variance bound test is expressed as:

$$\text{var}(H_t^*) \geq \text{var}(H_t) \quad or \quad VR \equiv \text{var}(H_t^*)/\text{var}(H_t) \geq 1 \quad (4.48)$$

It is worth stating that the variance inequality test (4.44) is also valid for a constant risk premium. Thus, the market efficiency test (4.44) applies not only to the expectations theory of the term structure of freight rates but also to the liquidity preference theory and the preferred habitat theory.

However, the variable bounds tests are only valid if the underlying time series are stationary (no underlying time trend or stochastic trend; see the Appendix for details). As the spread is stationary, the variance bounds tests in equation (4.44) are valid, but those on the time charter rate are not, as the series is non-stationary.

The variance bounds tests hold also if another specific expectations-generating mechanism than rational expectations is specified. Thus, if it is assumed that either an AR process, such as (4.15), or a VAR system, such as (4.18), is used to generate expectations, then we can generate historical values for the unobserved variable of the theoretical spread using equations (4.12a) or (4.12b). As before, the theoretical spread should equal the actual spread from which similar variance bounds tests to (4.44) are derived with the only difference that S is to be interpreted as the theoretical spread rather than the perfect foresight spread.

5.7 WEAK FORM OF MARKET EFFICIENCY – COINTEGRATION TESTS

So far, market efficiency has been viewed as an 'all or nothing' condition. Markets are either efficient or inefficient; there is no grey area. In a dynamically evolving world market efficiency should hold in every period of time. This implies that economic agents learn instantly all new information, absorb its implications for profit opportunity or loss and react instantly to take advantage of it. This reaction is not restricted to a few agents (the smart ones) in the market, but to the majority. Therefore, the market reaction is so sweeping that this new profit opportunity is eliminated instantly (that is, in the same period that the information arrives). To put in perspective the condition of instantaneous adjustment of market efficiency, consider the recent event of the Greek bailout by the troika in May 2010. The event marked the reversal of economic policy in the major economies from restoring growth and eliminating unemployment to the pre-crisis levels, to reining in public finances. It is the major cause of the distressed shipping markets from the spring of 2010 to the first half of 2013, as demand fell behind supply, predetermined from projections of rosy demand conditions before the crisis and in the aftermath following the swift recovery of demand until the Greek crisis. If markets were efficient in the short run, then time charter rates would have adjusted instantly to the low level of demand from this change of policy in such a way that in terms of expected profit loss would have made no difference in choosing to stay in the spot or the time charter market.

Therefore, the assumption of instantaneous adjustment is extreme and the notion that profit opportunities would gradually diminish as more and more economic agents follow the trend seems common sense and widely acceptable. This implies that market efficiency should be viewed as a long-term equilibrium condition rather than one that must be satisfied in every period of time. In the long run, *ex ante* excessive profits are zero, but in the short run some economic agents may be able to take advantage. Over time, however, as more and more agents learn about these profit opportunities the *ex ante* excess profitability is eliminated. In the context of our earlier example the adjustment of the time charter rates would have been gradual rather than instantaneous to the new long-run equilibrium of lower demand for shipping services. Accordingly, owners that would have acted swiftly and locked in three-year contracts in the second half of 2010 would have been better off than staying in the spot market. The condition of market efficiency in the long run implies that these profits would in time decline. Hence, in the real world there is a weak form of market efficiency.

A key question in this example is how long it takes for these profit opportunities to be eliminated: Would owners that locked in after the first six months following the Greek crisis have failed to capitalise from relatively robust time charter rates? In other words, how long does it take for disequilibrium to be corrected before the new long-run equilibrium (no profit opportunity) is attained? These questions can be tackled using the framework of cointegration. In the Appendix some effort is made to explain this framework intuitively and mathematically. Broadly speaking, two variables are cointegrated if they are driven together not because of a common trend but because they are moving in such a way as to restore a new long-run equilibrium. In the freight markets this means that the time charter and spot rates are moving in such a way that their spread is eliminated in the long run. This would be true if the spread in the long run is equal to a weighted average of expected changes in spot rates in the lifetime of the time charter contract. Therefore, the expectations theory of the term structure of freight rates is viewed as the long-term condition of zero excess profitability of staying in the time charter market over a strategy of ensuring rolling spot contracts for the duration of the time charter contract. This long-term equilibrium condition or cointegration relation is expressed as:

$$S_t = H_t - \phi - FR_t = 0 \tag{4.47}$$

If the pure expectations theory of the term structure of freight rates is valid, which means zero excess profits in the long run, then H_t and FR_t will be cointegrated with cointegrating coefficients $[1, 0, -1]$ in (4.47). The condition of cointegration gives rise to the requirements that the coefficient φ, which captures the risk premium, should be zero, whereas the coefficient of the spot rate should be equal and of opposite sign to the time charter rate (that is, –1). A non-zero φ, but with the other cointegrating coefficients $[1, -1]$, respectively means that the expectations rather than the pure expectations theory is valid. The weak form of market efficiency implies that the spread is zero in the long run, but not in the short run. A non-zero spread is evidence of disequilibrium and cointegration forces the spread to be zero in the long run.

The weak form of market efficiency can be tested within a VAR framework without the need to run extra regressions of the perfect foresight spread or the theoretical spread on the actual spreads or to run volatility tests. All that is needed for the validity of the weak form of market efficiency is to show that the time charter and spot rates are cointegrated with cointegrating coefficients $[1, 0, -1]$. Thus, the VAR model (4.18) can be reformulated as follows:

$$\Delta H_t = c_1 + \sum_{i=1}^{n} \lambda_i \Delta H_{t-i} + \sum_{i=1}^{n} \mu_i \cdot \Delta FR_{t-i} + a_1[H_{t-1} + \phi + \beta FR_{t-1}] + \eta_{1t} \tag{4.48a}$$

$$\Delta FR_t = c_2 + \sum_{i=1}^{n} \xi_i \Delta H_{t-i} + \sum_{i=1}^{n} \zeta_i \cdot \Delta FR_{t-i} + a_2[H_{t-1} + \phi + \beta FR_{t-1}] + \eta_{2t} \tag{4.48b}$$

If cointegration exists then the residuals in the VAR should be white noise processes. This condition arises from the fact that both the time charter and spot rates are non-stationary processes in the real world and need to be differenced once to become stationary. The stationarity concept is the key to these tests for valid statistical inferences. If the first differences of the time charter and spot rates are stationary, then the cointegrating vector should also be stationary for valid statistical inferences from the VAR. No wonder, the test for cointegration boils down to the cointegration vector being stationary (see the Statistical Appendix for the meaning and tests of stationarity and cointegration).

5.8 REMARKS

There are four remarks that merit some attention here. First, although only one equation of the VAR is really necessary for testing for cointegration, as there is only one possible cointegrating vector, there is statistical informational efficiency to be gained by including the second equation of the VAR. In other words, more accurate estimates of the underlying coefficients are obtained by including both equations of the VAR. Nonetheless, as there is only one cointegrating vector, (that is, equation 4.47), this appears in both equations of the VAR. Second, the coefficients a_1 and a_2 capture the speed of adjustment back to equilibrium. Therefore, they provide quantitative evidence to the key timing issue of how long it is profitable to lock in time charter contracts before the opportunity is eliminated. Third, the coefficient β in the cointegrating vector captures the long-term relationship of the time charter and spot rates. If the expectations theory of the term structure of freight rates is weakly valid (that is, it holds as a long-run but not short-run equilibrium condition), the coefficient β should be one. If the *pure* expectations theory is valid then the coefficient φ should be zero. For the expectations theory a non-zero φ is permissible. Fourth, the coefficients λ_i and μ_i capture the short-run response of the time charter rate to the new long-run equilibrium of the lower demand for shipping services (that is, the short-run dynamics). The coefficients ζ_i and ξ_i capture the short-run dynamics of spot freight rates to the new long-run equilibrium.

5.9 TIME-VARYING RISK PREMIA – AN EXPLANATION OF THE WEAK FORM OF MARKET EFFICIENCY

In the theory of the term structure of freight rates, as in the theory of finance, there is a simple reason why there is empirical evidence of weak rather than strong market efficiency. Many economic agents simply do not appreciate the implications of the new information. For example, one would not expect many owners to have appreciated the full implications of the Greek bailout for locking in time charter rates. Only the very well-informed owners (with good consultants) would have been able to grasp the significance of this event for shipping. Even from those owners that they did, only a few might have been prepared to act upon it. The reason may well be that owners are not always prepared to take the risk. Thus, in our example the average three-year time charter rate in the first half of 2010 for a 52,000 Handymax was $17,300, but the average in the spot market for the fist six months of 2010 was $18,800. An owner may have viewed the cost of moving from the spot to the time charter market as giving up $1,500 per day for an event that might have come to nothing at the end. With hindsight (*ex post*) this would have been a very good decision as an owner operating in the spot market in the three-year period from the second half of 2010 to the end of the first half of 2013 would have made for the same ship only $9,600 per day. Therefore, the strategy of moving to a three-year time charter contract would have resulted in a net gain of more than $7,600 per day. Although this example illustrates the *ex post* and not the *ex ante* failure of market efficiency, it suggest that the failure of market efficiency in the short run may be due to time-varying risk premia.

The excess profit to be made from the two alternative strategies (moving to the time charter market or staying in the spot) can be measured by the difference between the actual time charter rate, H, and the implied time charter rate if the joint hypothesis of the expectations theory of the term structure of freight rates and rational expectations was valid. The expectations theory is given by equation (4.45). Assuming that expectations are formed rationally and substituting out the expectations of spot freight rates from equation (4.17) and finally restoring the risk premium, φ, in equation (4.45), the excess profit, π, for choosing the time charter market for the three-year period is given by:

$$\pi_t = H_t - k\sum_{i=0}^{p-1}\delta^i FR_{t+i} + \phi + \eta_t, \quad \eta_t = \varepsilon_t + \sum_{i=1}^{p-1}\delta^i \varepsilon_{t+i}, \quad k = \frac{1-\delta}{1-\delta^p}, \quad \delta = \frac{1}{1+r} < 1 \quad (4.49)$$

Notice that all variables in the definition of excess profit are observables and that the error term η_t is a moving average of the rational expectations forecast errors, ε_t.

If an appropriate function for the risk premium φ can be found such that the *ex ante* excess profit is zero, then the expectations theory, appropriately adjusted by time-varying risk premia, would be valid, thereby confirming that freight markets are efficient. In theory, it is always possible to find such a function for φ. Therefore, the time-varying risk premia version of the expectations theory essentially begs the question of market efficiency by maintaining the hypothesis of market efficiency and asking the question of how it should be transformed so that it is valid. The contribution of this literature, therefore, is not in testing for market efficiency, but in defining and modelling risk and explaining how decisions are affected by risk. In this context the excess profit function can be reformulated as follows:

$$\pi_t = \phi + \eta_t \quad (4.50)$$

The starting point of this methodology is that high risk is rewarded with high return. As an example, consider the problem of deciding whether to stay in the time charter market or move to the spot market. The latter involves risk, whereas the former is relatively safe. The difference (or spread) between the time charter rate and the spot rate should be negative in normal conditions, as the owner must be compensated to take the risk of the spot market (see Chapter 2 for an explanation). As the risk in the spot market rises, so would the negative spread. This is the essence of the first proposition.

The second point of this methodology is that risk can be approximated with the conditional variance of past forecast errors. The conditional variance is defined as the variance with all information available at the time of the forecast. Its name derives from the fact that the estimate of the variance is conditioned on the available information set. The unconditional variance is the actual one where predictable and unpredictable events took place. So, the conditional variance is *ex ante*, whereas the unconditional variance is *ex post*. The two estimates would give most of the time very different results.[6] The notion that risk can be approximated by the conditional variance of past forecast errors is intuitively appealing. Consider an owner that takes seriously the advice of an economic consultant and assume that in the recent past the ex-post variance of the forecast errors is small (the consultant's forecasts are relatively accurate). In this case the owner's perception of risk is low (or is reduced), as the owner can rely more on the consultant's advice to formulate a strategy.

In formalising these ideas it is instructive to start with by assuming that the risk premium is constant through time and test the validity of this hypothesis.

A time-invariant risk premium means that investors demand at all times a fixed amount of money as a compensation to move from the time to the spot market. Taking expectations of both sides of (5.39) and remembering that the mean of the error term η is zero we have:

$$E_t \pi_{t+1} = \phi \tag{4.51}$$

This means that the expected excess profit is the constant risk premium, φ. Now assume that expectations are formed rationally and η is the rational expectations forecast error, which is distributed normally with zero mean and constant variance σ^2:

$$\pi_{t+1} = E_t \pi_{t+1} + \eta_{t+1}, \quad \eta_t \approx N(0, \sigma^2) \tag{4.52}$$

Substituting (4.52) into (4.51) gives the interpretation of (4.50) that the excess profit is equal to a constant risk premium plus the rational expectations forecast error:

$$\pi_{t+1} = \phi + \eta_{t+1} \tag{4.53}$$

It is important to distinguish between the unconditional and conditional variance of the forecast errors η. Under the assumption that the forecast errors are normally distributed, the unconditional variance, σ^2, is just a single number, whereas the conditional variance is the variance of the one-step ahead forecast error $E_t(\sigma^2_{t+1})$. The conditional variable therefore is a time series. This is related to the conditional variance of the excess profit with information available at time t, I_t, by the following equation:

$$E_t[\text{var}(\pi_{t+1}|I_t)] \equiv E_t[\pi_{t+1} - E_t(\pi_{t+1}|I_t)]^2 = E_t(\eta^2_{t+1}) = E_t(\sigma^2_{t+1}) \tag{4.54}$$

In deriving (4.54) use has been made of the square of (4.52). Equation (4.54) shows that the conditional variance of the excess profit is equal to the conditional variance of the rational expectations forecast errors (or so-called *innovations*).

The next step is the formalisation of the observation made by Mandelbrot (1963): Large changes in the variance of forecast errors tend to be followed by large changes of either sign; and conversely small changes tend to be followed by small ones. In modern terminology this means persistence in the conditional variance of the forecast errors (or otherwise, variance-clustering). In a pioneering paper Engle (1982) formulated this persistence as an autoregressive process of order one, AR(1):[7]

$$E_t \sigma^2_{t+1} = a_0 + a_1 \eta^2_t \tag{4.55}$$

This is known as an *autoregressive conditional heteroscedastic* (ARCH) process of order 1. The name comes from the fact that the conditional variance is not constant through time (homoscedastic) but time varying (heteroscedastic), and modelled as an autoregressive process. The companion equation to the ARCH model out of which the forecast errors η are generated, i.e. equation (4.53), is called the *mean function*. It is clear that if $a_0 > 0$ and $a_1 = 0$ in (4.55), the conditional variance is constant through time (homoscedastic). If a_1 is positive and close to one there is strong persistence in the conditional variance. It is worth noticing that while the *conditional* variance is heteroscedastic, the *unconditional* variance is time invariant (homoscedastic) and it is equal to: $a_0/(1 - a_1)$, provided $a_1 < 1$.

The ARCH(1) process can easily be generalised to order-p:

$$E_t(\sigma_{t+1}^2) = a_0 + a(L)\eta_t^2, \quad a(L) = a_1 + a_2 L^2 + \cdots + a_p L^p \quad (4.56)$$

where $a(L)$ is a polynomial in the lag operator of order-p. The ARCH(p) model introduces a long 'memory' of dependence on past forecast errors. Note that for the conditional variance to be positive in (4.56) both a_0 and the coefficients in $a(L)$ must be nonnegative.

A test of the hypothesis that the risk premium is time invariant and equal to a constant, φ, is to test the hypothesis that $a_i \neq 0$. for $i = 1,...,p$ in (4.56). The test procedure is to run the OLS regression of (4.53) and save the residuals. Regress the squared residuals on a constant and p-lags and test TR^2 as χ^2 with p degrees of freedom, where T is the sample size and R^2 is the squared correlation coefficient from (4.56).

However, direct estimation of (4.56) with p-lags may invalidate the non-negativity constraints on the conditional variance. Thus, in applications of the ARCH model, it is common to find that the long memory is approximated by an arbitrary linearly declining lag structure. For example, Engle (1982) imposes 0.4, 0.3, 0.2 and 0.1 coefficients on the first four lags of the squared residuals η. (Note the sum of the the coefficients should be 1.) To overcome this problem, Bollerslev (1986) suggested the *generalised autoregressive heteroscedastic* (GARCH) model of order (p, q):

$$E_t(\sigma_{t+1}^2) = w + a(L)\eta_t^2 + \beta(L)\sigma_t^2 \quad (4.57)$$

where $\beta(L)$ is a polynomial in the lag operator of order-q. The conditional variance specified in equation (4.57) has three terms: a constant, w; news about volatility from the past (the ARCH term); and past forecasts of the conditional variance (the GARCH term). The GARCH model includes the ARCH model as a special case. An ARCH model is a GARCH(p, 0). One interpretation of the GARCH(p, q) model is an infinitely long ARCH.[8] A second interpretation is an autoregressive moving average process (ARMA) of the forecast errors η (see Bollerslev, 1986). Although an ARMA interpretation is more appealing, from a practical point of view it is easier to work with (4.57). A GARCH(p, q) process with the mean function (4.53) is a general model for testing the hypothesis that the risk premium is time varying, but it does not provide for an explanation of the excess profit. This means that the mean function has to be reformulated. A convenient formulation is:

$$\pi_{t+1} = \phi_0 + \phi_1 E_t(\sigma_{t+1}^2) + \eta_{t+1} \quad (4.58)$$

The mean function (4.58) along with the general GARCH model (4.57) is called a GARCH in mean model (GARCH-M). The specification of (4.58) has the appealing interpretation that the excess profit is related to the variance of the forecast errors and hence to the 'reward' of moving from the time charter market to the spot. As uncertainty increases (that is, the variance of the forecast errors increases), the owner has to be compensated more to be convinced to move from the time charter to the spot market.

This interpretation follows from viewing (4.58) as the demand for spot contracts (a risky asset) against the demand for time charter contracts (the safe asset) in a 'mean-variance' model of asset demands with two assets. Engle et al. (1987) offer this interpretation of a GARCH-M model by noting that the demand for the risky asset in a 'mean-variance' framework, denoted A^d, is:

$$A_t^d = \frac{E_t \pi_{t+1}}{\gamma E_t(\sigma_{t+1}^2)} \qquad (4.59)$$

where γ is the coefficient of the owner's risk aversion. The demand for spot contracts depends positively on the expected excess return of spot over the time charter rate, but negatively on the risk associated with the spot market. The risk is captured by the degree of risk aversion of the owner and the conditional variance of the forecast errors in predicting the excess return. On the standard assumption that the supply of assets is fixed or slowly changing, equilibrium in the asset markets implies:

$$E_t \pi_{t+1} = A\gamma E_t(\sigma_{t+1}^2) \qquad (4.60)$$

Equation (4.58) is obtained from (4.60) by assuming $\varphi_0 = 0$ and $\varphi_1 = A\gamma$. Therefore, the excess profit is a function of risk, where the risk is measured by the degree of the owner's risk aversion, γ, and the variance of the forecast errors. The more risk averse an owner is and the less accurate the previous forecasts are, the greater is the required profit (the risk premium) needed to compensate the owner for moving from the time charter to the spot market.

Two more formulations of the GARCH model are useful in modelling the risk premium in the freight market. These are the exponential and absolute value forms. For example, the ARCH(1) version of these models is:

$$E_t(\sigma_{t+1}^2) = \exp[a_0 + a_1 \eta_t^2] \qquad (4.61)$$

$$E_t(\sigma_{t+1}^2) = a_0 + a_1 |\eta_t| \qquad (4.62)$$

The exponential form ensures that the variance is positive for all values of alpha, but has the drawback that the variance is infinite for any value of $a_1 \neq 0$. The absolute value form requires all values of both alpha coefficients to be positive to ensure a positive variance. The linear assumption may be preferable in many applications.

5.10 THE EMPIRICAL EVIDENCE

As we have seen so far, tests of the efficiency of freight markets can take several forms. One battery test is the informational efficiency or orthogonality property of the EMH, namely whether innovations can help in predicting tomorrow's freight rates. A second battery of tests is to formulate an explicit hypothesis of market behaviour and examine whether the theory is empirically validated. All theories of the term structure of freight rates agree that time charter rates are a weighted average of expected future spot rates. These theories only differ with respect to their treatment of the risk premium, which can be zero, positive, positive or negative, related to a portfolio of contracts, increasing with the term to maturity or finally time varying. The expectations theory of the term structure assumes that the risk premium is constant and has been the subject of investigation in many studies of the efficiency of freight markets. Such studies abound in the literature of maritime economics (see, for example, Zannetos, 1966; Glen et al. 1981; Strandenes, 1984). However, none of these studies tests for the validity of the theory.

The validity of the term structure of freight rates can be tested in a number of alternative tests. Assume that the expectations theory is valid, while postulating a mechanism for generating expectations of future spot rates and test whether the joint hypothesis is empirically valid. Such an approach would be valid if the underlying series are stationary. However, empirical tests show that the time charter rates and spot rates are non-stationary.[9] Campbell and Shiller (1991) have suggested a transformation of the original series in the form of the spread (or difference) between the time charter and the spot rate. The spread though should be compared with the *change* rather than the *levels* of spot rates. Both variables are now stationary and valid statistical inferences can be drawn from computing the theoretical spread or the perfect foresight spread with the actual one. If the expectations theory is correct, then the actual spread should be equal to the theoretical or perfect foresight spread. A simple regression can test the hypothesis. The constant in the regression should be zero and the coefficient of the spread should be one.

However, the regression tests the validity of the joint hypothesis of the expectations theory and the explicit expectations-generating mechanism. In empirical studies, AR, VAR and rational expectations have been used.[10] If the hypothesis is accepted by the data, then freight markets are efficient, but if the hypothesis is rejected, as is the case, for example, with Veenstra (1999a) and Kavussanos and Alizadeh (2002a), then it is not sure which part of the hypothesis is invalid. It could be that the EMH is correct but the specific expectations scheme is incorrect. This creates an ambiguity on the empirical results of the regression tests.

The VAR methodology gives rise to alternative tests of the EMH (see, for example, Glen et al., 1981; Veenstra, 1999a; Kavussanos and Alizadeh, 2002a; Wright, 2003). One test is to compare the statistical fit of the unrestricted with the restricted VAR. The restrictions on the coefficients of the VAR ensure the validity of the theory. But this method is cumbersome and has not been applied in the term structure of freight rates because it involves the estimation of two equations. An alternative much simpler test is to estimate the unrestricted VAR and then test for the validity of the restrictions. Veenstra (1999a) applies this test procedure, but there are methodological issues, which are corrected in Kavussanos and Alizadeh (2002a). The empirical evidence from this test does not support the EMH. But there are methodological issues, as the validity of the restrictions is tested through a Wald statistic, which has poor small sample statistical properties. As an alternative to the Wald test block exogeneity tests can be conducted. Although such tests have been applied extensively in financial markets, there has been little publicity in freight markets probably because they test the weak form of market efficiency.

Another test that arises out of the VAR methodology is that of variance inequality or variance bounds tests. But these tests are based on the same principles as the regression tests and therefore they are subject to the same criticism that rejection of the null hypothesis of market efficiency is a rejection of the joint hypothesis of market efficiency and the assumed expectations generation mechanism.

The conclusion is that none of these tests of the efficiency of freight markets are immune to criticisms. Although the empirical evidence is not overwhelming, on balance, it is not supportive of the EMH. Excellent surveys of this literature can be found in Glen and Martin (2005) and Glen (2006). Adland and Strandenes (2006) use technical analysis to illustrate that excess profits can be made, thereby rejecting market efficiency. It is worth noting that Adland and Cullinane (2005) claim on purely theoretical grounds that the risk premium must be time-varying and must vary with the level of freight rates and the duration of the contract.

Although the EMH, on balance, can be rejected if the requirement of efficiency is in every period of time, it cannot be rejected in its weak form when it is viewed as a long-term condition. Such tests involve the existence of a cointegrating relation between time charter and spot rates. There are many applications of this methodology (for example, Wright, 1999 and 2003; Kavussanos and Alizadeh, 2002a). Although cointegration is found, the results are mixed when the restrictions imposed by the expectations theory are explicitly tested.

The possible failure of the EMH may be due to a time-varying risk premium, which is the excess profit that has to be earned by an owner so that he can take the risk of moving from the safety of the time charter market to the risky spot market. This risk premium can be modelled with ARCH/GARCH in mean models. These models explain the risk as a function of the variance of forecast errors and the degree of risk aversion of the owner. The variance of forecast errors is modelled as a function of past variances and the squares of past innovations. The contribution of this methodology is not so much in testing the efficiency of freight markets. Rather, it is in defining and modelling risk and explaining how decisions of moving between the time charter and spot markets are affected by risk (see, for example, Kavussanos, 1996a, 1996b, 1996c, 1997; Glen and Martin, 1998; Kavussanos and Alizadeh, 2002a; Jing et al., 2008; Alizadeh and Nomikos, 2011; Wright, 2011).

6 THE EFFICIENCY OF SHIP PRICES

6.1 STATEMENT OF THE PROBLEM

The methodology that has been developed for testing the efficiency of freight rates has also been applied in testing the efficiency of ship prices (newbuilding and secondhand). The only difference is the theoretical price. In all theories of the term structure of freight rates the time charter rate is a weighted average of expected future spot rates during the lifetime of the time charter contract. According to the EMH this theoretical price should be equal to the actual price both in the long and the short run (that is, in every period of time). Similarly, the theoretical vessel price is equal to the discounted present value of expected future (operational) profits. As a vessel has a finite economically useful life (between 5 and 20 years) in computing the present value one should include along with profits the expected capital gains from the sale of the ship in the secondhand market or in the scrap market. In anticipation of time varying risk premia, it is useful to compute the present value with time varying discount rates. Thus, denoting by P^* the newbuilding price, the secondhand price by PS, the operational profit by Π, the economic life of the vessel by n and the interest rate used in discounting profits and capital gains by R, the theoretical newbuilding price is given by:

$$P_t^* = \sum_{i=0}^{n-1} \frac{1}{\prod_{j=0}^{i}(1+E_t R_{t+1+j})} E_t \Pi_{t+1+i} + \frac{1}{\prod_{j=0}^{n-1}(1+E_t R_{t+1+j})} E_t PS_{t+n} \quad (4.63)$$

To appreciate what is involved in (4.63) consider the simple case where the ship is bought as new, operated for two years (i.e. $n = 2$) and then sold in the secondhand market. Then (4.63) simplifies to:

$$P_t^* = \frac{E_t \Pi_{t+1}}{1+E_t R_{t+1}} + \frac{E_t \Pi_{t+2}}{(1+E_t R_{t+1})(1+E_t R_{t+2})} + \frac{E_t PS_{t+2}}{(1+E_t R_{t+1})(1+E_t R_{t+2})} \quad (4.63a)$$

Hence, the theoretical price is equal to the present value of the expected profit in period $t+1$ and $t+2$ plus the present value of the secondhand price that is expected to be fetched after two years. In computing the present value it is assumed that the discount (interest) rate would be different in each period. The product of the expected interest rates in the second period appears in the denominator of the last two terms.

Equation (4.63) can be used to test the efficiency of newbuilding as well as secondhand prices. For example, the theoretical price of a five-year-old ship operated for ten years and then sold can be computed through (4.63) by setting the five-year-old price on the left-hand side of (4.63) and the 15-year-old price. Similarly, the formula can be applied for an older ship, say 15-year, which is operated for ten years and then it is scrapped. The theoretical price is equal to the present value of expected profits for the 10-year period and the present value of the capital gains from the expected scrap price in the tenth year.

The present value of expected profitability captures the composite effect of two factors, profitability and interest rates. Similarly, the composite impact of expected capital gains and interest rates is embedded in the present value of capital gains. To assess the impact of risk in efficient ship pricing it is important to decompose this composite effect to its two constituent components. This is done by borrowing the methodology developed by Campbell and Shiller (1988) in financial markets. Campbell and Shiller apply equation (4.63) to a stock price, where dividends appear instead of profits and where there is no resale value in period n so that the present value is computed for an infinite horizon. The authors compute a linearized version of (4.63)) around the geometric mean of P, and Π, denoted by $\overline{P}$ and, $\overline{\Pi}$ respectively, using a first order Taylor series expansion:[11]

$$p_t^* = \sum_{i=0}^{n-1} \rho^i (1-\rho) E_t \pi_{t+1+i} - \sum_{i=0}^{n-1} \rho^i E_t r_{t+1+i} + \rho^n E_t ps_{t+n} + c \quad (4.64)$$

where lower-case letters denote natural logarithms; $E_t ps_{t+1} = \ln(E_t PS_{t+1})$ and similarly for Π; $E_t r_{t+1} = \ln(1 + E_t R_{t+1})$; $\rho = \overline{P}/(\overline{P} + \overline{\Pi})$; and c is a constant.

Note that (4.64) consists of three terms: the present value of expected profits; the present value of the expected interest rates; and the present value of the capital gains. Each present value though is computed with a constant discount rate, ρ, rather than time-varying discount rates. Therefore, the linearisation of (4.63) into (4.64) enables the separation of the impact of expected profits (and capital gains) from the time varying interest rates. This is very useful in evaluating the importance of risk premia as an independent factor of the theoretical price.

In analogy with the spread in the term structure Campbell and Shiller suggest subtracting π from both sides of (4.64). This transformation is necessary to make the underlying series stationary. As freight rates are non-stationary,[12] but can rendered stationary by taking their first difference, similarly vessel prices (new and secondhand) and profits are non-stationary, but can become stationary by taking their first difference. By taking the difference between two such non-stationary variables, the resulting variable is stationary.[13] The transformation enables (4.64) to be written as:

$$S_t^* = \sum_{i=0}^{n-1} \rho^i E_t e\pi_{t+1+i} + \rho^n E_t eps_{t+n} + d \quad (4.65)$$

where $S_t^* = p_t - \pi_t$, $e\pi_t = \Delta\pi_t - r_t$, $eps = ps_t - \pi_t$. In (5.54) all variables are stationary enabling valid statistical inferences.

Thus the theoretical spread of the (log) price from the (log) profit is equal to the present value of the spread of profits from the interest rate and the present value of the spread of the capital gains.

6.2 TESTS OF MARKET EFFICIENCY

Once the expectations of the variables are substituted out, the theoretical spread can be compared with the actual spread. If vessel prices are efficient, the theoretical spread should be equal with the actual spread. A test like (4.63) can be conducted to examine the EMH of vessel prices.

Similarly with the term structure of freight rates, three different expectations schemes can be considered: rational expectations, autoregressive and VAR.

The VAR approach opens up a battery of tests similar in nature to the efficiency of freight markets. The VAR model should explain the three state variables in (4.65):

$$S_t = k_1 + a(L)S_t + \beta(L)e\pi_t + c(L)eps_t + \varepsilon_{1t} \quad (4.66a)$$

$$e\pi_t = k_2 + d(L)S_t + e(L)e\pi_t + f(L)eps_t + \varepsilon_{2t} \quad (4.66b)$$

$$eps_t = k_3 + g(L)S_t + h(L)e\pi_t + m(L)eps_t + \varepsilon_{3t} \quad (4.66c)$$

where each coefficient is a polynomial in the lag operator of order p. For example:

$$a(L) = a_1 L + a_2 L^2 + \cdots + a_p L^p.$$

The VAR system can be expressed in the matrix notation of (4.19) and forecasts can be computed in the form of (4.30). These can be entered into (4.65) to compute the theoretical spread. If the EMH of vessel prices is valid, the non-linear restrictions (4.32) must hold. Thus, one test of the EMH is to compare the statistical fit of the constrained with the unconstrained VAR through log-likelihood statistic, as in equations (4.37)–(4.39). The second approach is to estimate the unrestricted VAR and test whether the restrictions (4.32) hold. The null hypothesis of efficient prices is that the restrictions are valid, which can be tested through a Wald statistic. As an alternative to a Wald-test block exogeneity tests can be conducted. Finally, the null hypothesis of efficient pricing can be tested through a variance ratio test of the form of (4.44).

6.3 WEAK FORM OF MARKET EFFICIENCY – COINTEGRATING TESTS

Efficient ship pricing may not hold in every period of time, but it may be true in the long run. In other words, there may be a weak form of market efficiency. As in the freight markets, a test of the weak form of efficient pricing is through cointegration tests. A possible cointegrating relation is between prices (newbuilding, secondhand and scrap) and profits.[14] The theory imposes no numerical restrictions on the coefficients of the cointegrating relation, other than a positive sign on the coefficient of profits. In other words, an increase in profitability should increase vessel prices. The cointegration test within a VAR framework takes the form:

$$\Delta p_t = \sum_{i=1}^{p} a_i \Delta p_{t-i} + \sum_{i=1}^{p} \beta_i \Delta \pi_{t-i} + \gamma_1 [p_{t-1} + \phi_0 + \phi_1 \pi_{t-1}] \quad (4.67a)$$

$$\Delta \pi_t = \sum_{i=1}^{p} c_i \Delta p_{t-i} + \sum_{i=1}^{p} d_i \Delta \pi_{t-i} + \gamma_{21} [p_{t-1} + \phi_0 + \phi_1 \pi_{t-1}] \quad (4.67b)$$

The VAR is specified with respect to the two possible variables that might form a cointegrating relation, namely the vessel price p_t and profit π_t. There is an equation for each one of the state variables. The order of the VAR is p; there are p-lags in each of the state variables. The order of the VAR is chosen with the Akaike Information Criterion or the Schwarz Information Criterion (see the Statistical Appendix). The existence of a cointegrating relation is tested through the Johansen procedure (see the Statistical Appendix).

6.4 TIME-VARYING RISK PREMIA

According to the EMH the profits from shipping should not exceed those from alternative investments. Efficient ship pricing implies that the ship price reflects the profits from operation and the capital gains, which is the essence of (4.63). But the variables in (4.63) are unobserved. An alternative way to postulating generation expectations mechanisms is to consider one-period holding yields. Let $H\pi_t$ denote the one-period holding shipping yield, which is defined by:

$$H\pi_{t+1} = \ln(P_{t+1} + \Pi_{t+1}) - \ln P_t \quad (4.68)$$

Recall that $r_{t+1} = \ln(1 + R_{t+1})$ denotes the return on money. Efficient ship pricing implies that the one-period holding shipping yield should be equal to the return on money for the same period:

$$H\pi_{t+1} = r_{t+1} \quad (4.69)$$

An alternative way of expressing efficient ship pricing is that the excess return from shipping over the return on money should be zero. Hence a test of efficient pricing is to examine whether the excess return is independent of the information available at time t. This gives rise to two complementary tests: testing whether the excess return is correlated with previous information and testing the predictability of the excess return. These procedures are testing the joint hypothesis of efficient pricing and rational expectations. But failure of the joint hypothesis may be due, as in the case of freight markets, to time varying risk premia. As a result by accounting for risk premia the excess return is zero. But as has been argued in the case of freight rates, one can always specify a function that renders the excess return equal to zero. Accordingly, the contribution of this literature lies in show showing how risk affects ship prices. As with freight rates one can specify a time invariant risk premium and examine whether the hypothesis is consistent with empirical evidence. If it is not, then a function explaining risk has to be postulated. The capital asset pricing model (CAPM) offers a theory of relating return with risk. According to CAPM, there is a positive relationship between risk and return: as risk rises, the excess return required by an owner to buy the ship is increased. Let XR denote the excess return of shipping over the return on money and define the *mean function* in the ARCH/GARCH-M class of models as an AR(m) process augmented by a time varying risk premium that depends on the conditional variance (or standard deviation) of forecast errors in predicting the excess return, σ^2.

$$XR_t = a_0 + \sum_{i=1}^{m} \beta_i XR_{t-i} + \phi\sigma_t^2 + \eta_t \quad (4.70)$$

In the mean function the second term involving the sum represents the AR(*m*) process; the coefficient φ measures the importance of risk, which according to CAPM should be positive; and η are the residuals, which should be a white noise process.

The forecast errors upon which risk is measured are obtained from an ARMA(*p*, *q*) process:

$$XR_t = c_0 + \sum_{i=1}^{p} b_i XR_{t-i} + \sum_{i=1}^{q} d_i \varepsilon_{t-i} + \varepsilon_t, \quad \varepsilon_t \approx IID(0, \sigma_t^2) \quad (4.71)$$

The first sum captures the AR(*p*) process and the second sum the MA(*q*) process. The ARMA specification is necessary if there is autocorrelation in the residuals.

The conditional variance of the forecast errors, which measure the time varying risk, can be modelled, for simplicity, as a GARCH process of order (1, 1):

$$\sigma_{t+1}^2 = f_0 + f_1 \varepsilon_t^2 + f_2 \sigma_t^2 \quad (4.72)$$

6.5 THE EMPIRICAL EVIDENCE

Early empirical work (for example, Beenstock, 1985) simply used the EMH to model vessel prices. Strandenes (1984, 1986) investigates the price formation in the dry bulk and tanker markets using the present value model. The empirical evidence shows that long term profitability is more important than current profits in explaining ship prices, which she interprets as support for the semi-rational expectations. Vergottis (1988) is one of the authors to test the efficiency of ship pricing by using the principles of rational expectations of unbiasedness and informational efficiency or orthogonality. Hale and Vanags (1992) test for market efficiency through the use of block exogeneity tests (Granger causality). But the procedure, as noted above, only tests for the weak form of market efficiency. These authors examine also the existence of cointegration relations between various vessel prices using the Engle–Granger two-stage procedure, but with mixed results. Glen (1997) employs the more powerful Johansen approach to test for cointegration in a multivariate setting. Veenstra (1999b) examines the existence of cointegration between secondhand prices, a time charter rate, newbuilding prices and scrap prices.

Kavussanos and Alizadeh (2002b) provide a thorough and exhaustive examination of efficient pricing using the elaborate testing procedures outlined above. Their empirical findings reject the EMH for newbuilding and secondhand prices. The authors attribute this inefficiency to time varying risk premia, showing that there is a positive relationship between risk and return in shipping in line with CAPM.

A common theme in all the abovementioned studies is that there is no theoretical basis for the cointegrating vector. Rather, the cointegrating vector is an empirical result based on intuition. For example, in the exemplary study of Kavussanos and Alizadeh (2002b) the cointegrating vector is based on the notion that prices are cointegrated with profits, where the profit is modelled as the spread between a time charter equivalent rate and operating costs. The latter are modelled as an exponential growth rate regression. A notable exception to this rule is the study by Tsolakis, Cridland and Haralambides (2003), where the cointegrating vector is the reduced form of a demand–supply framework in the secondhand market. The cointegrating vector to be tested empirically includes secondhand prices, a time charter rate, newbuilding prices, Libor and the orderbook to fleet ratio. Their empirical results suggest that newbuilding prices and time charter rates form a cointegrating vector with secondhand prices. Libor is only significant in the dry bulk market but not in the tanker market, a strange result. Haralambides et al. (2004) extend the above results in the newbuilding market. The possible cointegrating vector for newbuilding prices includes, in addition, cost variables, such as steel prices roll-plates in Japan. These theoretical approaches to the determinants of the cointegrating vector have their foundations on structural models of newbuilding and secondhand prices, such as Koopmans (1939), Hawdon (1978) and Jin (1993), among others. This is a trend in the right direction in establishing the importance of structural models in maritime economics. A notable study in this new trend is that of Jiang and Lauridsen (2012), which analyses the price formation of Chinese dry bulk carriers. The empirical evidence suggests that the time charter rate has the most significant positive impact on new prices followed by the cost of shipbuilding, the profit margin and the shipyard capacity utilisation.

7 CONCLUSIONS

This chapter has explained the EMH and the statistical tests of the efficiency of freight rates and ship prices. The methodology has been borrowed from the financial markets, but has been adapted to shipping mainly by correcting for the finite life of ships and of time charter contracts. The empirical evidence suggests that both freight rates and vessel prices are, on balance, inefficient. The literature on time-varying risk premia has shed light on the nature of this inefficiency and has highlighted models that explain risk and how it affects shipping decisions. The empirical evidence also suggests the advantage of structural models over purely statistical models of shipping as a guideline to the specification of the cointegrating vectors.

STATISTICAL APPENDIX

The literature on the efficiency of shipping markets involves many statistical and econometric concepts. This Appendix provides a summary of concepts and statistical tests which are used in this and other chapters. The objective of this Appendix is simply to make the book self-contained. Therefore the Appendix is not meant to replace econometrics textbooks, such as Hamilton (1994) and Greene (1997), which cover all the material covered here.

STATIONARY AND NON-STATIONARY UNIVARIATE TIME SERIES

All shipping variables, such as freight rates, can be thought of as statistical time series, as they assume a value for each period, be that a day, week, month, quarter or a year.

A very common statistical model of shipping and economic time series is the univariate (single variable) autoregressive model of order 1, denoted by AR(1). Thus

$$y_t = a + \beta \cdot y_{t-1} + \varepsilon_t \qquad |\beta| < 1 \qquad (A.1)$$

The concept of autoregression means that the variable y depends on past values of itself through an explicit linear function, such as (A.1). The function is linear because the terms are added and a and β are constant coefficients. The order of the autoregression is characterised by the maximum lag of y. Thus, an AR(2) process is written as

$$y_t = a + \beta_1 \cdot y_{t-1} + \beta_2 \cdot y_{t-2} + \varepsilon_t \qquad |\beta_i| < 1 \quad \text{for } i = 1, 2. \qquad (A.2)$$

The variable y depends linearly, through the constant coefficients a, β_1 and β_2, on two lagged values of itself. In statistical and econometric analysis the aim is to identify the exact functional form (A.1), (A.2) or another more complicated but unknown form) and the precise value of their coefficients.

In each autoregressive scheme a disturbance (error or residual) term, ε, is added. This means that the exact value of y in every period t is determined by a deterministic component captured by $a + \beta \cdot y_{t-1}$ in (A.1) and a stochastic (or random) component, ε. The disturbance term assumes random values in each period and therefore y differs in each period because of the stochastic nature of the disturbance term. The values of ε depend on events such as strikes, weather conditions, political events or the influence of other important variables which were unintentionally omitted from the specification of (A.1) or (A.2). Because of the stochastic nature of ε, the variable y is also stochastic following an autoregressive stochastic process.

The Identification of the 'true' (or population) values of the coefficients through statistical or econometric analysis depends on the properties of the disturbance term. From a sample of data on y the estimates of the coefficients (a and β) would approach the 'true' (population) values as the sample size tends to infinity if the disturbance term is purely random. In statistical analysis the properties of the estimates of the coefficients that must be satisfied so that they approximate the 'true' values are best-linear-unbiased estimators (BLUE). The 'best' property means that the estimator has a minimum variance, while the unbiasedness property means that the sample mean of the estimator is equal to the population value. If the random variable ε satisfies some properties then the estimate of the coefficients are BLUE. These properties of the ε can be summarised as follows.

$$E\varepsilon_t = 0 \quad \text{for all } t \qquad (A.3a)$$

$$\text{var}(\varepsilon_t) = E(\varepsilon_t^2) = \sigma^2 \quad \text{for all } t \qquad (A.3b)$$

$$\text{cov}(\varepsilon_t, \varepsilon_{t-j}) = 0 \quad \text{for all } j \neq 0 \text{ and all } t \qquad (A.3c)$$

Although in the real world we observe just one value of ε in each period, the stochastic nature of ε implies that there is an infinite number of observations that could have been observed forming an entire distribution. For each distribution in every period t the first property (A.3a) implies that the mean of the distribution is zero. The second property[15] (A.3b) implies that the variance of each distribution assumes the same constant value, σ^2. This property is usually referred to as *homoscedasticity*. When the variance is not constant through time the disturbance term is *heteroscedastic*. The third property (A.3c) is that the covariance[16] of any pair of ε, such as ε_t and ε_{t-1} is zero. For the covariance to be zero the correlation of ε with any past or future value of ε must be zero. These three properties of the disturbance term are usually referred to as identically and independently distributed and they are denoted as

$$\varepsilon \approx IID(0, \sigma^2) \tag{A.4a}$$

When the disturbance term satisfies these properties it is called a *white noise process*. If the distribution of the disturbance term is also normally distributed, then it is called a *Gaussian white noise process*, denoted as

$$\varepsilon \approx NIID(0, \sigma^2) \tag{A.4b}$$

When the error term follows a Gaussian white noise process the estimates of the coefficients of (A.1) or (A.2) obtained from a relatively large sample ($T > 30$) are BLUE. When the error term follows a white noise process, (A.4b), the estimates are asymptotically BLUE (i.e. they become BLUE as the sample size tends to infinity).

What is the meaning of an autoregressive process? The variable y oscillates in a random way around the population mean. The population mean can be calculated by assigning to ε its mean value of zero and assuming that y converges to its mean value (that is, when $y_t = y_{t-1} = y_{t-2}$). For convergence[17] the absolute value of β should be less than 1. Hence, the population (or unconditional) mean of y, Ey, is

$$Ey_t = \mu = \frac{a}{1-\beta} \tag{A.5}$$

It is clear from (A.5) that the absolute value of $\beta < 1$. If $\beta = 1$, then y does not converge to its mean value but diverges to infinity.

The population (or unconditional) variance of y can also be calculated as follows. First compute for (A.1) the deviation from the mean, which enters into the definition of the variance.

$$(y_t - \mu) = a + \beta \cdot y_{t-1} + \varepsilon_t - \mu = \beta \cdot (y_{t-1} - \mu) + \varepsilon_t$$

This relationship is obtained by substituting for $a = \mu(1-\beta)$ from the definition of the mean (A.5). Substituting this value into the definition of the variance of y we have

$$\begin{aligned} \text{var}(y_t) &= E(y_t - \mu)^2 = \beta^2 \cdot E(y_{t-1} - \mu)^2 + \text{var}(\varepsilon_t) \\ &+ 2\text{cov}(y_{t-1} - \mu, \varepsilon_t) = \sigma^2/(1-\beta^2) \end{aligned} \tag{A.6a}$$

In deriving (A.6) we have made use of (i) the var(y_t) = var(y_{t-1}) and (ii) the covariance of $(y_{t-1} - \mu)$ and ε_t is zero and (iii) the (absolute) value of $\beta < 1$.[18]

We can now define a *stationary* time series y, given by any stochastic process such as (A.1) or (A.2), as a series whose mean, variance and auto-covariance converge to a finite value.[19] The convergence means that for a series y the mean, the variance and the auto-covariance (the latter for a given lag length) are independent of time and are finite (that is, they do not tend to infinity). For a given lag length j, the auto-covariance of y_t and y_{t-j} is constant. The auto-covariance changes with the lag j. A series y is a stationary if the absolute value of *all* $\beta < 1$.

More formally a series is 'covariance' stationary if:

$$Ey_t = \mu, \quad \text{var}(y_t) = \sigma_y^2, \quad \text{cov}(y_t, y_{t-j}) = \gamma_j \quad (A.8)$$

In plain English a stationary variable is one that is trendless. It is not upward sloping, nor downward sloping against time. The process is mean reverting. A shock causes a deviation from the mean (or the equilibrium value) for a while, but the deviation peters out in the long run. In a graph the series frequently crosses the mean. The variability of the series around the mean is, on average, constant.

As the condition for stationarity is that *all* $\beta < 1$ (in absolute value), it follows that a series is *non-stationary* if at least one β is equal to or greater than 1. There are two widely used non-stationary models: a *random walk with drift* and a *random walk without drift*. Considering (A.1) and setting $\beta = 1$, we obtain a random walk with drift.

$$\Delta y_t = y_t - y_{t-1} = a + \varepsilon_t \quad (A.9)$$

The coefficient a is the drift. The interpretation of (A.9) is appealing if y is the natural logarithm of Y and therefore Δy is the rate of growth of Y. According to (A.9) the rate of growth is equal to a constant plus a white noise process. Therefore Y is non-stationary as its mean is increasing through time. A time varying mean is meaningless as a measure of central location of the distribution of Y.

The statistical properties of non-stationary variables are different from those that are stationary. To illustrate some of these differences it is convenient to define the lag operator L. For any variable y, $Ly_t = y_{t-1}$; $L^2 y_t = y_{t-2}$; and in general $L^k y_t = y_{t-k}$. Using the lag operator, a random walk without drift (that is, $a = 0$ in (A.9)) can be written as

$$y_t = (1-L)^{-1}\varepsilon_t = \varepsilon_t + \varepsilon_{t+1} + \cdots \quad (A.10)$$

The (unconditional) mean and (unconditional) variance of (A.10) are:

$$Ey_t = 0, \quad \text{var}(y_t) = E\left[\sum_{i=0}^{n-1}\varepsilon_{t+i}^2 + \sum_{t \neq s}\varepsilon_t \varepsilon_s\right] = n\sigma^2 \quad (A.11)$$

Therefore, the mean of a random walk without drift is zero, but the variance tends to infinity as n increases (that is, it is not independent of time). The difference between a random walk with and without drift is that with drift the mean is time varying, whereas without drift the variance is time varying.

Unlike the unconditional mean, the conditional mean uses the information in the time series to predict future values of y. For a random walk with and without drift the forecast of y 'm' periods ahead is

With drift: $E_t y_{t+m} = y_t + a \cdot m$ Without drift: $E_t y_{t+m} = y_t$ (A.12)

Another important difference between stationary and non-stationary variables is that the former may give rise to a *deterministic trend*, while the latter to a *stochastic trend*. To illustrate this difference consider again the random walk model with drift, equation (A.9), and assume that the initial value of y is y_0. Then (A.9) can be rewritten as:

$$y_t = y_0 + a \cdot t + \sum_{i=1}^{t} \varepsilon_i \qquad (A.13)$$

It can be seen from (A.13) that y_t does not return to the deterministic trend, defined by $(y_0 + a t)$, because of the accumulation of past random error terms. The variable y follows a 'stochastic-trend' because y will drift up or down depending on the sign of a. Note that the first difference of y_t (Δy_t) is stationary. For this reason y_t is referred to as 'difference-stationary'.

In contrast consider the following model:

$$x_t = \delta + a \cdot t + \varepsilon_t \qquad (A.14)$$

The variable x moves around a deterministic trend $(\delta + a t)$ by the disturbance term ε, which, by assumption, is stationary. Accordingly, x is said to be 'trend-stationary' because although it follows a trend the deviations from the trend are stationary. A comparison of (A.13) and (A.14) shows that both variables follow a linear trend, but the disturbance term in (A.13) is non-stationary, whereas the disturbance term in (A.14) is stationary. The variable y follows a stochastic trend and is difference-stationary, while the variable x follows a deterministic trend and is trend-stationary. The difference between difference-stationary and trend-stationary variables complicates the testing of unit roots, as we shall see later.

The different statistical properties between stationary and non-stationary variables have huge implications for statistical estimation and statistical inference. Estimation of non-stationary variables may give rise to 'spurious'[20] regression results. This means that the estimated regression gives the impression of good 'fit' (high R^2 and adjusted R^2) and statistically significant coefficients (high t-statistics), when there is no 'true' relationship between the variables. This would be the case if the variables for which a model is built are related through a common trend or a third variable that is omitted from the model. There are also huge problems with statistical inference, as the hypothesis underlying testing procedures (such as t-statistics χ^2- and F-statistics) assume that the variables are stationary.

STATIONARITY (UNIT ROOT) TESTS

Luckily, these problems can be resolved because any non-stationary variable can be transformed into a stationary one by differencing it a number of times. If a variable y becomes stationary when it is differenced once (i.e. Δy is stationary), then it is said that it is integrated of order 1, denoted by $I(1)$. If $\Delta^2 y$ is stationary, then y is integrated of order two, $I(2)$. In general, if $\Delta^k y$ is stationary then y is integrated of order k, $I(k)$.

Granger (1966) shows that most economic variables follow a stochastic process of the form

$$\Delta y_t = a + \varepsilon_t + \lambda \varepsilon_{t+1} \tag{A.15}$$

If y is again the natural logarithm of a variable Y, then (A.15) implies that Y grows at the constant rate a plus a moving average error term. This implies that all economic and shipping variables need to be differenced once or twice to become stationary, namely that they are $I(1)$ or $I(2)$.

A simple approach in detecting whether a variable is stationary or non-stationary, which is frequently referred to in the literature of the efficiency of shipping markets, is the *autocorrelation function* and the *correlogram*. The autocorrelation between y_t and y_{t-j}, denoted by $\rho(j)$, is defined as

$$\rho(j) = \frac{\text{cov}(y_t, y_{t-j})}{\text{var}(y_t)} \tag{A.16}$$

When $j = 0$, $\rho(0) = 1$. For a stationary variable $\rho(j)$ tends to zero as j increases, whereas for a non-stationary variable it remains significantly above zero. The correlogram simply plots the autocorrelation coefficient against j and provides a visual guidance to the stationarity or not of a variable. A quick drop to zero indicates a stationary variable, whereas a flat line above zero suggests a non-stationary variable.

But there are also formal tests of stationarity and we discuss the Dickey–Fuller (DF), the augmented Dickey–Fuller (ADF) tests and the Phillips–Perron test, as they are extensively used in the literature of the efficiency of shipping markets. Consider again the AR(1) process described by equation (A.1) and repeated here for convenience

$$y_t = a + \beta \cdot y_{t-1} + \varepsilon_t \tag{A.1}$$

We have established that if $\beta < 1$ (in absolute terms), y_t is stationary, $I(0)$, provided that ε_t is stationary, $I(0)$, as well.

The random walk model with drift is obtained by setting $\beta = 1$ in (A.1). Thus

$$\Delta y_t = a + \varepsilon_t \tag{A.17}$$

The right hand side of (A.17) is stationary, provided the error term is stationary. Hence, for $\beta = 1$, Δy_t is stationary and therefore y_t is $I(1)$. This reasoning suggests the following test of stationarity. Subtract y_{t-1} from both sides of (A.1)

$$\Delta y_t = a + \phi \cdot y_{t-1} + \varepsilon_t \quad \phi = \beta - 1 \tag{A.18}$$

If $\varphi = 0$, then $\beta = 1$. If $\varphi < 0$, then $\beta < 1$. Therefore, a test of stationarity (or unit root test) is whether $\varphi = 0$ (null hypothesis) against the alternative that $\varphi < 0$. If the null is accepted, then the right hand side of (A.18) is stationary, provided ε_t is also stationary. This means that Δy_t is stationary and therefore y is $I(1)$. The null hypothesis of non-stationarity (or unit root) can be tested by running the regression (A.18) and computing the t-statistic of the coefficient φ. This is the Fuller–Dickey test. The t-statistic of φ under non-stationarity does not follow a standard t-distribution but a Dickey–Fuller distribution. The authors compute Tables for the adjusted critical values of the t-statistic.

We can generalise the Dickey–Fuller test for an autoregressive process of order n, $AR(n)$. Assume that y is defined by

$$y_t = \beta_1 y_{t-1} + \beta_2 y_{t-2} + \cdots + \beta_n y_{t-n} + \varepsilon_t \quad \text{or} \quad \beta(L) y_t = \varepsilon_t \quad (A.19)$$

$$\text{where} \quad \beta(L) = 1 - \beta_1 L - \beta_2 L^2 - \cdots - \beta_n L^n \quad (A.20)$$

If the roots of the characteristic equation $\beta(L) = 0$ are all greater than unity then y_t is stationary. For the simple $AR(1)$ process, if the root of the characteristic equation $(1 - \beta_1 L = 0)$ is greater than unity in absolute value then y is stationary. This implies that $\beta_1 < 1$, because the root of L is $L = 1/\beta_1$. If $L = 1$, then $\beta_1 = 1$ (i.e. a unit root, which implies a non-stationary variable).

The Dickey–Fuller test is valid if the error term is white noise, namely if the 'true' model is indeed the one assumed. Thus, suppose that the true model is an $AR(2)$ process and that y depends in addition on x:

$$y_t = \beta_1 y_{t-1} + \beta_2 y_{t-2} + \beta_3 x_t + \varepsilon_t \quad (A.21)$$

But the modeller does not know the true model and uses instead an $AR(1)$ to test for stationarity (unit root). The omitted variables will cause autocorrelation in the ε residuals. Because the modeller does not know the 'true' model, and therefore which variables are omitted, a solution is to assume an n-order, $AR(n)$, process of the form (A.19). This implies adding lagged values of Δy to (A.18) until the error term ε becomes a white noise process. This is the augmented Dickey–Fuller test (ADF):

$$\Delta y_t = \phi y_{t-1} + \sum_{i=1}^{n} \phi_i \Delta y_{t-i} + \varepsilon_t; \quad \text{where } \phi = \beta_1 + \beta_2 + \cdots + \beta_n - 1 \quad (A.22a)$$

Solutions to the choice of the order-n of the AR process involve tests of significance of the additional lags (that is, if it increases the adjusted R^2, which for a linear model is equivalent to the Akaike Information Criterion (AIC)[21]).

We have deliberately chosen to include a constant in (A.18) and not in (A.22a) to illustrate a problem with the use of the DF- and ADF-tests. The true model may include a constant in (A.22) and a deterministic trend:

$$y_t = a + \mu t + \beta_1 y_{t-1} + \beta_2 y_{t-2} + \cdots + \beta_n y_{t-n} + \varepsilon_t \quad (A.24)$$

In testing for stationarity, the hypothesised model must 'nest' (i.e. include as special cases) both the null and alternative hypotheses. This implies that a constant or a trend or both should be included in (A.24) if the alternative hypothesis incorporates any single one of them or both. Thus, the ADF becomes:

$$\Delta y_t = \phi y_{t-1} + \sum_{i=1}^{n} \phi_i \Delta y_{t-i} + a + \mu t + \varepsilon_t \quad \text{where} \quad \phi = \beta_1 + \beta_2 + \cdots + \beta_n - 1 \quad (A.22b)$$

As the true model is unknown a good strategy is to use the methodology of moving from the 'general' to the 'specific'. Thus, one should start from (A.22b) and then move to simpler forms (i.e. dropping the time trend and then the constant, while ensuring in each case an order-n in the AR process so that the error term is a white noise process). In the first step we start with (A.22b) and test for stationarity (that is, $\varphi = 1$) with the adjusted t-statistic found in the Dickey–Fuller Tables. At the 5% level of significance the t-statistic varies between -3.60 and -3.45 depending on the size of the sample, T. If we fail to reject the null we test the joint hypothesis of a unit root and $\mu = 0$ through an F-test. The adjusted critical values of the F-distribution are obtained from the Dickey–Fuller Tables. If we fail to reject the null with the general specification of (A.22b), then sequentially we drop the trend and then the constant repeating the process. The critical values of the adjusted t-statistic when (A.22b) includes only a constant vary from -3.00 to -2.89 for small sample size ($T < 100$). This is the Perron sequential testing procedure.

But the general to specific methodology implies a bias towards accepting the null hypothesis of a unit root (that is, non-stationarity) when the true model is in fact stationary. An alternative approach is suggested by Phillips and Perron (1988), based on the Phillips (1987) Z-test, which involves a non-parametric correction to the t-test statistic to account for the bias introduced by the autocorrelation in the error term when an AR(1) is assumed.

Testing for non-stationarity involves more problems when there are structural breaks in the series, which lead to under-rejecting the null.

COINTEGRATION IN SINGLE EQUATIONS

For a statistical appendix we have devoted a great deal of effort into the concept and testing procedures of stationarity. The motivation of this big effort is that building models without checking for the stationarity of the underlying variables can lead to spurious correlation (regression) results with false inferences. We briefly touched upon the concept of spurious correlation in the last section, but it is important to dwell on it a bit more.

A spurious regression suggests that there is strong correlation between two (or more) variables and yet this does not imply a causal relationship between the variables. Spurious correlation arises when two variables are related through a common trend. But to start with assume that x is stationary but y is trend-stationary. There are two approaches that deal with a time trend. In the first case, the impact of the time trend can be removed from the data by regressing the variable in question on a time trend and computing the residuals or the deviations from the trend. The new variable is stationary, as the trend has been removed, and can be included in the model of the two variables (x and y). There is an alternative method in which a time trend is added in the model of the two variables. In this alternative case the residuals are again stationary. In both cases the standard regression model (estimated through ordinary least squares, OLS) is computed with stationary series which have constant means and finite variances. Statistical inferences based on t- and F-statistics are valid.

If one of the series in the model is non-stationary, then regressing it on a time trend does not yield a stationary variable because as we have seen a non-stationary variable includes a stochastic-trend (namely, the accumulation of past disturbances, see equation (A.13)). Accordingly, using OLS to estimate the true relationship between the two variables gives rise to spurious regression (correlation) results that lead to invalid inferences based on t- and F-statistics. This point can be further illustrated through the following model. Consider two non-stationary variables, each one following the random walk model without drift:

$$y_t = y_{t-1} + u_{1t} \qquad u_{1t} \approx NIID(0,1) \qquad (A.25)$$

$$x_t = x_{t-1} + u_{2t} \qquad u_{2t} \approx NIID(0,1) \qquad (A.26)$$

Suppose that the modeller does not know that the two variables follow the random walk model without drift and attempts to unravel their true relationship by running the model:

$$y_t = b_0 + b_1 x_t + \varepsilon_t \qquad (A.27)$$

Each variable includes a stochastic trend (the accumulation of past disturbances in u_1 and u_2) and therefore the modeller would not be able to reject the null hypothesis that $b_1 = 0$, when in fact the null hypothesis is valid. The spurious regression arises because of the common stochastic trend. The disturbance term ε in (A.27) will not be a stationary variable and therefore not a white noise process. The adjusted R^2 in (A.27) will also be high, suggesting a causal relationship, which does not exist. This would become apparent, if the modeller regressed Δy on Δx. The model would have a poor fit and the t-statistic on Δx would be insignificant. If y and x are the natural logs of Y and X, then Δy and Δx are the rates of growth of the two variables. Hence, their levels are correlated because of the existence of the common stochastic trend, but their rates of growth would not be correlated. The problem of spurious correlation is further complicated by the fact that t- and F-statistics do not have the standard distributions which are tabulated for stationary variables.

The spurious regression means that one should be careful in modelling and should first test the degree of integration of each variable to be included in the model. We have seen that if a variable is integrated of order d, $I(d)$, i.e. it needs to be differenced d-times to become stationary, then it has d unit roots. The residuals of a model with variables of different degree of integration would generally follow the highest order of integration. Thus, if y is $I(1)$ and x is $I(0)$, then the residuals $u = (y - bx)$ are $I(1)$. Therefore, statistical inference would only be valid if the model includes only $I(0)$ variables. Since most economic variables are $I(1)$ or $I(2)$, the above logic dictates that we should be modelling rates of growth or growth acceleration rather than levels of variables. But such a model would be poor because the information included in the levels is wasted.

So is there any way that $I(1)$ or $I(2)$ variables can be included in a model without producing spurious regression results? The short answer is yes, provided the variables included in the model are cointegrated. Starting with two variables, each one $I(1)$, the two variables are cointegrated if there is a linear combination between them that gives rise to $I(0)$ residuals. In general, if two variables are $I(d)$ and there exists a linear combination such that the disturbance term from the regression $(u_t = y_t - \beta x_t)$ is $I(d-b)$, where $b > 0$, then Engle and Granger (1987) define y_t and x_t as cointegrated of order (d, b), denoted by $CI(d, b)$.

This definition implies that only variables of the same order of integration might be combined to form a long-run equilibrium relationship. But this is not necessarily the case, if the number of variables is greater than 2. Consider three variables, where y and x are $I(2)$, whereas z is $I(1)$. If there exists a linear combination between y and x such that the residuals from this regression are $I(1)$, then these residuals can form another cointegrating vector with z, if the residuals of the last equation are $I(0)$. This allows variables with different order of integration to be combined to form cointegrating vectors, although in each vector only variables with the same degree of integration are allowed.

All this may sound very technical, but there is an economic logic behind it that appeals to common sense. The cointegrating vector simply summarises the long-term relationship between the two variables. In economic theory production, Y, and sales, X, cointegrate so that inventory adjustment $Z = Y - X$ is $I(0)$. In the context of this chapter a time charter rate of duration τ and the spot rate must be cointegrated as for a holding period yield of one period, say a year, the two are equal in equilibrium. The demand function for vessels derived in Chapter 3 (see equation 20 in the Appendix) must form a cointegrating vector with a constant reflecting technological factors, the demand for shipping services, Q, with elasticity of one, and the freight rate per unit of the user cost of capital, FR/UC, with elasticity equal to σ, the degree of substitution between speed and fleet.

The residuals of the cointegrating vector show the deviations from this long-term relationship. These deviations may persist in the short term, but there would be a tendency for the system to move back to equilibrium. Thus, in the term structure of freight rates there would be prolonged periods over which the time charter rate would deviate from the sum of expected future spot rates and possibly time-varying risk premia, but there would be a tendency for the system to return to equilibrium in the long run.

So, how this system is represented in a cointegrating framework? Let y denote the three-year time charter rate and x the spot freight rate. The equilibrium relationship of the term structure of freight rats states that the holding period yield of a time charter contract is equal to the holding period yield of a number of rolling spot contracts:

$$\Delta y_t = b_1 \Delta x_t - (1 - a_1)(y_{t-1} - \beta_0 - \beta_1 x_{t-1}) + u_t \qquad (A.28)$$

Engle and Granger call such a model an Error Correction Model (ECM).

The long-run equilibrium relationship, called the cointegrating relationship, between time charter and spot rates is given by the equation:

$$y_t = \beta_0 + \beta_1 x_t + \varepsilon_t \quad \text{or} \quad \varepsilon_t = y_t - \beta_0 - \beta_1 x_{t-1} \qquad (A.29)$$

This relationship appears in (A.28) as deviation from equilibrium. This deviation is equal to ε_t in (A.29), which is called the Error Correction (EC) mechanism because it measures the degree of disequilibrium at any point in time. When the EC is zero the system is in long run equilibrium; otherwise the system is in disequilibrium.

If y and x are cointegrated, the error term, ε, in (A.29) is stationary, $I(0)$. The error term, ε, is not a white noise process, otherwise there would be a random tendency for the system to come back to equilibrium. If instead ε is autocorrelated, then there would be a persistent tendency to return to equilibrium. The system is guaranteed to return to equilibrium (that is, it is stable) if the coefficient of the EC term in (A.28) is negative (that is, if $a_1 < 1$). The negative coefficient implies that if the time charter rate is higher than the spot rate (a positive deviation from equilibrium), then time charter rates will fall in the future (the change in the time charter rate will be negative – a move in the direction of equilibrium). Notice that the EC term appears with a lag and therefore previous disequilibrium leads to current and future changes in the time freight rate in the opposite direction. The coefficient $(1 - a_1)$ measures the speed at which the system returns to equilibrium. As it is apparent from (A.28) the dynamic adjustment to equilibrium is affected by the change in the spot rate. If spot rates rise, the dynamic adjustment would be slower than otherwise.

If y and x are $I(1)$ and they are cointegrated (i.e. ε_t is $I(0)$), then statistical inference based on t- and F-statistics computed from (A.28) is valid. This is so, because all variables appearing in (A.28) are $I(0)$ and therefore the equation can be estimated with OLS. Notice that both Δy and Δx are $I(0)$ because they are the first difference of y and x, which are $I(1)$ – they need to be differenced once to become stationary. If y and x are cointegrated then the residuals of (A.29) are also $I(0)$. Hence, the EC term in (A.28) is also $I(0)$. As all variables in (A.28) are $I(0)$, the residual u_t is also $I(0)$. If sufficient lags in Δy and Δx are allowed in the specification of (A.28) to eliminate any correlation in the residuals u, then the error term u_t will be a white noise process leading to valid statistical inferences.

The ECM methodology provides a solution to the problem of using levels or rates of change of variables.[22] The method shows that both levels and rates of change should be used in modelling, but in a consistent way that takes into account economic theory. The rationale of the ECM specification can be further elucidated by considering the unrestricted form of (A.28). Thus consider that the modeller starts with the notion that time charter and spot rates move together in the long run, but because of the inability of economic agents to adjust instantly to new information, the adjustment is spread over time. The gradual adjustment may be an optimal response to high costs of adjustment (as in the optimal fleet capacity expansion problem considered in the previous chapter). This suggests that the current time charter rate would depend not just on the current spot rate, but on lagged values of the time charter and spot rates. Therefore, the modeller may think that the true model of the term structure of freight rates is:

$$y_t = a_0 + \delta_1 x_t + \delta_2 x_{t-1} + a_1 y_{t-1} + v_t \quad (A.30)$$

This unrestricted model is related to the long-run model (A.29) through some restrictions on the coefficients of (A.30). If y and x are the natural logarithms of Y and X, then δ_1 measures the short run elasticity of the time charter rate with respect to the spot rate and therefore δ_1 should be positive ($\delta_1 > 0$). Similarly, δ_2 should be positive ($\delta_2 > 0$). For stability we also require that a_1 should be less than 1 ($a_1 < 1$). The long run elasticity of the time charter rate with respect to the spot rate is $(\delta_1 + \delta_2) / (1 - a_1)$. This should be equal to β_1 in (A.29). Moreover, the constant β_0 in (A.29) is related to the constant a_0 in (A.30). Thus the following restrictions should be imposed on the unrestricted model (A.30).

$$\beta_1 = \frac{\delta_1 + \delta_2}{1 - a_1} \quad \beta_0 = \frac{a_0}{1 - a_1} \quad (A.31)$$

The advantage of the ECM methodology is that it enables the estimation of (A.30) with the restrictions of (A.31) imposed. This is the rationale. The unrestricted model (A.30) model cannot be estimated to identify the population coefficients. The variables y and x are by definition $I(1)$, thus leading to spurious regression results with invalid statistical inferences. In addition, x_t and x_{t-1} are highly correlated (multicollinear). This gives rise to *multicollinearity* problems. Multicollinearity gives rise to high adjusted R^2, but to imprecise estimates of the coefficients δ_1 and δ_2. Most of the time one of the coefficients would be strongly positive (high t-statistic) while the other strongly negative. Both estimates are not unbiased, namely the means of the distributions are not equal to the population means. Therefore, the ECM specification enables a parameterisation of the unrestricted equation (A.30) with the restrictions in (A.31) imposed.

We have already mentioned that the error term, u_t, in (A.28) is $I(0)$ and can be made into a white noise process by adding lags in Δy and Δx to take account of the influence of omitted variables from the specification of (A.28). This generalises the ECM of two variables. Assume that p-lags should be added to the dependent variable Δy and q-lags to the independent variable Δx. Then, the general form of the ECM is:

$$A(L)\Delta y_t = B(L)\Delta x_t - (1 - \pi)[y_{t-p} - \beta_0 - \beta_1 x_{t-p}] + u_t \quad (A.32a)$$

$$A(L) = 1 - a_1 L - a_2 L^2 - \cdots - a_p L^p \quad B(L) = b_0 + b_1 L + b_2 L^2 + \cdots + b_q L^q \quad (A.32b)$$

$$\pi = (a_1 + a_2 + \cdots + a_p) \quad (A.32c)$$

This is the Engle–Granger *representation theorem*, which states that if y_t and x_t are cointegrated $CI(1, 1)$, then there must exist an ECM (and conversely, that an ECM generates cointegrated series).

Despite its many advantages equation (A.32) cannot be estimated directly through OLS in its current form because such estimation does not disentangle the product of coefficients $(1 - \pi)\beta_0$ and $(1 - \pi)\beta_1$. The long term coefficients of (A.29) and the speed of adjustment appear in a product form. One widely used method to estimate (A.32) is the two-stage estimation procedure suggested by Engle and Granger. This involves estimating first the long run equilibrium relationship (A.29), compute the residuals and then insert them in (A.28) or (A.32). OLS estimation of the cointegrating vector (A.29) is possible, despite the presence of $I(1)$ variables because of the *superconsistency* property of the OLS estimator. According to the superconsistency property, if y_t and x_t are both $I(1)$ and ε is $I(0)$, namely, y and x are cointegrated, then as the sample size, T, increases the OLS estimator of the vector-β converges to its true value at a much faster pace than the usual OLS estimator with $I(0)$ variables (Stock, 1987).[23] The residual, ε, in (A.29) will be autocorrelated, but this is not a problem because of the property of superconsistency.

Having estimated the cointegrating vector with OLS, then the next step in the Engle–Granger procedure is to test for the stationarity of the residuals of (A.29) and therefore test for cointegration. The DF-test or the ADF test and the Phillips–Perron methodology or the Z-test by Phillips discussed in this chapter can be employed for this purpose. If we denote the realisations of ε by e, then the cointegration test is based on:

$$\Delta e_t = a + \mu t + \phi e_{t-1} + \sum_{i=1}^{n} \phi_i e_{t-i} + u_t \qquad (A.33)$$

The null hypothesis of no-cointegration can be tested through the t-statistic of the φ-coefficient against the alternative hypothesis of cointegration. Thus, the computed t-statistic should, in absolute terms, be greater than the t-statistic from the Dickey–Fuller Tables for cointegration to exist.

The question of including a time-trend and/or a constant in (A.33) depends on whether the cointegrating vector (A.29) includes these terms or not. As the cointegration test must nest the alternative hypothesis, the deterministic components (that is, constant and time-trend) can be added to either the cointegration vector (A.29) or to (A.33), but not to both. This method is valid, if the β-vector in the cointegrating equation is known from economic theory and does not need to be estimated. For example, in the demand for new vessels the elasticity of the demand with respect to the demand for shipping services is one, while that for the freight rate per unit of the user cost of capital is σ. If σ is known, then the cointegrating equation need not be estimated; the coefficients can be imposed. Similarly, in the term structure of freight rates the coefficient relating the time charter and the spot is one, when each term is expressed as a one-year holding period yield. But if the coefficients in the cointegrating vector have to be estimated, then testing for cointegration through (A.33) is problematic for two reasons. First, the Dickey–Fuller tabulated t-statistics depend on the number of regressors in the cointegrating vector. Second, as the OLS estimator of the coefficients in the cointegrating vector would choose the minimum variance of ε, making the ε as stationary as possible, a test of non-cointegration through (A.33) would tend to over-reject the null and accept the alternative hypothesis of cointegration, even when in the true model there is no cointegration; and conversely, under-reject the null when it is false. These problems are compounded when the issue of adding deterministic components to (A.33) are also considered giving rise to a large number of permutations. Fortunately, MacKinnon (1991) has provided a criterion function and has tabulated the t-statistics for the null of no-cointegration in accordance to the inclusion of the deterministic components.

THE VAR METHODOLOGY

So far, we have dealt with the issues of stationarity and cointegration for two variables in a single equation and have generalised the results in the case of multi-variables but still in a single equation. The next step is to generalise these issues for a system of equations, each one containing an arbitrary number of variables. This involves the vector autoregressive (VAR) methodology. As a first step towards analysing VAR models we extend the AR process to other statistical processes: the moving average (MA) process, the moving average autoregressive process (ARMA) and the autoregressive integrated moving average (ARIMA) process and show their interrelationship.

In contrast to an AR process where the variable y depends on past values of itself, in a moving average process the variable y depends on past values of the disturbance term. Thus a moving average process of order 1, MA(1) is:

$$y_t = \varepsilon_t + \theta \varepsilon_{t-1} = (1 + \theta L)\varepsilon_t \quad \varepsilon_t \approx NIID(0,1) \quad (A.34)$$

If the y_t process is stationary, then (A.34) has an equivalent representation, which implies an infinite auto-regressive process.

$$(1 + \theta L)^{-1} y_t = \varepsilon_t \implies y_t = -\sum_{i=1}^{\infty}(-\theta)^i y_{t-i} + \varepsilon_t \quad (A.35)$$

The stationarity assumption is critical for this alternative representation. The term $(1 + \theta L)^{-1} = 1/(1 + \theta L)$ is the sum of a geometric series with declining weights at the rate of $-\theta L$, if and only if the absolute value of $\theta < 1$, that is, the y-process is stationary. It is this assumption that enables the representation of the solution of y in the right-hand side of (A.35). As this representation is obtained by inverting (A.34) into (A.35), the moving average process is said to be *invertible*, provided (absolute) $\theta < 1$. Hence, we have our first result: an MA(1) process may be represented as an infinite autoregressive process, provided the process is stationary.

But the reverse is also true. An AR(1) process, such as (A.1), reproduced here for convenience:

$$y_t = a + \beta \cdot y_{t-1} + \varepsilon_t \quad |\beta| < 1 \quad (A.1)$$

can be represented as an infinite MA process:

$$(1 - \beta L)y_t = \varepsilon_t \implies y_t = (1 - \beta L)^{-1}\varepsilon_t \implies y_t = \sum_{i=0}^{\infty}\beta^i \varepsilon_{t-i} \quad (A.36)$$

Again, the AR(1) process should be invertible for the equivalent infinite MA representation to be valid. Notice that the solution of y in (A.36) is only valid, if the process is stationary, (that is, if (absolute) $\beta < 1$). This establishes Wold's *decomposition theorem*, according to which any stationary stochastic process y may be represented as a univariate infinite moving average stochastic process of white noise disturbances (plus a deterministic component, a constant and/or a time trend, which for simplicity of exposition is ignored).

$$y_t = -\sum_{i=1}^{\infty}(-\theta)^i y_{t-i} + \varepsilon_t = \sum_{i=0}^{\infty} \beta^i \varepsilon_{t-i} \qquad (A.37)$$

The Wold's decomposition theorem enables any stationary stochastic process to be represented as an autoregressive moving average process of order (p, q), denoted by ARMA(p, q):

$$B(L)y_t = \Theta(L)\varepsilon_t \qquad (A.38)$$

$B(L)$ and $\Theta(L)$ are polynomials in the lag operator of order-p and order-q, respectively. Thus:

$$B(L) = 1 - \beta_1 L - \beta_2 L^2 - \cdots - \beta_p L^p \text{ and } \Theta(L) = 1 + \theta_1 L + \theta_2 L^2 + \cdots + \theta_q L^q \qquad (A.39)$$

Recall that $B(L)$ is the characteristic equation of (A.38), which for stability requires that all the roots lie outside the unit circle, which in turn means that all (absolute) β are less than 1. Therefore, any stationary ARMA process, such as equation (A.38), can be rewritten as a MA process or an AR process:

$$y_t = B^{-1}(L)\Theta(L)\varepsilon_t \quad \text{or} \quad B(L)\Theta^{-1}(L)y_t = \varepsilon_t \qquad (A.40)$$

Therefore, a stationary stochastic process has a number of equivalent representations.[24] By using a matrix formulation this enables a multivariate system to be expressed similarly in a number of different representations. We can generalise the single equation ARMA model to a system of equations. Consider, for simplicity, that there are only three variables, each with an arbitrary large number of lags:

$$\beta_{11}(L)y_{1t} = \beta_{12}(L)y_{2t} + \beta_{13}(L)y_{3t} + \theta_{11}(L)\varepsilon_{1t} + \theta_{12}(L)\varepsilon_{2t} + \theta_{13}(L)\varepsilon_{3t}$$

$$\beta_{22}(L)y_{2t} = \beta_{21}(L)y_{1t} + \beta_{23}(L)y_{3t} + \theta_{21}(L)\varepsilon_{1t} + \theta_{22}(L)\varepsilon_{2t} + \theta_{23}(L)\varepsilon_{3t}$$

$$\beta_{33}(L)y_{3t} = \beta_{32}(L)y_{2t} + \beta_{31}(L)y_{1t} + \theta_{31}(L)\varepsilon_{1t} + \theta_{22}(L)\varepsilon_{2t} + \theta_{33}(L)\varepsilon_{3t}$$

This system of equations is called vector-ARMA or simply (VAR) and can be written in matrix notation in the same form as (A.38):

$$\mathbf{B}(L)\mathbf{Y_t} = \Theta(L)\varepsilon_t \qquad (A.41)$$

where $\mathbf{Y} = [y_{1t}, y_{2t}, y_{3t}]'$, $\varepsilon = [\varepsilon_{1t}, \varepsilon_{2t}, \varepsilon_{3t}]'$ and $\mathbf{B}(\mathbf{L})$ and $\mathbf{\Theta}(\mathbf{L})$ are comfortable matrices of coefficients β_{ij} and Θ_{ij}. The coefficients β_{11}, β_{22} and β_{33} have a similar structure. For example, $\beta_{11} = (1 - \beta_{11}L - \beta_{11}^2 L^2 - ...)$.

Using Wold's decomposition theorem the above system of equations in the vector-ARMA representation (A.41) has two alternative representations, a moving average of current and past white noise terms or an infinite vector autoregression plus a linear combination of white noise error terms:

$$\mathbf{Y_t} = \mathbf{B}^{-1}(\mathbf{L})\mathbf{\Theta}(\mathbf{L})\varepsilon_t \quad \text{or} \quad \mathbf{\Theta}^{-1}(\mathbf{L})\mathbf{B}(\mathbf{L})\mathbf{Y_t} = \varepsilon_t \qquad (A.42)$$

A VAR model of an n-order can be reduced to order one. Thus, the above system of equations can be represented as a vector-AR(1) process:

$$\mathbf{Y_t} = \mathbf{A}(\mathbf{L})\,\mathbf{Y_{t-1}} + v_t \qquad (A.43)$$

In the above representation each variable y_{it}, such as y_{1t}, is of the form:

$$y_{1t} = a_1(L)y_{1t-1} + a_2(L)y_{2t-1} + a_3(L)y_{3t-1} + v_t \qquad (A.44)$$

This representation is particularly helpful in applying the stationarity methodology developed for a univariate AR(1) process to a multivariate context.

Moreover, a multivariable VAR model of m-variables can be reduced to a univariate AR process in terms of y_{1t}. This can be done by solving (or expressing) each of the last $m - 1$ endogenous variables as a function of y_{1t}.

THE JOHANSEN APPROACH TO COINTEGRATION

The Johansen approach generalises the cointegration tests of a single equation to a VAR system and overcomes some theoretical problems of the Engle–Granger two-stage estimation procedure for a single equation of many variables.

Using the methodology of the previous section it is assumed that a VAR model can be reduced to the following form:

$$\mathbf{z_t} = \mathbf{A}_1\,\mathbf{z}_{t-1} + \cdots + \mathbf{A}_k\,\mathbf{z_{t-k}} + \mathbf{u_t} \qquad \mathbf{u}_t \approx \mathrm{NIID}(0, \Sigma) \qquad (A.45)$$

Σ is the variance-covariance matrix of the residuals with the usual property of constant variance and zero serial correlation. The vector-z includes the total number of n-variables considered by a modeller, which may consist of y_i endogenous variables and x_j exogenous variables. The modeller is not certain whether the variables he regards as exogenous are truly exogenous and in order to test for their exogeneity he includes them in the vector-z. Thus the sum of y and x variables is n and they are all included in the vector-z. Thus, the vector-z includes all *potential* endogenous variables with a maximum lag-k. Each of the A_i for $i = 1,...,k-1$, is an $(n \times n)$ matrix of coefficients. A VAR in the form of (A.45) is called a reduced form because all variables on the right-hand side are predetermined (i.e. all endogenous variables on the right hand side appear with a lag). This means that the system is not simultaneous; none of the variables on the right hand side are endogenous at time-*t*. Hence, a 'reduced form' model is simply the solution of a system of simultaneous equations. The solution is an algebraic expression of past values of the endogenous and exogenous variables. The special case, in which the variables on the right-hand side are all lagged values of the endogenous variables, z, as in (A.45), is called the 'final form'.[25]

The final form VAR model (A.45) can be estimated with OLS if all variables in the z-vector are $I(0)$. But if some variables are $I(1)$, then the VAR model (A.45) can be reformulated as follows:

$$\Delta z_t = \Gamma_1 \Delta z_{t-1} + \cdots + \Gamma_{k-1} \Delta z_{t-k+1} + \Pi z_{t-k} + u_t \quad u_t \approx \text{NIID}(0, \Sigma) \quad (A.46)^{26}$$

$$\Gamma_i = (I - A_1 - A_2 - \cdots - A_{k-1}) \quad \Pi = (I - A_1 - A_2 - \cdots - A_k) \quad i = 1, \ldots, k-1$$

The critical issue for cointegration is what the *rank* of matrix Π is. The rank, *r*, of Π is defined as the maximum number of linearly independent columns in Π. If the matrix Π has full rank (i.e. there are $r = n$ linearly independent columns), then the variables in z are $I(0)$ and estimation of (A.46) can take place in levels through (A.45). If the rank is zero, then there are no cointegrating relations (i.e. a linear transformation of the columns of Π will result in an $(n \times n)$ matrix of zeros) and consequently (A.46) can be estimated in first differences, as in the Box–Jenkins approach.

The case for cointegration arises when Π has a reduced rank; that is, there are $r \leq (n-1)$ cointegrating vectors. In this case Π can be decomposed as:

$$\Pi = \alpha \beta' \qquad (A.47)$$

The matrix-β captures the coefficients of the long-run relationships among the variables in the vector-z, $\beta'z$. The matrix-α captures the speed of adjustment back to equilibrium. In this case the matrix-α is $(n \times r)$ and the matrix β is $(r \times n)$. In other words, there are r cointegrating relations. Therefore, the test of the rank of Π is equivalent to testing the number of columns in vector-α that are zero. The first *r*-columns of Π are non-zero, while the last $(n - r)$ columns are statistically equal to zero.

To sum up, if cointegration exists, then the long run relationships that bind together the variables in vector-z, are described by $\beta'z_{t-1}$ with β being the cointegrating vector of long run coefficients. The matrices Γ_i and Π describe the short-run and long-run adjustment to changes in z_t. If cointegration exists, all variables in (A.46) are $I(0)$; the residuals u_t are $I(0)$ white noise processes; the problem of spurious correlation disappears; and statistical inference on *t*- and *F*-statistics is valid.

An example would help to clarify what is involved. Assume that there only three variables in z_t (that is, $n = 3$) two of which are purely endogenous (y_1 and y_2) and one exogenous variable *x*, which the modeller includes in z to test whether it is really exogenous. Assume also that the VAR order is two (that is, only current and lagged once variables appear in z, $k = 2$). Since there are only three variables

($n = 3$), there can be at most two cointegrating vectors ($r = 2$). Then, (A.46) has the following form:

$$\begin{bmatrix} \Delta y_{1t} \\ \Delta y_{2t} \\ \Delta x_t \end{bmatrix} = \mathbf{\Gamma}_1 \begin{bmatrix} \Delta y_{1t-1} \\ \Delta y_{2t-1} \\ \Delta x_{t-1} \end{bmatrix} + \begin{bmatrix} a_{11} & a_{12} \\ a_{21} & a_{22} \\ a_{31} & a_{32} \end{bmatrix} \begin{bmatrix} \beta_{11} & \beta_{21} & \beta_{31} \\ \beta_{12} & \beta_{22} & \beta_{32} \end{bmatrix} \begin{bmatrix} y_{1t-1} \\ y_{2t-1} \\ x_{t-1} \end{bmatrix} \quad (A.48)$$

The first row of $\mathbf{\Pi}$ explains the relationship of the α- and β-coefficients for the first equation. Thus, denoting by $\mathbf{\Pi}_i$ (for $i = 1, 2, 3$) the ith row of $\mathbf{\Pi}$ we have:

$$\Pi_1 \mathbf{z}_{t-1} = (a_{11}\beta_{11} + a_{12}\beta_{12})y_{1t-1} + (a_{11}\beta_{21} + a_{12}\beta_{22})y_{2t-1}$$
$$+ (a_{11}\beta_{31} + a_{12}\beta_{32})x_{t-1} \quad (A.49a)$$

$$\Pi_2 \mathbf{z}_{t-1} = (a_{21}\beta_{11} + a_{22}\beta_{12})y_{1t-1} + (a_{21}\beta_{21} + a_{22}\beta_{22})y_{2t-1}$$
$$+ (a_{21}\beta_{31} + a_{22}\beta_{32})x_{t-1} \quad (A.49b)$$

$$\Pi_3 \mathbf{z}_{t-1} = (a_{31}\beta_{11} + a_{32}\beta_{12})y_{1t-1} + (a_{31}\beta_{21} + a_{32}\beta_{22})y_{2t-1}$$
$$+ (a_{31}\beta_{31} + a_{32}\beta_{32})x_{t-1} \quad (A.49c)$$

The two cointegrating vectors are given by $\boldsymbol{\beta}' \mathbf{z}_{t-1}$:

$$\beta_{11} y_{1t-1} + \beta_{21} y_{2t-1} + \beta_{31} x_{t-1} \quad (A.50a)$$

$$\beta_{12} y_{1t-1} + \beta_{22} y_{2t-1} + \beta_{32} x_{t-1} \quad (A.50b)$$

The two cointegrating relations (A.50a) and (A.50b) would appear in each of the three equations.

The Johansen approach identifies the exact number of cointegrating vectors r that maximise the log-likelihood function, which make the residuals $\mathbf{u}_t$ in (A.46) white noise processes. The maximisation process involves the use of the Full Information Maximum Likelihood (FIML) method. The Johansen approach can be described as follows.

Let $\mathbf{z}_t$ denote an $(n \times 1)$ vector of $I(1)$ non-stationary variables. The maintained hypothesis is that $\mathbf{z}_t$ follows a VAR(k) in levels. Recall from equation (A.45) that any k-order VAR can be written in the form (A.46). The Johansen approach assumes that although each element in $\mathbf{z}_t$ is $I(1)$, there are r linear combinations of z that are stationary, that is, they are cointegrated. This means that they can be expressed in the form of (A.47), where the matrix-a is $(n \times r)$ and the β-matrix is $(r \times n)$. The FIML maximises the log-likelihood function with respect to the coefficients

$$\Gamma_i = -(\mathbf{I} - \mathbf{A}_1 - \mathbf{A}_2 - \cdots - \mathbf{A}_{k-1}) \quad \mathbf{\Pi} = -(\mathbf{I} - \mathbf{A}_1 - \mathbf{A}_2 - \cdots - \mathbf{A}_k) \quad i = 1,\ldots,k-1$$

subject to the constraint that $\mathbf{\Pi}$ can be written in the form of (A.47).

The Johansen approach involves three steps. First, it estimates a set of auxiliary regressions. The first set estimates a VAR($k - 1$) for $\Delta \mathbf{z}_t$. In other words, each element of Δz_t is regressed on $\Delta z_{t-1}, \Delta z_{t-2}, \ldots, \Delta z_{t-k+1}$ with OLS to obtain estimates of $\mathbf{P}_1, \mathbf{P}_2, \ldots, \mathbf{P}_{k-1}$. All these OLS regressions for $i = 1, \ldots, n$ are collected in a vector form:

$$\Delta \mathbf{z}_t = \mathbf{P}_1 \Delta \mathbf{z}_{t-1} + \mathbf{P}_2 \Delta \mathbf{z}_{t-2} + \cdots + \mathbf{P}_{k-1} \Delta \mathbf{z}_{t-k+1} + \mathbf{\eta}_t \quad (A.51)$$

The second set of auxiliary regressions estimates again through OLS each element of z_{t-k} on the same regressors to obtain estimates of $R_1, R_2, \ldots, R_{k-1}$. All these regressions for $I = 1, \ldots, n$) are stacked together in a vector form:

$$\mathbf{z}_{t-k} = \mathbf{R}_1 \Delta \mathbf{z}_{t-1} + \mathbf{R}_2 \Delta \mathbf{z}_{t-2} + \cdots + \mathbf{R}_{k-1} \Delta \mathbf{z}_{t-k+1} + \mathbf{\varsigma}_t \quad (A.52)$$

The second step involves the calculation of the variance-covariance matrices of the OLS estimates of the residuals η and ζ from (A.51) and (A.52):

$$\Sigma_{\eta\eta} = (1/T) \sum_{t=1}^{T} \eta_t \eta_t' \quad (A.53a)$$

$$\Sigma_{\varsigma\varsigma} = (1/T) \sum_{t=1}^{T} \varsigma_t \varsigma_t' \quad (A.53b)$$

$$\Sigma_{\eta\varsigma} = (1/T) \sum_{t=1}^{T} \varsigma_t \eta_t' \quad (A.53c)$$

$$\Sigma_{\eta\varsigma} = \Sigma_{\varsigma\eta}' \quad (A.53d)$$

The first two equation denote the variance of the η-vector and ζ-vector, whereas the last two their covariance. These are the observed variance-covariance matrices from the sample.

From equations (A.53) find the eigenvalues of the following equation:

$$\left| \Sigma_{\eta\eta}^{-1} \Sigma_{\eta\varsigma} \Sigma_{\varsigma\varsigma}^{-1} \Sigma_{\varsigma\eta} \right| = 0 \quad (A.54)$$

and order the eigenvalues in descending order: $\lambda_1 > \lambda_2 > \cdots > \lambda_n$. The third step involves maximising the log-likelihood function by choosing $\mathbf{\Pi}$ subject to the constraint (A.47).

Johansen proposes two types of tests for r: the trace statistic and the maximum eigenvalue statistic. In the trace statistic the null hypothesis (H_0) is there are exactly r cointegrating relations among the elements of $\mathbf{z}_t$ in (A.46). The VAR is restricted by the requirement that $\mathbf{\Pi}$ can be written in the form (A.47) with α ($n \times r$) and β ($r \times n$) matrices. Under the alternative hypothesis (H_A) there are n cointegrating relations, where n is the number of the endogenous variables included in $\mathbf{z}_t$. This amounts to the claim that every linear combination of $\mathbf{z}_t$ is stationary.

A likelihood ratio test of the null that there are exactly r cointegrating relations against the alternative that there are n can be computed by comparing the maximum value, L_0, that can be achieved for the log likelihood function that satisfies the constraints with the maximum value, L_A, for the log-likelihood function without the constraints. The trace statistic is a likelihood ratio (that is, the log difference) test based on:

$$LR_{tr}(r/n) = 2(L_A - L_0) = -T \sum_{i=r+1}^{n} \log(1 - \lambda_i) \qquad (A.55)$$

Twice the log-likelihood ratio (A.55) is asymptotically distributed as χ^2. The test is repeated sequentially for $r = 0, 1, \ldots, n-1$. Thus if there are five endogenous variables ($n = 5$), then in the first step the null of no cointegration against 5 is tested. The χ^2 under the null hypothesis of no cointegration should be close to zero. Accordingly, we will reject the null if the calculated χ^2 is larger than some critical value, for example, the value at the 5% level of significance. If the null ($r = 0$) is rejected, the next step is to test that there is one cointegrating vector ($r = 1$) against the alternative that there are 4. If the null is again rejected the next step is to test the null that there are two cointegrating vectors ($r = 2$) against the alternative that there are three. The process is repeated two more times, as the maximum number of possible cointegrating vectors is ($n - 1$).

In the simple case where the vector **z** is a scalar (that is, only one variable is included in $\mathbf{z}_t$ ($n = 1$)), the null hypothesis of no cointegration amounts to $\mathbf{\Pi} = 0$ in (A.46) or that Δz follows an AR($k - 1$). Hence, Johansen's procedure provides for an alternative approach to testing for unit roots in univariate series.

Johansen has suggested an alternative test of the null hypothesis of r cointegrating relations against the alternative of $r + 1$. Twice the log-likelihood ratio for this case is:

$$\begin{aligned} LR_{max}(r/r+1) &= 2(L_A - L_0) = -T \log(1 - \lambda_{r+1}) \\ &= LR_{tr}(r/n) - LR_{tr}[(r+1)/(n-1)] \end{aligned} \qquad (A.56)$$

This is known as the maximum eigenvalue statistic. In a similar way to trace statistic the process is repeated: we first test whether the null hypothesis $r = 0$ against the alternative $r = 1$, then $r = 1$ against $r = 2$ and so on. In practice, it is common to find that the two tests provide contradictory results. Juselius (1995) suggests that in practice the unrestricted VAR in levels (i.e. A.45) is used to calculate the roots of the characteristic equation and select the ones that are near to one as possible candidates of unit roots.

What is the intuition of the Johansen approach? Since by definition all elements of $\mathbf{z_t}$ are $I(1)$, $\Delta \mathbf{z}$ is $I(0)$. Thus the approach boils down to finding the linear combination of $\mathbf{z_t}$ that produces $I(0)$ and maximises the correlation with $\Delta \mathbf{z}$, while making the residuals white noise processes. In other words, the method chooses the cointegrating relations that maximise the log-likelihood function. It does so, by separating the unit roots from the entire set of roots.

The Johansen approach provides an advantage over the Engle–Granger two-stage estimation procedure even when estimating a single equation model of more than two variables. To illustrate, consider that a modeller believes that there is a single cointegration relation among three variables and assume that this is indeed the case in the real world. Assume that this cointegration relation is represented by (A.50a). This implies that the second column of matrix-*a* in (A.48) is zero. The Engle–Granger procedure is valid only when $a_{21} = a_{31} = 0$ (i.e. only when y_{2t} and x_t are weakly exogenous). It is only then that the estimates of β are efficient (that is, they have minimum variance); otherwise there is an information advantage to be gained by estimating the entire model (A.48) and testing whether there is indeed one cointegration relation. If y_{2t} and x_t are not weakly exogenous, then the estimates of β with a single equation, β_s, have a larger variance than the β obtained from the full three equation system (A.48), β_F. Johansen (1996) shows that the variance of $\beta_s > \beta_F$ (that is, the system estimates are more efficient than the single equation estimates). The Engle–Granger approach provides an accurate test and estimation of cointegration when there are only two variables and the hypothesis to be tested is the existence of one cointegrating relation against none. Although this is the verdict from theory, in practice the Engle–Granger procedure may be valid more often than not.

In addition to the trace and maximum eigenvalue tests of cointegration, Johansen suggests a likelihood ratio test of parametric restrictions on β of the form: $\beta = H\varphi$, where H is a given $(n \times s)$ matrix of rank $s \leq r$ that expresses the restrictions and φ is an unrestricted $(s \times r)$ matrix. For example, if $r = 1$ and $n = 2$, one might wish to test whether the coefficients in β are bound together by a strict proportionality $H = [1, -1]$. These restrictions can be tested through the likelihood ratio statistic

$$-2\ln[L(r, H\phi)/L(r, \beta)] \qquad (A.57)$$

which has a limiting χ null distribution with $r(n - s)$ degrees of freedom.

5 BUSINESS CYCLES

EXECUTIVE SUMMARY

From a statistical point of view fluctuations in economic activity, called business cycles, are generated by random shocks, such as abrupt and sustained oil price changes or productivity improvements. These shocks can be either transitory or permanent. If all shocks are transitory, then business cycles can be decomposed into a trend and a cycle. Transitory shocks give rise to fluctuations around the trend. In the absence of such shocks the economy would be growing on the stable path of the trend, which is also called potential output. The path of potential output may not necessarily be linear; it could be cyclical. The shocks cause a transient, but *persistent*, deviation from potential output. Business cycles are somewhat predictable and there is room for stabilisation policies, as any deviation from potential output results in economic inefficiencies.

Although the traditional view was that business cycles are caused by random transitory shocks, in the 1980s this view was challenged. An alternative explanation was offered in which all shocks are permanent. If all shocks are permanent, business cycles cannot be decomposed into a trend and a cycle. Permanent shocks give rise to a trend, which in itself may be cyclical or purely random (non-linear). If such shocks are random, then business cycles are unpredictable and there is no room for stabilisation policies. The school of economic thought that advocates this approach argues that these fluctuations in economic activity do not result in economic inefficiency. Instead, they are equilibrium outcomes and therefore should not be corrected by stabilisation policies.

Now, the consensus accepts that both permanent and transitory shocks hit the economy. Permanent shocks are usually coming from the supply side of the economy (demographic factors, technological factors causing productivity improvement) and hence affect the rate of growth of potential output. On the other hand, most transitory shocks come from the demand side (fiscal and monetary policies, consumption, investment – driven by consumer tastes and confidence, and what Keynes called 'animal spirits' – and world trade). A few transitory shocks may also come from the supply side (commodity prices, oil price). These shocks give rise to business cycles. The persistency of transitory shocks is due to multiplier (or magnifying) effects that arise from the interaction of key macroeconomic variables.

The New Consensus Macroeconomics (NCM) model provides a framework to analyse the effects of such shocks. Demand shocks cause output and inflation to move in the same direction (they both rise or fall). Thus, a positive shock increases output and inflation. Supply shocks cause output and inflation to move in opposite direction to each other. Thus a rise in the price of oil reduces output, but fuels inflation (stagflation) in the short to medium term. Even In the absence of external shocks the economy tends to become overheated as confidence in the personal and corporate sectors grows over time. Thus, the usual reason for the end of a business cycle is that the policymakers react to the consequences of overheating by adopting tight monetary policy and/or tight fiscal policy.

The NCM model provides an explanation of how expectations in shipping are generated. Inflation depends on the expected path of future output (real GDP). Output, in turn, depends on current and future monetary policy. Through this causal relation both inflation and output depend upon monetary policy. The expected path of nominal interest rates is determined by the current state of the economy and the objectives of the central bank (where the economy is and where the central bank wants to steer it). This provides a coherent framework for understanding how expectations are formulated.

Monetary policy is conducted with the aim of achieving the targets of a central bank. These are inflation and, in most cases, growth or employment. These targets have been insufficient to stabilise the economy along the potential output path in the new millennium because of excessive liquidity. This liquidity has financed a series of bubbles in the past, such as the internet, housing, commodities and shipping and continues to finance new bubbles now, which pose a threat to future financial and economic stability.

Potential output is important for defining when the economy is overheated or operates with spare capacity. However, the rate of growth of potential output cannot be measured accurately. Trend estimates provide a good explanation of what happened in the past, but are of limited success in predicting the future business cycle especially if the determinants of potential output change. A structural approach based on the production function (the Solow approach) seems to provide a reasonable estimate of potential output as it depends on demographic factors, which are already known. However, the most important determinant of potential output is multi-factor productivity and this cannot be measured accurately. If there are large and unpredictable productivity changes between cycles then the estimate of potential output is very unreliable. Econometric estimates of potential output, although theoretically more sound, have huge problems and so far have not provided more accurate estimates.

This chapter is organised as follows. The first section offers a statistical explanation of business cycles, which helps to put in perspective the issues that arise in business cycle analysis. Section 2 provides an economic interpretation of business cycles. It traces the development of economic thought regarding business cycles so that the New Consensus Macroeconomics (NCM) model, which forms the basis for analysing business cycles, can be analysed in Section 3. Section 4 describes how monetary policy is conducted by major central banks using the NCM model. Section 5 discusses the long-term policy implications of the NCM model. It also uses the model to show how business cycles are generated by random shocks and how central banks react to them, shaping together the business cycle. Section 6 offers a reformulation of the NCM model so that it can account for the impact of liquidity and suggests that central bank should include asset price inflation to their traditional targets of inflation in goods and services and growth or employment. Section 7 discusses the role of fiscal policy and its limitations. Section 8 provides a case study in the practice of fiscal policy by considering the US President's fiscal budget for fiscal year 2013. Section 9 analyses the factors that determine potential output, while the last section concludes.

1 A STATISTICAL EXPLANATION OF BUSINESS CYCLES

Fluctuations in economic activity (usually called business or economic cycles) have invariably been costly in terms of employment, profits, the income of households and the distribution of income and wealth. In a recession, people lose their jobs, companies go bust and company profitability and incomes are eroded with the poor being hurt more than the rich. On the other hand, in a period of boom but rising inflation, company profitability may increase, self-employed and even some private sector employees may become richer while others, like pensioners and government sector employees, may become poorer. Hence, large fluctuations in activity, which accentuate these phenomena, are considered as socially undesirable.

For this reason economists have devoted a great deal of effort to understanding what causes fluctuations in economic activity. Government popularity fluctuates in the course of business cycles, as voters blame the government for not taking the appropriate measures to deter them. Governments have invested a lot of effort in designing policies that minimize fluctuations in economic activity (dampening or reducing the amplitude of business cycles).

From a purely statistical point of view business cycles are caused by shocks that hit the economy. These shocks may be considered as random and transitory, but their effect is usually long lasting (**persistent**) because of the dependence of economic activity on its own past values. The persistency is due to slowly changing habits, uncertainty and costs of adjustment that make it optimal for households and businesses to delay decisions. The dependence of economic activity on its own past values, which gives rise to the persistency effect, can be written as

$$Y_t = f(Y_{t-i}, u_t) \tag{5.1}$$

where Y is the general level of economic activity (GDP) and u are random shocks which hit the economy.[1]

The shocks arise from the demand or the supply side of the economy. If the shocks are coming from the demand side the business cycle is said to be demand-led, whereas if the shocks emanate from the supply side the business cycle is termed supply-led. Shocks can be further distinguished as to whether they arise from the domestic economy or from the rest of the world. Typical domestic demand shocks occur from a change in monetary policy, or a change in fiscal policy or from the private sector. For example, a change in the Fed funds rate signifies a shock from monetary policy. A change in tax rates (personal tax or corporate income tax or expenditure tax, like a sales tax or VAT) or a change in public consumption or public investment represents a shock from fiscal policy.

Similarly, shocks from the private sector arise from changes in the propensity to save of the personal sector, as for example the effect of the baby-boomers, which resulted in initially a drop in the saving ratio or from the corporate sector – an increase in business investment as a result of animal spirits (becoming more or less confident about the future). Shocks from the rest of the world can also be characterised as demand or supply shocks. For example, an increase in world trade represents a demand shock while a surge either in commodity prices or the price of oil represents a supply shock.

The shocks can be permanent or transitory. Prime examples of permanent shocks are improvements in productivity or demographic changes, whereas changes in the money supply or government spending are transitory shocks. The distinction between permanent and transitory shocks depends on the relevant time horizon dictated by the problem at hand. In financial markets a transitory shock may last from a few days to several months, depending on the investor's horizon. In economics, a transitory shock may last from a few months to a year or two. From the energy conservation point of view the relevant horizon is very long as it takes a long time for production methods to be changed and consumers to adapt to alternative cheaper sources of energy. Hence, a transitory shock may last for ten years or more. Figure 5.1 shows the nominal and real price of oil (West Texas Intermediate). For financial markets and economics the quadrupling in the price of oil in 1973–74 (known as OPEC-I) is permanent. The doubling in the price of oil in 1978–79 (known as OPEC-II) is also permanent, but with a few transitory shocks. The collapse of OPEC in 1985 is also a permanent shock, but with a few transitory shocks. The Gulf crisis in 1990 is certainly a transitory shock

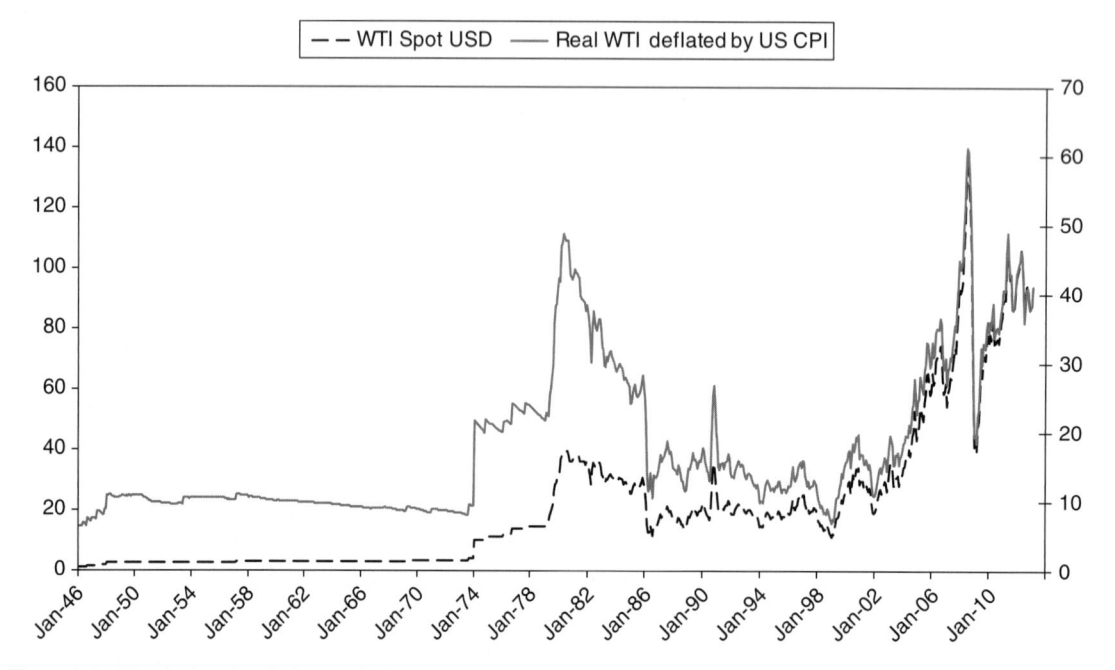

Figure 5.1 Nominal and real oil price (WTI)
Source: Federal Reserve Bank of St Louis and authors' calculations.

as it lasted for approximately two quarters. For energy policy all shocks from 1973 to 2003 can be considered as transitory as the real price of oil fluctuated around a mean value of $15 per barrel in inflation-adjusted (real) 1983 prices.

The distinction between permanent and transitory shocks is important in separating the trend from the cycle in equation (5.1). Permanent shocks give rise to a trend, whereas transitory shocks give rise to cyclical fluctuations. It is easy to understand why this is the case. The trend is a non-stationary series. Permanent shocks shape the trend, as they are also non-stationary. On the other hand, transitory shocks are stationary.[2] By construction they cannot affect the trend, hence they give rise to transient fluctuations around the trend – namely, they give rise to cyclical fluctuations. Demand shocks are usually transitory, whereas most supply shocks are permanent.

As an example of generating business cycles consider a specific form of (1) estimated using US data over the period 1978Q1–1997Q1

$$YGAP = 0.83 * 10^{-4} + 1.29 YGAP_{t-1} - 0.41 YGAP_{t-2} - 0.32 u_{t-8} - 0.39 u_{t-12} + u_t \quad (5.2)$$

where $YGAP$ is the deviation of the logarithm of the level of real GDP from a cubic trend and u are the residuals (shocks) from this equation.[3]

In this formulation the level of GDP is equal to its trend (or potential output) and random shocks cause it to deviate from this value. The trend in GDP is defined, for simplicity, by a cubic trend. Figure 5.2 shows the level of potential output and actual GDP. When the level of GDP exceeds its potential the economy

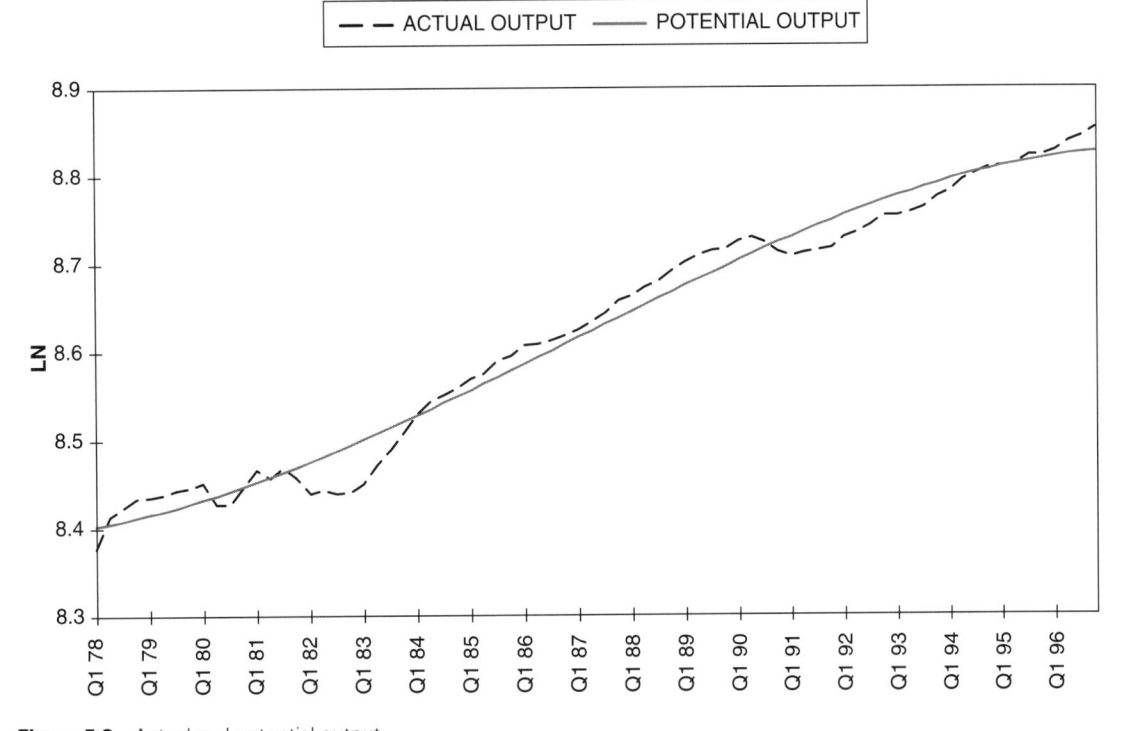

Figure 5.2 Actual and potential output

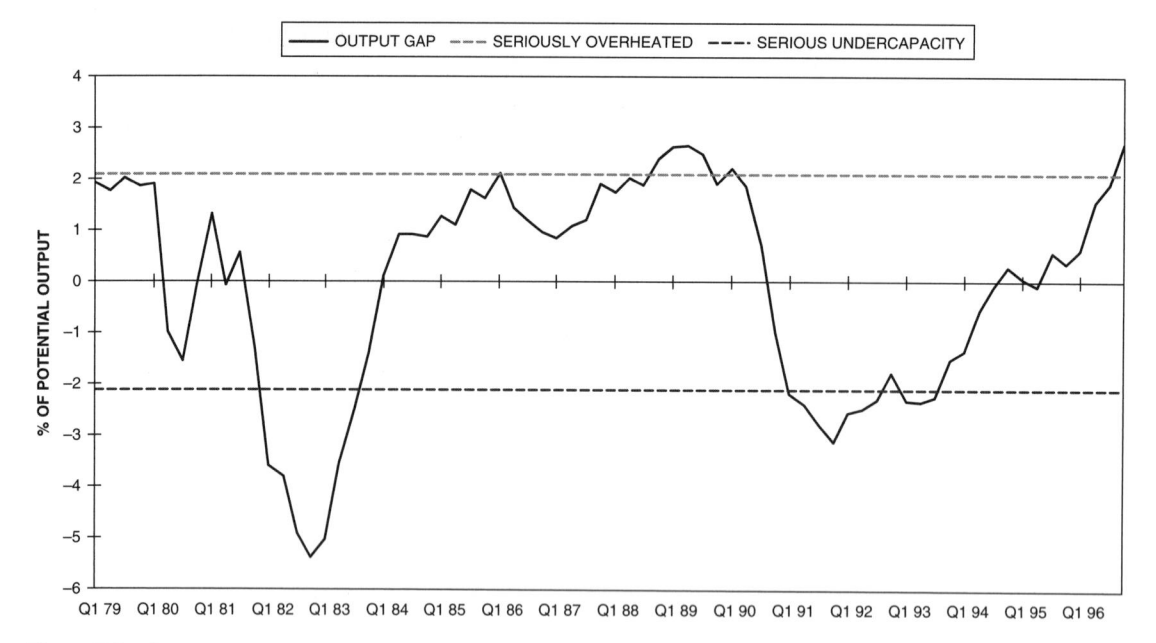

Figure 5.3 Output gap

is overheated, and when it falls short the economy operates with spare capacity. The deviation of GDP from its potential is plotted in Figure 5.3. The early 1980s recession was deep, but relatively short. In contrast, the early 1990s recession was shallow, but relatively long, as the recovery was anaemic. On this interpretation business cycles are caused by random shocks. The shocks cause a transient, but *persistent,* deviation from potential output, thereby explaining how business cycles are generated. Although many shocks may be random, there are other shocks, which might arise out of a systematic effort on the part of the policymakers to stabilise the economy along the path of potential output.

However, the view that business cycles are caused by transitory shocks and that in their absence the economy would simply grow smoothly on its trend (or potential output) is not the only one. This orthodox view has been challenged by Prescott (1986). If all shocks are permanent then there is no distinction between trend and cycle. All fluctuations are reflecting changes in trend caused by permanent shocks. As an example of this alternative decomposition consider the following equation estimated over the same period

$$y_t = 1.13 + 1.1 y_{t-1} - 0.28 y_{t-2} - 0.14 y_{t-4} - 0.085 y_{t-12} \\ + 0.64 u_{t-2} - 0.19 u_{t-12} + u_t \quad (5.2a)$$

where y = US GDP year-on-year (yoy) growth rate and u = residuals from this equation.[4]

Figure 5.4 illustrates the fit of this equation, the residuals or errors along with its post-sample forecasting performance for 1997. The equation provides a close fit; it explains 90 per cent of the total variation of the year-on-year growth rate over the period 1983–96. However, its predictive power is not impressive

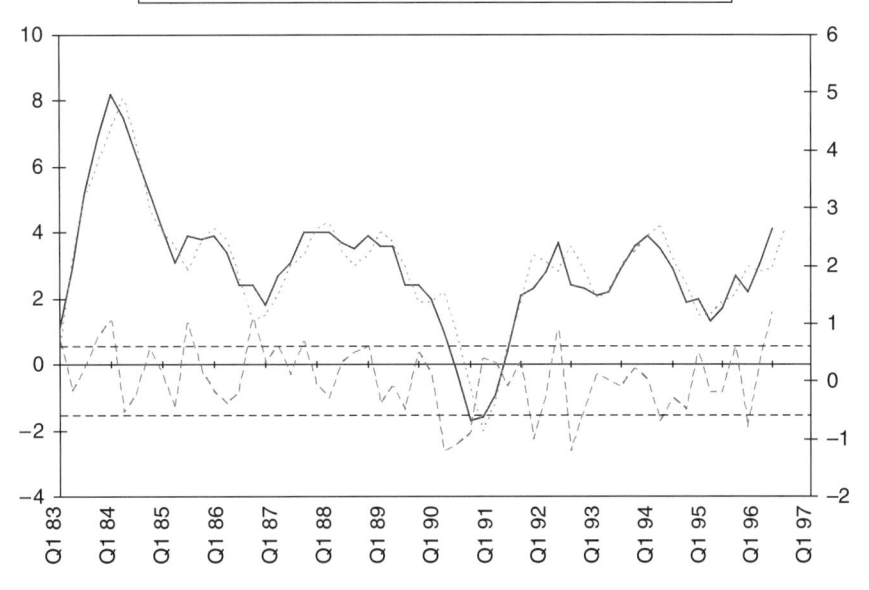

Figure 5.4 US real GDP %YOY

as the standard error of the equation is 0.6 per cent. Nonetheless, the equation exhibits the long lag structure inherent both in the dependent variable and the residuals, up to 12-quarters, which is characteristic for all major industrialised economies.[5]

The implications of this equation can be examined by computing the underlying steady state rate of growth. This is found by setting

$$y_t = y_{t-1} = y_{t-n} = \overline{y}$$
$$u_t = u_{t-1} = u_{t-n} = 0 \tag{5.3}$$

The equilibrium (steady-state) rate of growth is

$$\overline{y} = \frac{1.13}{1 - 1.1 + 0.28 + 0.14 + 0.085} = 2.8\% \tag{5.4}$$

This compares with 2.9 per cent for the average rate of growth over the historical period 1983–96 and 2.7 per cent for the last business cycle 1978:Q4–1987:Q4 measured from peak to peak on the year-on-year rate.

Figure 5.5 illustrates how business cycles can be generated using this equation. We start from a steady state with growth in the absence of shocks at 2.8 per cent. Let us now consider the effects of a positive shock, for example an increase in labour productivity by 1 per cent. This shock generates business cycles in terms of the year-on-year rate of growth, which ultimately die out. However, it leads to a permanent shock in terms of the level of GDP.

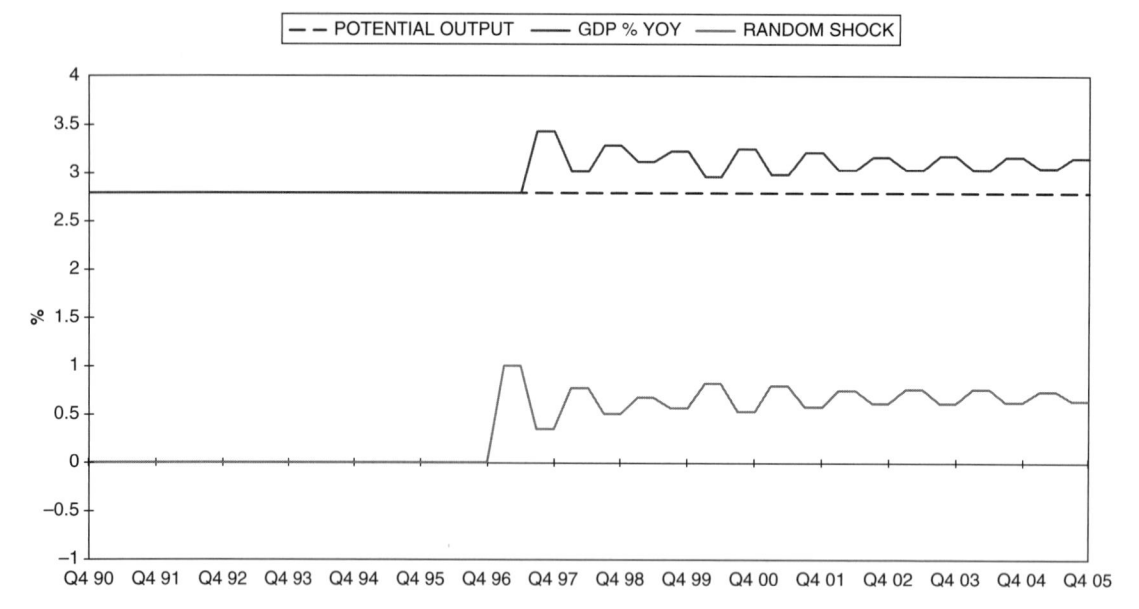

Figure 5.5 Generating business cycles using US GDP ARMA (12, 12) permanent shock = +1

Which method of decomposition is correct is therefore important for understanding both what causes business cycles and also the role of economic policy. If business cycles are caused by random transitory shocks then it makes sense for policymakers to pursue stabilisation policies. If, on the other hand, cycles are caused by permanent shocks, then it makes little sense to pursue such policies, as the chances of success are limited. Unfortunately, the statistical evidence to date has been unable to discriminate between the two rival approaches (see Campbell and Mankiw, 1987). Both fit the data reasonably well. Therefore, without theory we have reached a deadlock. The only approach forward is to specify a theoretical model for each of the two schools of economic thought and test which model can account better for the real world. This is the task of the next section.

2 AN ECONOMIC INTERPRETATION OF BUSINESS CYCLES

An economic interpretation of business cycles aims to explain not only fluctuations in output (real GDP), but also its relationship to other key macroeconomic variables. The stylised facts of this interrelationship are:

- Changes in nominal GDP are strongly correlated with changes in output (real GDP), but have little correlation with inflation (measured for example by the GDP-deflator). The implications are, first, that shocks in nominal demand have persistent effects on output and, second, that prices are sticky.
- Innovations in money supply (i.e. forecast errors) are positively correlated with innovations in output. This suggests that changes in money supply may have large effects on output.
- There is very little correlation between real wages and output or employment and this is pro-cyclical rather than counter-cyclical.

- Changes in the demand for labour produce large fluctuations in employment and small changes in the real wage rate. This means that wages are sticky.
- Unemployment is largely involuntary and unemployed are unhappier than those employed.

These stylized facts differ across industries (see Bresnahan, 1989; Carlton, 1986; Cecchetti, 1986; Stigler and Kindahl, 1970), across different historical periods and across countries (see, for example, Gordon, 1982, 1983).

In this section we utilise an economic model, which is accepted by a wide range of economists, to show how business cycles are generated. The model employed is based on the New Consensus Macroeconomics (NCM), which emerged in the 1980s and the 1990s. To put the NCM model in perspective it is instructive to trace the development of economic thought on the role of economic policy.

The field of macroeconomics has evolved rapidly since the publication of Keynes' (1936) *General Theory*. Prior to Keynes, macroeconomics had very little to contribute to the practice of economic policy. According to classical economists, the *real* sector of the economy, namely the goods and labour markets, are always in equilibrium because of an assumed infinitely elastic response of wages and prices to imbalances in perfectly competitive goods and labour markets. Perfect wage flexibility ensures that the labour market is always in full employment. Perfect price flexibility ensures that the demand in the goods market is equal to supply (output) and therefore to the full employment level of output. In other words, perfect wage and price flexibility ensures that demand in the labour and the goods markets are always equal to the corresponding supply. This principle was expressed in Say's Law that the supply of output creates its own demand. Say's Law was thought to be valid both in the short and the long run by classical economists making redundant any government intervention to boost output or employment. The economy was assumed to be self-regulated, as prices in each perfectly competitive market in the economy, namely goods, labour and money, move infinitely fast to clear any disequilibrium.

With output effectively being fixed by supply considerations in the Classical system, the goods market determines the interest rate via the equilibrium condition that saving is equal to investment. With output and the interest rate fixed in the goods market, the money (or assets) market determines the average level of prices. Accordingly, in the Classical system there is a 'dichotomy' between the real and monetary sectors of the economy. With output being fixed at the full employment level, any increase in the supply of money triggers a 'proportional' increase in the average level of prices. This is the essence of the quantity theory of money. Inflation is a monetary phenomenon caused by 'too much money chasing too few goods'. Money is 'neutral' in that it cannot affect output and the entire real system of the economy, but just prices.

Keynes made demand the centre of macroeconomics by arguing that there is some inertia or stickiness in wages and prices to imbalances in the labour and the goods markets.[6] Wages and prices do not move fast enough to clear instantly the labour and the goods markets. Therefore, in the Keynesian system disequilibrium in the real sector is a typical state of affairs. Each market may have a different response speed, with some markets or economies moving faster to equilibrium than others. As the response of wages and prices may vary from very sluggish to infinitely fast Keynes claimed that the General Theory is indeed the general theory and the Classical system (infinitely fast response of wages and prices) is obtained as a special case. Once disequilibrium is regarded as the norm, output is demand, and not supply, determined. Accordingly, Say's law (one of the pillars of the Classical system) does not hold. The level of output (supply) responds to the level of demand – whatever is demanded is supplied, provided there is spare capacity. There is no point in producing up to the point of potential capacity because the excess output would pile up in stocks of unsold goods, thus hurting company profitability. In the Keynesian system, the primary determinant of the demand for goods and services (for example, consumption) is the *current* income, while the interest rate plays a secondary role. The demand for labour would be such as to produce the output that is demanded. The demand for labour is obtained from a cost minimisation of a given output that is demand determined. If there in not sufficient demand in the economy, the demand for labour would fall short of supply and unemployment would emerge. Thus, shocks in the goods market are transmitted to the labour market. Disequilibrium in the goods market would spill over to the labour market in a way that the original shock is magnified, as there is a feedback effect from the labour market to the goods market. The lower demand for labour emanating from a negative shock in the goods market would feed back to the goods market via lower income, thus accentuating the original drop in demand – a multiplier effect.

In the Keynesian system, there is room for policy intervention. Indeed, fiscal policy can take the slack of demand in the economy to restore full employment by spurring the demand for goods and services. The government can do that directly by spending more on goods and services or indirectly by cutting taxes or increasing its subsidies to the private sector, so that private demand is restored to the level that is needed for full employment. In this theorising the Great Depression of the 1930s was the result of insufficient (or deficient) demand in the goods markets. Fiscal policy had a major role in boosting demand and lowering unemployment.

In the Keynesian system, the interest rate is primarily determined in the money (assets) market by the demand for and the supply of money. But the demand for money depends, in addition to the interest rate, on income, as households and businesses need money to finance a given volume of transactions in the economy. In this framework, output (or income) and the interest rate are determined simultaneously by the interaction of the goods and money markets. The 'dichotomy' of the real and monetary sectors of the economy, one of the pillars of the Classical system, is no longer valid. The money market cannot be in equilibrium unless the goods market is also in equilibrium. Thus, shocks in one market spill over to the other market. A decrease in demand in the goods market would lower output (or income) and employment. As a result, it would also reduce the demand for money enabling the interest rate to decline. This will spur demand in the goods market, thus mitigating the original shock of a drop in demand. An increase in the supply of money would lower the interest rate thus boosting demand in the goods market and therefore output. In the Keynesian system the 'neutrality' of money, another pillar of the Classical system, no longer applies. Monetary policy, therefore, can affect the real economy and not just prices by changing not just the nominal but also the real (inflation adjusted) interest rate. Only if output is at the full employment level, an increase in the supply of money will affect the general price level. Inflation is thus not a monetary phenomenon and the quantity theory of money is no longer valid. Inflation in the economy rises when the demand in the goods market exceeds potential (that is, there is an excess demand for goods and services). This puts the macroeconomy on par with microeconomics. As the price of any good would rise when demand exceeds supply, so does the general level of prices in the economy. An increase in the general price level would turn into inflation only if a wage-price spiral is set in motion.

In the next forty years following the publication of the *General Theory* in 1936 Keynesianism was in ascendancy epitomised in the cover page of *Time* magazine in 1965: 'We are all Keynesians now'. The article concluded: "the modern capitalist economy does not automatically work at top efficiency, but can be raised to that level by the intervention and influence of the government." 'Neo-Keynesian Economics', an attempt to formalise Keynes' General Theory in mathematical terms by economists renowned as John Hicks, Franco Modigliani and Paul Samuelson, came to dominate macroeconomic thought until the mid-1970s. The truth is that the 1950s and 1960s, the heyday of Keynesian economics, appear in retrospect as the 'Golden Age of Capitalism', to use a well-known phrase attributed to Professor Gordon Fletcher.

However, even in the heyday of Keynesian economics cracks began to appear in the underpinnings of the theory. The dependence of consumption on *current* income was questioned empirically by Kuznets and new theories were advanced by Franco Modigliani and his associates (the Life-Cycle Hypothesis) and Milton Friedman (the Permanent Income Hypothesis) that related consumption either to income and wealth or just permanent income – an average of expected income. The nearly zero interest elasticity of investment, a cornerstone of the supremacy of fiscal over monetary policy in demand management in the Keynesian system, was challenged theoretically and empirically by Jorgenson's Neoclassical theory of investment. The 'liquidity trap' that played a vital role in Keynes' theory of the demand for money (Liquidity Preference) in claiming the impotence of monetary policy in conditions of huge excess supply, such as the Great Depression of the 1930s, and therefore the overpowering influence of fiscal policy in such conditions, was interpreted as a theoretical curiosum. The interest elasticity of the demand for money was shown to be nowhere near infinity (liquidity trap), but less than one by considering money as an asset competing with other higher-yielding risky assets, such as bonds and stocks, within a portfolio approach. Even the income elasticity of the transactions demand for money was thought on theoretical grounds that it cannot be as high as one, thus limiting the spillover effect from the goods market to the money market. In the mid-1950s, Friedman also restated the Quantity Theory of Money, thus offering an alternative to the liquidity preference as the demand for money. The verdict of the so-called 'Neoclassical Synthesis', which aimed to combine Keynes' General Theory with Neoclassical economics, was that Keynes won the policy war, while Neoclassical economics won the theoretical battlefield. The General Theory became the special case of Neoclassical Economics, but Keynesian policy prescriptions remained valid because wages and prices could not be trusted to clear markets fast enough to prevent the emergence of unemployment or inflation. Demand management through fiscal and monetary policies continued to be the norm. But another policy message was also clear. If the market mechanism was enhanced, by increasing the speed at which wages and prices move to clear markets, then ultimately policy intervention would not be necessary. This was a recipe that trade unions, minimum wage laws and the welfare state were obstacles to the market mechanism. Weakening all these distortions or imperfections and move closer to perfectly competitive markets would enhance the ability of the economy for self-correction and self-regulation.

The message that Keynes had won the policy war but the Neoclassical Synthesis the theoretical war called into question by Friedman in the late 1960s in his presidential address to the American Economic Association. Friedman challenged the stability of the Phillips curve, an empirical relationship between wage inflation and unemployment, which gave the impression that policymakers can choose between inflation and unemployment. Friedman argued that a persistent attempt to lower unemployment below the 'natural' rate of unemployment would result in 'stagflation' (that is, the simultaneous occurrence of recession and inflation) or simply accelerating inflation at the unchanged natural rate of unemployment, as rising expected inflation would shift the short-run Phillips curve upwards, resulting in a vertical Phillips curve in the long run. In the 1970s, Friedman's prediction of a vertical Phillips curve became a reality, although the evidence was circumstantial because for the first time since the Great Depression of the 1930s, the shocks that caused business cycles were no longer demand driven but supply driven, emanating from the two oil shocks in the early and late 1970s and monetary policy in the US and the UK was accommodating. As a result, fiscal policy came out of fashion and monetarism with its emphasis on control of inflation via monetary policy went to ascendancy.

In the 1980s, the efficacy of demand management according to Keynesian principles was dealt another blow through the Rational Expectations Revolution (see Chapter 4, Section 2 for more details). Forward-looking expectations gained ground over backward-looking ones because they avoid systematic errors, especially in forecasting turning points in business cycles. Rational expectations generated as predictions of economic models (model consistent expectations) came to be interpreted as denying any room for policy intervention. But as it transpired from later research the policy ineffectiveness was not caused by rational expectations, but by assumptions in the underlying economic model. It was clear to most economists that the debate could only be settled by setting up models on common assumptions and testing which of these models can explain better the real world. Neoclassical economics provided the common background, as the principles that people make rational choices and use all available information to optimise utility or profits are acceptable to most economists. Thus, economists engulfed in building the microfoundations of macro-economics. Any theorising not founded on microfoundations was expelled from mainstream economics as heresy.

This led to the development of Dynamic Stochastic General Equilibrium (DSGE) models, which are immune to the Lucas critique.[7] Two strands stand out. Real Business Cycles (RBC) models build on the Neoclassical growth model under the assumption of perfect wage and price flexibility and competitive equilibrium.[8] In the RBC models business cycles are generated by productivity shocks, which are non-stationary. As such, productivity shocks shift the growth trend up or down for a while. Two mechanisms propagate the random and temporary shocks giving rise to the persistency effect (the serial correlation of output, which is otherwise called business cycle). The first mechanism works via the consumption–investment decision. A positive but temporary shock in productivity increases the effectiveness of workers and capital and raises output today. But because of consumption 'smoothing' an individual household would defer some consumption for tomorrow. The output not consumed today would be invested in capital and increase production tomorrow. The Life-Cycle Hypothesis with its emphasis on consumption smoothing would make sure that in periods of high income (output) households will save and invest for periods of low income (output), thus propagating the shock to a series of periods. The second mechanism that propagates the temporary shock in productivity is the labour-leisure decision. Higher productivity today encourages substitution of current work for future work since workers will earn more per hour today compared to tomorrow. More labour and less leisure results in higher output today, greater consumption and investment today. But there is also an income effect, which mitigates the substitution effect. As workers are earning more today, they may not want to work as much today and in future periods. In the real world the substitution effect dominates the income effect as labour moves pro-cyclically. In RBC models business cycles are the optimal choice to no business cycles at all. This does not mean that people prefer to be in recession. But recessions are preceded by an undesirable productivity shock that constrains the actions of households and firms. So, a recession is the best possible response to a negative productivity shock. It follows that laissez-faire (non-intervention) is the best macroeconomic policy in RBC models.

The second strand of DSGE models is based on New-Keynesian principles. These models build on a similar structure to RBC models with rational expectations, but instead of perfectly flexible prices they assume that prices are set by monopolistically competitive firms, which cannot be adjusted instantly and without cost. The New-Keynesian DSGE models attempt to explain the wage and price rigidity that was simply assumed in the original Keynesian models.[9] They derive wage or price stickiness as the optimal response of households or firms under imperfect market competition and rational expectations. Three approaches have been advanced to account for real wage rigidity, which are not mutually exclusive. The first falls under the general heading of 'implicit contracts', according to which firms act as if they are offering insurance contracts to employees against income uncertainty, thereby producing a relatively stable real wage rate with large fluctuations in employment. Layoffs act as the insurance premium that workers pay for the stability in the insurance schedule in the long run. The second approach, which falls under the heading of 'unions' or 'insider–outsider' models, assumes that employed workers have more bargaining power than the unemployed in wage negotiations because of turnover costs, such as the costs of hiring, firing and training. The implication for employment is that there is absence of downward wage pressure even when there is high unemployment. The third approach falls under the heading 'efficiency wages', according to which the productivity of labour depends on the real wage rate. In this case it is firms rather than unions or insiders that induce real wage resistance and therefore large fluctuations in employment.

Price stickiness is partly the result of coordination problems. "Price setters" in imperfectly competitive markets may find that, given other prices, not changing their own prices or changing them only infrequently, may cost them relatively little. But the macroeconomic implication may be slow changes in the price level, large effects of aggregate demand on output, and large output fluctuations. Wage and price stickiness, and the other market failures present in New Keynesian models, imply that the economy may fail to attain full employment. Therefore, New Keynesians argue that macroeconomic stabilization by the government (using fiscal policy) or by the central bank (using monetary policy) can lead to a more efficient macroeconomic outcome than a laissez faire policy would.

3 THE NEW CONSENSUS MACROECONOMICS OR NEO-WICKSELLIAN MODEL

The macroeconomic model derived from New-Keynesian DSGE models has been called the "New Consensus Macroeconomics" (NCM) or Neo-Wicksellian Model (Woodford, 2003, 2009). It reconciles Keynesian with Neoclassical macroeconomics by building models on common micro principles with rational expectations. Galí and Gertler (2007) suggest that the New Keynesian paradigm, which arose in the 1980s, provided sound micro-foundations along with the concurrent development of the real business cycle framework that promoted the explicit optimising behaviour on the part of household and firms. Although the NCM model is at the other polar of RBC models in terms of policy implications it endorses the policy implications of monetarism and therefore achieves the greatest possible consensus among macroeconomists. The birth of the NCM was made possible after the collapse of the Grand Neoclassical Synthesis in the 1970s (Galí and Gertler, 2007). It draws heavily on the so-called New Keynesian economics (Goodfriend and King, 1997; Clarida et al., 1999; Woodford, 2003; Meyer, 2001; Carlin and Soskice, 2005, 2006). The NCM model encapsulates certain features that previous paradigms lacked, such as the long run vertical Phillips curve and a monetary-policy rule (Woodford, 2009). Blanchard (2009) summarises this development when he suggests that 'there has been enormous progress and substantial convergence ... The state of Macro is good' (p. 210).[10]

The starting point of the NCM models is the adoption of the notion advanced by DSGE models of the RBC type, namely that an equilibrium condition is expressed as a stochastic process for all endogenous variables in the economic system. This is consistent with optimal intertemporal decisions by households and firms. However, the NCM models depart from RBC ones in that they assume monopolistic competition in the goods market and reject the hypothesis of perfectly competitive markets; and, secondly, they accept nominal rigidities in product and labour markets, which result in sticky wages and prices. There is ample evidence from micro surveys to justify nominal rigidities. Taylor (1999) finds that firms adjust prices once a year. Nakamura and Steinsson (2006) find that prices included in the US CPI are changed between 8 and 11 months. Services are adjusted less frequently. There is also evidence of wage rigidity.

Taylor (op. cit.) finds that wages are adjusted once a year. As a result of these nominal rigidities, the impact of monetary policy is non-neutral in the NCM models, which by construction are immune to the Lucas' critique (see footnote 7).

The NCM model can be expressed by the following five equations:

$$x_t = E_t(x_{t+1}) - \sigma^{-1}[r_t - E_t(p_{t+1}) + u_{xt}] \tag{5.5}$$

$$x_t = y_t - \bar{y} = 0 \tag{5.6}$$

$$p_t = \beta E_t(p_{t+1}) + \kappa x_t + u_{pt} \tag{5.7}$$

$$u_{xt} = \rho_x u_{xt-1} + \varepsilon_{xt} \tag{5.8a}$$

$$u_{pt} = \rho_p u_{pt-1} + \varepsilon_{pt} \tag{5.8b}$$

Lower-case letters are the natural logarithms of the underlying variables; x is the output gap, i.e. the deviation of actual (log) of output (real GDP) from its potential (or natural) rate, $\bar{y}$; r is the nominal interest rate; r^n is the natural rate of interest; p is the inflation rate; E is the expectations operator with information available at time t, so that $E_t(p_{t+1})$ is expected inflation in period $t+1$ as with information at t; u_i is a disturbance term, which follows an autoregressive process of order one, AR(1); ε_i are random disturbance terms with zero mean and constant variances.

The intertemporal optimisation of households and firms gives rise to an equilibrium condition for the goods market, equation (5.5), and an inflation equation, equation (5.7).[11]

Equation (5.5) makes the output gap a negative function of the expected real interest rate via the elasticity of substitution between current and future consumption, σ. The equation is derived by log-linearizing the optimality condition (that is, the first-order condition or Euler equation) around the long-run equilibrium with constant rates of inflation and consumption growth. The equation describes the optimal consumption decision of the representative household. It is sometimes called the dynamic or 'new' IS curve. The similarity with the traditional IS curve is obvious, as equilibrium in the goods market depends on output and the expected real interest rate. But there is a difference. In the traditional IS curve the equilibrium in the goods market depends on *current* output and the expected *current* real interest rate. In the dynamic IS curve equilibrium in the goods market depends on the expected *future* output and expected *future* real interest rate. The emphasis on the future as opposed to current values has huge policy implications (see below). It can intuitively be seen why the intertemporal optimisation of households would lead to the dependence on future as opposed to current values of output and the real interest rate. In the Life Cycle Hypothesis and Permanent Income Hypothesis of consumption a representative household would smooth his consumption over a planning horizon, which can be as long as his lifetime or even infinite, if he also cares about his heirs, in the face of volatile current income.

It is intuitively appealing that the optimal plan consists of maintaining a smooth (or steady) level of consumption amidst income volatility. But the ability to do so depends on subjective and objective factors. How much smoothness does a household wish? Can he achieve this smoothness and is it worthy? The subjective factors depend on the preferences of the individual household, which are expressed in his utility function, and in particular in the elasticity of substitution between current and future consumption, σ. The appearance of the interest rate in this choice reflects the income that would be received by investing in an asset (for example, a riskless discount bond), namely the income that is not consumed today and saved for future consumption – an objective factor. Therefore, it is intuitive that higher expected future output raises current output, as households spend more on current consumption in order to smooth consumption patterns over a planning horizon.

More formally, a representative household makes a twin decision on how much to consume and how much to work (supply of labour) taking as given the prices of goods and services P, and the wage rate, W. Assume that the preferences of the representative household are described by the following utility function of consumption, C, and hours worked or employment, N:

$$U(C_t, N_t) \; U_{ct} > 0, \; U_{cct} < 0, \; U_{nt} < 0, \; U_{nnt} < 0 \tag{5.9}$$

where U_{xt} is the first-order partial derivative of the utility with respect to consumption and employment in period t; and U_{xxt} is the second-order partial derivative. The sign of the derivatives imply that more consumption increases utility, but at a decreasing pace. Working harder (more hours worked) decreases utility at an increasing rate.

Denoting the price of one-period discount bond by Q and the amount of bonds by B, the budget constraint of the representative household is:

$$P_t C_t + Q_t B_t = B_{t-1} + W_t N_t \tag{5.10}$$

The budget constraint states that the nominal value of the current consumption and the amount of bonds purchased today must be equal to the investment income (that is, the amount of money from the holdings of bonds that matured in the previous period) plus the nominal value of the wages by working N hours.

The representative household maximises the expected utility over an infinite planning horizon discounting the future by δ:

$$E_0 \sum_{t=0}^{\infty} \delta^t U(C_t, N_t) \tag{5.11}$$

subject to the budget constraint (5.10).

Consider a departure from the optimal plan that consists of an increase in consumption, dC_t, and an increase in the supply of labour (hours worked), dN_t with everything else unchanged. The optimal plan implies that the utility derived remains unchanged, $dU = 0$, to the increase in consumption and hours worked. This implies:

$$U_{ct}\, dC_t + U_{nt}\, dN_t = 0 \qquad (5.12)$$

This choice must also satisfy the budget constraint:

$$P_t dC_t = W_t dN_t \qquad (5.13)$$

Combining equations (5.12) and (5.13) gives the first-order condition for the optimal plan:

$$-\frac{U_{nt}}{U_{ct}} = \frac{W_t}{P_t} \qquad (5.14)$$

The left-hand side of (5.14) measures the rate of substitution between consumption and supply of labour. Hence, the optimality condition implies that the representative household would increase consumption and the supply of labour up to the point where the rate of substitution between consumption and hours worked is equal to the real wage rate.

The second optimality condition relates to a reallocation of consumption between periods t and $t+1$, with all other variables remaining unchanged. Consider a departure from the optimal plan through a reallocation of consumption. The optimal plan implies that the utility derived remains unchanged, $dU=0$, to the reallocation of consumption. Thus:

$$U_{ct}\, dC_t + \delta E_t[U_{ct+1}\, dC_{t+1}] = 0 \qquad (5.15)$$

If the household defers some consumption today, $(-P_t dC_t)$, and invests it in a bond, then the amount earned is:

$$-P_t dC_t = Q_t B_t \qquad (5.16)$$

Since all other variables remain unchanged, the money available for next period consumption is equal to B_t. Using the above equation next period's consumption is:

$$P_{t+1} dC_{t+1} = B_t = -\frac{P_t}{Q_t} dC_t \qquad (5.17)$$

Solving this equation for dC_{t+1}/dC_t and substituting into (5.15) the second optimality condition is obtained:

$$E_t\left[\frac{U_{ct+1}}{U_{ct}} \frac{P_t}{P_{t+1}}\right] = \frac{Q_t}{\delta} \qquad (5.18)$$

For an additive and separable utility function of the form:

$$U(C_t, N_t) = \frac{C_t^{1-\sigma}}{1-\sigma} - \frac{N_t^{1+\varphi}}{1+\varphi} \qquad (5.19)$$

where σ is the elasticity of substitution between current and future consumption and φ is the elasticity of substitution of current and future labour supply, the optimality condition (5.14) can be written in logs as:

$$w_t - p_t = \sigma c_t + \varphi n_t \qquad (5.20)$$

where lower-case letters denote natural logs. This equation can be solved for n to derive the supply of labour

$$n_t = \frac{1}{\varphi}[(w_t - p_t) - \sigma c_t] \qquad (5.21)$$

Equation (5.21) states that the supply of labour depends positively on the real wage rate and negatively on consumption. The dependence of the labour supply on the real wage rate is through the elasticity of substitution between hours worked today and tomorrow, φ. The higher the elasticity of substitution is, the lower the dependence of the current labour supply on the real wage rate.

The log linearization of the second optimality condition (5.18) for the utility function (5.19) gives equation (5.5) using the market-clearing condition $Y = C + G$, where G is government expenditure, and assuming that G evolves exogenously.

Equation (5.6) defines the output gap, namely the deviation of output from the potential (or natural) rate. In equilibrium the output gap is zero. The rate of potential (or natural) output is defined as the level of output obtained when wages and prices are fully flexible. This implies that the level of output deviates from the level of potential because of nominal rigidities. As the level of potential output is assumed to be fixed in the NCM model, the level of output varies one-to-one with the level of the output gap. Accordingly, the output gap is used interchangeably with the level of output in the text.

Equation (5.7) is often called the New Keynesian Phillips Curve (NKPC). It describes the optimal behaviour of firms in setting prices. It is derived from staggered nominal price setting based on the work of Fischer (1977), Taylor (1980) and Calvo (1983). The Calvo approach has been adopted as the standard model. According to this model, each firm resets its price with probability $1 - \theta$ in any given period, independent of the time elapsed since the last adjustment. Therefore, θ is a measure of price stickiness, as in each period a fraction θ of firms reset their prices, while $(1 - \theta)$ leave them unchanged. For example, if $\theta = 0.75$ and the data are quarterly, then the average time with unchanged prices is $1/(1 - \theta) = 4$ quarters; namely prices are changed once a year. The resemblance of equation (5.7) to the standard expectations augmented Phillips curve is obvious. Both relate current inflation to the output gap and expected inflation, but with two important differences. First, in the traditional version it is expected *current* inflation, $(E_{t-1} p_t)$, whereas in the New

Keynesian version it is expected *future* inflation, $(E_t p_{t+1})$. Second, the coefficient κ of the output gap in equation (5.5) is a decreasing function of θ, namely the degree of price rigidity, see Gali (2007). This formulation implies that the longer prices remain unchanged, the less sensitive is inflation to the output gap.

The last two equations of the NCM model, (5.8a) and (5.8b), introduce persistence to any shock. In the goods market, such shocks can be considered to take the form of fiscal policy, or shocks in demand from the rest of the world. Because of the autoregressive nature of equation (5.8), the shocks once they occur, continue to impact on the goods market for a number of periods. In the inflation equation, the shocks capture 'cost-push' inflation, emanating for example from imported inflation. The effects of these shocks are again persistent, that is, they last for more than one period. Shocks to the inflation equation are also introduced from the deviation of marginal cost from its equilibrium level. In the Calvo model, equation (5.7) is expressed as

$$p_t = \beta E_t(p_{t+1}) + \kappa m c_t \qquad (5.22)$$

where mc is the marginal cost. To derive equation (5.7), the literature of the NCM model invokes a relation of proportionality between marginal cost and the output gap, on assumptions about technology, preferences, and the structure of the labour market. Thus,

$$mc_t = \mu x_t \qquad (5.23)$$

Substituting (5.23) into (5.22) equation (5.7) is obtained.

The policy implications of the NCM model emerge by solving forward the first order differential equations (5.5) and (5.7). The solution of equation (5.5) is:

$$x_t = E_t \sum_{i=0}^{\infty} [-\sigma(r_{t+i} - p_{t+1+i}) + u_{1t+i}] \qquad (5.24)$$

This equation states that the current output (or output gap) depends on the path of expected future real interest rates and unexpected shocks, which leads to the first Proposition.

Proposition 1: If monetary policy can affect the entire path of current and future real interest rates, then output depends on current and future monetary policy.

The solution of equation (5.7) leads to the second Proposition:

$$p_t = E_t \sum_{i=0}^{\infty} \beta^i [\kappa x_{t+i} + u_{2t+i}] \qquad (5.25)$$

Proposition 2: Inflation today depends on the path of current and expected future output gaps. As these depend on current and monetary policy, inflation also depends on current and future monetary policy.

---

4 MONETARY POLICY IN THE NCM MODEL

The NCM model, summarised in equations (5.5)–(5.8), is incomplete, as there is no equation determining the nominal interest rate. In the traditional IS-LM model the nominal interest rate (or the money supply) is determined by the central bank exogenously of developments in the economy. In the NCM model, the central bank decides on the nominal interest rate by maximising an objective function that summarises the statutory targets of monetary policy. In the US, the Fed has the twin objective of controlling inflation and promoting maximum employment. In Europe, the European Central Bank (ECB) has the sheer target of controlling inflation. In the NCM model, the central bank sets the nominal interest rate by maximising the objective function (26) subject to the way the nominal interest affects the economy, described by equations (5.5)–(5.8). Thus,

$$\max\left\{(-1/2)E_t\left[\sum_{t=0}^{\infty}\beta^i[a\,x_{t+i}^2 + p_{t+i}^2]\right]\right\} \qquad (5.26)$$

In this specification, the central bank has two targets: the output gap and inflation. The central bank penalises deviations of each target from each corresponding target level. These target levels are assumed to be zero for the output gap and inflation. The target level of zero for the output gap implies that the central bank aims to minimise fluctuations of the level of output around the exogenously given level of potential (or natural) output. The inflation target of zero is not restrictive, as inflation in the NCM model is expressed in deviation from the trend. The inflation trend can be chosen at any arbitrary rate, such as the 2 per cent inflation target, which is adopted by many central banks, like the Fed, the ECB or the Bank of England.

The central bank penalises the *squared* deviations of the output gap and inflation from zero, instead of the levels. This means that the central bank cares equally about positive and negative deviations. In other words, the central bank has an equal dislike of both inflation and deflation. It also dislikes equally a positive or negative output gap (overheating of the economy or spare capacity). As the objective function penalises the squared deviations of each target from its target value, it is also called a loss function. The maximisation of a negative function means that in essence the central bank minimises the loss function. Therefore, the optimisation is referred to as the maximisation of the objective function or the minimisation of the loss function. It is obvious that the unconstrained minimum of (26) is obtained at the point where the output gap and inflation are zero.[12]

The importance of each target in the objective function of the central bank is measured by the corresponding weight. In (5.26) the weight on inflation is normalised at unity. Therefore, the weight a on the output gap highlights the importance of this target relative to that on inflation.

The central bank wants to achieve the targets of inflation and the output gap in the current and all subsequent periods, which theoretically means up to infinity. But it cares more about the present and the near future and less about the distant future. This is captured by applying the discount factor β into the future that declines with time. As the future realisations of inflation and the output gap are unknown, the central bank penalises the expected values of these variables. Expectations about future inflation and the future output gap are generated by the model of equations (5.5)–(5.8).

Although an objective function like (5.26) has the advantage of being pragmatic, it has attracted some criticism as it raises the issue of how it relates to a welfare function. In particular, one issue is that (5.26) ignores the impact of the variability of the target variables, inflation and the output gap, on financial planning and hence the welfare cost of this variability (for example, De Long, 1997). However, Rotemberg and Woodford (1998) have shown that an objective function like (5.26) can be obtained as a quadratic approximation of the utility based welfare function.

The optimal monetary policy (the expected nominal interest rate path) is obtained by minimising the loss function (5.26) subject to the constraints summarised by equations (5.5)–(5.8), which describe how the nominal interest rate affects the output gap and inflation.

The optimal policy depends on whether the central bank makes a binding commitment over the course of its future monetary policy. A commitment would restrict the options of the central bank in the future, but it may have the advantage that it favourably affects private sector expectations, thereby potentially reducing the cost of disinflation, namely the output loss per unit reduction of inflation. As in the real world no central bank makes such commitment, the optimal policy discussed here is without commitment.[13]

Without commitment, the central bank optimises the objective function (5.26) by taking as given private sector expectations. Under rational expectations, the private sector forms expectations by taking into account that the central bank adjusts policy, namely that the central bank is free to re-optimise every period. In this environment, the optimal policy can be computed in two steps. First, determine the output gap and inflation (x and p) that maximise the objective function (5.26) subject to the inflation equation (5.7) and given private sector expectations. Second, determine the value of the nominal interest rate from equation (5.5), the dynamic IS curve, which is consistent with (and therefore conditional upon) the optimal values of x and p. This process yields the following optimality conditions:[14]

$$x_t = -\frac{\kappa}{a}p_t; \quad r_t = \gamma_p E_t(p_{t+1}) + \sigma u_{xt}; \quad \gamma_p = 1 + \frac{(1-\rho_p)\kappa\sigma}{\rho_p a} > 1 \quad (5.27)$$

The first condition relates the output gap to inflation. If inflation is above the central bank target (say 2 per cent), the optimal policy is to cause a negative output gap that may result in recession. This type of policy is usually called 'leaning against the wind'. The extent of the negative output gap depends on two parameters, κ and a. The first parameter, κ, is a structural coefficient of the economic system. It measures the output loss per unit reduction in inflation, which as we have seen depends on the degree of price rigidity in the economy. The more rigid prices are the deeper the required recession. But this is counterbalanced by the willingness of the central bank to tolerate a negative output gap. This is measured by the coefficient a in the objective function (5.26), which reflects the importance the central bank assigns to the output gap relative to inflation. The more the central bank cares about the output gap relative to inflation, the shallower the required recession, but the longer the process of converging back to the target inflation.

The second optimality condition gives the optimal feedback law for the interest rate. This depends on expected future inflation and the shocks to the real economy, u_x. Assume for the moment that such shocks are zero. Next period's expected inflation can be computed from equation (5.8b):

$$E_t(p_{t+1}) = \rho_p p_t \qquad (5.28)$$

The second optimality condition shows that the interest rate is related to expected inflation via the coefficient γ_p, which is greater than one. This yields the first result on optimal monetary policy.

To analyse the more general case in which shocks in the real economy and in inflation are present it is useful to compute the reduced form of the economic system (5.5)–(5.8). This means expressing x and p in terms of the exogenous variables of the system. The reduced form equations can easily be computed and they are:[15]

$$x_t = \kappa q u_{pt} \qquad p_t = a q u_{pt} \qquad q = \frac{1}{\kappa^2 + a(1 - \beta \rho_p)} \qquad (5.29)$$

Using the above equation (5.28) becomes:

$$E_t(p_{t+1}) = \rho_p a q u_{pt} \qquad (5.30)$$

This suggests that the central bank in setting the nominal interest rate through the feedback law (5.27) should respond to shocks in the output gap, u_x, and shocks to inflation, u_p. The latter has been interpreted as 'cost-push inflation', such as imported inflation, triggered by the price of oil or imported raw industrials and even imported final goods and services. The response to unexpected shocks introduces a trade-off between the output gap and inflation. Following Henry, Karakitsos and Savage (1982), this trade-off can measured by constructing the 'efficient policy frontier'. This is the locus of points that characterise how the unconditional standard deviation of the output gap and inflation under the optimal policy, s_x and s_p, vary with the central bank priority of the output gap relative to inflation, a. In the (s_x, s_p) space the efficient frontier is negatively sloped and convex to the origin. Points above the frontier are inefficient, while points below are infeasible. Along the frontier there is a trade-off. There are two polar cases, when a tends to zero and when a tends to infinity, worth mentioning:

$$\lim_{a \to 0}(s_x) = \frac{\sigma_p}{\kappa} \quad \lim_{a \to 0}(s_p) = 0 \quad \lim_{a \to \infty}(s_x) = 0 \quad \lim_{a \to \infty}(s_p) = \frac{\sigma_p}{1 - \beta \rho_p} \qquad (5.31)$$

Where σ_p is the standard deviation of the inflation shocks, described by equation (5.8b). If the central bank does not care about the output gap, $a = 0$, then the unconditional standard deviation of inflation tends to zero, but the unconditional standard deviation of the output gap tend to σ_p / κ. The larger the volatility of the inflation shocks, the deeper the recession. The opposite polar case arises when the central bank wants to avoid a negative output gap at any cost, a tends to infinity. In this case the volatility of inflation tends to the last expression in (5.31). The volatility of inflation increases with the volatility of the inflation shocks, σ_p, the degree of dependence of inflation to expectations of inflation, β, and the persistence of cost-push inflation, ρ_p.

The above analysis leads to two important results in the design of monetary policy, which are robust to a variety of models:

Result 1: Inflation above the central bank target requires a negative output gap, which may lead to recession. The central bank should lift the nominal interest rate more than one for one in response to any change in inflation, so that the real interest rate rises and causes a negative output gap. The required recession depends on price rigidity, a structural characteristic of the economy. The process of convergence back to the inflation target depends on the importance the central bank attaches to the output gap relative to inflation.

Result 2: In the presence of exogenous cost push inflation (for example, imported inflation) there is a trade off between inflation and output variability. The convergence of inflation to the target should be gradual.

5 THE INTERACTION OF SHOCKS AND MONETARY POLICY IN GENERATING BUSINESS CYCLES

To work out the dynamics of the interaction of shocks and the reaction of monetary policy in generating realistic business cycles we need to make an adjustment to the NCM model. In the basic NCM model the dynamics of inflation and the output gap (or simply output) arise from purely *exogenous* shocks, which are autocorrelated, thereby creating persistent effects, see equations (5.8a) and (5.8b). But there is strong empirical evidence of *endogenous* persistence in inflation and output. It is easy to reformulate the NCM model to embed endogenous inflation and output persistence. This is done by introducing the lagged dependent variable in the dynamic IS curve and the NKPC. The general NCM model can be summarised by the following set of equations:

$$x_t = a_1(G-T) + a_2 x_{t-1} + (1-a_2) E_t(x_{t+1}) + a_3[r_t - E_t(p_{t+1})] + u_{1t}$$
$$a_1 > 0, \quad 0 < a_2 < 1, \, a_3 < 0 \tag{5.32}$$

$$p_t = b_0 + b_1 x_t + b_2 p_{t-1} + (1-b_2) E_t(p_{t+1}) + u_{2t} \tag{5.33}$$

$$r_t = (1-c_0)[RN + E_t(p_{t+1}) + c_1(Y_{t-1} - \bar{Y}) + c_2(p_{t-1} - p^T)] + c_0 r_{t-1} + u_{3t}$$
$$c_1, c_2 > 0, \quad 0 < c_0 < 1 \tag{5.34}$$

$$x_t = Y_t - \bar{Y} = 0 \qquad (5.35)$$

$$E_t(X_{t+1}) = X_{t+1} + \varepsilon_{6t}; \quad \lim_{T \to \infty} X_T = X_{T-1} \qquad (5.36)$$

The justification of the lagged output gap, x_{t-1}, in equation (5.32) can easily be justified by invoking costs of adjustment. The justification of lagged inflation, p_{t-1}, in equation (5.33) is more problematic. Nonetheless, the model of equations (5.32)–(5.36) nests as special cases some models that have attracted a lot of attention in the literature. With $a_2 = 0$ and $b_2 = 0$, we obtain the basic NCM model. With $a_2 = 1$ and $b_2 = 1$, the model becomes backward looking (Ball, 1997 and Svenson, 1997). Therefore, the model with $a_2 < 1$ and $b_2 < 1$ provides a general framework. In this general model, fiscal policy is allowed to have a direct impact on the output gap via the fiscal budget, $G - T$, where G is government expenditure and T is taxes. The coefficient a_1 captures the traditional balanced budget multiplier (see below). The disturbance terms are now assumed to be white noise processes. We depart from the traditional model in one more important respect: the conduct of monetary policy.

The central bank operates monetary policy via a simple feedback rule that relates the level of the nominal interest rate to the output gap and the deviation of observed inflation from its target (see equation 5.34). Such simple feedback rules have been popularised in the literature by Taylor (1993, 1999) and Svenson (1999, 2003), although their appeal in conducting credible monetary policy that affects favourably inflation expectations and the derivation of the optimal parameters had already been demonstrated by Artis and Karakitsos (1983) and Karakitsos and Rustem (1984, 1985). In the long-run equilibrium, when inflation is equal to the central bank target and the output gap is zero, the nominal short-term interest rate is equal to the natural interest rate and expected inflation. The lagged interest rate in equation (5.34), often ignored in the literature, represents interest rate 'smoothing' undertaken by the monetary authorities (see, for example, Rotemberg and Woodford, 1997; Woodford, 1999; Clarida, Galí and Gertler, 1998, 2000). It actually reflects the willingness of the central bank to persevere for a period of time with the pursuit of its targets by implementing systematically and consistently a one-way directional conduct of monetary policy. In other words, the central bank avoids stop–go policies (that is, swings in interest rates), which confuse economic agents as to what the central bank is trying to achieve.

It can be shown (for example, Clarida, Gali and Gertler, 1999) that the guidelines to monetary policy summarised as Results 1 and 2 for the basic NCM model apply to the more general model of (5.32)–(5.36). In this section, we first discuss the policy implications of the basic and general NCM model. These are analysed with the help of Figure 5.6.[16] The dynamic IS curve is negatively sloped in the (r, Y) space, as an increase in the *nominal* interest rate leads to higher *real* interest rates because of wage and price rigidity. Higher real interest rates reduce aggregate demand in the economy, hence the negative slope of the dynamic IS curve. An increase in expected inflation or a higher expected income (assuming for

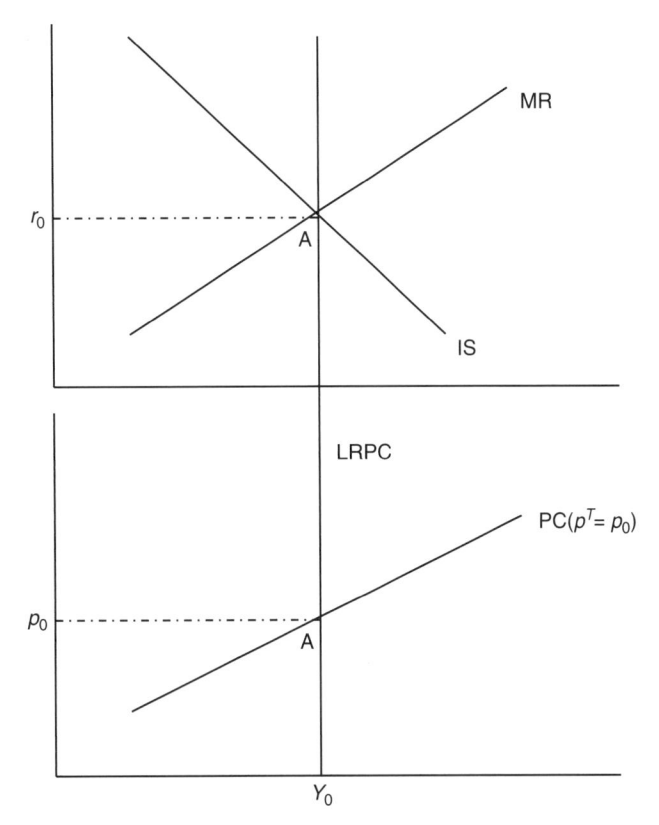

Figure 5.6 Long-run equilibrium in the NCM model

simplicity that expected income is an average of past and future income) shifts the dynamic IS curve to the right. The monetary rule is portrayed as the positively sloped MR curve with slope equal to the central bank priority on growth. As growth increases the central bank raises nominal interest rates. Higher inflation shifts the MR curve up. The short-run Phillips curve is obtained for a given level of expected inflation. It is an upward sloping curve in the (p, Y) space, as higher output (income) leads to higher inflation. The short-run Phillips curve shifts up as expected inflation increases.

In the long run, output converges to the exogenously given level of potential output Y_0 in Figure 5.6. This implies the validity of Say's Law in the long run – supply determines the level of output. But in the short run output is demand determined, as in the Keynesian system. Say's Law does not hold true in the short run. Fiscal policy can influence output (and hence employment and unemployment) in the short run, but not in the long run. Similarly, monetary policy can affect output and unemployment in the short run, but not in the long run. Hence, demand management (fiscal and monetary policy) has only transient macroeconomic effects. In the long run policy has a zero effect on real variables – long-run policy neutrality. The long-run policy neutrality property of the NCM model reflects the Classical system and achieves maximum consensus in the profession. In contrast to output, inflation is under the sole control of the central bank, in the long run. The central bank can choose as its target any rate of inflation, as this is independent of the level of potential output. This property results from a vertical long-run Phillips curve that passes through Y_0 in the lower panel of Figure 5.6. The central bank can choose the level of the short-run Phillips curve by adopting and announcing an explicit inflation target p^T. In the long run, the economy would achieve this inflation target as the central bank influences expected inflation. Thus, if the inflation target is p_0 in terms of Figure 5.6 then the central bank can choose the short-run Phillips curve along which expected inflation is equal to the target. In terms of Figure 5.6, this short-run Phillips curve is labelled $PC(p^T = p_0)$ and passes through the vertical long-run Phillips curve at point A.

With output fixed in the long run by supply factors, the dynamic IS curve helps to determine the level of the real interest rate. This long-run real interest rate is what Wicksell called the natural interest rate. It is the level of the real interest rate that would prevail in a fictitious economy where there are no nominal rigidities, that is, in an economy in which nominal adjustment is complete. This can be verified from the long-run solution (steady state) of the monetary rule:

$$r = RN + p^T \tag{5.37}$$

To show how business cycles are generated consider the interaction of a negative demand shock and the reaction of the central bank. To simulate the dynamic adjustment path of key macroeconomic variables we use a numerical analogue of equations (5.32)–(5.36). Figure 5.7 shows the dynamic adjustment path of the output gap to a negative demand shock under the optimal monetary policy response of the feedback law (5.34). Figure 5.8 shows the same response for inflation and the interest rate. Assume that prior to the shock the economy was in long-run equilibrium with zero output gap, inflation at 2 per cent and nominal interest rate at 4.5 per cent. The negative demand shock reduces output and creates a negative output gap. This reduces inflation, as a demand shock drives output and inflation in the same direction. The central bank cuts the nominal interest rate to fight back the negative demand shock. Given the lags with which monetary policy affects inflation and output, the central bank cannot immediately offset the impact of the negative demand shock; it can only alleviate it and accelerate the speed of adjustment back to the long-run equilibrium. To make the example realistic we have assumed the characteristics of the US economy and we have assumed that the demand shock occurs in the fourth quarter of 2007, when the last recession started. The negative demand shock is assumed to be 1 per cent of real GDP. The economy falls into recession with a negative output gap of more than 3.5 per cent. Inflation declines by one-half of one per cent. The central bank cuts the nominal interest rate by nearly one and a half percent to alleviate the recession and steer the economy back to the path of potential output. The adjustment back to the long run equilibrium is oscillatory and takes a few years, although the shock was only transient lasting for one year only.

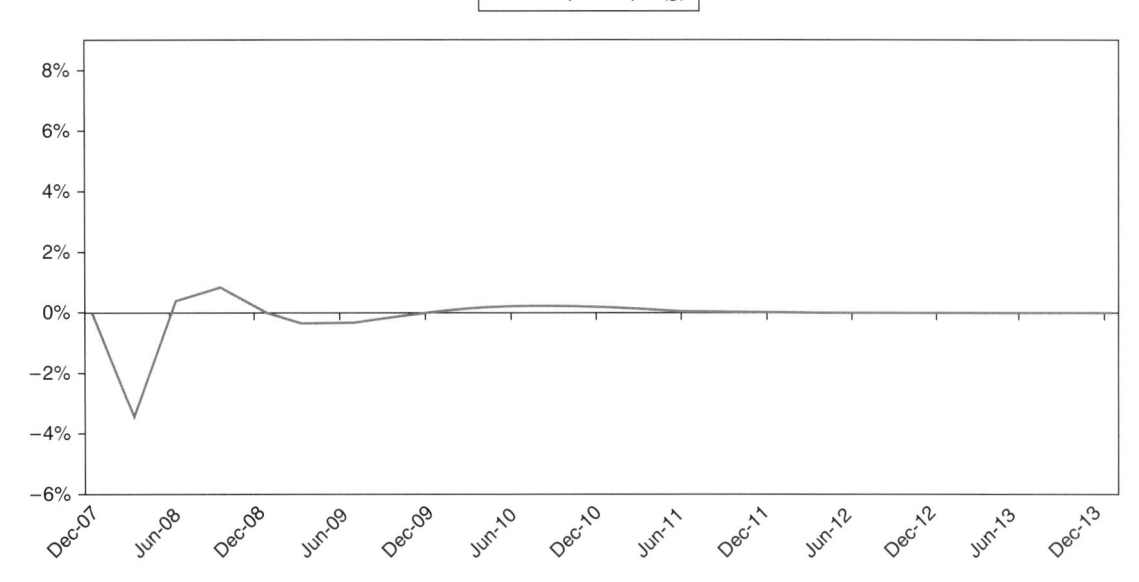

Figure 5.7 Output gap response to a negative demand shock under optimal policy

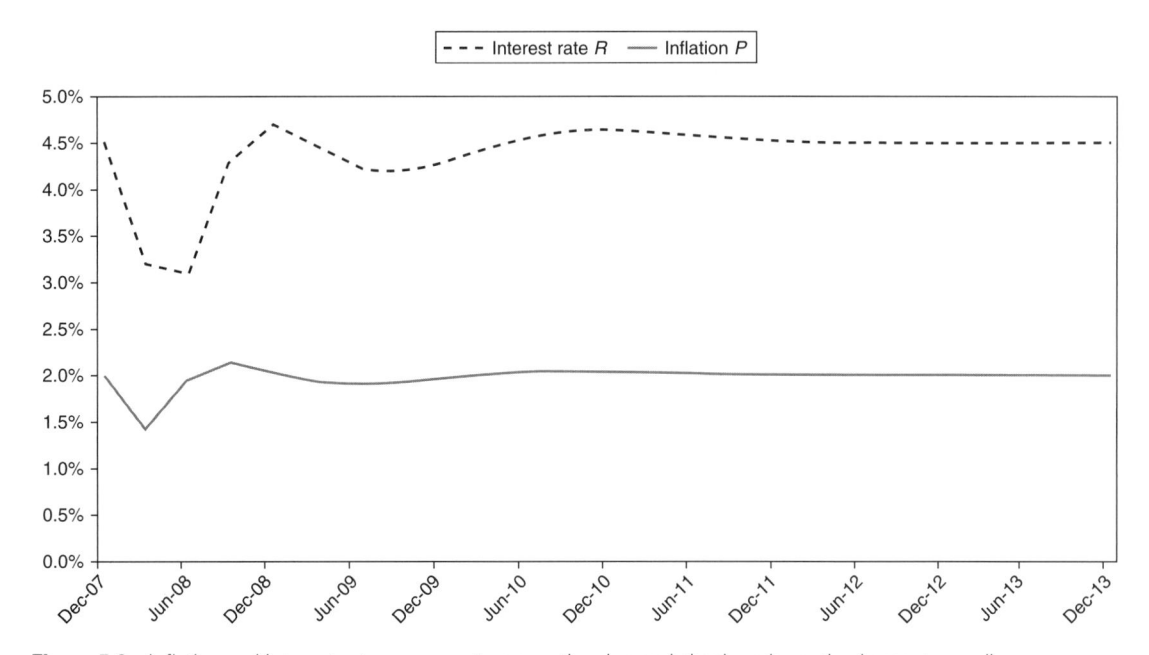

Figure 5.8 Inflation and interest rate response to a negative demand shock under optimal monetary policy

6 A REFORMULATION OF THE NCM MODEL

Despite the popularity of the NCM model and its acceptance by major central banks, monetary policy has not been successful in shielding the financial and economic system from the excessive liquidity that has financed a series of bubbles in the new millennium.[17] The biggest testament of this failure is the credit crisis of 2007–08 and the resulting Great Recession. Central banks have been unable to detect and monitor this liquidity, which has developed in a shadow banking system outside the regulatory control of the monetary authorities. In the aftermath of the crisis central banks have extended their regulatory umbrella to shadow banking. But this may not be sufficient to deter another crisis in the future.[18] There are two reasons for this pessimism. The first is that central banks have not changed their practices. Following Greenspan, they are trying to deal with the consequences of the burst of a bubble rather than preventing bubbles from ballooning in the first instance. The second reason is that central banks continue to rely on the NCM model, which provides the theoretical basis for inflation targeting. In the NCM model inflation targeting is sufficient to stabilise the economy, but as the experience of the new millennium has shown this is not true in the real world. The monetary sector of the economy, captured by the traditional LM curve, does not play a role in determining inflation or output in the NCM model; it has been replaced by the feedback law that determines the nominal interest rate. Therefore, only the interest rate matters in the NCM model, not the money supply (liquidity) in the economy. This liquidity is channelled into real and financial assets and therefore it is reflected in the wealth of the private sector. In the NCM model wealth does not play an independent role in consumption–saving decisions. The natural rate of interest, which in Wicksell's original work is the real profit rate, is just a constant in the NCM model. Potential output is exogenous in the NCM model, whereas in the real world it is endogenously determined. These deficiencies of the NCM model can be corrected (see, for example, Arestis and Karakitsos, 2004, 2007, 2010, 2013 and Karakitsos 2008 and 2009). The reformulated NCM model introduces a wealth effect in consumption; it endogenises wealth, potential output and the natural interest rate. As inflation targeting in the extended model is not sufficient to stabilise the economy, it includes as an additional target of economic policy real private sector wealth. This bypasses the problem of having an explicit target for housing prices or for equities, which would simply be unacceptable. The reduced form of the reformulated model consists of six equations that determine the output gap, inflation, the nominal interest rate, the natural interest rate, potential output and net wealth, NW.

$$x_t = a_1(G-T) + a_2 x_{t-1} + (1-a_2)E_t(x_{t+1}) \\ + a_3[r_t - E_t(p_{t+1}) - RN_t] + a_4 NW_t + u_{1t} \\ a_1 > 0, \quad 0 < a_2, a_4 < 1, \quad a_3 < 0 \tag{5.38}$$

$$p_t = b_0 + b_1 x_t + b_2 p_{t-1} + (1-b_2)E_t(p_{t+1}) + u_{2t} \tag{5.39}$$

$$r_t = (1-\gamma_0)[RN_t + E_t(p_{t+1}) + \gamma_1 Y^g_{t-1} + \gamma_2(p_{t-1} - p^T) \\ + \gamma_3(NW_t - NW^T)] + \gamma_0 r_{t-1} + u_{3t} \tag{5.40}$$

$$RN_t = q + f_1[P_t - E_t(P_{t+1})] + f_2 y_t + f_3 r_t + f_4 x_t + u_{4t} \tag{5.41}$$

$$\bar{y}_t = q + \lambda y_t + \mu RN_t + u_{5t} \tag{5.42}$$

$$NW_t = \Omega_1 R_t + \Omega_2 RN_t + \Omega_3 RC_t + u_{6t} \tag{5.43}$$

$$\Omega_1, \Omega_3 < 0, \quad \Omega_2 > 0$$

The extended model nests the NCM model as a special case, which is obtained by setting $a_4 = 0$ in (5.38), $\gamma_3 = 0$ in (5.40) and ignoring the last three equations (5.41)–(5.43), which endogenise the natural rate of interest, potential output and real net wealth. The natural interest, equation (5.41), is the real profit rate, which depends on the profit margin, the excess of price over unit labour cost, output and the nominal interest rate, see Arestis and Karakitsos (2013) for a derivation of the structural equation and the reduced form presented here. The rate of potential output, equation (5.42), depends on the capital accumulation process, which in turn, is based on the joint optimising behaviour of firms and households (see Arestis and Karakitsos, 2007). The final equation is the reduced form of a system of equations that determine housing and financial wealth in a rudimentary way. A proper specification is provided in Arestis and Karakitsos (2004). In the reduced form real net wealth depends on the nominal interest rate, the natural rate and on credit risk, the spread between corporate bonds and US Treasuries of the same maturity, RC. This variable is important in simulating the causes of the recent credit crisis. Credit risk soared during the crisis, resulting in a widening of this spread and other spreads in the financial system. The presence of a real net wealth target is sufficient to stabilise the economy even in the presence of shocks resulting from widening credit spreads (see Karakitsos, 2009).

7 FISCAL POLICY IN BUSINESS CYCLES

Fiscal policy is an arm of demand management policies,[19] which aim to reduce the amplitude of business cycles by controlling the level of demand in the economy, as opposed to supply.[20] The emphasis on demand rather than supply management is because demand changes more rapidly than supply. The underlying assumption in demand management is that if it wasn't for unexpected transitory shocks the economy would have been growing on a smooth (not necessarily linear) trend, equal to the rate of growth of potential output.[21] For demand management purposes the rate of growth of potential output does not change in each business cycle; it is treated as a constant. It can be estimated empirically after the cycle is completed by the trend measured from peak to peak in the last business cycle. Thus, the rate of growth of potential output in the current business cycle is equal to that of the last business cycle plus a change that is assumed to take place in all supply factors that shape the rate of growth of potential output (see below).

Fiscal policy aims to influence the level of demand in the economy so that it can reduce the amplitude of the business cycles. In particular, it aims to reduce demand when the economy is overheated and stimulate demand when the economy operates with spare capacity. Therefore, fiscal policy should be tightened, become contractionary or restrictive when the economy is growing faster than potential (that is, when GDP growth exceeds the rate of growth of potential output) and expansionary or easy when the economy is in recession or operates with spare capacity (GDP growth less than potential).

Fiscal policy impacts the rate of growth of GDP through the following channels: (a) the level of government expenditure used for consumption. This includes the wages and salaries of all public sector employees and procurement by all government departments, such as computers, public health care and defence expenditure; (b) the amount of public investment for infrastructure, such as ports, airports, bridges, railways, roads, schools, hospitals, sewage; (c) subsidies to the personal and corporate sectors of the economy, such as unemployment and housing benefits, provision of housing for the poor, school meals, training of the labour force and direct subsidies to particular industries; (d) tax rates for the personal sector and corporate sector; and (e) income tax allowances for the personal and corporate sectors. Each channel has a different multiplier.[22]

Public investment and public procurement have the biggest multiplier effects on the economy. In other words, these instruments of fiscal policy have the maximum impact on GDP, because every dollar that is earmarked for spending is actually spent, nothing is saved. This means that if government spending, G, goes up by $1 billion the demand in the economy would also go up by $1 billion. In other words, the initial multiplier is one. This is not true about allowances and taxes, as how much of the money that is earmarked for stimulating the economy would actually be spent depends upon the saving attitude of the personal and corporate sectors. Thus, $1 billion of personal tax cuts (either through a cut of the marginal tax rate or an increase in allowances) would stimulate demand by $(1 - S_y)$, where S_y is the marginal propensity to save or the marginal propensity to consume $C_y = 1 - S_y$. If the marginal propensity to consume is 0.8, then for every $1 billion of tax cuts, only $800 million would be spent and therefore stimulate demand in the economy; the remaining $200 million would be saved.

The total boost to demand would be even greater than the $1 billion increase in public investment, as in every subsequent period there would be a diminishing stimulus (that is, there is a multiplier effect). The reason of this multiplier effect stems from the fact that out of every $1 billion increase in demand, output would also increase by $1 billion initially and this would be distributed as income to households and profits to companies. Assuming that all profits are distributed to shareholders (households) and a marginal propensity to consume equal to 0.8, consumption in the second period would be boosted by $1000 × 0.8 = $800 million. This would increase demand and income by $800 million. In the third period consumption would be boosted by $800 × 0.8 = $640 million. In the fourth period the stimulus to demand and income would be $640 × 0.8 = $512 million and so on. The total stimulus is equal to the sum of the stimuli:[23]

Total stimulus = $1000 \times (1 + 0.8 + 0.8^2 + 0.8^3 + \cdots) = 1000 \times (1/(1 - 0.8)) = 5$
In general, the total stimulus to GDP is: $(1 + C_Y + C_Y^2 + C_Y^3 + \cdots) \cdot$
$$dG = \frac{1}{1 - C_Y} \cdot dG = 5 \cdot dG$$

Accordingly, each successive stimulus is smaller as the marginal propensity to consume is less than 1. As the marginal propensity to consume is equal to 1 less the marginal propensity to save (that is, $C_Y = 1 - S_Y$), in each period there is a leakage in demand because of saving. In every period consumers keep $(1 - C_Y)$ of the increase in income in the form of saving and consume the remainder.

The total stimulus to GDP can alternatively be derived by computing the partial derivative of GDP with respect to government expenditure, in the equilibrium GDP relationship.[24] The multiplier is equal to

$$\frac{dY}{dG} = \kappa = \frac{1}{1-C_Y} = \frac{1}{0.2} = 5 \qquad (5.44)$$

This means that out of every $1 billion increase in government expenditure, GDP would increase by $5 billion.

Saving is not the only leakage in the multiplier process. Households would also pay taxes out of the extra income they have generated. The marginal tax rate (t_y) shows the extra taxes that would be paid by households out of every extra dollar of income they generate. In this case the multiplier is reduced in value by two leakages: saving and taxes. Assuming a marginal tax rate at 0.25, namely 25 cents is paid in the form of taxes to government out of every extra dollar of income, the total stimulus is:

Total stimulus: $(1 + [0.8 \cdot (1-0.25)] + [0.8 \cdot (1-0.25)]^2 + [0.8 \cdot (1-0.25)]^3 + \cdots) \cdot 1000$

$$\omega = C_Y \cdot (1-t_Y) = 0.8 \cdot (1-25) = 0.6$$

In general

$$dY = \{1 + [C_Y \cdot (1-t_Y)] + [C_Y \cdot (1-t_Y)]^2 + [C_Y \cdot (1-t_Y)]^3 + \cdots\} \cdot dG$$
$$= \frac{1}{1 - C_Y \cdot (1-t_Y)} \cdot dG$$

In every round demand and income increases by $\omega = 0.6$. This is less than $C_Y = 0.8$. Hence, the leakage when there is taxation is bigger (the stimulus is smaller: 0.6 compared to 0.8).

$$\frac{\partial Y}{\partial G} = \kappa = \frac{1}{1 - C_Y \cdot (1-t_Y)} = \frac{1}{1 - 0.8 \cdot (1-0.25)} = \frac{1}{0.4} = 2.5 \qquad (5.45)$$

Hence, the impact of taxation is to lower the multiplier from 5 to 2.5. Still, 2.5 is a large multiplier signifying that fiscal policy is very powerful (that is, it has a big impact on the economy).

Although the government would spend $1 billion the budget would not be in deficit by $1 billion after the full impact of the multiplier is felt. Thus, in equilibrium the budget deficit out of $1 billion extra spending would increase by only $375 million:

$$\Delta(T-G) = t_Y \cdot \Delta Y - \Delta G = 0.25 \cdot 2.5 - 1 = -0.375 \qquad (5.46)$$

The $625 million would be self-financed. This is the difference between microeconomics and macro-economics.

If the budget deficit is large and public debt is uncomfortably high, fiscal policy can still be used to get the economy out of a recession. This can be done by spending and taxing at the same time, so that the budget does not increase even in the short run. Thus, assume that the government increases spending by $1 billion, but at the same time it increases taxes by an amount to be determined below so that the budget would be balanced. The impact on the economy would not be zero, but positive. This is so because the value of the so-called 'balanced budget multiplier' is still positive and not zero.

An income tax cut of dt_0 would have the following impact on GDP:[25]

$$dY = \frac{-C_Y}{1 - C_Y \cdot (1 - t_Y)} dt_0 \quad (5.47)$$

The impact on GDP from a change in government spending, G, and taxes, T is given by the total differential:

$$dY = \frac{\partial Y}{\partial G} \cdot dG + \frac{\partial Y}{\partial t_0} \cdot dT = \frac{1}{1 - C_Y (1 - t_Y)} \cdot dG + \frac{-C_Y}{1 - C_Y \cdot (1 - t_Y)} dt_0 \quad (5.48)$$

For simplicity, it is assumed that the tax increases take place by reducing tax allowances (tax thresholds), which are denoted by t_0. To derive the amount by which tax allowances should be reduced we take the total differential of the tax equation and solve it for dt_0

$$T = t_0 + t_Y \cdot Y \Rightarrow dT = dt_0 + t_Y \cdot dY \Rightarrow dt_0 = dT - t_Y \cdot dY \quad (5.49)$$

By substituting (5.49) into (5.48) and setting $dG = dT$, because of the requirement that the budget remains balanced, we get the total impact on GDP:

$$dY = \frac{1}{1 - C_Y \cdot (1 - t_Y)} \cdot dG + \frac{-C_Y}{1 - C_Y \cdot (1 - t_Y)} \cdot (dG - t_Y \cdot dY) = dG \quad (5.50)$$

Thus, the balanced budget multiplier is one, which implies that fiscal policy can still stimulate the economy without causing a budget deficit.

But there are further leakages to the multiplier process when money is introduced into the economy. This is because the demand for money increases with higher income, as households need to finance the extra consumer purchases in every round of the multiplier process (purchases must be backed by money). With unchanged supply of money the interest rate would move up to equilibrate the demand for and supply of money. The extra demand for money requires portfolio rebalancing. If there are two assets, money that is kept in the current account and savings accounts (or time deposits), then portfolio rebalancing requires moving money from the savings account to the current account. In response to this drain of liquidity banks would raise the interest rate on savings accounts. If the two assets are money and government bonds, an increase in the demand for money would have to be met by selling government bonds. Selling bonds would lower their price, as there is an excess supply of bonds, and the yield on bonds (which moves inversely with the price) will rise.

But higher interest rates would curtail investment expenditure by firms for plant and equipment, residential investment and the demand for consumer durables, such as cars and furniture. The impact on investment capturing all interest rate sensitive components of demand is calculated as follows. Let $M_Y = \partial M/\partial Y$ denote the income sensitivity of the demand for money (how much the demand for money would increase out of one extra dollar of income). Let $M_r = \partial M/\partial r$ denote the interest rate sensitivity of the demand for money (how much the demand for money would decrease by one percent rise in the interest rate). The ratio M_Y/M_r measures the increase in the interest rate required for a one dollar increase in income. Let I_r denote the interest rate sensitivity of investment, consumer durables and residential investment (how much investment would fall for one percent increase in the interest rate. The product $M_Y \cdot I_r/M_r$ measures the impact on investment of a dollar increase in income. For example, assume that $M_Y = 0.9$, $M_r = -0.5$ and $I_r = -0.2$. Then the impact of the money market on the multiplier can be computed as follows. First, we need to compute by how much the demand for money should increase to finance consumer purchases when income is increased by \$1000. This is: $M_Y \times dY = 0.9 \times 1000 = \$900$ million. Households need to sell \$900 million of bonds at \$100 face value to raise \$900 million in cash. By selling \$900 million of bonds the price of bond would fall and the interest rate would rise. The increase in the interest rate is equal to: $-M_Y/M_r = -0.9/(-0.5) = 1.8$ per cent. But as the interest rate increases by 1.8 per cent, investment in housing, consumer durables and plant and machinery would be cut back. The impact on investment and hence on GDP is: $I_r \times M_Y \times (dY)/M_r = -\$360$ million.

The leakages in every round are given by:

$$\omega = C_Y \cdot (1 - t_Y) - M_Y \frac{I_r}{M_r} = 0.8 \cdot (1 - 0.25) - 0.9 \frac{(-0.2)}{(-0.5)} = 0.24$$

The total GDP stimulus is: $[1 + \omega + \omega^2 + \omega^3 + \cdots] \cdot dG = \frac{1}{1-\omega}$.

$$dG = \frac{1}{1 - 0.24} \cdot dG = 1.32 \cdot dG$$

The fiscal multiplier is accordingly reduced by this factor

$$\frac{\partial Y}{\partial G} = \kappa = \frac{1}{1 - C_Y \cdot (1 - t_Y) + M_Y \frac{I_r}{M_r}} \tag{5.51}$$

With three leakages (savings, taxes and money) the multiplier is reduced to 1.32. Thus, for every billion increase in government spending GDP increases by 1.32 billion.

The above formula (5.51) shows the factors that affect the size of the multiplier: the marginal propensity to consume, the marginal tax rate, the income sensitivity of the demand for money, the interest sensitivity of investment and the interest sensitivity of the demand for money. Each one affects the multiplier in the following way. First, the higher the marginal propensity to consume, the higher the fiscal multiplier, because the marginal propensity to save is lower and hence the smaller the leakage. Second, the higher the marginal tax rate is, the lower the multiplier, because less money is available to households to purchase consumer goods (that is, the leakage in the multiplier process is bigger). Third, the higher the income sensitivity of the demand for money, the lower the multiplier. This is because the higher the income sensitivity of the demand for money, the higher the demand for money. People need more money to finance a given volume of transactions and hence higher interest rates to equilibrate the money market. Fourth, the higher the interest sensitivity of investment, the lower the multiplier, because more investment would be crowded out. The higher the interest sensitivity of the demand for money, the lower the multiplier, because interest rates would need to rise less to equilibrate the demand and supply of money. In the limit cases in which the interest sensitivity of investment becomes zero or the interest sensitivity of the demand for money becomes infinity (what is called liquidity trap), the multiplier is given by the big impact of (5.45).

There is one final leakage in the multiplier process, which stems from imports. For each dollar increase in income a small portion stimulates foreign demand and not domestic demand. By denoting the marginal propensity to import by $Q_Y = 0.1$ (the increase in imports for each dollar increase in income) the stimulus ω in each round of the multiplier process is equal to:

$$\omega = C_Y \cdot (1 - t_Y) - M_Y \frac{I_r}{M_r} - Q_Y = 0.8 \cdot (1 - 0.25) - 0.9 \cdot \frac{(-0.2)}{(-0.5)} - 0.1 = 0.14$$

The total GDP stimulus is: $\frac{1}{1 - 0.14} \cdot dG = 1.16 \cdot dG$.

The multiplier is still positive and above unity. Thus, for every billion increase in government spending GDP increases by 1.16 billion. Formula (5.51) is accordingly amended:

$$\frac{\partial Y}{\partial G} = \kappa = \frac{1}{1 - C_y \cdot (1 - t_Y) + M_Y \frac{I_r}{M_r} + Q_Y} = 1.16 \qquad (5.52)$$

The value of the multiplier has huge implications for the deficit that easy fiscal policy creates. The lower the multiplier, the bigger the deficit created by easy fiscal policy. A deficit can be financed in the short run by tapping capital markets (issuance of government bonds). But in the long run, a deficit would have to be financed by increasing taxes or cutting spending. In our previous example, the government would have to cut the budget by $375 million in the long run by increasing taxes or cutting spending. This suggests an optimal pattern in the use of fiscal policy. In the downswing of the business cycle (recession or simply spare capacity) fiscal policy should be easy; but once the economy starts growing at a higher rate than potential output, fiscal policy should become tight to eliminate the $375 million increase in the deficit. It is obvious that the degree of tightness of fiscal policy in the upswing of the cycle need not match the degree of easiness in the downswing because part of the stimulus is self-financed. Thus, in our previous example fiscal policy was easy in the downswing by $1 billion, but in the upswing of the cycle it should be tightened by only $375 million. But the lower the multiplier, the larger the fiscal tightness in the upswing. For as long as the multiplier is greater than unity, the fiscal tightness in the upswing of the cycle is smaller than the stimulus in the downswing of the cycle. A small part of the fiscal stimulus is self-financed making the whole exercise worthwhile.

7.1 CONCLUSIONS ON FISCAL POLICY

There are five implications from this optimal pattern of fiscal policy. First, policymakers should never attempt to balance the budget (through austerity measures) when the economy is in recession or with spare capacity because they would negate their own actions of stimulating the economy. Unfortunately, this is exactly how Europe has responded to the credit crisis. Second, the fact that easy fiscal policy would increase the budget deficit and public debt is not prima facie evidence against using fiscal policy when the economy is in recession. It simply becomes imperative to tighten fiscal policy when the economy is booming, as otherwise public debt would be increasing forever in successive business cycles. At some point in time, when public debt is high capital markets would refuse to provide any additional borrowing to the government making it insolvent, such as Greece and the other periphery countries in Europe. Third, even if the government does not become insolvent for a long time, the economy is operating with smaller efficiency the higher the share of government in GDP. Therefore, it is imperative that fiscal deficits created in the downswing are corrected in the upswing of the cycle. Otherwise, the debt would continue to soar, thereby making insolvency unavoidable in the long run. Fourth, an inefficient economy would result even if the government always uses a balanced budget stimulus, because on each stimulus the share of the public sector in GDP increases. Fifth, there is a trade off between the short to medium term impact and the long-term effects of fiscal policy. Easy fiscal policy boosts output in the short to medium term, but has a negative impact in the long run by cutting the rate of growth of potential output (see below for more details).

The above analysis points to a serious limitation in the use of fiscal policy in demand management. It is politically easier to implement easy than tight fiscal policy. Democrats in the US or the Labour Party in the UK or a socialist government in Europe would oppose spending cuts, as they would hurt the poor. Republicans in the US or the Conservative Party in the UK would oppose tax increases, as they would hurt the rich. Therefore, governments find it easier to leave the task of tightening fiscal policy to the next government. This creates an upward trend in public debt, which at some point like the severe recession of 2008–09 creates an insolvency problem for the government. Because of this drawback of fiscal policy, many countries have opted for the use of monetary policy in demand management. The task of tightening monetary policy rests not with the government, but with an independent central bank. This makes it politically easier to reverse policies when the economy is booming.

8 FISCAL POLICY IN PRACTICE: THE US PRESIDENT'S BUDGET FOR FY 2013

The US President's fiscal stimulus for fiscal year (FY) 2013 is $1.5 trillion (or 1 per cent of GDP) and consists of $1 trillion in tax cuts and $0.5 trillion of higher government outlays (see CBO, 2012). The changes in spending would consist of an increase in transfer payments; a permanent extension of the tax credit for research and experimentation; an allowance for companies to immediately deduct 100 per cent of new investments in equipment and certain shorter-lived structures; and reductions in purchases of goods and services, such as military spending. The changes in taxes consist of lower tax rates for incomes below $200,000 for individuals; and for incomes below $250,000 for married couples; and indexation of the Alternative Minimum Tax (AMT) for inflation. The total stimulus package is $1.5 trillion (or 1 per cent of GDP).

8.1 THE MACROECONOMIC EFFECTS OF THE PRESIDENT'S BUDGET

For the economy and consequently for shipping what matters are the macroeconomic effects of the budget. Fiscal policy can affect the economy's actual output as well as its potential output, which is defined as the level that is consistent with the normal rate of utilisation of labour and capital and hence with steady inflation. In the short to medium term (up to a five-year period) fiscal policy affects the actual output of the economy mainly through the aggregate demand for goods and services. In the long run (from year six to ten) the impact of fiscal policy is increasingly felt through potential output. The short- to medium-term impact is positive, whereas the long-term impact is negative. Therefore, fiscal policy involves a trade-off between boosting output in the short to medium term and reducing output in the long run. When the net impact of these two opposing forces shifts from positive to negative depends on various factors, including the impact of increased aggregate demand on output and the effect of deficits on consumption and investment and the long-run supply of labour and capital.

8.1.1 SHORT-TO-MEDIUM-TERM ECONOMIC EFFECTS

The impact of fiscal policy on the economy in the first few years stems from its influence on aggregate demand for goods and services, but at a diminishing pace. In time, the impact of fiscal policy on the economy works at an increasing pace via its impact on potential output. According to the CBO (2012), the fiscal stimulus of $1.5 trillion (or 1 per cent of GDP) would increase real GDP by between 0.6 per cent and 3.2 per cent in 2013. For the 2013–17 period, the CBO estimates that GDP might increase by 1.4 per cent or, even, fall by a sheer –0.2 per cent. Although the CBO estimates are wide varying, we can narrow them down with the help of the K-model[26] and the particular circumstances of the economy.

As the recent severe recession and ongoing slow recovery have shown, the nation's output can deviate for a long time from its potential level in response to a drop in demand for goods and services by consumers, businesses, government and foreigners, which is what has happened as a result of the credit crisis of 2008. Although GDP has now (2013) surpassed its pre-recession level, output remains well below its potential, and unemployment remains high. When there is spare capacity in the economy, as is now the case, tax cuts and increases in government spending can boost demand and hence hasten a return to the potential level of output, thus justifying the need for a fiscal stimulus. Increases in government's outlays in the first year would directly boost the demand for goods and services by that amount (first-round effects). Hence, the $0.5 trillion stimulus from increased government outlays would boost GDP by at least $0.5 trillion in the first year (2013).

Tax cuts and increases in transfers would boost the disposable income of households, increasing consumer demand. Increases in disposable income are likely to boost purchases more for lower-income households than for higher-income households, as the former save a smaller proportion of their income than the latter. As the President's budget aims primarily at incomes of less than $250,000 for married couples, the impact of the tax cuts on the economy is also likely to set up a virtuous circle between demand and output, leading to a multiple of the fiscal stimulus. However, if someone receives a tax cut of a dollar and spends 90 cents (saving the other 10 cents, which is usually called the marginal propensity to save) then provided there is spare capacity in the economy, as there is now, production would increase by 90 cents in the first year. The fact that the marginal propensity to save is positive, but less than unity, implies that the $1 trillion of tax cuts would increase GDP by less than $1 trillion in the first year. This would increase in subsequent years, as the stimulus is likely to set up a multiplier process between demand and output, because there is spare capacity. In the current economic environment, therefore, the fiscal stimulus of 1 per cent of GDP is more likely to increase output by more than 1 per cent in the first few years. Hence, we can eliminate the low end of the CBO range and concentrate on estimates higher than 1 per cent, but less than 2 per cent as consumers are involved in deleveraging (paying back debts).

These first-round effects may be enhanced or mitigated in subsequent periods. If there is spare capacity in the economy, the higher level of demand for goods and services is likely to be met by companies increasing output. This would induce companies to hire and invest to expand capacity. Such a response would multiply the first-round effects of fiscal policy. On the other hand, these first-round effects are muted, if greater government borrowing caused by tax cuts or spending increases leads to higher interest rates that discourage spending by households and businesses. An expansionary fiscal policy by increasing output would lead to higher short- and long-term interest rates, as the Fed removes the accommodation bias. These higher interest rates would discourage (crowd-out) spending by households and businesses, thereby mitigating the positive impact of fiscal policy on output in subsequent years. Nonetheless, in the current environment in which there is huge excess capacity, the extent to which interest rates would rise is minimal. The Fed does not foresee interest rates rising before the end of 2014. Therefore, the crowding-out effect on investment is likely to be minimal in the first two years of the stimulus (2013–14). According to the K-model the impact on the economy in 2013 would increase the nation's output by $0.6–0.7 trillion from the $0.5 trillion purchases of goods and services and by $0.9–1.1 trillion from the tax cuts. The K-model first year multipliers are 1.1–1.2 for government outlays and 0.8–0.9 for tax cuts. Accordingly, the President's stimulus package of $1.5 trillion would increase GNP in 2013 and 2014 by 1.0–1.3 per cent, according to the K-model. The impact of fiscal policy will fade in later years (in 2015 and beyond) as interest rates begin to rise. In the current economic environment the impact of fiscal policy peaks at the end of 2014 and tapers off after from 2015 onwards; it becomes zero in 2016 or 2017.

8.1.2 LONG-TERM ECONOMIC EFFECTS

The negative impact of fiscal policy in the long run stems from its influence on potential output. The latter depends on the size and quality of the labour force, on the stock of productive capital (such as factories, vehicles and computers) and on the efficiency with which labour and capital are used to produce goods and services – otherwise called multifactor productivity, which is the joint productivity of labour and capital. Fiscal policy affects potential output by influencing the amount of national saving and hence the supply of capital in the long run. A federal deficit represents negative public saving, but it can also influence private saving. Larger deficits would imply less public saving, but that would induce a small increase in private saving, as a result of higher interest rates and increases in disposable income, which can boost both spending and saving. This positive increase in private saving is not sufficient to offset the reduction in public savings and therefore national saving declines. The CBO estimates vary from small to medium and large effects. In the scenario of a small impact of deficits, every additional $1 of deficit is assumed to lead people to increase their private saving by 68 cents and thus to reduce national saving by 32 cents. In the scenarios of medium and large impact of deficits, private saving increases by 45 cents and 29 cents. This leads to a decline of 55 cents and 71 cents, respectively in national saving. An overall decline in national savings reduces the capital stock owned by US residents over time through a decrease in domestic investment, an increase in net borrowing from abroad, or both. In the CBO estimates, every $1 of deficit is assumed to lead to a 10 cent, 30 cent and 50 cent decline in investment. Moreover, changes in tax rates affect the long-run supply of labour, thus again influencing the level of potential output.

But the President's measures, by altering tax rates on capital and labour, would have a direct impact on potential output in addition to that achieved through national saving. According to CBO estimates, the President's fiscal measures reduced the marginal tax rate on capital income by 1.7 percentage points in 2012 from 14.5 per cent to 12.8 per cent and by an estimated −0.2 percentage points in 2013. From 2014 to 2022, the marginal tax rate on capital income is expected to increase by 0.4 to 0.8 percentage points. The President's proposals to cap at 28 per cent the rate on tax deductions, such as mortgage interest payments and property taxes, would generate the largest increase in the marginal tax rate on capital income. The CBO estimates that the President's policies would reduce the effective marginal rate on labour by 1.5 to 1.6 percentage points over the period 2013–22. The measures would affect the total hours worked (supply of labour) by increasing both people's total income (triggered by the effect of changes in average tax rates) and the marginal after-tax income (that is, the extra income for each additional hour of work triggered by changes in the marginal tax rate). Those changes would have opposing effects on the supply of labour. Employees would be encouraged to work longer hours because they would earn more for each extra hour of work. But at the same time the same workers would be encouraged to work fewer hours because they earn more at their current working hours. Thus, the responsiveness of the supply of labour, which is expressed as the total wage elasticity (the change in total labour income caused by a 1 per cent change in after-tax wages), has two components: a substitution elasticity (which measures the effect of changes in marginal tax rates) and an income elasticity (which measures the effect of changes in average tax rates). The CBO assumes that the total wage elasticity ranges from a low of −0.05 (composed of a substitution elasticity of 0.15 and an income elasticity of −0.20) to a high of 0.35 (composed of a substitution elasticity of 0.35 and an income elasticity of 0.0).

For 2013, the CBO assumes that the impact of fiscal policy would be felt entirely through aggregate demand. For 2014, 2015 and 2016, the estimates are based on a combination of effects on demand and effects on potential output. In particular, the blend for 2014 weights the effects on demand at 0.75 and weights the effect on potential output at 0.25; for 2015, those weights are 0.5 and 0.5; and for 2016, 0.25 and 0.75. The estimate for 2017–22 is based on effects on potential output. With these assumptions the CBO concludes that the President's fiscal measures would reduce GDP growth by as little as −0.5 per cent and as high as −2.2 per cent.

These estimates of the long-term effects of fiscal policy are also wide-ranging, as they encompass a broad spectrum of economists' views. Under more normal assumptions the negative impact on the capital stock would be mitigated by the effect of income tax cuts on labour. According to the K-model the negative impact on GDP through lower potential output would probably be between −0.5 per cent and −1 per cent.

9 POTENTIAL OUTPUT

The prevailing view among economists is that business cycles are caused by transitory but persistent random shocks.[27] This view entails that there are different forces that shape the trend in GDP and the cyclical fluctuations. Whereas the theory of business cycles deals with how transitory shocks propagate through the decisions of households and businesses to the entire economy for a number of years, making the shocks persistent, the theory of potential output relies entirely on the supply side of the economy. In economic theory the trend in GDP that matters is the one that measures the growth in the productive capacity of the economy. This is called potential output. It is loosely defined as the growth rate at which capital and labour are fully utilised. But this is not a technical ceiling. It is rather a rate of resource capacity utilisation at which inflation is steady. Thus actual GDP may grow faster than potential, but then bottlenecks arise in the production process that demand higher marginal costs and therefore cause higher inflation. Hence, when the economy grows faster than potential, there is overheating which is associated with accelerating inflation. On the other hand, if actual growth is less than potential, there is spare capacity in the economy, marginal costs are falling and therefore inflation abates. So, the rate of growth of potential output helps to differentiate when the economy is overheated and when it operates with spare capacity. Accordingly, potential output is properly defined as the maximum rate of growth consistent with steady (or non-accelerating) inflation or as simply the maximum *sustainable* growth of output. A steady rate of inflation means neither accelerating nor decelerating. It does not imply a constant rate of inflation through time. In different business cycles the steady rate of inflation may assume different values, for example 1 per cent, 2 per cent or 3 per cent.[28]

Although the conceptual definition of potential output is straightforward, it is not an observable variable in the real world. Hence, empirical counterparts to this theoretical construct must be provided. This is where the difficulty lies. A number of approaches have been suggested on how to measure potential output. These are reviewed under two headings: statistical estimates and supply-side determinants.

9.1 STATISTICAL ESTIMATES

One of the most commonly used methods for estimating potential output is smoothing of actual output. The smoothing method consists of fitting a trend either to actual output or through its peaks. The former takes the form of a linear, quadratic or a cubic spline.[29] Non-linear trends attempt to capture the fact that the long-term determinants of potential output may be changing constantly, but slowly through time. On the other hand, fitting a linear trend through peaks recognises that each business cycle may be different, and hence that the rate of growth of potential output may vary between cycles, but for each cycle the rate of growth of potential output is regarded as fixed. This implies a discontinuity in potential output growth between business cycles. A non-linear trend bypasses the discontinuity problem but it does not offer an explanation of why the trend is changing, despite being a smooth function. The degree of a non-linear trend is also arbitrary as it is assumed. Thus, a cubic spline implies an arbitrary choice of the various segments through which a separate curve is to be fitted.

These statistical estimates of potential output can be useful benchmarks for when inflationary pressures may emerge in the economy. The attraction of the smoothing approach lies in its simplicity. Its drawback, on the other hand, lies in the absence of information regarding the determinants of potential output. In particular, there is no benchmarking to inflation. Thus, a smooth trend may capture accurately the underlying trend in GDP, but it does not necessarily imply that this trend is associated with steady inflation. It provides, therefore, a measure of the trend in GDP, but it is not a measure of the underlying trend of the productive capacity of the economy because there is no benchmarking of the trend output to steady inflation. Moreover, although historical trend estimates of potential output may be reasonably good, ex-ante forecasting of the current or future business cycle may be poor if the determinants of potential output are likely to change.

9.2 SUPPLY-SIDE DETERMINANTS

The standard theory of the supply side of the economy is based on the Solow growth model (1956), with a Neoclassical production function. The production function for the overall economy describes how capital and labour are combined to produce a composite good (the goods and services in the economy as measured by GDP), for a given technology. But technology improves through time, thereby increasing output multiplicatively for a given stock of labour and capital. In this framework, labour productivity, which is defined as output per head (or output per hour worked), depends on the capital stock available to labour (the capital–labour ratio) and the state of technology.[30] Hence, improvements in labour productivity occur either through capital deepening (an increase in the capital–labour ratio) or through technological progress.

The Solow growth model enables the empirical calculation of potential output, by decomposing output to its conceptual determinants. Under the assumption of constant returns to scale the rate of growth of potential output can be obtained from a Cobb–Douglas production function

$$\bar{y} = a \cdot l + (1-a) \cdot k + q \qquad (5.54)$$

where $\bar{y}, l, k$ are the rates of growth of potential output, labour and capital, respectively; a and $(1-a)$ represent the contributions that the growth of labour and capital make to the growth of potential output; and q is total-factor productivity growth (TFE) or simply the Solow residual.[31] It is easy to see how the name Solow residual is derived. The Solow residual accounts for the proportion of growth that cannot be explained either by the growth of labour or capital. Depending on the decomposition of potential output, total-factor productivity would assume different values. If capital is substituted out, total-factor productivity would be higher. The coefficients a and $(1-a)$ can be approximated by the shares of labour compensation and capital income in the value of output.[32] As the share of capital income in the value of output has been constant at 30 per cent since 1947, it follows from equation (5.54) that potential output growth can be estimated by using the growth rates of labour, capital and total-factor productivity or the Solow residual.

Once installed, capital adds to the productive capacity of the economy, without affecting the non-accelerating inflation rate irrespective of the degree of its utilisation. On the other hand, the potential of labour force to increase the productive capacity of the economy depends on whether any additions to the labour force cause inflation to accelerate. As a result, it is that rate of growth of the labour force that is consistent with steady (or non-accelerating) inflation that comes into the computation of potential output. This is called the natural, normal or potential employment level of the labour force. Any labour input in excess of this natural rate produces acceleration in inflation. The natural rate of employment is associated with the natural rate of unemployment. In a similar fashion, the natural rate of unemployment is the rate of unemployment consistent with stable inflation. It is also called the non-accelerating inflation rate of unemployment (NAIRU). Hence, the rate of potential output and the natural rate of unemployment are similar measures of the productive capacity of the economy. The two are related to each other through Okun's Law (1962). This law links the goods with the labour market. In particular, it establishes a relationship in which shocks that emanate from the goods market are transmitted to the labour market. Thus, if there is a negative output gap (actual growth falls short of potential) it will lead to higher unemployment. If the goods market is in equilibrium then the labour market will also be in equilibrium. These concepts are summarised in Okun's Law

$$y - \bar{y} = g(U - \bar{U}) \quad g < 0 \tag{5.55}$$

where Y is the growth of GDP, $\bar{Y}$ is the rate of growth of potential output, U is the unemployment rate, is the natural rate of unemployment and g is the Okun coefficient. Thus, if the economy is overheated then unemployment is below its natural rate. If the economy is operating with spare capacity, then unemployment is higher than the natural rate. If the economy is growing at the rate of potential output then unemployment is equal to the natural rate. The Okun coefficient shows how much the economy should grow above potential so that unemployment can fall by 1 percent. The natural level of unemployment is a cyclically adjusted measure of unemployment. The impact of fluctuations in demand has been removed. The natural level of unemployment consists of structural and voluntary unemployment. The former arises because of the coexistence of vacancies and unemployment – a mismatch between the skills required in vacant jobs and the skills of the unemployed. Structural unemployment can decline if the labour market becomes more efficient in matching unemployment to vacancies by quick training of the unemployed. Voluntary unemployment arises when employees quit their jobs in order to further their careers by undertaking studies or because they feel that can look for a higher paid job more efficiently if they take time off from job or simply because they do not want to work at the current wage rate (for example mothers that are paid less than the cost of caring for their children).

The natural rate of unemployment can also be computed statistically in a similar fashion to potential output or through structural models that benchmark it to steady inflation. Benchmarking is achieved through the Philips curve, which connects the change in inflation to unemployment and other variables, including productivity, the price of oil and wage and price controls.

Obviously, given an estimate of the natural rate of unemployment the rate of growth of potential output is determined through (5.55). This implies that potential output can be computed without having to measure the capital stock in the economy. Such an approach is advantageous for those economists that believe that capital cannot be measured accurately because of problems in aggregating different vintages of capital stock, each vintage characterised by different technology. In this approach, potential output depends on labour and total-factor productivity. The contribution of the labour input to potential output requires some deeper analysis. At the simplest level demographic changes (birth rate) affect the rate of growth of labour force. Social and economic changes affect the labour participation rate. For example, adverse changes in the income distribution may force a higher participation of the labour force. The participation pace of women in the labour force depends partly on the child dependency ratio (the number of children per woman aged 20 to 54), which in turn depends on prior birth rates and hence on demographic factors. The intensity with which employers use the labour force, measured as hours per week, also affect the labour input of the productive capacity in the economy.

As an example of this method of computing potential output consider Table 5.1, which accounts for the rate of growth of US potential output by concentrating on the labour productivity growth accounting. In particular, it shows the contribution of population, labour force participation and total-factor productivity to output growth both historically and as projected by the Council of Economic Advisers (1997). The last line shows the rate of growth of potential output. This is obtained as the sum of the figures in all other lines. The first period 1960 III to 1973 IV covers two business cycles (from 1960 to 1970 and from 1970 to 1973). Since the supply-side components do not differ drastically and as in addition these are demand-led cycles they can be grouped together. The period from 1973 IV to 1990 III also covers two business cycles (from 1973 to 1980 and from 1980 to 1990). These are supply-led cycles and can also be grouped together. The third period is the business cycle which started in 1990 III and which the US Administration expected to last until the year 2003.

The striking feature of Table 5.1 is the progressive slowdown of the rate of growth of potential output from 4.2 per cent in the 1960s to 2.3 per cent in the 1990s. This is due to the slowing of two components of labour force growth, the working-age population and the labour force participation rate. Fluctuations in the participation rate are mainly accounted for by women and reflect probably long-term demographic factors. The child dependency ratio fell between the late 1960s and early 1980s as a result of a previous slowdown in the birth rate. This enabled women to increase their participation rate during most of the supply-side business cycles. However, the child dependency ratio flattened subsequently and this explains why the participation rate of women fell again in the 1990s.

A number of features seem to emerge from Table 5.1. First, most of the trends in the constituent components of the supply-side factors have been offsetting each other in these three periods. Second, the slowdown of the rate of potential output, from 4.2 per cent in the demand-led cycle of the 1960s to 2.7 per cent in the supply-led cycles of the 1970s and the 1980s, is due mainly to falling productivity, since the decline in the population growth was offset by an increase in the participation rate. Third, as a result of major reforms in the 1980s the labour market has become more flexible. This has halted the decline in the growth of average weekly hours in the 1990s.

An alternative method to computing potential output is to use all the components of equation (5.2), namely labour, total-factor productivity and capital. This method requires the cyclical adjustment of both hours worked and total-factor productivity so that potential rates can be computed for these two variables. This is achieved either by a statistical model or by use of the Okun's Law. Projections are generated by setting the unemployment and output gaps to zero. These potential rates are then substituted into equation (5.54), along with the capital input to compute potential output. Unlike the labour input and total-factor productivity, the capital input does not require cyclical adjustment, as the unadjusted measure already represents its potential contribution to output.

As an example of this method consider Table 5.2, which accounts for the rate of growth of potential output in the non-farm business sector and the overall economy using the supply determinants of equation (5.54). These are CBO estimates based on annual data until 2000. Potential output for the overall economy is derived by summing up the potential output of five sectors: non-farm business,

Table 5.1 Accounting for growth in US potential output
Average annual percent change

Item	1960 III to 1973 IV	1973 IV to 1990 III	1990 III to 2003
Working age population (16 and over)	1.8	1.5	1.0
Labour force participation Rate	0.2	0.5	0.1
Average weekly hours	−0.4	−0.3	0
Productivity (Output per hour)	2.8	1.1	1.2
Share of non-farm business output in GDP	−0.2	−0.1	0
Potential output growth	4.2	2.7	2.3

Source: Adapted from the 'Economic Report of the President' (1997).

government, households and non-profit institutions, residential housing and farm. The non-farm business sector accounts for more than three-quarters of GDP, the government for 11 per cent, and the other three sectors for the remaining 14 per cent. Equation (54) is used only for the non-farm business sector, in which capital is a major input to output and data for capital stock exist. For the other sectors a different approach is used partly because there are no data on capital and partly because a Neoclassical production function is not appropriate in the production method. A noteworthy point is that productivity is higher in Table 5.1 than in Table 5.2. As an example, for the period until 1973 productivity is only 2 per cent in the growth model, but 2.8 per cent in the productivity growth accounting method. This is because in the former capital has an explicit role, whereas in the latter the role of capital is in subsumed in productivity.

Although there are small differences in the estimates of potential output in the various subperiods between Tables 5.1 and 5.2, the striking contrast is that in the 1990s. The Council of Economic Advisers of the US President estimated the growth of potential output at a mere 2.3 per cent, whereas the CBO estimate is markedly higher (3.5 per cent) for the second half of the 1990s.

Progressively in the course of the 1990s there was a heated debate whether the decline in productivity of the 1970s and the 1980s was reversed in the 1990s. Anecdotal evidence in the second half of the 1990s suggested that productivity growth could have been higher. The advocates of higher potential output argued that the explosive growth in many areas of information technology (IT) – including telecommunications, personal computers and the Internet – combined with more flexible labour markets resulted in substantial gains in total-factor productivity. The implication was that the US economy could grow much faster without endangering a rekindling of inflation. Others, like the US Administration, argued that there was no structural change in the US economy with the implication that there was a need for tight monetary policy to contain inflationary pressures. Greenspan,

Table 5.2 Accounting for growth in US potential output – CBO method

Contributions to growth of potential output (Percentage points)	1951 to 1973	1974 to 1981	1982 to 1990	1991 to 1995	1996 to 2000	2000 to 2011	Average 1951–2000
Potential hours worked	0.9	1.5	1.1	1.1	1.0	0.8	1.1
Capital input	1.1	1.3	1.1	0.8	1.5	1.5	1.1
Potential TFP	2.0	0.8	1.0	1.1	1.5	1.4	1.5
Potential Output (Non-farm business sector)	4.0	3.6	3.2	2.9	4.0	3.7	3.7
Potential Output (Overall economy)	3.9	3.2	3.0	2.6	3.5	3.3	3.5

Source: Adapted from Table 2 of 'CBO's Method for Estimating Potential Output' (2001).

the then Fed chairman, sided with the structural change camp and did not tighten monetary policy; and yet higher inflation never materialised. The US economy grew much faster and inflation remained subdued.

The revision of the National Income and Product Accounts (NIPA) in September 1999 confirmed the structural change in the economy. The more accurate estimate of potential output by the CBO is due not only to the analysis being conducted four years later than the Administration, but also to a different methodology. The US Administration estimate is based on the labour productivity growth accounting, whereas the CBO figure is based on all three factors, including capital. The extensive revision of NIPA, dating back to 1958, simply accounted for expenditure on software and computer hardware not as consumption but as investment. Although from a purely accounting point of view this doesn't change the overall picture since higher investment is offset by lower consumption, it does have implications from an economic point of view. Higher investment rates mean that the capital stock in the economy was larger than previously estimated and this implies that the capital–labour ratio was higher throughout the period. The CBO attributes more importance to capital than to productivity. It estimates that the structural change in the economy is due to 0.7 per cent faster capital growth and to a 0.4 per cent improvement in total-factor productivity. Moreover, the faster capital growth is not only due to information technology, but also to policies that produced surpluses in the federal budget and reduced federal debt. A reduction in the federal budget deficit increases national savings and this stimulates private investment. The stagnation of private investment in the first half of the 1990s, along with efforts to curb the federal deficit in the second half of the 1990s, led the CBO to conclude that the structural change in the economy took place in the second half of the 1990s.

9.3 CONCLUSIONS ON POTENTIAL OUTPUT

The concept of potential output plays a vital role in the demand management of the economy and in defining neutral fiscal policy (the stance of fiscal policy consistent with the economy growing at the rate of potential output). Potential output helps to distinguish when the economy is overheated and therefore when inflation is on the rise and when there is spare capacity in the economy, which leads to higher unemployment. As a result, potential output serves as a natural target for economic policy, and particularly for monetary policy, which is more flexible than fiscal policy in demand management. When there is persistent overheating monetary policy should be tightened and when there is ample spare capacity monetary policy should be loosened.

Statistical estimates of potential output may explain the past well, but have a large margin of error in forecasting future business cycles. Statistical models provide accurate measures of trend GDP, but not necessarily of potential output because there is no benchmarking of the trend in GDP with stable inflation. Models of potential output based on the supply-side of the economy take care of the benchmarking problem. The growth model that decomposes output to its three constituent components (labour, capital and total-factor productivity) may, on occasions (like the second half of the 1990s), be superior to the productivity growth accounting method because it takes explicit account of the impact of investment on the capital stock. Spending on IT and fiscal policy with its impact on national savings are factors, which explicitly affect the level and quality of capital stock. Hence, the growth model provides more insights about the sources of change in potential output than the labour productivity growth accounting method.

But there are also clear disadvantages with supply-side models. These models require cyclical adjustment (the removal of the cyclical component caused by changes in demand) for all determinants of potential output. For some of these variables, like the natural rate of unemployment, the benchmarking to steady inflation can be achieved through the use of Okun's Law and the Phillips curve, but for others the same statistical techniques that are used for potential output may be necessary. Okun's Law has a number of drawbacks for calculating potential output. First, g may be time varying between cycles and within a business cycle. Secondly, the estimate of potential output depends on the natural rate of unemployment creating a conceptual circularity. Thirdly, the natural rate of unemployment may be more unstable than potential output if the distinction is made between cyclical and structural unemployment. Hence, if the estimate of potential output is based on a prior estimate of the natural rate of unemployment then it may be subject to unwarranted fluctuations. Statistical estimates of NAIRU are highly uncertain. Few models can claim a margin of error within one percentage point. Despite these drawbacks careful statistical estimation may overcome or mitigate these problems in estimating potential output.

10 CONCLUSIONS

In theory, it is the response of central banks to exogenous shocks that generates business cycles. But in practice most shocks have an insignificant effect on output, invoking no response from central banks. Shocks which necessitate central bank response are increases in commodity prices, such as a significant and long-lasting surge in the price of oil, provided that such an increase sets up a wage–price spiral; and shocks in world trade, which spill over recession from one economy to another. The common cause of business cycles is central bank response to rampant inflation as a result of overheating in the economy. This arises naturally as confidence in the private sector grows over time.

Most central banks have adopted inflation targeting, which may be supplemented with a zero output gap target. This practice stems from the policy implications of the NCM model, where control of inflation results in zero output gap in the long run. In the new millennium, inflation targeting has failed to shield the financial and economic system from the shocks created by excessive liquidity. As a result, the amplitude of business cycles has increased with dire consequences for unemployment and income inequality. In the extended NCM model, presented in this chapter, these policy properties of the basic model break down. Accordingly, central banks are advised to extend inflation targeting to include growth (or employment) and a target on real wealth to deter the repeat of financial crises and severe recessions. Unfortunately, central banks are moving in a different direction – stricter regulation of traditional banking and extension of the regulatory umbrella to shadow banking. The disadvantage of this approach is that regulation is backward looking; it closes the loopholes of the past, but cannot anticipate future practices. Any given regulation has loopholes, which can be exploited. Moreover, central banks instead of encouraging deleveraging (the drain of liquidity) inject new liquidity with the aim of avoiding deflation. This stems from the practice of central banks in dealing with the consequences of the burst of a bubble rather than trying to prevent the ballooning of the bubble in the first instance. From this point of view the adoption of three targets (inflation, output gap and real wealth) is preferable to regulation, without this meaning that regulation should be eased. Unless central banks change their practices, the future entails more of the same.

The analysis of business cycles is a prerequisite for understanding shipping cycles, as the former are the primary cause of the latter. The interrelationship of business and shipping cycles is taken up in the next chapter.

APPENDIX: NOMINAL RIGIDITIES

In this Appendix we present a summary of models that attempt to explain nominal rigidities in the economy. The approach is neither thorough nor exhaustive. Nonetheless, for the sake of completeness and of a self-contained book it may be helpful to the reader to appreciate the causes of these rigidities, which justify the use of monetary and fiscal policy in aiming to reduce the amplitude of business cycles. These models are discussed under three headings: implicit contracts, union or insider–outsider models and efficiency wage models, which attempt to explain real wage rigidity. Real price rigidity models are discussed next and finally models of nominal price and wage rigidity.

IMPLICIT CONTRACTS

The 'implicit contract' theory was developed simultaneously by Azariadis (1975), Baily (1974) and D. Gordon (1974) and aims to explain nominal wage rigidity. The original models could not explain greater volatility in employment and unemployment than the perfectly competitive market (that is, the Walrasian model) (see for details Akerlof and Miyazaki, 1980 and Grossman and Hart, 1983). This result is consistent with the findings of Arrow and Debreu that insurance contracts improve the functioning of competitive economies by making them Pareto efficient, see Azariadis (1975) and Azariadis and Stiglitz (1983) for surveys of this literature. If, however, firms are risk averse rather than risk neutral as was assumed in the original version of the implicit contract theory and they are better informed about the state of the world than workers (i.e. there is asymmetric information), then the implicit contract model can explain 'non-Walrasian' unemployment, see Azariadis (1983) and Grossman and Hart (1983); for a survey of the asymmetric information version of the implicit contract theory see Hart (1983). However, this result is not robust. The asymmetric information theory of implicit contracts can result over-employment or underemployment depending on the nature of the utility function and the degree of risk aversion of firms, see Stiglitz (1986). Despite the popularity of implicit contract theory in the 1980s, applications of the implicit contracts theory in labor economics has been in decline since the 1990s. The theory has been replaced by 'search' and 'matching theory' to explain labor market imperfections (see Pissarides, 2000, and Mortensen and Pissarides, 1994).

UNIONS OR INSIDER–OUTSIDER MODELS

It is a common belief among many academic economists, practitioners, politicians and the general public that real wage rigidity is caused by or is, at least, connected with labour unions. Models that explore the role of labour unions in real wage rigidity are based on bargaining between workers and firms over wages and employment. In some of these models the labour union is assumed to maximise the expected utility of its 'representative member' on the implicit assumption that all members are equally treated by the firm and the union. In other models the union maximises the expected utility of the 'median voter' on the assumption that workers differ in terms of seniority, see Oswald (1985) and Pencavel (1985) for some of these problems, surveys of the literature and extended bibliographies. Although bargaining can be over wages and employment, in reality actual labour contracts appear to set a wage and to leave the employment decision to the firm.

The most influential model in this category is the 'right-to-manage', according to which the firm and the union bargain over the real wage rate and then employment is chosen by the firm so as to maximise profits (Nickell, 1982). An extreme version of this model is the early approach of the 'monopoly union' according to which the union chooses the real wage rate unilaterally and the firm maximises profits taking as given that wage rate, Dunlop (1938). A monopoly union that maximises the expected utility of its representative member would choose that level of the real wage rate at which the highest indifference curve of the union between real wages and employment is tangent to the firm's labour demand curve. The reason is that the labour demand curve can be thought of as the locus of maximum points of the firm's isoprofit curves between real wages and employment for given values of the real wage rate (that is, the locus of points along which the marginal revenue product of labour is equal to the wage).

However, as Leontief (1946) first pointed out, this optimal combination of the real wage rate and employment chosen by a monopoly union is inefficient, in the sense that both the firm and the union can be better off (that is, there can be a Pareto improvement) by a lower real wage rate that leads to higher employment. Thus, there is no incentive for the union to choose a point on the labour demand curve either in conjunction with the firm (the right-to-manage approach) or unilaterally (the monopoly union approach). Both would be better off by bargaining over both the real wage rate and employment. In this framework McDonald and Solow (1981) show that large fluctuations in employment can be associated with small variations in the real wage rate on the assumption that product market conditions are more sensitive to the business cycle than the reservation wage. The importance of this assumption can be seen as follows. Suppose that economic conditions worsen (that is, the economy enters the downswing of the business cycle). Then for any given real wage rate firms would choose lower employment. However, at the same time the labour market conditions deteriorate and workers should lower their reservation wage. The assumption implies that the former effect dominates the latter, thereby resulting in smaller variation in the real wage rate and larger fluctuations in employment. If the labour market is segmented into a unionized part and a competitive one, then aggregate demand shocks will have different effects on the two sectors. Obviously, a negative demand shock would force wages down more in the competitive than the unionized sector. Hence the differential wage between the two sectors would widen during recessions, as would the fluctuations in employment. The unionized sector would experience smaller wage variability but larger employment fluctuations than the competitive sector (McDonald and Solow, 1985). However, there are a number of problems with the explanation of the union's approach to real wage rigidity. First, the real wage rigidity result is not robust, since the firm and the union are involved in repeated bargaining, reputation considerations may force the firm to give up short run profits for better contracts in the future thereby increasing real wage flexibility (Espinosa and Rhee, 1987). Second, membership considerations imply a dynamic model and this may induce more real wage flexibility than the static model (Gottfries and Horn, 1987). Third, as Blanchard and Fischer (1990, Ch. 9) argue, if the union cares about employment, why does it bargain only over the real wage rate and not also over employment? Fourth, at least in the US there is evidence of price stickiness before unions became important (Gordon, 1990). Finally, there is the issue of what Gordon calls the 'indexation puzzle': why don't unions agree contracts with full indexation of the growth rate of the nominal wage rate with the growth rate of nominal GNP?

If the bargaining power of workers does not come from powerful trade unions, where does it come from? An insight into this problem comes from the 'insider–outsider' models; see Lindbeck and Snower (1986, 1987, 1988) and Solow (1985). The labour market consists of incumbent employees (insiders) and unemployed workers (outsiders). In these models the determination of wages is considered as the outcome of a bargain between insiders and employers, while outsiders are unable to influence either the wage rate or the level of employment. Insiders have bargaining power that arises from turnover costs, such as the costs of hiring, firing and training. Incumbent employees exploit these turnover costs in bargaining with employers without taking into consideration the interests of outsiders and raise wages above the clearing market level, thereby creating unemployment. The bargaining power of insiders makes it possible for them to extract a share of the product market rents earned by firms. Real wage rigidity is an optimal behaviour in these models because turnover costs make it costly for employers to replace their incumbent employees with unemployed workers. Outsiders are unable to gain employment by offering their labour services at a lower wage because insiders cooperate amongst themselves and threat firms with collective action, harass new entrants and make it, in general, unpleasant for outsiders to underbid and gain jobs. Insiders can maximise their bargaining power by forming unions. The implications of insider–outsider models for wages and employment are similar to those obtained under the theory of unions. Insider–outsider models can generate 'hysteresis' effects and have been used by Blanchard and Summers (1986) and Layard and Nickell (1986) to explain high and persistent unemployment in Europe in the 1980s.

EFFICIENCY WAGES

The most promising theory of unemployment is the 'efficiency wage hypothesis' developed in the 1980s. The main idea is that the productivity of workers depends on the wage paid. In this case firms may be unwilling to lower the real wage rate in the face of excess supply of labour if the increase in costs due to falling productivity exceeds the gains from lower wages. It is therefore possible that profit-maximising firms may pay a real wage rate that is higher than the competitive (Walrasian) equilibrium and that would explain real wage stickiness, unemployment and large fluctuations in employment in the presence of unexpected shocks. The basic efficiency wage result can be illustrated in a perfectly competitive environment in which the production function depends on the number of workers and their effort, which, in turn, is a positive function of the real wage rate. The optimal wage is called the 'efficiency wage' because it minimises labour costs per efficiency unit. Each firm would hire workers up to the point where the value of its marginal product equals the efficiency wage, see Solow (1979). This basic model has been extended by adding in the effort function of the average wage rate, the unemployment rate and unemployment benefits; see Katz (1986), Stiglitz (1986), Weiss (1980) and Yellen (1984) for surveys of the literature.

The explanation of wage stickiness is important, but it rests on the assumption that productivity depends on wages (the effort function), which is postulated rather than derived. There are at least five versions of the efficiency wage hypothesis, which differ in terms of the explanation they offer for the relationship between productivity and wages. The original idea is usually attributed to Leibenstein (1957) in the theory of development economics who postulated a relationship between, on the one hand, the level of nutrition and productivity and, on the other hand, the level of wages. The second theory emphasizes labour turnover costs which are wholly or partly borne by firms as long as workers are more risk averse than firms, see Arnott and Stiglitz (1982), Hashimoto (1981), Phelps (1970) and Stiglitz (1982). In this case, the lower the wage rate, the higher the rate of labour turnover and the lower the labour productivity. The rationale can be seen as follows. If a firm is paying a wage rate that is lower than that paid by other firms (the equilibrium wage), then this firm would experience a higher quit rate by workers who are searching for higher wages. Since total labour cost depends on the wage bill (the product of the wage rate times employment) and the training cost which, in turn, is a positive function of the quit rate, it follows that these costs are minimised at the equilibrium wage rate. In this economy unemployment will be created by any shock in the economy that requires a reduction of the real wage rate as no firm has an incentive to adjust first its real wage rate, since to do so would invite a higher quit rate. Hence, unemployment follows from a coordination failure. The third version of the efficiency wage hypothesis (the quality-efficiency wage model) assumes that labour is not homogeneous and because of imperfect information, firms are unable to attract the right quality of labour force unless they pay higher wages, see Stiglitz (1976) and Weiss (1980). The fourth version of the theory is based on 'shirking', according to which, because of imperfect monitoring regarding the actions of a non-homogeneous labour force, firms pay higher wages to induce workers not to shirk, see Calvo (1979), Salop (1979) and Shapiro and Stiglitz (1984). The fifth explanation of the positive relationship between productivity and wages is based on sociological factors; see Akerlof (1984) and Akerlof and Yellen (1987). In these models a decrease of the real wage rate is considered as 'unfair' among workers who as a result reduce their effort. Accordingly, firms do not lower the real wage rate in the face of excess supply of labour.

The advantage of the efficiency wage hypothesis is not only that it can explain real wage rigidity, unemployment and large fluctuations in employment in the presence of unexpected shocks, but also queues for high-paid jobs, the pro-cyclical fluctuation of the quit rate, segmented labour markets, why unemployment may hit some groups more than others. The latter, for example, can be explained by invoking that various groups differ in their relationship between wage and productivity. The theory, therefore, can explain patterns of observed unemployment and many aspects of micro labour market behaviour. Although all versions of the theory can explain the above results, their policy implications would differ substantially. For example, in the shirking version an increase in unemployment benefits would result in higher wages and higher unemployment as the opportunity cost of being fired when caught shirking is reduced and, therefore, firms have to pay even higher wages to induce workers not to shirk. However, in the quality-efficiency version the effect of unemployment benefits on wages and unemployment is ambiguous and depends on parameter values. The main objection to the efficiency wage hypothesis is based on the argument that more elaborate pay schemes could be designed to overcome the coordination failure implicit in all versions of the theory. In the labour-turnover version workers could be forced to pay all of their training costs and therefore receive only their reservation wage, which would lead to Walrasian equilibrium. Similarly, in the case of imperfect monitoring, workers could be forced to post a bond, which they would forfeit if they were found shirking. In the case of quality-efficiency version, contracts could be tied with performance bonds which the workers could forfeit if it turned out that they were not as good as they claimed. These issues have given rise to more elaborate models, like the 'moral hazard' in which firms have an incentive to charge individuals for training that they do not provide. Other models are based on the observation that individuals do not have sufficient wealth to post bonds or that they are more risk averse than firms. These issues are not yet resolved because critics counter-argue that some effective bonding is observed in the real world in the form of low-wage apprenticeships and performance related paying schemes in which a proportion of wages-salaries is related to the company's earnings. Similar criticisms have been raised to sociological theories of the efficiency wage, which are based on the premise of 'fairness'. Why is it fair for some workers to keep their jobs at the same wage rate, while others are losing their jobs? The challenge for the proponents of the sociological version of the efficient wage theory is to explain rather than assume 'fairness'. Finally, as with all theories of real wage rigidity the puzzle is why workers do not agree contracts with full GNP indexation?

REAL PRICE RIGIDITY MODELS

Real price rigidity models attempt to explain why firms would choose to keep their profit margins (that is, the relationship between price and cost) constant in the face of unexpected shocks, or to keep constant the prices of their products relative to those of their competitors or to the general price level (that is, a form of relative prices) (see Okun, 1975). These models draw a distinction between auction and customer markets and originate from the analysis of Alchian (1969) and Phelps and Winter (1970). In auction markets prices would change to equate demand and supply at any point in time, whereas customer markets are characterised by price stickiness. The presence of search costs explains why firms build long-lasting relationships with customary suppliers and are willing to pay a premium over the competitive price. Similarly, search costs discourage customers from identifying whether intertemporal price changes apply to all firms or only to their preferred ones. These costs make customers willing to pay a premium above the competitive price. On the other hand, for the sake of preserving such long-lasting relationships firms are prepared to absorb transient increases in costs and apply mark-up pricing in which temporary increases in demand do not lead to frequent price adjustment. Moreover, Okun argues that customers would accept as fair any increase in prices due to a permanent increase in costs, but not to changes in productivity or changes in demand, which are generally regarded as transient. This sort of behaviour would explain price stickiness. The attraction of this theory is that it is intuitively appealing.

However, this approach suffers from a number of drawbacks. First, the argument of customer dissatisfaction is not sufficient to justify deviation from competitive pricing. Customer dissatisfaction arising in periods of boom would, on average, be offset in periods of recession. Second, the theory leaves unexplained what is regarded as 'fair'. Third, why firms do not revert to pricing practices which are based on full indexation of their costs to the nominal GNP?

NOMINAL PRICE AND WAGE RIGIDITY MODELS

Recently, the attention has switched from the labour to the goods market with the emphasis being placed on nominal price rigidity. If price rigidity is to be explained from optimising behaviour then the framework of monopolistic competition is more appropriate than perfect competition since in the former firms are allowed to set up prices. However, the mere introduction of imperfect competition is not sufficient to explain price stickiness and, moreover, money is neutral as under perfect competition, see Blanchard and Kiyotaki (1987), Fischer (1988), Kiyotaki (1988) and Rotemberg (1982). For an imperfectly competitive firm, shifts in demand can produce two alternative extreme combinations of price–output response. At one end of the spectrum, the price can remain unchanged with the whole adjustment borne by output under the assumption of constant marginal cost (that is, a flat marginal cost curve) and iso-elastic demand. At the other end of the spectrum, a decrease in demand would leave the profit-maximising level of output unchanged with the whole adjustment borne by the price if the marginal cost is reduced in proportion to the marginal revenue.

In between these two extremes both the price and output would change in response to shifts in demand, but whether most of the adjustment would fall on price or output depends on the following factors. The relaxation of the assumption of either an iso-elastic demand function or a flat marginal cost curve would introduce some degree of price adjustment. An upward-sloping marginal cost curve would allow for some price adjustment, even with an iso-elastic demand curve. On the other hand, a linear demand curve, even with a flat marginal cost curve, would also allow for some price adjustment. In the Blanchard and Kiyotaki model the marginal cost curve is flat under the twin assumptions of a constant marginal disutility of work and constant returns to labour in production. If either the marginal disutility of work increases with the amount of work or there are decreasing returns to labour in production, or both, then the marginal cost curve is upward sloping, implying some price adjustment. Furthermore, even with an iso-elastic demand curve and a flat marginal cost curve there would be some price adjustment if the marginal cost decreased (that is, when the curve shifts down). Thus, the justification of price stickiness suggests that the explanation ought perhaps to be sought in the sticky adjustment of marginal cost and this, in turn, points to factors in the labour market and material prices. Furthermore, in the Blanchard and Kiyotaki model, money is neutral because there is complete symmetry across producers. In response to a decrease in money, which decreases aggregate demand and, therefore, the demand of each firm an attempt by each producer to reduce its relative price leads to unchanged relative prices in the new equilibrium, because of the assumption of symmetry. Hence, money is neutral and, therefore, there is no explanation of price stickiness despite the framework of imperfect competition. Nevertheless, output is lower and the price is higher relative to perfect competition.

To make further progress in explaining aggregate price stickiness we must introduce barriers to nominal price adjustment at the micro level through adjustment costs, 'menu costs' (see Akerlof and Yellen, 1985 and Mankiw, 1985). These menu costs include not only the costs of listing new prices, but also the costs of informing customers, customer annoyance of changing prices and the administrative cost of taking the decision to change prices. In the presence of such menu costs it is optimal for firms not to change prices and allow output and employment to fluctuate in response to shifts in demand. But these costs by their very nature must be small and therefore the question arises as to how they can account for the large fluctuations in output and employment observed in industrialised countries such as the US or the UK. The important point is that even small menu costs can produce large fluctuations in output and employment and create nominal aggregate price rigidity. Firms in a monopolistic environment who face a fall in demand would only cut prices if the extra profit exceeds the menu cost. However, any cut in prices would lead to a welfare gain, in terms of the sum of the consumer and producer surplus, exactly because output under imperfect competition is lower than perfect competition. Thus, small menu costs deter firms from cutting their prices although doing so would be socially optimal. Moreover, if some firms do not adjust their prices then it is likely that other firms would not adjust too, thereby spreading the price rigidity. However, the menu cost approach suffers from three main criticisms and, therefore, cannot provide an adequate explanation of price stickiness. First, for small menu costs to generate large fluctuations in output and employment the condition that the marginal disutility of work is small is equivalent to the condition that the labour supply is very flat (see Blanchard and Fischer, 1990, Ch. 8). This, in turn, requires very strong intertemporal substitution effect of leisure, which is counterfactual. Second, if there are social losses during recessions because of a failure to reduce prices, symmetry suggests that these losses should be offset during booms because of a failure to increase prices. Thus, since the mean-output remains unchanged, the welfare cost of price rigidity is an increase in the variance of output, which surely must be a second-order effect. Hence, as Ball and Romer (1989a) and Gordon (1990) have pointed out the menu cost approach to price rigidity cannot conclusively reduce welfare and, therefore, explain large fluctuations in output and employment as a deviation from perfect competition because it involves the comparison of two second-order effects. Third, price rigidity is explained by postulating costs of adjustment of changing prices. If, however, there are costs in adjusting output, as in the theory of investment or inventory, then it is not sure whether the balance of adjustment falls on prices or output.

The above analysis of menu costs is static. The introduction of dynamics means that the issue is no longer whether but rather how often prices would adjust. This depends on the functional form of the price rules used by firms. These models fall within two broad categories: state-dependent and time-dependent rules. The former indicate that the price changes whenever it deviates by a particular percentage from the desired price, while the latter that price changes occur at fixed intervals because of explicit or implicit contracts with suppliers, the labour force and clients.

The first state-dependent rules were derived for a monopolist who faces demand shocks that take the form of a random walk without drift under fixed costs of adjusting prices (Barro, 1972). In this case the optimal policy is to set the price that maximises profits for a given value of the demand shock and then readjust it whenever future realisations of shocks (or, to be precise, the change in a shock) exceed a certain 'floor' or 'ceiling'. The determination of the lower and upper bounds, called S-s from which this literature derives its name, depends on a comparison of the opportunity cost of not adjusting the price (that is, having the 'wrong' price) with the cost of changing it. Explicit analytic solutions have not been derived except under very restrictive assumptions such as the one made in the Barro model that demand shocks follow a random walk, instead of serial correlation as observed in business cycles.

However, the Barro model cannot explain aggregate price stickiness and the stylised facts because money is neutral. Although this is the result of the particular assumptions made, their relaxation makes the S-s approach analytically intractable and therefore there are as yet not any further firm results, see also Blanchard and Fischer (1990, Ch. 8), Caplin and Spulber (1987) and Sheshinki and Weiss (1983) for some attempts of extending this approach. The alternative approach of explaining aggregate price stickiness is to explore time-dependent rules based on explicit or implicit contracts. Initial attempts emphasized overlapping staggered contracts in the labour market (Fischer, 1977b; Phelps and Taylor, 1977; Taylor, 1980). A contract specifies a fixed nominal wage for the duration of the contract. Staggered labour contracts reflect the absence of synchronization in the renewal of these contracts and therefore imply that in different firms wages change at different times. In these models prices in the goods market are flexible, but wages are sticky. The labour market fails to clear because at any point in time wages are given and, therefore, employment is determined by fluctuations in the demand for labour. In the Fischer model the nominal wage is set for the duration of the contract with the aim of maintaining a constant real wage rate on the basis of price expectations, which are formed rationally. In the Taylor model the wage rate is affected by past wages because of an overhang of unexpired contracts and expectations of both future wages and demand conditions in the labour market. Again expectations are formed rationally. The attraction of models on staggered contracts is that their assumptions are realistic reflecting observed behaviour in developed economies.

The contribution of these models was to clarify the role of rational expectations in producing the 'policy irrelevance proposition' in models of market clearing. Sargent and Wallace (1975) had previously shown that if expectations are formed rationally in a model of market clearing, then output is invariant to the money supply rule chosen by the monetary authorities, 'policy irrelevance proposition'. However, this conclusion results from the assumption of continuous market clearing rather than the assumption of rational expectations. In the Fischer and Taylor models in which there is wage stickiness monetary policy affects output in the short run despite the presence of rational expectations.

More recently, Blanchard (1986) generalised these models by extending the staggering to price decision too. In a monopolistic environment in the goods and the labour market workers attempt to maintain their real wage rate, while firms attempt to maintain their mark-ups of prices over wages. Expectations are formed rationally, but under staggering of price and wage decisions shifts in aggregate demand, caused, for example, by changes in nominal money supply, have long-lasting effects on output. The aggregate price level responds slowly to these nominal shocks, thereby producing a wage–price spiral in the adjustment to long-run equilibrium. Furthermore, on the assumption that the economy is hit predominantly by aggregate demand shocks, there is no correlation between the real wage rate and output. In this model, therefore, there is scope for stabilisation policy.

In the Blanchard–Fischer–Taylor models staggering is not explained and the timing of price changes is exogenous. Therefore the question arises as to why firms and unions adopt staggering while they would be better off under synchronization. Ball and Romer (1989b) investigate the conditions under which synchronisation and staggering is optimal and in equilibrium. They assume that synchronisation is socially optimal, but market failure leads to staggering. The model is an extension of Blanchard's in which the timing of staggering is made endogenous and there is an incentive for staggering because it is assumed that there are firm-specific shocks that arrive at different times for different firms. Ball and Romer (op. cit.) reach two major conclusions. First, staggering is inefficient, but it can be an equilibrium outcome provided there are firm specific shocks of any size. The reason is that 'a firm's decision to adjust at different times from others contributes, through its effect on the behaviour of the price level, to movements in relative prices and real aggregate demand that harm all firms. In a large economy, each price setter ignores these effects because it takes the behaviour of the price level as given' (Ball and Romer, 1989b, p. 180). Second, multiple equilibria are possible and therefore both synchronization and staggering can be an equilibrium.

There are three drawbacks with the nominal wage stickiness based on staggered contracts. First, these models cannot explain the diversity of stylized facts across time, across industries and across countries. Second, the approach lacks microfoundations. Staggered price and wage decisions are unexplained and they are not derived from optimising behaviour. As Barro (1977) has pointed out, this approach leaves open the question as to why, if fluctuations in output and employment are costly, firms and unions do not write contracts that avoid such costs. Third, in most of these models the predicted behaviour of the real wage rate is not consistent with the stylised facts described above. In particular, with a fixed nominal wage rate a negative disturbance in demand lowers the price level and raises the real wage rate, thereby suggesting a counter-cyclical behaviour of the real wage rate. However, as we have seen the evidence suggests that the real wage rate fluctuates pro-cyclically.

It follows from the above survey that there is not yet available a satisfactory theory of wage–price stickiness that can explain all stylised facts in the labour and the goods markets. Nevertheless, these theories are useful in clarifying the issues and suggesting the ways for future research.

6 THE THEORY OF SHIPPING CYCLES

EXECUTIVE SUMMARY

The theory of shipping cycles so far has been shaped primarily by two models, the Tinbergen–Koopmans model and the Beenstock–Vergottis model.

The fundamental contribution of the Tinbergen–Koopmans model is that shipping cycles arise even if the demand for shipping services is not cyclical. Shipping cycles are caused by cyclical fluctuations in the supply of vessels (that is, the net fleet). The cyclical behaviour of supply, on the other hand, is due to the lag between placing orders for ships and the ability of shipyards to deliver (the so-called delivery lag).

The Beenstock–Vergottis model is the first systematic approach to explain the interaction of the freight, time charter, secondhand, newbuilding and scrap markets under the twin assumption of rational expectations and market efficiency. It is a landmark because it treats ships as assets and applies portfolio theory to assess their values. As asset prices depend on expectations, Beenstock and Vergottis introduce rational expectations to account for the impact of expected and unexpected changes in key exogenous variables, such as the demand for dry, interest rates and bunker costs.

The BV model suffers from a major drawback, which has gone largely unnoticed and unchallenged in the maritime economics literature in the last twenty years. This is that the microfoundations of the BV model involve decisions intended to maximise short-term profits (that is, profits in every single period of time) instead of maximising long-term profits (that is, profits over the entire life of the vessel). Short-term profit maximisation is imposed either explicitly (as in the freight market) or implicitly by invoking market efficiency (as in the secondhand and newbuilding markets). The combination of short-term profit optimisation and market efficiency destroys the simultaneity of the BV model. Decisions in the four shipping markets (freight, secondhand, newbuilding and scrap) are not jointly determined. Rather the decisions can be arranged in such a way so that one follows from the other. This has serious implications for fleet expansion strategies.

To illustrate this point, consider an owner that maximises profits in each period of time, say a month, by choosing both the average fleet speed and the size of the fleet, so as to equate the return on shipping, adjusted for a variable risk premium, with the return on other competing assets, such as the short- or long-term interest rate. The fleet is adjusted monthly to reach the optimum via the secondhand and scrap markets. An owner adjusts his actual to the optimal fleet on a monthly basis by considering whether to buy or sell additional vessels in the secondhand market or scrap existing vessels according to the principle of monthly profit maximisation. Thus, an owner operating in the BV framework may expand the fleet one month and contract it the next month. In general, unanticipated random fluctuations in any exogenous variables, such as the demand for shipping services, interest rates and bunker costs, would trigger oscillations in the owner's fleet. Therefore, the owner in the BV model is myopic in that he ignores the consequences of his actions today for the lifetime of the vessel despite forming rational expectations.

The unsatisfactory features of the Beenstock–Vergottis model are dealt with in this book. In Chapter 3 it is shown that the appropriate framework for fleet expansion strategies is long-term profit maximisation. In Chapter 4 it is shown that shipping markets are inefficient for a horizon relevant to decision making, although shipping markets may be asymptotically efficient (that is, as the investment horizon tends to infinity). In this chapter we analyse the Tinbergen–Koopmans model in section 1 and the Beenstock–Vergottis model in section 2. In section 3 we reap the benefits of the painstaking path we have followed so far in this book in reaching the point where we can put together a complete shipping model capable of explaining shipping cycles in a way that overcomes the problems encountered in the studies reviewed so far. This is a model that integrates the main features of other shipping cycle models. In particular, it integrates the Tinbergen–Koopmans model of supply-led shipping cycles with the Beenstock–Vergottis model of expectations-driven shipping cycles. As the empirical evidence of shipping markets is that they are inefficient in the short run, but asymptotically efficient in the long run, the integrated model breaks away from the Beenstock–Vergottis model of assuming that shipping markets are efficient. The implication is that the arbitrage conditions between newbuilding and secondhand prices, and between the return of shipping and alternative assets, are removed. Instead, demand and supply factors in newbuilding and secondhand markets are allowed to interact in the determination of prices. This has the implication that all shipping markets interact with each other. As a result, a fleet capacity expansion strategy involves expectations of future freight rates, newbuilding, secondhand and scrap prices and the net fleet, which are jointly determined.

The integrated model also includes the business cycle model developed in the previous chapter. In the Beenstock–Vergottis model expectations are rational and drive the dynamics of the shipping model, along with the fleet accumulation dynamics, but the demand for dry is exogenous to the model. In the integrated model, by contrast, the demand for dry is endogenous. The implication of extending the model to cover the interactions between business and shipping cycles does not simply provide a more realistic explanation of this interaction. It also explains how expectations in shipping are formulated by economic policy; and in particular by how central banks react to economic conditions. As central banks choose their policies with the view of achieving their statutory targets, by observing current inflation and the output gap and knowing the central bank's targets one can deduce the future path of nominal interest rates. This provides a consistent explanation of how expectations in shipping are formed, which is analysed in section 4.

Section 5 illustrates how the integrated model explains the interaction of business and shipping cycles, while section 6 concludes. A temporary drop in aggregate demand in the economy lowers inflation and causes recession, which reduces the demand for shipping services. As supply is predetermined by past expectations of current demand, the fleet capacity utilisation drops, leading to lower freight rates, secondhand and new prices. The central bank, by cutting nominal interest rates, fosters the stabilising forces and the economy returns to its initial long-run equilibrium faster than if left to its own processes. This resuscitates the demand for shipping services and as a result freight rates, secondhand and newbuilding prices return to their long-run equilibrium. Therefore, a transient drop in aggregate demand causes both business cycles and shipping cycles. Accordingly, it can be concluded that shipping cycles are caused by business cycles. In the simulation results of the integrated model reported in section 5, shipping and business cycles are synchronised. The delivery lag of shipyards can distort this synchronisation. A two-year delivery lag can cause shipping cycles to follow with a lag business cycles largely because supply is predetermined by past expectations of current demand. The Tinbergen–Koopmans model is instructive of the implications of the delivery lag. Depending on parameter values the shipping cycles can appear out of phase with business cycles, thereby giving the impression that shipping cycles move counter-cyclically to business cycles (that is, the two cycles move in opposite directions). Such behaviour does not change the direction of causality. Business cycles cause shipping cycles.

Shipping cycles are also caused by expectations formed by charterers and owners in bargaining over current freight rates. Such expectations affect the demand for vessels, whether for newbuilding or secondhand prices. Rational expectations in shipping markets imply the discounting of future economic fundamentals in shipping variables. But such expectations depend on macro-economic fundamentals and, in particular, on how central banks would respond to the current business cycle. It is shown in this chapter that expectations in shipping are shaped by expectations of the future path of real interest rates and, consequently, on monetary policy. Beenstock and Vergottis emphasised the importance of these expectations in generating shipping cycles. Unfortunately, this was done in a model that assumed that shipping markets are efficient, an assertion which is not supported by empirical evidence.

1 SUPPLY-LED SHIPPING CYCLES: THE TINBERGEN–KOOPMANS MODEL

The seminal work of Tinbergen (1934) and Koopmans (1939) on shipping cycles has shaped all subsequent work. The fundamental concept behind these analyses is that shipping cycles arise even if demand for shipping services is not cyclical. Shipping cycles are caused by cyclical fluctuations in the supply of vessels (that is, the net fleet). The cyclical behaviour of supply, on the other hand, is due to the lag between placing orders for ships and the ability of shipyards to deliver (that is, the delivery lag).

The mechanics of this model can be summarised as follows. The demand for shipping services is considered as being perfectly inelastic to freight rates and it is assumed to be constant through time so that we can abstract from its influence in shipping cycles. Thus, at any point in time, t, the demand for shipping services, Q_t^d, is equal to a constant Q

$$Q_t^d = Q \qquad (6.1)$$

The supply of shipping services, Q^s, on the other hand, depends positively on the size of the fleet, K, and freight rates, F, but negatively on the price of bunkers, P_b.

$$Q^s = K^\alpha F^\beta P_b^\gamma \qquad \alpha, \beta > 0, \quad \gamma < 0 \qquad (6.2)$$

An increase in the fleet size increases the supply of shipping services almost proportionately. Accordingly, the elasticity of supply of shipping services to the size of the fleet should be equal to one (that is, $\alpha = 1$). An increase in freight rates for a given fleet would encourage fast steaming and thus again increase the supply of shipping services. Hence, the elasticity of supply with respect to freight rates is positive (that is, $\beta > 0$). On the other hand, an increase in the price of bunkers for given freight rates leads to slower steaming and hence a decrease in the supply of shipping services. Accordingly, the elasticity of supply of shipping services to bunker costs is negative (that is, $\gamma < 0$).

By equating the demand for and supply of shipping services, log-linearising equation (6.2), denoting by lower-case letters the natural logs of the variables and solving for freight rates, the reduced form of the model is obtained.

$$f = \frac{1}{\beta}[q - \alpha \cdot k - \gamma \cdot p_b] \qquad (6.3)$$

Assuming that in addition to the demand for shipping services, the price of bunkers does not change through time and that it is a constant, the freight rate equation is an instantaneous relationship and therefore does not give rise to cycles. Nonetheless, if the fleet fluctuates around its equilibrium, freight rates would also fluctuate with the same periodicity, but counter-cyclically (that is, in reverse order to the fleet).

To investigate the dynamic evolution of the fleet, consider the identity of the orderbook. At the beginning of period $t + 1$ the orderbook, O, is equal to its value in the previous period, t, plus the new contracts in period t, C, less order cancellations in period t, CL. Thus,

$$O_{t+1} = O_t + C_t - CL_t \qquad (6.4)$$

In this static framework, where demand for shipping services and the price of bunkers are constant, it is reasonable to assume that there are no order cancellations. This enables the study of fleet dynamics when everything else is held constant. Therefore, the change in the orderbook is equal to the demand for vessels, DV, which is assumed to be a positive function of freight rates. The demand for vessels by owners increases when freight rates improve.

$$\Delta O_{t+1} = DV_t + \lambda \cdot F_t \qquad \lambda > 0 \qquad (6.5)$$

In addition, it is convenient to abstract from scrapping on the basis that we are interested in just the dynamic evolution of the fleet with everything else being unchanged. Under these simplifying assumptions, the net fleet (gross fleet less scrapping and losses) is equal to the accumulation of fleet deliveries, DL.

$$K_t = \sum_i DL_{t-i} \qquad (6.6a)$$

Equivalently, the change in the fleet in period t is simply equal to the deliveries in the period

$$\Delta K_t = DL_t \qquad (6.6b)$$

Assume that shipyards deliver ships at time t from orders placed $t-\theta$ periods earlier. These orders were based on the estimate of demand for vessels at one period earlier than $t-\theta$. Let the expectations operator be denoted by E. The equation for deliveries is expressed as follows

$$DL_t = E_{t-\theta-1} DV_{t-\theta} = \Delta O_{t-\theta+1} \qquad (6.7a)$$

The term $E_{t-\theta-1} DV_{t-\theta}$ denotes the expectation of demand in period $t-\theta$, given the information available in the previous period. Under perfect foresight expected and actual demand are equal (an assumption relaxed later on). Finally, the freight equation (6.3) can be simplified as the demand for shipping services and the price of bunkers are constant

$$F_t = \rho \cdot K_t \qquad \rho = -\frac{1}{\beta} < 0 \qquad (6.7b)$$

Lagging equation (6.5) by θ and inserting it into (6.7a) and the resulting equation into (6.6b) equation (6.8) is obtained.

$$\Delta K_t = \lambda \cdot F_{t-\theta} \qquad (6.8)$$

Lagging equation (6.7) by θ and inserting it into equation (6.8), the equation that describes the dynamics of the fleet is finally obtained

$$K_t = K_{t-1} + \lambda \rho \cdot K_{t-\theta} \qquad \lambda \rho < 0 \qquad (6.9)$$

If it is assumed that the shipyard delivery lag is two years (that is, $\theta = 2$), which coincides with the historical average, then equation (6.9) is a second-order difference equation, which can be solved through time with two initial conditions. Equation (6.9) gives the dynamic adjustment of the fleet when all other variables are held constant. The dynamic adjustment of the fleet is taken as deviation from its equilibrium value, K^e, and depends on the product of the two coefficients, λ, and ρ, which is negative.[1] The first coefficient, λ, measures the response of the fleet to freight rates (how quickly owners adjust the fleet to a permanent change in freight rates), which depends on the elasticity of the demand for vessels to freight rates. The second coefficient, ρ, measures the response of freight rates to the fleet, which depends on the elasticity of supply of shipping services to freight rates. Thus, $-\lambda\rho$, which measures the intensity of reaction, is equal to the product of the elasticity of demand for ships to freight rates times the inverse of the elasticity of supply of shipping services to freight rates. Thus

$$-\frac{\partial K}{\partial F}\cdot\frac{\partial F}{\partial K} = -\frac{\partial VD}{\partial F}\cdot\left(\frac{\partial Q^s}{\partial F}\right)^{-1} \quad (6.15)$$

It is clear from (6.15) that the intensity of reaction $(-\lambda\rho)$ depends on the interaction of two markets, namely the newbuilding market, where the demand for vessels is equal to the supply; and the freight market, where the demand for shipping services is equated to the supply of shipping services.

Table 6.1 expresses the roots of equation (6.9) for different values of $-\lambda\rho > 0$. The fleet exhibits oscillatory (cyclical) adjustment when $-\lambda\rho > 0.25$ and monotonic adjustment when $-\lambda\rho < 0.25$. The system is stable when $-\lambda\rho < 1$ and unstable when $-\lambda\rho > 1$.

In the simulations reported below the two initial conditions of equation (6.9) are the actual values of the dry fleet in 2011 and 2012. Thus, equation (6.9) describes the dynamic evolution of the fleet from 2013 onwards. Figure 6.1 plots the dynamic adjustment of fleet and freight rates when $-\lambda\rho = 1$. The fleet and freight rates move symmetrically but in opposite cycles (one rises and the other falls). Each cycle lasts for six years, but the system never converges to equilibrium. It exhibits oscillations of constant amplitude of six-year cycles around the equilibrium.

Figure 6.2 plots the dynamic adjustment of fleet and freight rates when $-\lambda\rho = 0.7$. Recall from Table 6.1 that the system converges to equilibrium in an oscillatory way with damped cycles. The first cycle lasts for seven years and the second one for six years, but with much smaller amplitude. Each subsequent cycle is shorter and of decreasing amplitude. For practical purposes the first two cycles are important. Figure 6.3 plots the dynamic adjustment of the fleet and freight rates when $-\lambda\rho = 0.25$, the cut-off point between monotonic (real roots) and oscillatory (cyclical) adjustment (complex roots). For values of $-\lambda\rho < 0.25$ the dynamic adjustment is slower, the slower the coefficient.

The Tinbergen–Koopmans model may be rudimentary, but it captures a very important aspect of shipping cycles, namely the shipyard delivery lag. The model links the shipyard and freight markets in explaining shipping cycles by invoking

Table 6.1 Dynamic fleet adjustment

$-\lambda\rho < 0.25$	Real roots. Monotonic convergence to equilibrium.
$-\lambda\rho = 0.25$	Two equal roots. Fast monotonic convergence to equilibrium.
$-\lambda\rho > 0.25$	Complex roots. Oscillatory (cyclical) adjustment.
$-\lambda\rho < 1$	Damped oscillations. Stable system.
$-\lambda\rho = 1$	Regular oscillations around equilibrium.
$-\lambda\rho > 1$	Explosive oscillations. Unstable system.

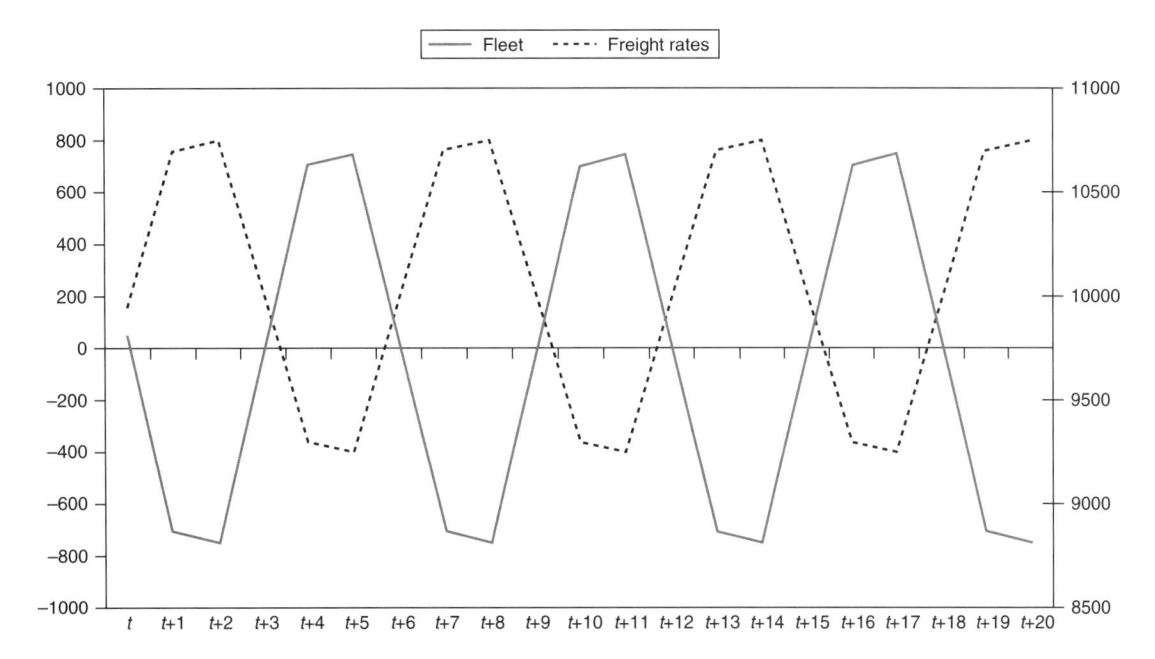

Figure 6.1 Fleet and freight rates adjustments when $-\lambda\rho = 1$

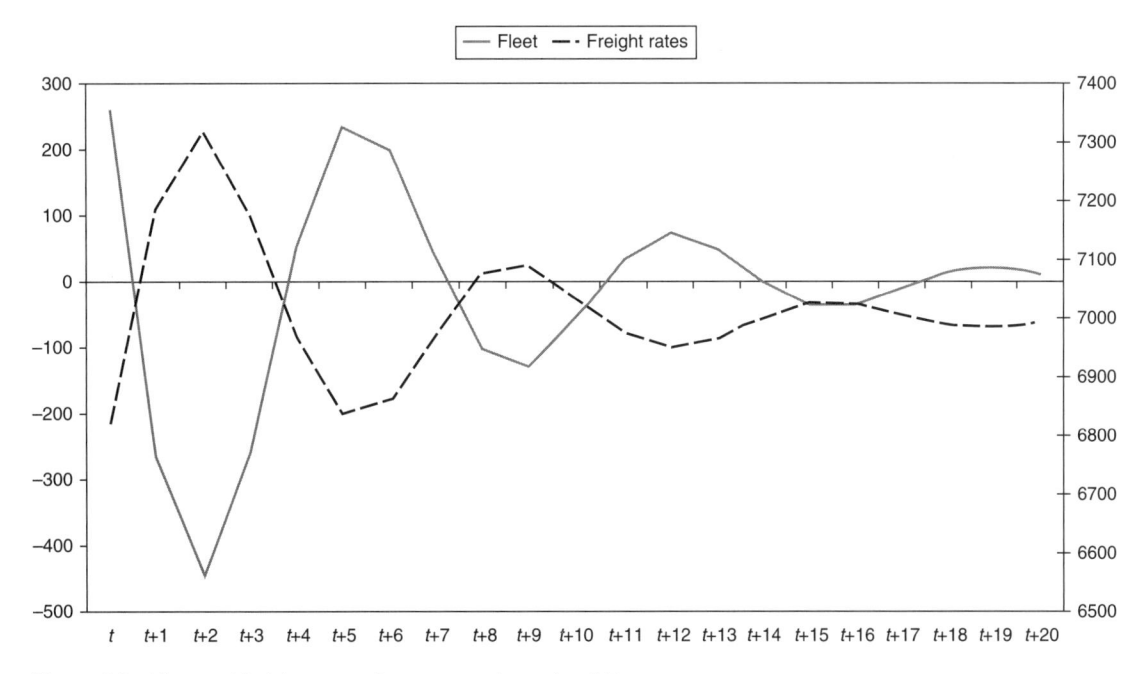

Figure 6.2 Fleet and freight rates adjustments when $-\lambda\rho = 0.7$

the delivery lag between orders for new ships and the ability of shipyards to deliver. In this model, even when the demand for shipping services, the price of bunkers and scrapping are constant, the fleet and freight rates exhibit oscillations, which are moving in reverse order. The freight market clears instantly, whereas the

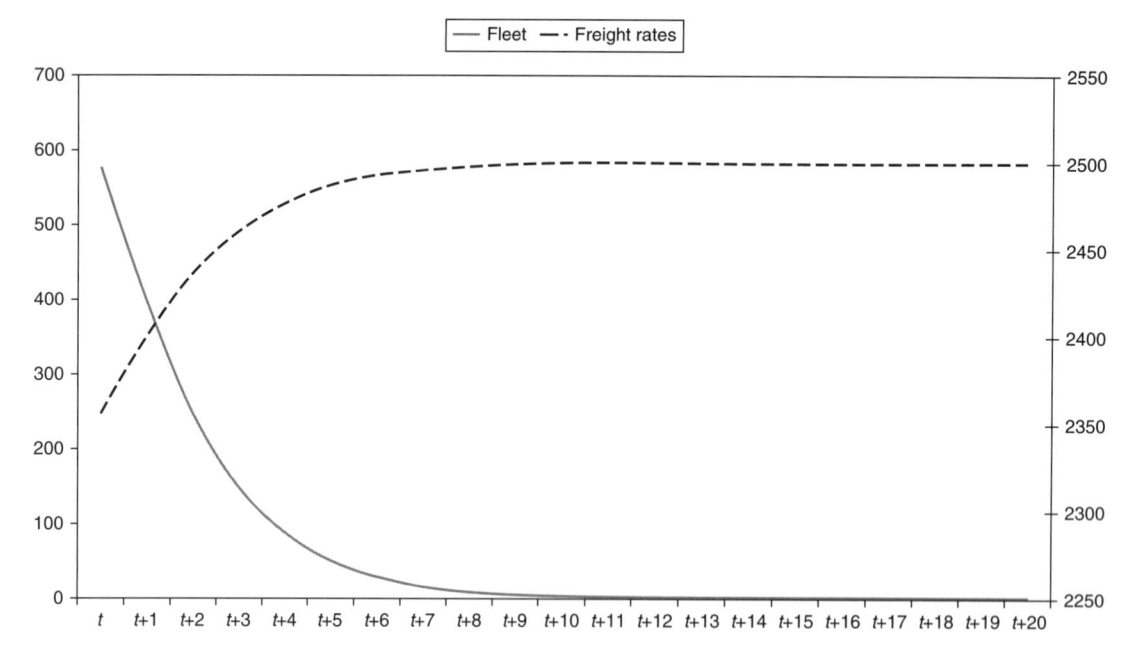

Figure 6.3 Fleet and freight rates adjustments when $-\lambda\rho = 0.25$

shipyard market responds sluggishly to demand for vessels because of the gestation lag of producing ships. The model is rudimentary because the demand for shipping services is perfectly inelastic to freight rates and the demand for vessels does not depend on vessel prices, the cost of capital, the availability of credit, the shipyard excess capacity and the fleet size. Nonetheless, the model has shaped all subsequent work on shipping cycles.

2 EXPECTATIONS-LED SHIPPING CYCLES: THE BEENSTOCK–VERGOTTIS (BV) MODEL

In an excellent review of the state of the art Glen (2006) argues convincingly that Beenstock and Vergottis set a high-water mark in shipping market modelling. In Glen's own words, "[t]he models developed and presented in Beenstock and Vergottis (1993) are a 'high-water mark' in the application of traditional econometric methods. They remain the most recent published work that develops a complete model of freight rate relations and an integrated model of the ship markets. It is a high-water mark because the tide of empirical work has turned and shifted in a new direction. This change has occurred for three reasons: first, the development of new econometric approaches, which have focused on the statistical properties of data; second, the use of different modelling techniques; and third, improvements in data availability have meant a shift away from the use of annual data to that of higher frequency, i.e. quarterly or monthly" (Glen 2006, p. 433). The BV model is the first systematic approach to explain the interaction of the freight, time charter, secondhand, newbuilding and scrap markets under the twin assumptions of rational expectations and market efficiency. The model is a landmark because it treats ships as assets and applies portfolio theory to assess their values. As asset prices depend on expectations Beenstock and Vergottis introduce rational expectations to account for the impact of expected and unexpected changes in key exogenous variables, such as the demand for dry, interest rates and the price of bunker costs. Rational expectations require model consistent expectations, and Beenstock and Vergottis make sure that these expectations comply with the steady-state properties of the model. The individual relationships in each market are derived from explicit micro-foundations (profit maximisation or market efficiency[2]).

The BV model suffers from a major drawback, which has gone unnoticed and unchallenged in shipping research in the last twenty years.[3] This is that the microfoundations of the BV model involve decisions so as to maximise short-term profits (that is, profits in every single period of time otherwise called static optimisation) instead of maximising long-term profits (that is, profits over the entire life of the vessel otherwise called dynamic optimisation). Short-term profit maximisation is imposed either explicitly as in the freight market or implicitly by invoking market efficiency, as in the secondhand and newbuilding markets. Therefore, the BV model rests on static rather than dynamic optimisation, which is analysed in Chapter 3.

In the BV framework an owner is maximising profits each period of time, say a month, by choosing both the average fleet speed and the size of the fleet, inorder to equate the return on shipping, adjusted for a variable risk premium, with the return on other competing assets, such as the short- or long-term interest rate. The fleet is adjusted monthly to reach the optimum via the secondhand and scrap markets. An owner adjusts his actual to the optimal fleet on a monthly basis by considering whether to buy or sell additional vessels in the secondhand market or scrap existing vessels according to the principle of short-term profit maximisation (that is, for each time period). Thus, in the BV framework an owner may expand the fleet in one month and contract it the next month. In general, unanticipated random fluctuations in any exogenous variables, such as the demand for shipping services, interest rates and bunker costs, would trigger oscillations in the owners' fleet. Therefore, the owner in the BV model is myopic in that he ignores the consequences of his actions today for the lifetime of the vessel despite forming rational expectations. In the BV model it is immaterial that such actions cannot affect the total fleet in the market, but only the secondhand price. At the aggregate level (that is, the shipping market as a whole), the fleet is determined by the level of demand for shipping services through the optimising behaviour of shipyards.

The condition of market efficiency implies instantaneous adjustment in demand to minute price changes so that returns of alternative assets, including appropriate risk premia, are equal in every time period. From this angle market efficiency implies short-term profit maximisation. The combination of static optimisation and market efficiency destroys the simultaneity of the BV model. Decisions in the four shipping markets (freight, secondhand, newbuilding and scrap) are not jointly determined. The decisions can be arranged in such a way so that one follows from the other. This has huge implications for the interaction of the secondhand and newbuilding markets. In the BV model the major asset market is the secondhand market. The price of secondhand vessels is obtained from an arbitrage condition that equates the returns, allowing for a risk premium, between shipping and alternative assets, on the assumption that all asset markets (and ships are such an asset) are efficient. The price of newbuilding vessels is simply equal to the expected value of secondhand prices at the time of the delivery of the new vessel. Secondhand and new vessels are perfect substitutes, except for age, and therefore market efficiency entails that the two are equal between two different time periods, where the difference is the delivery lag.

Market efficiency (arbitrage) would make sure that the newbuilding price at the time of delivery is equal to the expected secondhand price, plus or minus a risk premium reflecting such factors as vessel technological improvement, depreciation and demand uncertainty. The demand for new vessels by owners is infinitely elastic at the equilibrium price (that is, a horizontal demand curve for new ships). The only reason that the fleet is determinate in the BV framework is because it is not feasible or optimal for shipyards to respond to price fluctuations. The total fleet in the market is adjusted by the optimising behaviour of shipyards. In maximising profits, shipyards increase the supply of vessels as the price of new vessels increases relative to the cost of raw materials. Thus, in the BV model owners determine the price of new vessels and shipyards the fleet in the short run. In the long run, the fleet is determined by the exogenously given demand for shipping services and the price of secondhand and new vessels (since the two are equal to each other) by shipyards.

2.1 A SUMMARY OF THE BV MODEL

The BV model can be summarised as follows.

$$q_t^s = k_t + \gamma \cdot f_t - \gamma \cdot pb_t \tag{6.16}$$

$$\pi_t = (1 + \gamma) \cdot f_t - \gamma \cdot pb_t \tag{6.17}$$

$$q_t = q_t^s \tag{6.18}$$

$$ps_t = a_1 \cdot \pi_t + a_2 \cdot E_t\, ps_{t+1} - a_3 \cdot r_t \tag{6.19}$$

$$p_t = E_{t-1}\, ps_t \tag{6.20}$$

$$dl_t = \mu_1 \cdot p_t - \mu_1 \cdot pt_t \tag{6.21}$$

$$dm_t = \mu_2 \cdot psc_t - \mu_2 \cdot ps_t \tag{6.22}$$

$$\Delta k_t = dl_t - dm_t \tag{6.23}$$

The symbols have the usual interpretation, which for convenience is summarised below. Low-case letters denote the natural logarithms of the variables. For example, $q^s = \log(Q^s)$.

q^s = supply of shipping services, measured in tonne-miles;
k = net fleet, measured in tonnage (after taking account of scrapping and slipping).
The net fleet is predetermined at the beginning of each time-period;

f = freight rate in units of currency per tonne-mile;
pb = price of bunkers per tonne-mile;
π = profit per ship per time period;
r = the short or long term interest rate;
q = demand for shipping services, measured in tonne-miles;
ps = the price of a secondhand ship;
p = the price of a newbuilding vessel;
dl = deliveries of new vessels by shipyards, measured in tonnage;
dm = demolition of ships, measured in tonnage;
pt = price of steel;
psc = price of scrap.

Equation (6.16) specifies the supply of shipping services in the freight market. The supply is proportional to the fleet and depends positively on prevailing freight rates and negatively on bunker costs. The supply function is derived from explicit profit maximisation. Assuming speed is the only variable input in the production of shipping services and the fleet is predetermined in each time period, the supply of shipping services is proportional to the speed, S

$$Q_t^s = K_t \cdot S_t \tag{6.24}$$

Using equation (6.24), and ignoring operating costs, as they are treated as fixed and therefore do not affect the decision on the optimal speed, the profit per ship per unit of time is defined as revenue per ship less the fuel bill.[4] The profit function is defined by the equation

$$\Pi = F \cdot S - PB \cdot S^a \tag{6.25}$$

The parameter a measures fuel efficiency and the lower a is, the greater the fuel efficiency of the ship. However, the fuel consumption of a ship is a non-linear function of speed. For $a > 1$, as speed increases the fuel consumption increases proportionately more than the speed.

The optimal speed is the one that maximises short-term profits. This is obtained by setting the derivative of (6.25) with respect to speed equal to zero, and solving for S:

$$S = \left(\frac{F}{a \cdot PB} \right)^{\gamma} \quad \gamma = 1/(a-1) > 0 \tag{6.26}$$

By substituting the optimal speed to equation (6.24) and taking logs, equation (6.16) is obtained. Close inspection of (6.26) shows that the optimal speed depends on the ratio of freight rates to bunker costs, adjusted for the fuel efficiency of the vessel. The optimal speed depends positively on freight rates and negatively on the price of bunkers. Hence, higher freight rates induce owners to fast steaming, whereas higher bunker costs to slow steaming.

By substituting the optimal speed to equation (6.25) and taking logs, equation (6.17) is obtained. The profit function (6.25) implies that an owner maximises short-term profits (profits per time period) rather than long-term profits (profits over the lifetime of the vessel). As is shown in Chapter 3, this piecemeal approach is valid for speed decisions, but not for investment in buying ships. The reason is that a decision to buy a ship would have consequences for profitability across the entire lifetime of the vessel. Accordingly, the appropriate criterion is long-term profit maximisation. By contrast, the decision on the speed can be split down into decisions for each separate time period. This is a valid approach because in each time period the owner observes the price of bunkers and chooses the speed that maximises profits for that period only. In the next period, a different price of bunkers would entail a different optimal speed. The validity of the approach is clear from the analysis in Chapter 3, where it is shown that the optimal speed is derived by equating the marginal product of the speed at time t with the price of bunkers in the same period. This is not true with regard to choosing the fleet size, as the optimal fleet depends on expectations about key variables, such as the demand for shipping services, the price of vessels, freight rates and interest rates that affect future profitability throughout the lifetime of the vessel.

Equation (6.18) is the equilibrium condition in the freight market, which is regarded as perfectly competitive. Beenstock and Vergottis follow the tradition in maritime economics of treating the demand for shipping services as exogenous and perfectly inelastic with respect to freight rates.

As the fleet is predetermined in every time period, the freight market can be isolated from the rest of the system of equations (6.16)–(6.23). The system is not really simultaneous but recursive. This means that the system can be split into two separate subsystems and solved in a particular order without the need to solve all equations simultaneously. Equations (6.19)–(6.23) describe the interactions in the shipyard, secondhand and scrap markets and can be solved first to determine the fleet in each period of time. Then the fleet is substituted in equations (6.16)–(6.18) to determine freight rates. In plain English this means that freight rates must adjust in every period to make sure that the demand for shipping is equal to the predetermined stock of fleet. In every period owners choose a speed to maximise profits in that same period, given the price of bunkers.

Equation (6.19) describes the secondhand market, which is the cornerstone of the BV model. A secondhand ship is regarded as an asset and its price is determined using portfolio theory. The pillar assumption of portfolio theory is that asset markets are efficient and assets are perfect (or nearly perfect) substitutes, except in some features, like risk. Accordingly, the returns on alternative assets, allowing for risk, are equalised in every time period by corresponding changes in the demand for these assets. The demand for assets is perfectly elastic to the equilibrium price (or return). The demand becomes either infinitely large or zero in response to a slight divergence of the price (or return) from equilibrium. The adjustment to equilibrium, when exogenous variables change, is instantaneous. In other words, the equilibrium condition is enforced within the period of observation. Denoting by r^s the return on shipping and r the return on alternative assets, market efficiency (arbitrage) ensures the equality of returns in every time period, allowing for a risk premium, v.

$$r^s = r + v \qquad (6.27)$$

The return of a secondhand ship involves profits resulting from operating the vessel for T periods and capital gains or losses when the vessel is sold in period T in the secondhand market or scrapped. As the return is expressed as a percentage of the purchase price the two components of profits of the per period return are defined by the following equation

$$r^s = \frac{E_t \Pi_{t+1}}{PS_t} + \frac{E_t PS_{t+1} - PS_t}{PS_t} \qquad (6.28)$$

In the analytical work, Beenstock and Vergottis ignore the risk premium in equation (6.27). Then equating (6.27) with (6.28) and solving for PS_t equation (6.29) is obtained which relates the current price to its economic fundamentals

$$PS_t = \frac{E_t \Pi_{t+1} + E_t PS_{t+1}}{1 + r} \qquad (6.29)$$

Taking logs of both sides results in equation (6.19), on the assumption that the expected profit tomorrow is equal to today's profit. The BV framework is unrealistic, as owners are dedicated to their business and do not look to equalise the returns on shipping with those on financial assets. There seems to be little empirical support of high correlation between shipping profits and those in financial markets. The boom of 2003–08 attracted some outside investors (for example, private equity funds and hedge funds), but the return on shipping does not show any convergence to those of financial assets, such as stocks, corporate or sovereign bonds. In the boom and bust phases of the recent shipping cycle liquidity was channelled mainly to commodities. This liquidity distorted the prices of commodities and accentuated the boom and bust of shipping returns instead of enhancing convergence (see Chapter 8).

Putting aside the unrealistic nature of the secondhand market, equation (6.29) makes clear that the current price of a secondhand ship depends on the present value of expected profits in the next time period, discounted back to today by the current risk-free interest rate. These expectations turn the static nature of the arbitrage condition (6.27) into a dynamic one. Therefore, Beenstock and Vergottis, by replacing expected profits with current profits in equation (6.19), turn a dynamic optimisation problem into a static one with the implication that their model becomes recursive.

Beenstock and Vergottis invoke rational expectations to explain how expectations in their model are formed. It could be argued that the use of rational expectations make the BV model immune to the criticism that it involves static optimisation (profit maximisation per time period). The use of rational expectations per se is not objectionable, although many economists find it difficult to swallow the assumptions upon which rational expectations are based.

Rational expectations force the expected values of profits and secondhand prices to be equal to the predicted values of the model. These model consistent expectations, assume that all owners and charterers form the same expectations (uniform expectations) and use this particular model to form such expectations. Moreover, this model is further assumed to be the 'true' model of the shipping market because it is clear that if every economic agent uses a different model, then market expectations would not be uniform. But if this model happens to be the true model, then in time agents would learn from their mistakes and converge to the expectations generated by this model, thereby forming uniform expectations in the long run. Putting aside these objections, rational expectations require also some terminal conditions to be imposed – what happens in the final period T. One approach is simply to assume that after many periods of time these expectations make no difference to current decisions, namely that expectations converge to each other as T approaches infinity.

$$\lim_{T \to \infty} \left(E_{T-2} X_{T-1} \right) = E_{T-1} X_T \qquad (6.30)$$

The approach adopted by Beenstock and Vergottis, not uncommon in the economics literature, is somewhat different. They assume that the expectations in the final period T are equal to the steady-state properties of the model. In this context the choice of the model affects the rational expectations solution. If the model implies static optimisation in the final period T, then working backwards rational expectations force the same principle today. Therefore, the use of rational expectations does not make the BV model immune from the criticism that it implies static optimisation. This is evident from (6.17), in which the profit today depends on current values of freight rates and bunker costs.

Equation (6.20) describes how newbuilding prices are determined. In the BV model, newbuilding prices are set as if there was a fully efficient shipyard futures market, in which shipyards, owners and speculators could buy and sell new building contracts. This would entail an arbitrage equation in which the price of a newbuilding ship today is equal to the expectation formed with information today of the secondhand price m periods ahead, where m is the delivery lag, minus a risk premium that reflects the accuracy of the forecast and other risks. Thus,

$$P_t = E_t PS_{t+m} + \text{risk premium} \qquad (6.31)$$

In the BV framework, the importance of the newbuilding market is downgraded; it is regarded as being subservient to the secondhand market. In this model the demand for newbuilding vessels is perfectly elastic at a price fixed by the arbitrage equation (6.31). A horizontal demand curve for vessels means that owners do not have an optimal fleet, as we have shown in Chapter 3, where the optimal fleet and the speed at which it is acquired maximises long-term profits for the entire life of the fleet. Whereas in Chapter 3, newbuilding prices and the fleet are jointly determined by the demand for ships by owners and the supply of ships by shipyards, in the BV model, prices are determined by owners and the fleet by profit maximising shipyards.

Equation (6.21) describes the supply function of shipyards, the deliveries of vessels. This is a positive function of the newbuilding price and a negative function of the price of variable inputs necessary for the production of ships, such as steel and labour. Equation (6.21) is a standard specification of a normal upward-sloping supply curve. At any point in time a shipyard has an orderbook, which reflects the stock of orders for ships. The change in the orderbook from the previous to the current time period is the demand for new vessels in the current period, assuming no cancellations. But this demand would be met with a delivery lag of m time periods. The length of the delivery lag is shaped by technological but also economic factors. With a given shipyard capacity a ship would take an m number of years to produce. But the delivery lag can be shortened or lengthened by economic considerations. A shipyard may employ more labour to speed up the delivery or even expand the capacity if the orderbook is large. The economic decision of a shipyard, therefore, is how quickly to deplete a given orderbook taking into account that some features in the production process cannot be squeezed beyond limits or lie outside the control of shipyards, such as the capacity of other industries to deliver necessary equipment. The economic decision depends on expectations of demand for ships in the future, namely on projections of how quickly the orderbook would grow. Accordingly, the economic decision of a shipyard on the pace of deliveries would have to be decided by maximisation of long-term profits. In the BV model the supply function of vessels is derived from short-term rather than long-term profit maximisation. Nonetheless, the BV choice can be justified for the sake of simplicity. But it should be borne in mind that the shipyard decision would affect the evolution of future newbuilding prices. For example, in the boom years of 2003–08 lack of shipyard capacity exacerbated the upside of newbuilding prices. Shipyard capacity was expanded with a lag when the orderbook as a percentage of the fleet began to decline. This has acted as a drag to the recovery of newbuilding prices in the recovery phase, which probably started in 2013.

In the BV model equation (6.21) can easily be derived from short-term profit maximisation. Assume for simplicity that there is only one variable factor of production, namely steel, denoted by X, and that the price of steel is denoted by PT. Then the shipyard production function is

$$DL = \gamma \cdot X^b \qquad (6.32)$$

The parameters b and γ are technological constants with $b < 1$, reflecting diminishing returns to scale, and $\gamma > 0$ reflecting technological progress.

Then the short-term profit function of a typical shipyard is

$$\Pi = P \cdot DL - PT \cdot X \qquad (6.33)$$

Substituting the production function in (6.33) and setting the derivative of profit with respect to X equal to zero gives the optimal level of steel. Then substituting this optimal level of steel to the production function gives the level of deliveries (shipyard output) that maximises short-term profits.

$$DL = \gamma \cdot (b \cdot \gamma)^{1/(1-b)} \cdot \left(\frac{P}{PT}\right)^{1/(1-b)} \qquad (6.34)$$

With the restrictions imposed on b and γ, deliveries are a positive function of the newbuilding price and a negative function of the price of steel. Taking logs on both sides of (6.34) yields equation (6.21). Equation (6.34) can easily be generalised for more than one variable input. Thus, for two variable inputs, steel and labour equation (6.34) involves the ratio of the newbuilding price to the price of steel and the ratio of the newbuilding price to wages.

Equation (6.22) explains the decision of owners to scrap a vessel. A ship is scrapped when its scrap value exceeds its trading value, namely what it can fetch in the secondhand market. The scrap value is simply the scrap price per tonne (PSC) times the tonnes of metal contained in the ship. If both the secondhand price and the price of scrap are expressed per dwt then the decision of owners to supply ships for scrap depends on the ratio of the two prices. This expression of the relative price of scrap suggests that the supply of vessels for scrap should be expressed as a proportion of the fleet for scrap. The demand for scrap is decided by scrapyards and the equilibrium condition that demand must equal supply determines the scrap price and the vessels for scrap as a proportion of the fleet. Thus, denoting the vessels offered for demolition by DM, measured in dwt, and the fleet as K, measured also in dwt, then the supply of vessels for demolition is specified by equation (6.35)

$$\frac{DM}{K} = \mu \cdot \frac{PS}{PSC}, \quad \mu < 0 \tag{6.35}$$

The vessels offered for demolition is a positive function of the relative price of scrap. An increase in the price of vessels in the secondhand market relative to the scrap price reduces the supply of vessels for demolition. In contrast, an increase in the price of scrap relative to the secondhand price increases the supply of vessels for demolition. Taking logs of both sides of (6.35) yields equation (6.22).

The final equation (6.23) in the BV model is an identity that states that the fleet today is equal to the fleet of yesterday plus the deliveries of new vessels today less the demolition of ships today.

$$K_t = K_{t-1} + DL_t - DM_t \tag{6.36}$$

Dividing both sides of (6.36) by K_{t-1} and taking logs yields equation (6.23). Therefore, the proper interpretation of equation (6.23) is that the rate of growth of net fleet is equal to the proportion of new deliveries in the fleet less the proportion of the fleet for scrap.

2.2 THE STEADY-STATE PROPERTIES OF THE BV MODEL

In the long-run equilibrium, all variables converge to steady values (hence the name steady state) and expectations are realised. The latter means that expected values are equal to actual ones. Denoting the long-run equilibrium values of the variables with the same symbols but without the time subscript the long-run equilibrium (or steady state) is described by the following equations.

$$\Delta q = \Delta k = dl - dm = \mu_1 \cdot ps - \mu_1 \cdot pt - \mu_2 \cdot psc + \mu_2 \cdot ps \tag{6.37}$$

$$cu = q - k = \gamma \cdot f - \gamma \, pb \tag{6.38}$$

$$\pi = (1+\gamma)\cdot f - \gamma \cdot pb \qquad (6.39)$$

$$ps = \pi - r \qquad (6.40)$$

In the steady state the system of eight equations (6.17)–(6.23) has been reduced to four equations. It can be seen from equation (6.20) that newbuilding prices are equal to secondhand prices in the steady state because the expectation of the secondhand price is equal to the actual one. The equilibrium condition (6.18) in the freight market can be expressed in rates of growth. This means that in the long-run equilibrium the rate of growth of fleet must be equal to the rate of growth of demand (the first part of equation (6.37)). By substituting deliveries and demolition (equations (6.21) and (6.22)) into equation (6.23) and replacing newbuilding prices with secondhand prices, as in the steady state they are equal, the second part of equation (6.37) is obtained.

It is also clear that in the steady-state equation (6.19) implies that secondhand prices are equal to the difference of the per vessel profit less the interest rate. This can be seen from equation (6.29), which in the steady state implies that

$$PS = \frac{\Pi}{r} \qquad (6.41)$$

Taking logs of (6.41) yields equation (6.40).

Defining the fleet capacity utilisation, cu, as the (log) difference of demand and the fleet (or demand as percent of the fleet), then equation (6.16) can be solved for cu, yielding equation (6.38). Finally, equation (6.39) is simply equation (6.17).

The steady-state system of (6.37)–(6.40) is not really simultaneous. The four equations do not have to be solved simultaneously to obtain the steady-state values of the four endogenous variables. The system is recursive, meaning that it can be solved sequentially if arranged in a particular order. Notice that as Δq, pt and psc are exogenous variables equation (6.37) can be solved for ps, the only endogenous variable in the steady state. Thus

$$ps = a \cdot [\Delta q + \mu_1 \cdot pt + \mu_2 \cdot psc], \quad a = 1/(\mu_1 + \mu_2) \qquad (6.42)$$

Once the secondhand price, ps, is obtained, equation (6.40) can be solved for π. Thus,

$$\pi = a \cdot [\Delta q + \mu_1 \cdot pt + \mu_2 \cdot psc] + r \qquad (6.43)$$

Once π is obtained from (6.43), equation (6.39) can be solved for the freight rate, f. Thus,

$$f = \frac{1}{1+\gamma}\{(a \cdot [\Delta q + \mu_1 \cdot pt + \mu_2 \cdot psc] + r) + \gamma \cdot pb\} \qquad (6.44)$$

$$cu = q - k = \gamma \cdot f - \gamma \cdot pb \qquad (6.45)$$

Finally, substituting the value of freight rates from (6.44) into (6.38) the steady-state value of cu is obtained.

Equation (6.42) implies that in this uniquely defined steady-state vessel prices (newbuilding and secondhand ones, as the two are equal in the long run) are shaped by purely exogenous factors, namely the rate of growth of demand for shipping services, the price of steel and the price of scrap. An increase in any one of these factors would raise vessel prices. This is the level of prices that would make sure that the rate of growth of fleet is equal to the pace of demand. Given this level of prices and the exogenous level of the interest rate, equation (6.40) or (6.43) determines the owners' profit rate for running this fleet. The profit rate is equal to the price of vessels plus the interest rate. Then equation (6.44) determines the level of freight rates that would produce this profit rate. Freight rates, in the steady state, are determined by the profit rate and the price of bunkers. Finally, equation (6.45) defines the fleet capacity utilisation that is consistent with this level of freight rates, given bunker costs.

The BV structure is peculiar as it implies that vessel prices are determined by the optimising behaviour of shipyards and the decision of owners to scrap vessels. Owners and shipyards make sure that for given steel and scrap prices the fleet is growing at the same pace as the demand for shipping services. To a large extent it is the shipyards that make sure that the fleet grows at the same pace as the demand for shipping services, as deliveries are a much larger component of the fleet development than demolition. Therefore, it is mainly the profit-maximising behaviour of shipyards that makes sure that the fleet grows at the same pace as demand. This is counterintuitive as the fleet capacity utilisation is a residual rather than a choice variable. The owners do not have an optimal fleet and place sufficient orders to make sure that the profit rate is equal to prices plus the interest rate. The demand for fleet is perfectly elastic to this arbitrage relationship, which follows from treating ships as a purely financial investment. The shipping market attracts a sufficient number of investors to make sure that the return on shipping is equal to that on alternative assets.

This is a counterintuitive structure because in Chapter 3 it is shown that the owners would decide on the optimal fleet level, given the existing level of capacity utilisation (the level of demand as a percentage of the existing fleet) and their expectations of future fleet capacity utilisation. The optimal fleet would also depend on current and expected values of freight rates, ship prices, and bunker costs. In this framework all endogenous variables, fleet, freight rates, newbuilding prices and secondhand prices are determined simultaneously and not recursively.

The recursive nature of the BV model is due to a number of factors. The first, which is acknowledged by Beenstock and Vergottis, is the assumption that the demand for shipping services is perfectly inelastic to freight rates. As is shown in Chapter 2 charterers are not price takers in the freight market. Instead, they bargain over the freight rate with owners, where each party's bargaining strength depends on actual and expected market conditions (cu_t and $E_t cu_{t+1}$). This would destroy the recursive nature of the BV model and make it simultaneous. The second factor that makes the BV structure recursive is the arbitrage equation that equates returns on shipping and alternative assets. This means that the demand for ships by owners simply follows from the demand for shipping services. The demand for ships is perfectly elastic to the level of prices which ensures that prices are such as to make the fleet grow with the exogenously given level of demand. In reality, an excess demand for shipping services induces owners to place orders for ships if they believe this would be profitable. Thus, an excess demand for shipping services would not necessarily spill over to an excess demand for ships. As we have seen in Chapter 3, this depends on the degree of uncertainty. With a high degree of uncertainty it is optimal for owners to wait before they invest to expand the fleet. But even when the extra demand for shipping services translates into a corresponding demand for ships, the fleet may not grow as fast as demand. Apart from the delivery lag, the pace at which shipyards would meet this demand depends on their existing capacity. An excess capacity would force profit-maximising shipyards to compete for orders and this would act as a drag to higher newbuilding prices. On the other hand, lack of spare capacity would lead to fast increases in newbuilding prices. The third factor that makes the BV model recursive is the hypothesis that secondhand and new ships are perfect substitutes, except for age; and therefore efficient futures markets would force equality of secondhand prices with newbuilding ones. As we have seen in Chapter 3, because of uncertainty secondhand prices may be at a discount or a premium over newbuilding prices even when allowance is made for the degree of depreciation. For example, if owners are uncertain as to whether the increase in demand for shipping services is permanent, they would opt for secondhand vessels instead of new ones, because they can take immediate advantage of the transient increase in demand and can commit a smaller amount of capital for fleet expansion. This extra demand for secondhand ships would increase their prices creating a premium over newbuilding prices, and vice versa. The fourth factor that makes the BV model recursive is the assumption that the return on shipping depends on current rather than expected profit. This is apparent by comparing equation (6.29) with (6.19). This simplifying assumption is responsible for the recursive nature of the model. Dependence on expected profit does not allow the freight market to be isolated from the rest of the system, namely freight rates to be determined independently of developments in the shipyard, secondhand and scrap markets.

2.3 THE DYNAMIC ADJUSTMENT IN THE BV MODEL

Beenstock and Vergottis work out the dynamic adjustment of their system in both discrete and continuous time. As the two methods are equivalent we present here only the continuous time analysis. In continuous time the Δ operator is replaced with the time derivative denoted by a dot over a variable. Thus, the continuous time equivalent of the discrete time operation $\Delta X_t = X_{t-1}$ is simply $\frac{d}{dt}(X)$ or $\dot{X}$.

The continuous time version of the BV model can be described by the following four equations, which determine secondhand pieces, fleet capacity utilisation, freight rates and profits.

$$cu = \gamma \cdot f - \gamma \cdot pb \qquad (6.46)$$

$$\pi = (1 + \gamma) \cdot f - \gamma \cdot pb \qquad (6.47)$$

$$ps = a_1 \cdot \pi + a_2 \cdot \dot{ps} - a_3 \cdot r \qquad (6.48)$$

$$\dot{k} = \mu \cdot ps - \mu_1 \cdot pt - \mu_2 \cdot psc, \quad \mu = \mu_1 + \mu_2 > 0 \qquad (6.49)$$

In deriving the laws of motion to the long-run equilibrium use is made of the property that in the steady state secondhand and newbuilding prices are equal. The supply of shipping services, equation (16), has been expressed in the form of (6.46) making use of the definition of fleet capacity utilisation ($cu = q - k$), as in equation (6.38). Equation (6.47) is the same as equation (6.17). Equation (6.48) is the same as (6.19), which is the log-linear version of (6.29), but with the assumption that next period's expected profitability is equal to that of today. This assumption is analytically necessary to keep the laws of motion to just two equations. Without this assumption the dynamics would have been characterised with three rather than two equations. But this assumption, as we have seen, is responsible for the recursive nature of the BV model. Equation (6.49) is simply the continuous time equivalent of (6.37).

In the BV framework the four equations (6.46)–(6.49) can be reduced to just two equations that describe the dynamic evolution to the long-run equilibrium in terms of cu and ps. This is done by eliminating freight rates and profits from the system. Solve equation (46) for f and substitute into (6.47). The resulting equation for π is then substituted into equation (6.48). Thus

$$ps = a_1 \cdot \delta \cdot cu + a_1 \cdot pb + a_2 \cdot \dot{ps} - a_3 \cdot r, \quad \delta = (1 + \gamma)/\gamma > 0 \qquad (6.50)$$

Subtracting the rate of growth of demand from both sides of (6.49) and then solving for the time derivative of the fleet capacity utilisation, yields equation (6.51).

$$\dot{cu} = \dot{q} - \dot{k} = \dot{q} - \mu \cdot ps + \mu_1 \cdot pt + \mu_2 \cdot psc \qquad (6.51)$$

Equations (6.50) and (6.51) form a system of two first-order differential equations that describe the dynamic adjustment of secondhand prices and fleet capacity utilisation to their long-run equilibrium values. These are obtained from (6.50) and (6.51) by setting the time derivatives equal to zero.

$$\dot{cu} = 0 \Rightarrow \dot{q} - \mu \cdot ps + \mu_1 \cdot pt + \mu_2 \cdot psc = 0 \qquad (6.52)$$

Thus, the long-run equilibrium value of secondhand prices, ps^*, is obtained by solving (6.52) with respect to ps.

$$ps^* = \frac{1}{\mu} \cdot [\dot{q} + \mu_1 \cdot pt + \mu_2 \cdot psc] \qquad (6.53)$$

In the (cu, ps) space equation (6.53) is represented as a vertical line passing through ps^* as secondhand prices do not vary with cu. This is a representation of the locus along which the time derivative of fleet capacity utilisation is zero, namely where the fleet capacity utilisation is constant.

The long-run equilibrium value of fleet capacity utilisation, cu^*, is obtained from equation (6.50) by setting the time derivative of secondhand prices equal to zero and substituting the long-run equilibrium value of secondhand prices, ps^*.

$$\dot{ps} = 0 \Rightarrow cu^* = \frac{1}{a_1 \cdot \delta} \cdot [ps^* - a_1 \cdot pb + a_3 \cdot r] \qquad (6.54)$$

This is an upward-sloping line in the (cu, ps) space and shifts downward for an increase in bunker costs and upwards for an increase in the interest rate. The two equations are plotted in Figure 6.4 and their intersection (point A) defines the long-run equilibrium values of secondhand prices and fleet capacity utilisation given by equations (6.53) and (6.54).

The vertical curve $\dot{cu} = 0$ is responsible for the recursive nature of the model. In the long-run equilibrium secondhand prices are shaped exclusively in the shipyard and scrap markets; there is no interaction with the freight market. The common-sense idea that owners decide on the size of the fleet by projecting what would happen to freight rates in the long run is lost in the BV model. This enables the reduction of the system of four equations (6.46)–(6.49) into two equations (6.50) and (6.51).

To work out the dynamic adjustment to the long-run equilibrium consider the reaction of the fleet capacity utilisation when SH prices are not equal to their long-run equilibrium value, ps^*, namely to points on the right or the left of the $\dot{cu} = 0$ locus. When SH prices are higher than ps^* then the fleet capacity utilisation is decreasing and vice versa.[5] For example, when SH prices are higher than ps^*, such as at point C, then the fleet capacity utilisation is decreasing; notice the downward direction of the arrows at points on the right of the $\dot{cu} = 0$ locus in

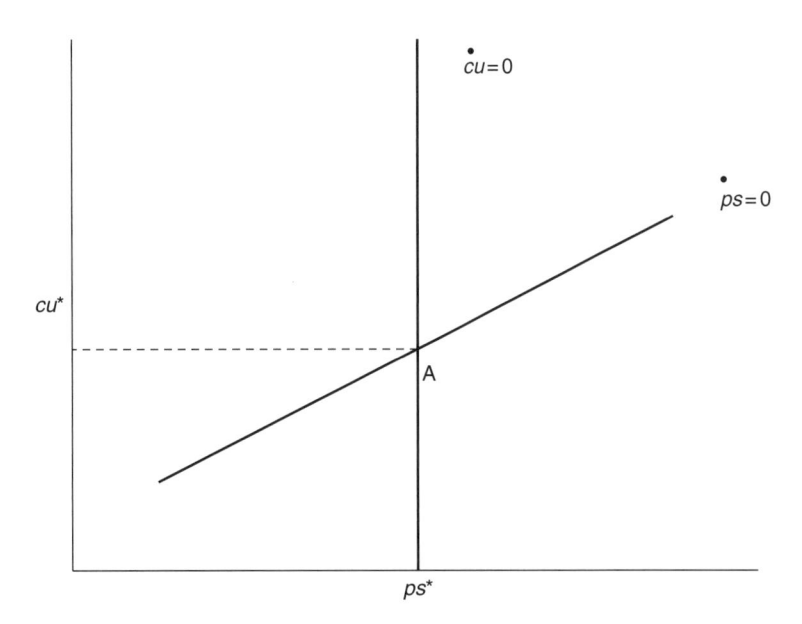

Figure 6.4 Long-run equilibrium in the BV model

Figure 6.5. Similarly, when prices are lower than ps^*, such as point B, then the fleet capacity utilisation is increasing; notice the upward direction of the arrows to the left of the $\dot{cu} = 0$ locus Figure 6.5.

Next consider the dynamic adjustment of SH prices for a given level of the fleet capacity utilisation when the system is at points that lie on the left or the right of the $\dot{ps} = 0$ locus (see Figure 6.6).

It is obvious from equation (6.54) that at points on the right-hand side of the $\dot{ps} = 0$ locus, such as E in Figure 6.6, SH prices are increasing for a given value

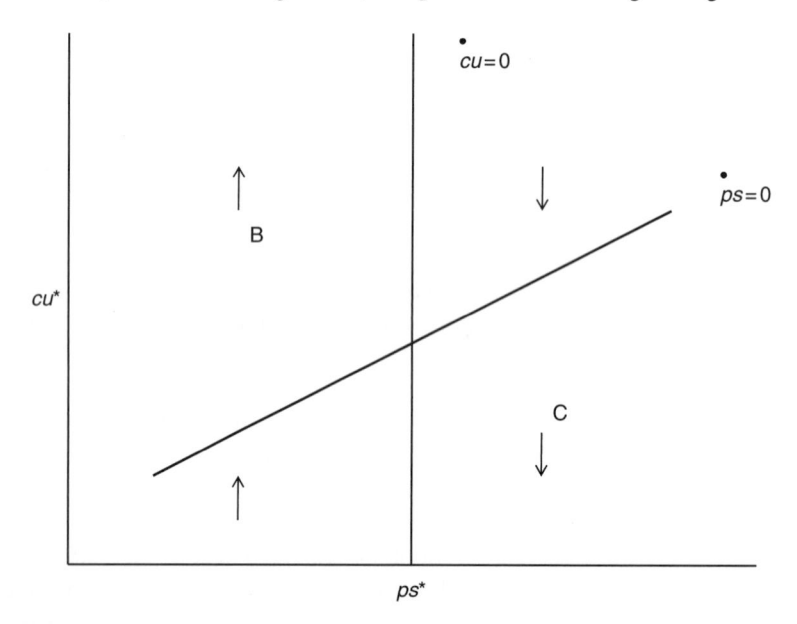

Figure 6.5 Adjustment of capacity utilisation

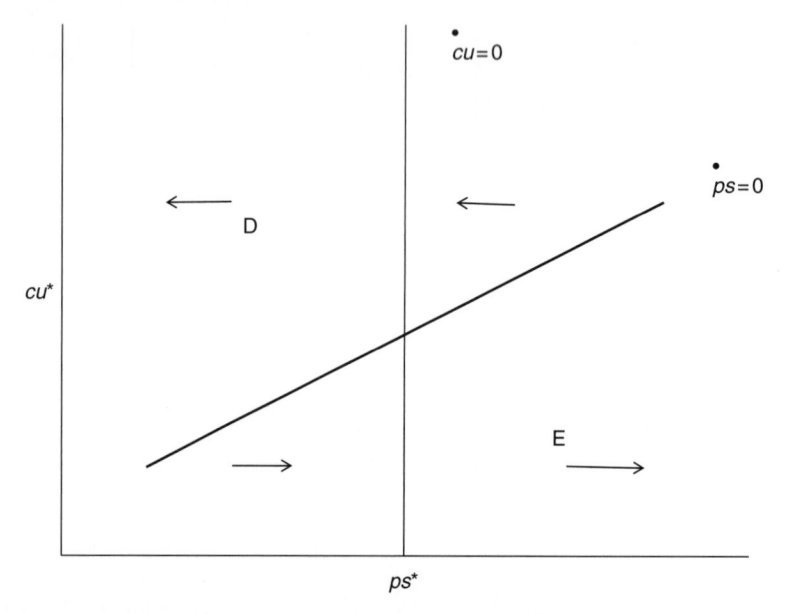

Figure 6.6 Adjustment of SH prices

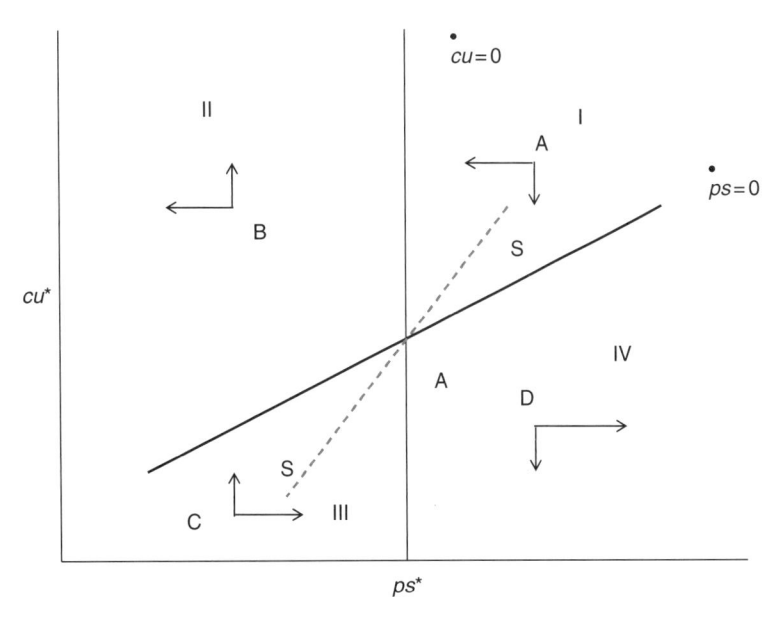

Figure 6.7 Adjustment to long-run equilibrium

of fleet capacity utilisation. Notice that the rightward direction of the arrows at points on the right-hand side of the $\dot{ps} = 0$ imply instability, as any perturbation from equilibrium moves the system further away from it. Owners buy ships when prices are increasing. This destabilising behaviour of owners was also observed in Chapter 3, but with the only difference that there the signal was provided by newbuilding prices rather than secondhand prices. Similarly, SH prices are decreasing at points on the left-hand side of the $\dot{ps} = 0$ locus, such as point D in Figure 6.6. The leftward direction of the arrows again implies instability, as any perturbation from equilibrium that implies lower SH prices than consistent with equilibrium imply even lower prices; owners are selling ships when prices are falling.

Figure 6.7 combines the dynamic adjustment of the fleet capacity utilisation and SH prices in all quadrants formed by the two lines. The system is unstable in points like B and D, which lie in quadrants II and IV, but it is stable in points such as A or C, which lie in quadrants I and III. The instability in quadrants II and IV is associated with bubbles. Convergence to the long-run equilibrium A is achieved along the unique saddle path SS.

It can be seen from equation (6.53) that a permanent increase in the growth rate of demand would shift the $\dot{cu} = 0$ locus to the right, but would leave the $\dot{ps} = 0$ locus in the same position (see equation (6.54)). Figure 6.8 portrays the dynamic adjustment of SH prices and the fleet capacity utilisation rate in response to a permanent increase in the rate of growth of demand.[6] Initial long-run equilibrium is at A with SH prices at ps^* and fleet capacity utilisation at cu^*. Now assume that the rate of growth of demand increases from a to b and this shifts the $\dot{cu} > 0$ locus to the right. Final equilibrium would be attained at B with higher fleet capacity utilisation at cu^{**} and higher SH prices at ps^{**}. The dynamic adjustment of SH prices involves an initial jump from A to C at the initial fleet capacity utilisation

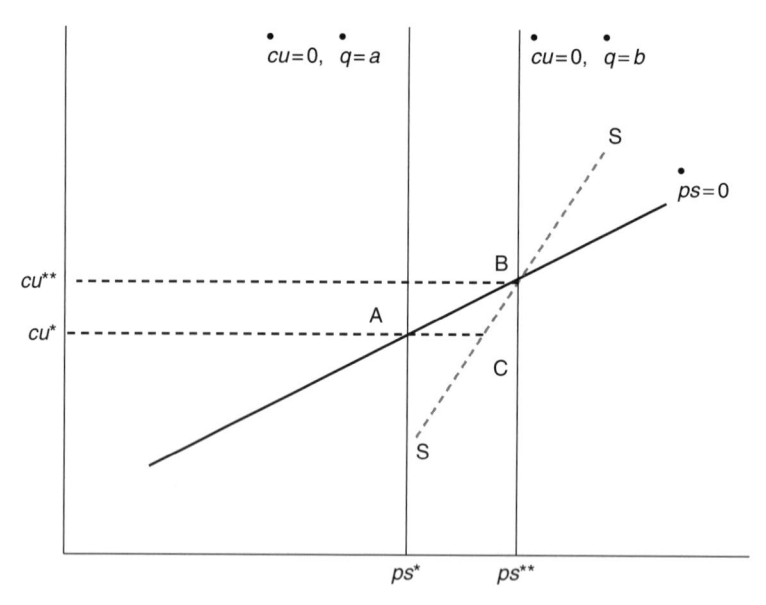

Figure 6.8 Adjustment for a permanent increase in demand

rate cu^*. Therefore, before conditions in the demand–supply balance in the freight market had a chance to change, anticipatory behaviour leads to higher SH prices. The jump in SH prices is necessary for the system to reach the new unique saddle path to the new equilibrium B. Notice that the increase in demand also shifts the saddle path to the right. The new saddle path SS passes through the new equilibrium at B. After point C is reached, both SH prices and the fleet capacity utilisation rate increase gradually to their new long-run values. It can be seen from equation (6.46) that this implies a gradual increase in freight rates in line with the improvement in fleet capacity utilisation. The profit rate also increases gradually in line with higher freight rates (see equation 6.47). Therefore, the BV model explains the cyclical adjustment of prices, fleet capacity utilisation, freight rates and profit rate in response to a permanent increase in demand. However, there is a drawback in the detailed dynamic adjustment. The only asset variable (jump variable) is vessel prices. Freight rates do not jump; rather they increase gradually as the fleet capacity utilisation adjusts. In our model, freight rates are also an asset variable. Freight rates jump in line with prices because it is not only the current fleet capacity utilisation that affects freight rates, but also expectations of the future fleet capacity utilisation. Moreover, the profit rate in the BV model also adjusts gradually in line with freight rates. In our model, the profit rate also jumps. Notice that the profit rate in the BV model would have jumped if it wasn't for the assumption that future profits are equal to current ones.

3 AN INTEGRATED MODEL OF BUSINESS AND SHIPPING CYCLES

We have followed a painstaking path in reaching the point where we can put together a complete shipping model capable of explaining shipping cycles in a way that overcomes the problems encountered in the studies reviewed so far. This is a model that integrates the main features of other shipping cycle models. In particular, it integrates the Tinbergen–Koopmans model of supply-led shipping cycles with the Beenstock–Vergottis model of expectations-driven shipping cycles, which is analysed, however, under the hypothesis that shipping markets are efficient. As the empirical evidence of shipping markets is that they are inefficient in the short run, but asymptotically efficient in the long run, the integrated model breaks away from the Beenstock–Vergottis model of assuming that shipping markets are inefficient. The implication is that the arbitrage conditions between newbuilding and secondhand prices, and between the return of shipping and alternative assets, are removed. Instead, demand and supply factors in newbuilding and secondhand markets are allowed to interact in the determination of prices. This has the implication that all shipping markets interact with each other. As a result, a fleet capacity expansion strategy involves expectations of future freight rates, newbuilding, secondhand and scrap prices and the net fleet, which are jointly determined.

The integrated model also includes the business cycle model developed in the previous chapter. In the Beenstock–Vergottis model expectations are rational and drive the dynamics of the shipping model, along with the fleet accumulation dynamics, but the demand for dry is exogenous to the model. In the integrated model the demand for dry is endogenous. The implication of extending the model to cover the interactions between business and shipping cycles does not simply provide a more realistic explanation of this interaction. It also explains how expectations in shipping are formulated by economic policy; and, in particular, by how central banks react to economic conditions. As central banks choose their policies with the view of achieving their statutory targets, this provides a consistent explanation of how expectations in shipping are formed.

The integrated model consists of 11 equations, where all symbols have their usual meaning with the exception of the newbuilding price, which in this chapter is denoted by PN instead of the usual P; in line with Chapter 5 P denotes the Consumer Price index and p the inflation rate; lower-case letters denote the natural logarithms of the underlying variables. The integrated system is:

$$fr_t = fr^e + b_1 E_t(fr_{t+1}) + b_2 E_t(cu_{t+1}) + b_3 E_t(z_{t+1}) \qquad (6.56)$$

$$0 < b_1 = \frac{1}{1+r} < 1, \quad b_2, b_3 > 0$$

$$k^d = A_1 + q + \sigma[fr - uc] \quad UC_t = (r_t + \delta) \cdot PN_t + G_K(I_t, K_t) - [E_t(PS_{t+1}) - PS_t] \quad (6.57)$$

$$k^s = c_1\, pn + c_2\, px \quad c_1 > 0, \quad c_2 < 0 \qquad (6.58)$$

$$k^d = k^s = k \qquad (6.59)$$

$$ktr^d = A_2 + q + \sigma[fr - ucs]$$
$$UCS_t = (r_t + \delta) \cdot PS_t + G_K(I_t, K_t) - [E_t(PSC_{t+1}) - PSC_t] \qquad (6.60)$$

$$ktr^s = f_1 A + f_2[s_t - E_{t-1} s_t] + f_3 E_t(q_{t+1}) + f_4 E_t[fr_{t+1} - ucs_{t+1}] + f_5\, ps_t \quad (6.61)$$
$$f_1 > 0, f_2 < 0, f_3 < 0, f_4 < 0, f_5 > 0$$

$$ktr^d = ktr^s = ktr \quad (6.62)$$

$$x_t = E_t(x_{t+1}) - \phi[r_t - E_t(p_{t+1})] \quad \phi = 1/\sigma \quad (6.63)$$

$$p_t = \beta\, E_t(p_{t+1}) + \kappa\, x_t + u_{pt} \quad (6.64)$$

$$r_t = r^n + E_t(p_{t+1}) + \gamma_p(p_{t-1} - p^T) + \gamma_x x \quad (6.65)$$

$$q = \tau\, y = \tau(x + \bar{y}) \quad \tau > 1 \quad (6.66)$$

Equation (6.56) is the game-bargaining approach to determining freight rates, which is equation (2.7) in Chapter 2 with the only difference that the equilibrium freight rate is explicitly introduced, whereas in Chapter 2 the freight rate is expressed in terms of deviation from equilibrium. Equation (6.56) shows that in bargaining over the current freight rate, charterers and owners take into account current as well as expected developments in freight rates, the fleet capacity utilisation and exogenous variables included in the vector z, such as bunker costs and port congestion. This specification of the freight rate equation destroys the recursive nature of the BV model. The freight market cannot be isolated from the rest of the shipping model; all shipping variables are simultaneously determined.

Equations (6.57)–(6.59) describe the newbuilding market. The demand for new vessels is given by equation (6.57), which is a log-linear version of equation (3.27) in Chapter 3. The demand for new vessels is derived from the first-order condition for long-run maximum profits. Owners decide on the optimal fleet and the speed at which to arrive from the current fleet to the optimal level. The first-order condition for long-run profit maximisation gives rise to a present value rule found in empirical work on models of newbuilding prices, but with one important difference. The present value rule in the integrated model is based on an explicit model founded on economic theory, whereas the present value rule in empirical work is without any theory. In these models the present value rule is derived from the definition of profits, as cash flows from the operation of the fleet and the definition of capital gains, as in equation (6.29). There is no model to explain the determinants of operational profits and expected capital gains. Moreover, whereas in traditional models the present value rule determines the newbuilding price, because shipping markets are assumed to be efficient, in the integrated model this gives rise to the demand price. The newbuilding price is determined by the equilibrium condition that demand should equal supply in the shipyard market.

The demand for new vessels is a function of the demand for shipping services, q; relative prices (the freight rate relative to the user cost of capital), uc, and technological factors, which are suppressed in the constant A_1. The elasticity of the demand for vessels relative to the demand for shipping services has an elasticity of one, whereas with respect to relative prices the elasticity is equal to the elasticity of substitutions between fleet and average fleet speed, σ. Under normal conditions $\sigma < 1$, and this implies that the demand for shipping services is more important than the relative freight rate. An increase in the price of new ships decreases the demand for new vessels in a non-linear way through the user cost of capital. The non-linear dependence is through the interest rate. As the interest rate increases so does the price elasticity of demand. Note that the user cost of capital is the same as equation (3.29) in Chapter 3 with the only difference being that expected capital gains in the secondhand market are expressed here in discrete rather than continuous time.

The supply of new ships is derived from the short-run optimising behaviour of shipyards.[7] It is a log-linear version of equation (3.29) in Chapter 3 for one factor of production. The single factor of production can be viewed as a composite of the shipyard cost of steel, labour and equipment with per unit cost denoted by *px*. Supply increases with the price of new ships and decreases with the shipyard cost, *px*. Equilibrium in the newbuilding market requires that the demand should be equal to supply, (6.59). The equilibrium condition determines the net fleet (that is, the accumulation of deliveries less the accumulation of demolition) and the price of newbuilding prices. In deriving the demand for and supply of new vessels in Chapter 3, it is assumed that scrapping is proportional to the existing net fleet. Therefore, in line with Chapter 3 equilibrium in the shipyard market determines the net fleet and the newbuilding price.

At this point it is worth noting the interdependence of all shipping markets in the determination of the net fleet. Deliveries depend on the newbuilding price of ships, where the latter is the equilibrium price in the shipyard market rather than the price determined by a net present value rule, which lacks theoretical foundations. The newbuilding price depends through the user cost of capital on the demand for vessels, which in turn depends on freight rates. The user cost of capital is also a function of the expected capital gains in the secondhand market. Therefore, the net fleet requires the simultaneous equilibrium of all three markets: freight, newbuilding and secondhand.[8] The recursive nature of the BV model is the result of special assumptions regarding the efficiency of shipping markets. As we have seen in Chapter 4, the empirical evidence is not in favour of the assumed efficiency. Hence, the integrated model is a general model that nests other models as special cases.

Equations (6.60)–(6.62) define the secondhand market. The demand for secondhand ships is a negative function of their price. An increase in the secondhand price reduces the demand for secondhand vessels in a non-linear way. The negative effect works via the user cost of capital in the secondhand market, denoted by *ucs*. The latter is affected by the secondhand price through the interest rate. The demand for secondhand vessels is a positive function of the demand for shipping services with unitary elasticity. Relative prices (that is, the freight rate relative to the user cost of capital) affect the demand for secondhand vessels via the elasticity of substitution between fleet and average fleet speed in supplying shipping services, σ. As in most cases $\sigma < 1$, the demand for shipping services is more important than relative prices – a result that is common in the newbuilding and secondhand markets. The demand for secondhand ships (6.60) is a log-linear version of equation (3.28) in Chapter 3.

The supply of secondhand ships is given by equation (6.61), which is a log-linear version of equation (3.30) in Chapter 3. Equilibrium in the secondhand market requires that demand should be equal to supply, equation (6.62). The equilibrium condition determines the traded fleet in the secondhand market and the secondhand price.

Equations (6.63)–(6.66) explain the demand for shipping services, Q, by linking shipping with the world economy. In the real world, modelling of the world economy requires the specification of a macro model for the US, China, Europe and Japan, which account for almost all world trade. Hence, the model of (6.63)–(6.65) can be thought of as a representative macro model for each of these four economic regions. This macro model, for simplicity, is the basic NCM model analysed in Chapter 5. Equation (6.63) explains the output gap, x, the deviation of real GDP from its potential, which is equation (5.5) in Chapter 5. In the macro model σ denotes the elasticity of substitution between current and future consumption. In shipping σ denotes the elasticity of substitution between the fleet and average fleet speed in providing shipping services. We have retained σ in shipping, while we have termed the inverse of the macro elasticity as φ to avoid notational confusion.

Equation (6.64) explains inflation in the macro-economy, which is equation (5.7) in Chapter 5. The output gap and inflation are the two targets of monetary policy. Equation (6.65) shows how the central bank sets nominal interest rates. Equation (6.65) is a simplified form of (5.34) in Chapter 5. In this section we abstain from 'interest rate-smoothing', which is pursued by major central banks in the real world, as they want to avoid stop-go-policies that would confuse financial markets and other economic agents (for example, employers and employees in bargaining over wages and employment) as to what the central bank is trying to achieve. Accordingly, equation (6.65) is obtained by setting $c_0 = 0$ in (5.34) in Chapter 5.

Equation (6.66) relates the demand for shipping services to real GDP, which acts as a proxy for world trade. For simplicity, it is assumed that the demand for shipping services is a constant multiple of real GDP, $\tau > 1$. In the real world τ is time varying, which depends on the degree of spare capacity in each of the four major economic regions and the degree of synchronisation of their business cycles. As the output gap is the deviation of real GDP from potential, the demand for shipping services is a multiple of the rate of growth of potential output and the output gap, which is treated as a fixed constant. With this specification even when the output gap is zero, the demand for shipping services is increasing at the rate of potential output. When the economy becomes overheated (positive output gap), the demand for shipping services increases more than proportionately because $\tau > 1$. When the economy operates with spare capacity or is in recession (negative output gap), the demand for shipping services falls proportionately more than the economy's output. Therefore, the amplitude of shipping cycles is bigger than that of the world economy. This is a stylised fact as we shall see in Part III Ch. 9.

The system of 11 equations (6.56)–(6.66) determines the following nine variables: freight rates in the spot market; the newbuilding price and the net fleet (deliveries less scrapping and losses); the secondhand price and the volume of traded ships; the demand for shipping services; the output gap; inflation; and the nominal interest rate. The freight period market, analysed in Chapter 2, is obtained as a residual (recursively) after the system of equations (6.56)–(6.66) has been solved. Hence, the freight period market does not impact directly on the rest of the system, but the other way round. In this sense it is determined as a residual (recursively).

4 THE PROPERTIES OF THE INTEGRATED MODEL

In working out the dynamic properties of the integrated system, it is convenient to solve first the macro model, which is an input to the shipping model. The macro environment affects shipping through the demand for shipping services and by shaping expectations of key variables in the shipping market. These links are highlighted through the use of the following equations.

The inflation equation (6.64) is a first-order difference equation. Ignoring for simplicity the exogenous shocks, the solution of this difference equation is given by equation (6.67), which is equation (5.25) in Chapter 5.

$$p_t = E_t \sum_{i=0}^{\infty} \beta^i (\kappa\, x_{t+i}) \qquad (6.67)$$

Therefore, as is shown in Chapter 5, inflation depends on the expected future path of the output gap.

The output gap equation (6.63) is also a first-order difference equation. The solution is given by equation (6.68), which is equation (5.24) in Chapter 5.

$$x_t = E_t \sum_{i=0}^{\infty} \left[-\frac{1}{\phi}(r_{t+i} - p_{t+i+1}) \right] \qquad (6.68)$$

Therefore, the output gap depends on the expected future path of real interest rates, as shown in Chapter 5. By combining the last two equations it follows that central banks by controlling nominal interest rates affect real interest rates, which in turn shape the future path of the output gap; through the latter central banks control inflation. Therefore, by observing current inflation and the output gap and knowing the central bank's targets one can deduce the future path of nominal interest rates. These results are captured in Propositions 1 and 2 in Chapter 5.

It is shown in Chapter 2 that the spot freight rate equation (2.9), reproduced here as equation (6.56), has a solution of the following form (ignoring for simplicity the exogenous shocks):

$$fr_t = \frac{1}{1-b_1} fr^e + E_t \left\{ b_2 \sum_{j=0}^{\infty} b_1^j cu_{t+j} + b_3 \sum_{j=0}^{\infty} b_1^j z_{t+j} \right\} \qquad (6.69)$$

Equation (6.69) shows that in bargaining over the current freight rate, charterers form expectations of current and future economic fundamentals. These expectations relate to the fleet capacity utilisation and exogenous variables, such as bunker costs and port congestion. Expectations of fleet capacity utilisation require separate expectations for the demand for shipping services and the fleet. Assuming a shipyard delivery lag of two years and that scrapping is proportional to the existing fleet, the fleet today is equal to expectations of demand formed three years ago, which is equal to the change of the orderbook two years ago. Thus

$$k_t = E_{t-3} K^d_{t-2} = E_{t-3} K^d_{t-2} \qquad (6.70)$$

The demand for shipping services or the change in the orderbook is a multiple of the output gap in the economy. Using equations (6.66) and (6.70), the fleet capacity utilisation is:

$$cu_t = q_t - k_t = \tau(\bar{y} + x_t) - \tau E_{t-3}(\bar{y} + x_{t-2}) \qquad (6.71)$$

Substituting (6.71) into (6.69), the negotiated freight rate becomes:

$$fr_t = \frac{1}{1-b_1} fr^e + E_t \left\{ b_2 \sum_{j=0}^{\infty} b_1^j [\tau(\bar{y} + x_{t+j}) - E_{t-3} \tau(\bar{y} + x_{t-2+j})] \right. \\ \left. + b_3 \sum_{j=0}^{\infty} b_1^j z_{t+j} \right\} \qquad (6.72)$$

Finally, using equation (6.68) and suppressing unnecessary terms the negotiated freight rate is given be the following equation:

$$fr_t = E_t \left\{ \sum_{j=0}^{\infty} v^j [(r_{t+j} - E_t p_{t+j+1}) + E_{t-3}(r_{t+j-2} - E_{t-2} p_{t+j-1})] \right. \\ \left. + b_3 \sum_{j=0}^{\infty} b_1^j z_{t+j} \right\} \qquad v < 1 \ (6.73)$$

The last equation enables the formulation of the following proposition.

Proposition 1: the bargaining of charterers and owners over the current freight rate depends mainly on current and past expectations of future real interest rates and, consequently, on the future conduct of monetary policy. Expectations of higher real interest rates (tightening of monetary policy) imply lower demand for dry and, consequently, lower freight rates at present; and vice versa. Therefore, by observing current inflation and the output gap and knowing the central bank's targets one can deduce the future path of nominal interest rates.

In working out the properties of the newbuilding and secondhand prices it is convenient to assume a specific function for the costs of adjusting the fleet, equation (A4) in the Appendix of Chapter 3. The requirement of (A4) is that costs of adjustment are convex. The convexity assumption implies that the cost is increasing at an accelerating pace (first- and second-order derivatives with respect to net investment are positive). When net investment is broken down to gross investment, I, less replacement investment due to depreciation, the cost is increasing at an accelerating pace with respect to gross investment, but decreases with respect to the fleet (as the fleet increases, it creates economies of scale). These requirements are fulfilled by a quadratic function. Thus, the cost of adjusting the fleet is assumed to be given by

$$G(I - \delta K) = 1/2[I - \delta K]^2 \quad G_I = I - \delta K > 0, \\ G_{II} = 1, \quad G_K = -\delta[I - \delta K] < 0, \quad G_{KK} = \delta^2 > 0 \tag{6.74}$$

It is worth recalling from Chapter 3 that the demand for and supply of new vessels are derived from profit optimisation. The demand function for new vessels is obtained from the first-order condition for maximum profits. This is presented here as

$$\Delta PN_t = (r_t + \delta) \cdot PN_t - a / A^p FR_t CU^{(1/\sigma)} - \delta(I_t - \delta K_{t-1}) \tag{6.75}$$

The above equation is the discrete time version of equation A.10 in the Appendix of Chapter 3. It arises by replacing the marginal product of fleet, F_k, by equation (A.19), Q/K by the fleet capacity utilisation and G_K by the third expression in (6.74). This equation drives the dynamics of the demand price. The demand function is obtained by setting (6.75) equal to zero and solving for pn.

The dynamics of the supply side are described by the evolution of the net fleet:

$$\Delta K_t = I_t - \delta K_{t-1} \tag{6.76}$$

In the real world, owners pay in three instalments the full price of a ship according to the progress made towards completion. Hence, gross investment expenditure is proportional to the progress made towards completion. If we assume that shipyard production is smooth, then gross investment expenditure should be approximately equal to the shipyard deliveries – the supply of vessels. Inserting (6.58) into (6.76), the net fleet dynamics are obtained:

$$\Delta K_t = c_1 PN + c_2 PX - \delta K_{t-1} \tag{6.77}$$

Equations (6.75) and (6.77) form a system of two simultaneous equations. This system drives the dynamics of the equilibrium newbuilding price (not the demand price) and the net fleet, for a given freight rate. As the freight rate is an endogenous variable, the proper system should include the freight rate equation (6.56). As it is more difficult to work out an analytic solution for a three-equation system, the dynamics of the newbuilding price and the net fleet are analysed for a given freight rate. The solution for the newbuilding price is:

$$PN_t + (I_t - \delta K_{t-1}) = E_t \sum_{s=t}^{T} \frac{1}{(1 + r_s + \delta)^{s-t+1}} \\ \cdot \left(\frac{a}{A^p} FR_{s-t+1} \cdot CU^{1/\sigma}_{s-t+1} + \delta(I_{s-t+1} - \delta K_{s-t+1}) \right) \tag{6.78}$$

The second term on the left-hand side is the cost of adjusting the fleet and the last term on the right-hand side are the economies of scale produced by adding one more vessel to the owner's fleet. According to equation (6.78), the marginal cost of one extra ship, which includes the price and the cost of adjusting the fleet, is equal to the present value of the expected marginal benefits accruing throughout the entire useful life of the vessel. These benefits depend on expectations of freight rates, fleet capacity utilisation and economies of scale. As the current freight rate and the fleet capacity utilisation depend on current and past expectations of future real interest rates (see Proposition 1), the newbuilding price is mainly a function of the expected conduct of monetary policy. It is worth noting that the dependence on monetary policy is non-linear. Thus, equation (6.78) implies

$$PN_t + (I_t - \delta K_{t-1}) = E_t \mu^j [(r_{t+j} - E_t p_{t+j+1}), \\ E_{t-3}(r_{t+j-2} - E_{t-2} p_{t+j-1}), \\ \delta(I_{s-t+1} - \delta K_{s-t+1})] \quad \mu < 1 \tag{6.79}$$

The last equation enables the formulation of the following proposition.

Proposition 2: The cost of a new vessel (price plus cost of fleet adjustment) depends in a non-linear manner on current and past expectations of future real interest rates and economies of scale. As the latter is not very sensitive to shipping cycles, it follows that in the main expectations in the shipyard market are formed by the future conduct of monetary policy. If monetary policy is expected to be tightened, newbuilding prices are expected to fall, and vice versa.

The dynamic optimisation problem of fleet capacity expansion, outlined in Chapter 3 and its Appendix, gives rise to the demand for ships whether they are newbuilding or secondhand and derives their corresponding demand prices. The demand for secondhand ships is given by equation (6.60). This is a function of the same variables as the demand for new vessels, namely the demand for shipping services and relative prices. But relative prices are computed as the freight rate relative to the user cost of capital in the secondhand market. The latter is a function of secondhand prices. The capital gains are for a ship of older age or the scrap value. In addition, there is no cost of adjustment, or at least the cost is much smaller than a new vessel. A secondhand ship requires an inspection and evaluation of its true value so that the price paid is fair for the owner and the seller. Nonetheless a secondhand ship might still involve economies of scale. Bearing these differences in mind the secondhand demand price is given

$$PS_t = E_t \sum_{s=t}^{T} \frac{1}{(1+r_s+\delta)^{s-t+1}} \cdot \left(\frac{a}{A^p} \cdot FR_{s-t+1} \cdot CU_{s-t+1}^{1/\sigma} + \delta(I_{s-t+1} - \delta K_{s-t+1}) \right) \tag{6.80}$$

Equation (6.80) states that the secondhand demand price should be equal to the marginal profits that accrue from one extra vessel in its entire useful life. Evaluation of the secondhand price involves expectations of the entire path of freight rates, capacity utilisation and economies of scale. As the current freight rate and the fleet capacity utilisation depend on current and past expectations of future real interest rates (see Proposition 1), the secondhand price is mainly a function of the expected conduct of monetary policy. It is worth noting that the dependence on monetary policy is non-linear. Therefore, the conclusion regarding newbuilding prices is also valid for secondhand prices. Accordingly, this enables us to arrive at a new proposition.

Proposition 3: The price of a secondhand vessel depends in a non-linear manner on current and past expectations of future real interest rates and economies of scale. As the latter is not very sensitive to shipping cycles, it follows that in the main expectations in the secondhand market are formed by the future conduct of monetary policy. Expectations of higher real interest rates (tightening of monetary policy) imply lower demand for dry and, consequently, lower freight rates at present; and vice versa.

In conclusion, the time path of all main shipping variables, namely freight rates, fleet capacity utilisation, secondhand and new prices, are determined by expectations of future real interest rates, and, consequently, on the future conduct of monetary policy. Therefore, by observing current inflation and the output gap and knowing the central bank's targets one can deduce the future path of nominal interest rates.

5 THE INTERACTION OF BUSINESS AND SHIPPING CYCLES

In working out the interaction of business and shipping cycles it is useful to relax the simplifying assumption made in the previous section that scrapping is proportional to the existing net fleet. In this section scrapping becomes an economic decision: if a ship is economically viable, it remains in the active fleet. This economic decision is described by equation (3.23) in Chapter 3, where demolition (DM), expressed as a proportion of the existing net fleet is a function of the scrap price relative to the secondhand price and the age of the fleet, A. In this chapter equation (3.23) in Chapter 3 is written as equation (6.81). Taking the total differential of the discrete time, log-linear version of (3.23) in Chapter 3 gives:

$$\Delta dm_t = h_1 \Delta psc_t - h_1 \Delta ps_t \qquad (6.81)$$

The symbols in the above equation have their usual meaning. Lower-case letters are the natural logarithms of the underlying variables. Δ is the first difference operator, which, when applied to the log of a variable, represents the compound percentage change over the previous period. Thus, equation (6.81) states that the rate of growth of fleet demolition is a positive function of the rate of growth of scrap prices (psc) and a negative function of the rate of growth of secondhand prices. The elasticity of demolition with respect to scrap and secondhand prices is equal to h_1, but with opposite signs. A one percent increase in scrap prices, other things being equal, leads to an h_1 percent increase in demolition, while a one percent increase in secondhand prices leads to an h_1 decrease in demolition. The equal but opposite impact of scrap prices and secondhand prices on demolition is the logical implication of owners comparing the price of scrap relative to the secondhand price in deciding whether to scrap a vessel or not.

As a result of this change, the shipyard market in the integrated system of equations (6.57)–(6.59) determines the newbuilding price and the gross fleet, KG, rather than the net fleet, K. The net fleet is determined by an identity that links the shipyard with the demolition market. The change of the net fleet between two consecutive periods is equal to the deliveries that took place in period t, DL, less the demolition, DM, in the same period. The deliveries in period t are equal to the change in gross tonnage, ΔKG. Thus,

$$\Delta K_t = DL_t - DM_t = \Delta KG - DM_t \qquad (6.82)$$

Dividing both sides of (6.82) by K_{t-1} the rate of growth of the net fleet is defined as

$$g_t \equiv \frac{K_t - K_{t-1}}{K_{t-1}} = \frac{DL_t}{K_{t-1}} - \frac{DM_t}{K_{t-1}} = \frac{\Delta KG}{K_{t-1}} - \frac{DM_t}{K_{t-1}} \qquad (6.83)$$

This equation can be restated in logs as

$$\Delta k_t = \Delta kg_t - \Delta m_t \qquad (6.84)$$

The equilibrium relationship, (6.59), in the shipyard market now becomes

$$k^d = k^s = kg \qquad (6.85)$$

Taking the total differential of (6.57) and applying the equilibrium condition (6.85) gives the equation for the rate of growth of gross fleet

$$\Delta kg_t = \Delta q_t + \sigma [\Delta fr_t - \Delta uc_t] \qquad (6.86)$$

The equation for the rate of growth of newbuilding prices is obtained by equating the demand for new vessels, (6.57), with the supply of new vessels, (6.58), solving for the price and taking the total differential

$$\Delta pn_t = \frac{1}{c_1}[\Delta q_t + \sigma(\Delta fr_t - \Delta uc_t) - c_2 \Delta px_t] \quad c_1 > 0, \quad c_2 < 0 \qquad (6.87)$$

The rate of growth of new prices depends positively on the rates of growth of demand for shipping services, freight rate and the cost of building a ship, px, but negatively on the user cost of capital, uc. The output elasticity of price is equal to the inverse of the slope of the supply curve $(1/c_1)$. The elasticity would be 1, if instead of assuming a log-linear supply function we had started with a linear function.[9]

In deriving the equation for secondhand prices it is convenient to simplify the supply of vessels by owners in the secondhand market. In particular, it is assumed that in equilibrium all expectation errors in the supply function, (6.61), are zero and that the supply is a positive function only of the secondhand price. By equating the demand for secondhand vessels, (6.60), with the new simplified supply, solving for *ps* and taking the total differential, the equation for the rate of growth of secondhand prices is derived

$$\Delta ps_t = \frac{1}{f_s}[\Delta q_t + \sigma(\Delta fr_t - \Delta ucs_t)] \tag{6.88}$$

The coefficient f_s measures the supply elasticity with respect to the secondhand price, which is assumed to be positive for simplicity (but see section 3.9 in chapter 3 for a discussion when the elasticity is negative).

The rate of growth of secondhand prices is a positive function of the rates of growth the demand for shipping services and freight rates, but a negative function of the user cost of capital, *ucs*, in the secondhand market. The output elasticity of the secondhand price may also be one (see footnote 9).

With these changes the integrated model consists of the following equations:

$$fr_t = fr^e + b_1 E_t(fr_{t+1}) + b_2 E_t(cu_{t+1}) + b_3 E_t(z_{t+1}) \tag{6.89}$$

$$\Delta pn_t = \frac{1}{c_1}[\Delta q_t + \sigma(\Delta fr_t - \Delta uc_t) - c_2 \Delta px_t] \quad c_1 > 0, \quad c_2 < 0 \tag{6.90}$$

$$\Delta ps_t = \frac{1}{f_s}[\Delta q_t + \sigma(\Delta fr_t - \Delta ucs_t)] \tag{6.91}$$

$$\Delta k = \Delta kg - \Delta m = \Delta q + \sigma[\Delta fr - \Delta uc] - (h_1 \Delta psc_t - h_1 \Delta ps_t) \tag{6.92}$$

$$\Delta cu = \Delta q - \Delta k \tag{6.93}$$

$$UC_t = (r_t + \delta) \cdot PN_t + G_K(I_t, K_t) - [E_t(PS_{t+1}) - PS_t] \tag{6.94}$$

$$UCS_t = (r_t + \delta) \cdot PS_t + G_K(I_t, K_t) - [E_t(PSC_{t+1}) - PSC_t] \tag{6.95}$$

$$x_t = E_t(x_{t+1}) - \phi[r_t - E_t(p_{t+1})] \quad \phi = 1/\sigma \tag{6.96}$$

$$p_t = \beta E_t(p_{t+1}) + \kappa x_t + u_{pt} \tag{6.97}$$

$$r_t = r^n + E_t(p_{t+1}) + \gamma_p(p_{t-1} - p^T) + \gamma_x x \tag{6.98}$$

$$q = \tau y = \tau(x + \overline{y}) \quad \tau > 1 \tag{6.99}$$

This system determines the (spot) freight rate, the rates of growth of newbuilding prices, secondhand prices and the net fleet; the fleet capacity utilisation, the user cost of capital in the newbuilding and secondhand markets, the output gap in the macro-economy, inflation, the nominal interest rate and the demand for shipping services. The system of equations (6.89)–(6.99) is recursive. The macro model can be solved first to determine the demand for shipping services. With demand determined, the shipping model determines the rest of the shipping variables. As we have seen in section 6.4, the importance of the integrated model lies in explaining the demand for shipping services and how expectations in the shipping market are formulated.

It is worth noting that the equations for newbuilding and secondhand prices are non-linear difference equations through the dependence of the corresponding user cost of capital on the level of prices. Hence, it is not only the dimensionality of the integrated model that makes the calculation of an analytic solution very difficult, if not impossible, but also the non-linear nature of the two price equations. Accordingly, in working out the interaction of business with shipping cycles we use simulations of a calibrated model.

In the calibrated model, the demand for shipping services in the long-run equilibrium is growing at twice the pace of potential output growth in the macro-economy, namely $\tau = 2$ in equation (6.99). The rate of growth of potential output is assumed at 3 per cent per annum. Inflation is growing at 2 per cent and the nominal interest rate is 4.5 per cent. The freight rate in equilibrium is $25, expressed in thousands, the newbuilding price is $100, expressed as an index of the prices of the various vessel types, and the secondhand price is $80, again expressed as an index. This assumes a depreciation factor of 4 per cent with the economic life of a vessel at 25 years. Deliveries are growing at 10 per cent, while demolition at 4 per cent. The fleet capacity utilisation is assumed to be at 90 per cent, very near to the long-term historical average. The elasticity of substitution between fleet and average speed is assumed at 0.1. The impact of fleet capacity utilisation on freight rates is assumed at 20, very near to empirical estimates. The output elasticity of new and secondhand prices is assumed to be 1.

It is assumed that both the macro-economy and shipping markets are in long-run equilibrium and that there is an unexpected shock triggering a widening of credit spreads, in the first year, which is assumed to take place in 2007. The widening of credit spreads results in a drop of 2 per cent in the aggregate demand of the economy in 2008. This provides a realistic simulation of the events leading to the last financial crisis of 2007–08 and the resulting 'Great Recession'.

In the first set of simulations it is assumed that the rate of growth of fleet (the supply) is predetermined by previous expectations of demand two years ago (that is, assuming a two-year delivery lag by shipyards). As the economy and shipping markets are in equilibrium prior to the shock expectations held two years earlier (that is, in 2005 and 2006) for demand growth are 6 per cent. Therefore, in this set of simulations it is assumed that the fleet continues to grow at 6 per cent.

The results of this unexpected shock are studied with the help of Figures 6.9–6.12. The drop in aggregate demand of 2 per cent leads to a recession in the economy with a negative output gap of −3.5 per cent in 2008 (see Figure 6.9). As a result of the recession the demand for shipping services drops by 7 per cent; the growth in demand falls from 6 per cent, its equilibrium growth, to −1 per cent (see Figure 6.9). With the economy in recession inflation decreases from 2 per cent to 1.5 per cent in 2008 (see Figure 6.10).

With the economy in recession and inflation falling, the central bank cuts the nominal interest rate from 4.5 per cent to 3 per cent in 2008 (see Figure 6.10). The nominal interest rate cut is sufficient to decrease real interest rates, a necessary

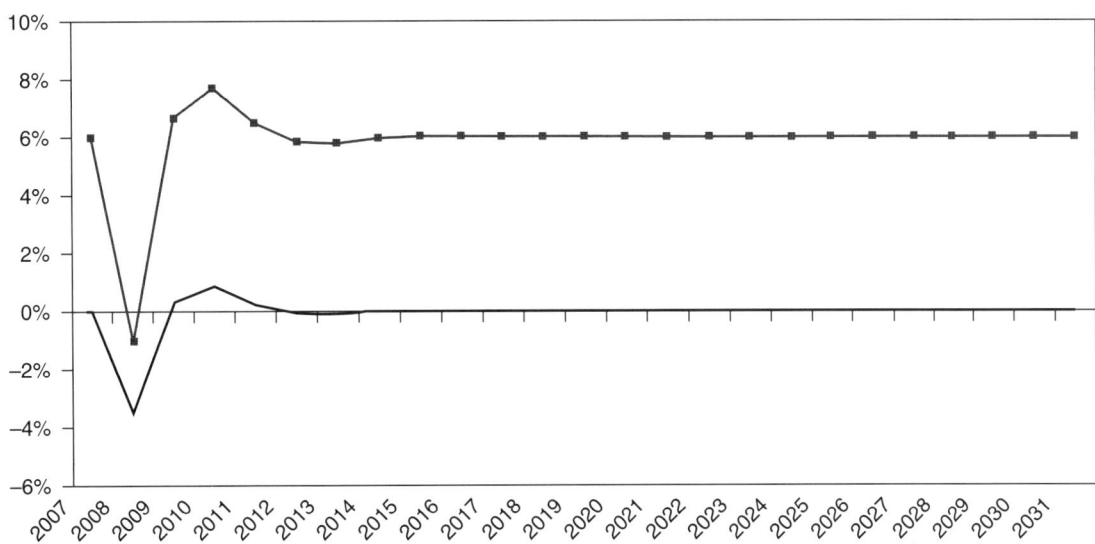

Figure 6.9 Output gap and demand for shipping

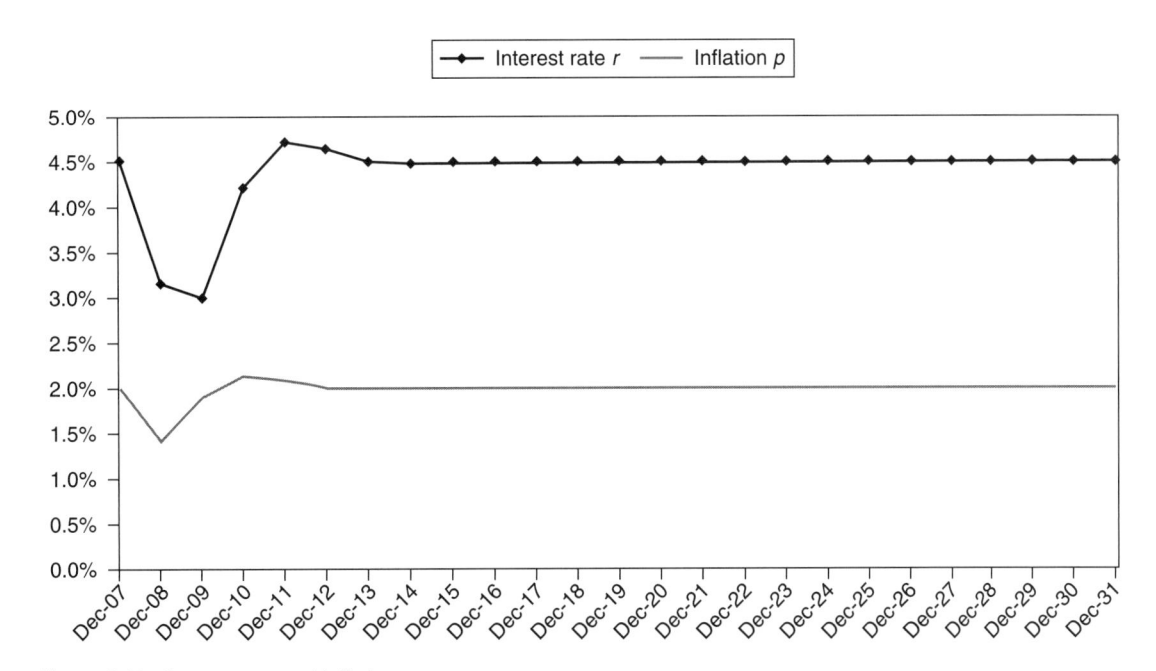

Figure 6.10 Interest rate and inflation

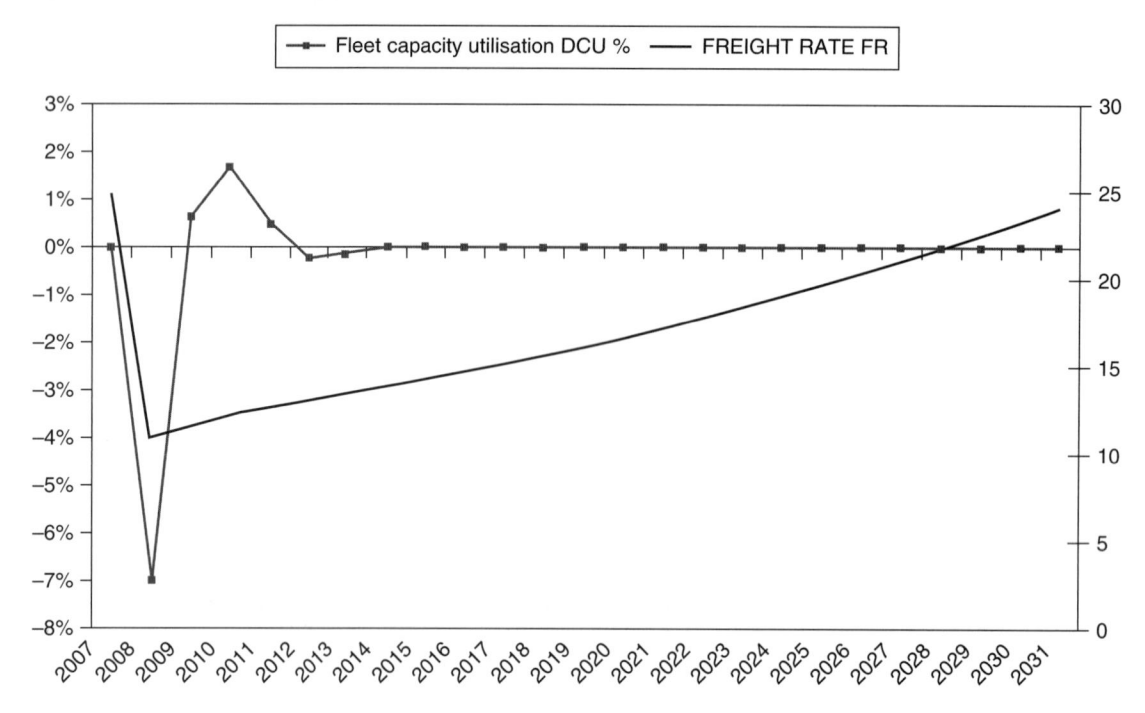

Figure 6.11 Freight rate and fleet capacity utilisation

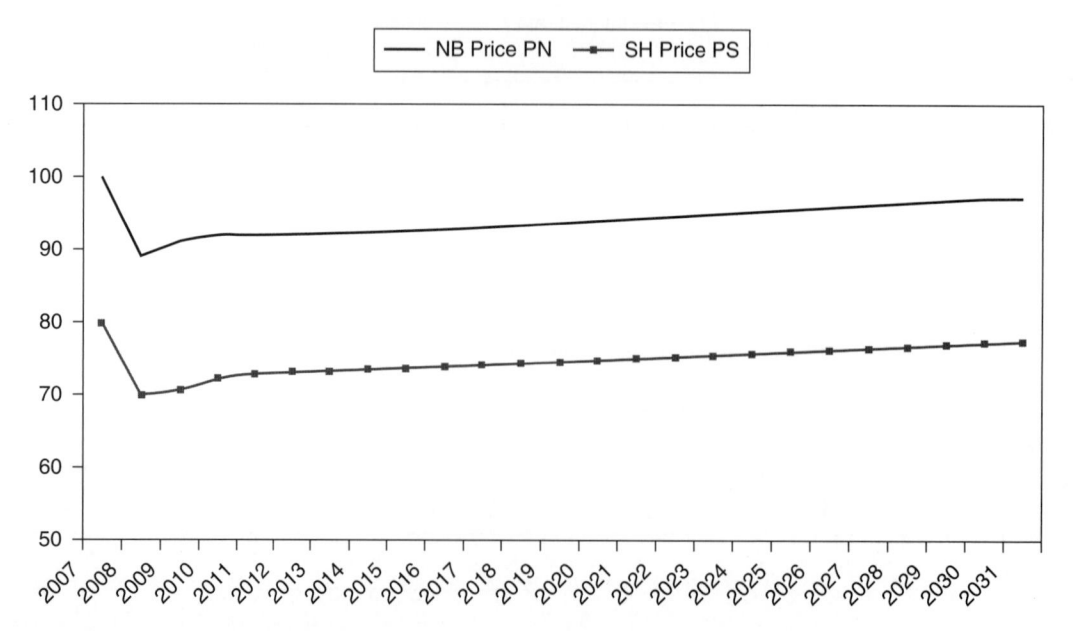

Figure 6.12 NB and SH price adjustment

condition for the effectiveness of monetary policy (see Chapter 5). The drop in demand for shipping services lowers the fleet capacity utilisation rate from its equilibrium rate of 90 per cent to 83.7 per cent in 2008 (see Figure 6.11). As a result, the spot freight rate drops from $25 to $11, equivalent to a fall of −56.4 per cent (see Figure 6.11). Newbuilding prices slide from 100 to 89 in 2008, while secondhand prices diminish from 80 to 70 (see Figure 6.12).

The central bank reaction fosters the economy's stabilising forces back to equilibrium and the economy recovers in 2009 with the output gap becoming positive (see Figure 6.9). The recovery gathers momentum in 2010 with a positive output gap of 0.8 per cent. This ultimately peters out and the economy returns to the initial equilibrium (prior to the shock) in 2012 with zero output gap. The adjustment of the output gap is not a gradual return to equilibrium (monotonic) but it is cyclical, involving a small overshooting of the long-run equilibrium. Therefore, the unexpected shock generates a five-year cycle in the output gap.

The adjustment of the demand for shipping services mimics the output gap. The recovery of the economy in 2009 helps the demand for shipping services to rebound from −1 per cent in 2008 to nearly 8 per cent in 2009. Demand returns to its long-term rate of growth of 6 per cent in 2014. The adjustment path of the demand growth rate also involves an overshooting of the long-run equilibrium. The adjustment of demand mimics the output gap because of the assumption that τ is time invariant; there is no persistency effect in the demand for shipping services; and there is no lag between the output gap and the demand for shipping services – the two are moving together. In the real world, τ is time varying and although demand responds to the output gap more or less at the same time, there is a strong persistency effect that makes the stimulus in demand last a little longer than the output gap. The shock generates a seven-year cycle in the demand for shipping services.

The adjustment of freight rates, newbuilding and secondhand prices back to equilibrium is gradual because of the persistence of adverse expectations. This is a self-fulfilling prophecy, which is enforced through rational expectations because of expected capital losses in selling vessels in the secondhand market or for scrap. This is apparent by inspecting the definition of the user of capital in the newbuilding and secondhand markets, which involve expected capital gains or losses (see equations (6.94) and (6.95)). The adjustment of freight rates, newbuilding and secondhand prices back to equilibrium is extremely lengthy in this rudimentary model. In the real world it is much faster.

To get more insight into the dynamic adjustment path of freight rates, a second set of simulations is carried out in which the fleet is allowed to respond. It is hypothesised that shipyards respond in any given period of time to a proportion of the orders placed by owners. This is captured by introducing the following equation:

$$k_t = \gamma k^* + (1-\gamma)k_{t-1} \quad 0 < \gamma < 1 \quad 0 < \gamma < 1 \quad (6.100)$$

The desired fleet by owners is k^*. But shipyards can only deliver a proportion γ of the desired fleet in a period of time. The proportion γ varies between zero and one. Equation (6.100) nests the previous model when $\gamma = 0$, as the fleet

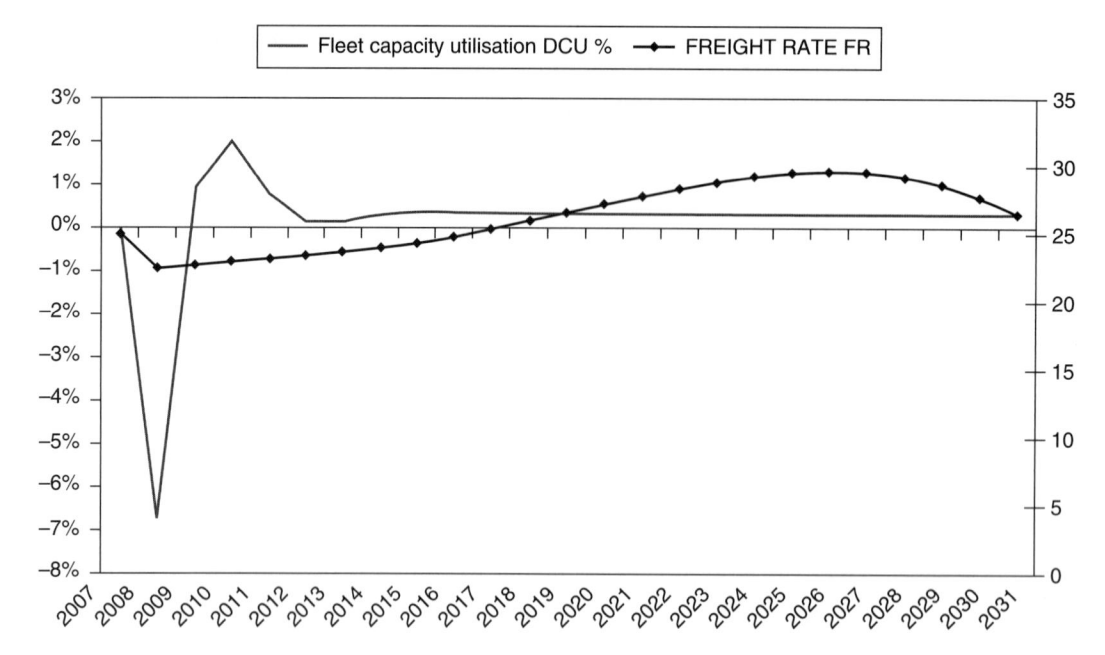

Figure 6.13 Adjustment with small response of freight rates to fleet capacity utilisation

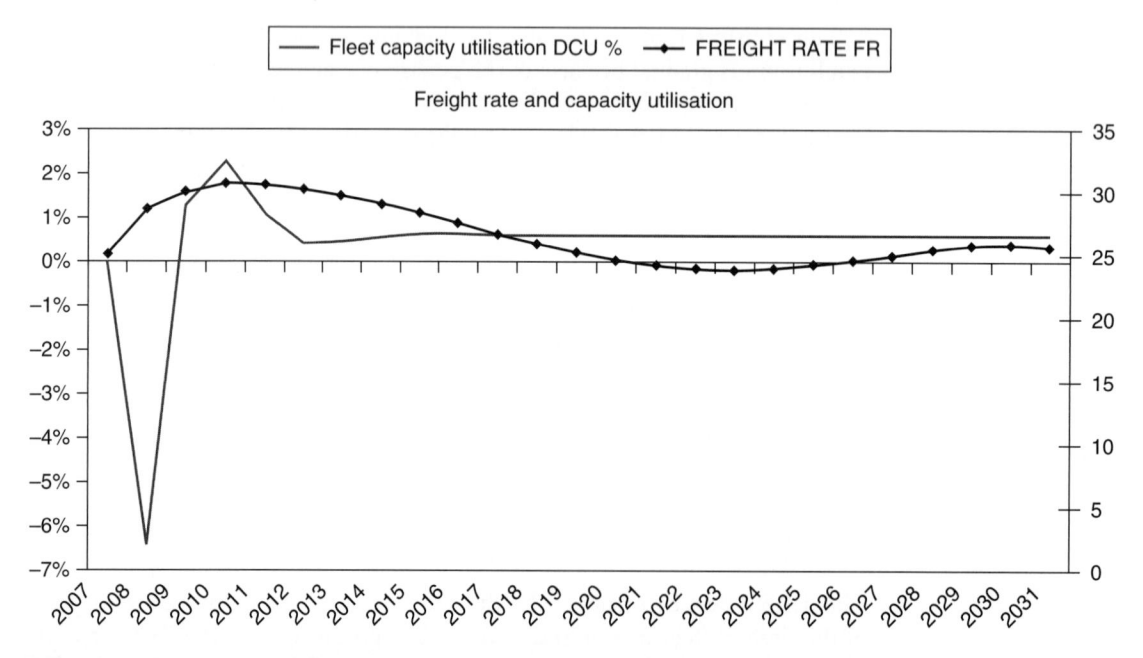

Figure 6.14 Adjustment with large response of freight rates to fleet capacity utilisation

grows at the same pace as the last period. Starting from the long-run equilibrium with fleet growing at 6 per cent and introducing an unexpected shock, the fleet would continue to grow at 6 per cent throughout the adjustment. Therefore, the higher the value of γ, the more flexibility is introduced in the adjustment process. When the unexpected shock occurs owners would adjust the desired fleet down. If $\gamma = 1$, the rate of growth of the fleet would adjust immediately down.

Assuming $\gamma = 0.2$, the dynamic adjustment path of freight rates depends on the elasticity of freight rates to the fleet capacity utilisation rate, the coefficient b_2 in equation (6.56). In the first set of simulations this elasticity is assumed to be 20, very near to empirical estimates. In this set of simulations this elasticity takes two different values, representing a small response ($b_2 = 15$) and a large response ($b_2 = 40$). The results are studied with the help of Figures 6.13 and 6.14.

Figure 6.13 plots the dynamic adjustment path of freight rates with a small response to fleet capacity utilisation. The freight rate falls on impact, but the drop is very small compared to the normal case. In this case the drop is just –12 per cent, whereas in the normal case it is –56 per cent. The adjustment back to the long-run equilibrium takes an equally long time as in the normal case and it is oscillatory rather than monotonic (compare Figures 6.11 and 6.13).

Figure 6.14 plots the dynamic adjustment path of freight rates with a large response to fleet capacity utilisation. The freight rate does not fall even on impact. Due to rational expectations and the large response to future rates of fleet capacity utilisation, freight rates begin to climb immediately because of discounting of future economic fundamentals by owners and charterers when bargaining over current freight rates. This is an extreme case, not likely to be experienced in the real world. Nonetheless, the simulation highlights the importance of discounting future economic fundamentals in freight rate bargaining, which can generate perverse results under extreme assumptions.

6 CONCLUSIONS

All shipping variables exhibit cyclical fluctuations around the long-run equilibrium in response to unexpected shocks in the economy. Demand shocks in the economy, such as a temporary drop in aggregate demand, cause cyclical fluctuations in the economy. Thus, a temporary negative demand shock causes recession in the economy and lowers inflation. The recession triggers a fall in the demand for shipping services. The fleet is largely predetermined by past expectations of current demand. In the light of the unexpected adverse shock, the past expectations of demand for shipping services are now considered to be overoptimistic and lead to a lower fleet capacity utilisation. Freight rates, newbuilding and secondhand prices fall on impact.

After the initial impact, which depends on the intensity and durability of the shock, the economy would tend to return back to long-run equilibrium. These stabilising forces are reinforced by central banks. The central bank reacts to such an adverse demand shock by cutting the nominal interest rate (the overnight rate, such as the Fed funds rate) sufficiently to engineer a drop in real interest rates, as it is that these rates affect the spending decisions of households and firms. Lower real interest rates expedite the adjustment of the economy back to long-run equilibrium. The shipping market reacts to *actual* and *expected* developments in the economy. The actual demand for shipping services rebounds in response to the recovery of the economy, thereby triggering improvements in the fleet capacity utilisation rate. Expectations of such developments also affect current freight rates, thereby speeding up the process of adjustment. The actual and expected developments reverse the decline in freight rates, newbuilding and secondhand prices. After some time all shipping markets return to long-run equilibrium. Therefore, the negative demand shock in the economy generates business and shipping cycles.

Shipping cycles are caused by business cycles. In the simulation results of the integrated model, shipping and business cycles are moving together. This synchronisation can be distorted by the delivery lag of shipyards. A two-year delivery lag can cause shipping cycles to follow with a lag business cycles largely because supply is predetermined by past expectations of current demand. The Tinbergen–Koopmans model is instructive of the implications of the delivery lag. Depending on parameter values the shipping cycles can appear out of phase with business cycles, thereby giving the impression that shipping cycles move counter-cyclically to business cycles (in an opposite way to each other). But such behaviour does not change the direction of causality. Business cycles cause shipping cycles.

It is worth stressing the role of expectations in generating shipping cycles. Rational expectations imply discounting of future economic fundamentals and, in particular, market conditions (the demand–supply balance) in shipping. Expected market conditions shape the result of the bargaining of charterers and owners over current freight rates, which, in turn, affect the demand for vessels, whether for newbuilding or secondhand prices and, consequently, shipping cycles. But such expectations depend on macroeconomic fundamentals and, in particular, on how central banks would respond to the current business cycle. It is shown in this chapter that expectations in shipping are shaped by expectations of the future path of real interest rates.

Beenstock and Vergottis emphasised the importance of these expectations in generating shipping cycles. Unfortunately, this was done in a model that assumed that shipping markets are efficient. The empirical evidence presented in Chapter 4 shows that shipping markets are inefficient for a horizon relevant to economic decisions in the shipping market. Nonetheless, it should be borne in mind that market efficiency is asymptotically valid. The integrated model presented here, which creates a synthesis of the whole effort of all previous chapters, corrects for the assumption of shipping market efficiency. This chapter shows how expectations affect shipping cycles when shipping markets are inefficient.

PART III

FROM THEORY TO PRACTICE

7 THE MARKET STRUCTURE OF SHIPPING AND SHIP FINANCE

EXECUTIVE SUMMARY

In Part III of the book we move from theory to practice. In this Chapter we analyse the market structure of shipping, the key players and basic facts related to earnings and asset values. The capital market structure of shipping is investigated in section 2. We then examine in Section 3 the role of finance in shaping shipping cycles. The constraints of shipping finance and its future prospects are analysed in section 4. It is argued that the availability of ship finance can have a significant impact on shipping cycles

In every severe recession the pessimistic view that shipping would never recover gains ground. This is based on the well-known conspiracy theory that it is in the interest of the country that depends most on world trade, such as Japan in the last twenty years of the twentieth century or China today, to increase the fleet in order to keep freight rates low. In section 5 we deal with this issue and we conclude that on a cost–benefit analysis it does not make any sense for China to do so; the benefits of lower freight rates are offset by the cost of extra investment required to increase the fleet.

1 THE MARKET STRUCTURE

Shipping is one of the last large perfectly competitive markets where the laws of supply and demand control the market and prices. In particular, freight rates are set by the equilibrium price of demand and supply.

Let us examine again the key players, namely the shipowners and charterers, which in turn represent the supply and demand in shipping. Other key players include the shipyards, again a component of supply, and the brokers who, in effect, can be seen as a 'catalyst' bringing the various parties together.

In this market the shipowners are perhaps the single most important component. In June 2008 the value of the world commercial fleet exceeded $1 trillion, showing that the industry is undoubtedly of major significance. The total global marine-related market was estimated to be valued at over US$3trillion in 2013 and continues to grow. In the EU, 90 per cent of external trade is carried by sea.

Greek owners represent around 20 per cent of the world fleet in dwt terms, but of the approximately 800 owners, only around 200 are of much global significance. The largest owners are now found in the Far East, including Cosco with approximately 1000 vessels, and the two large Japanese carriers, NYK and MOL.

In the last decade of the twentieth Century Marsoft calculated that the return on capital to owners was 12% in containers, 8 per cent in tankers and 4 per cent in bulk. The return to bulk owners was hardly satisfactory at a time of higher interest rates. Matters changed significantly in 2003, following China's accession to the World Trade Organisation (WTO). This led to the most sustained boom ever known in shipping, especially in the dry bulk markets. Utilisation rates approached 100 per cent, which simply led to no more short-run supply but an increase in freight rates which resulted in capesize vessels in June 2008 earning US$300,000 per day. Although markets fell sharply in October 2008, following the collapse of Lehman Brothers, freight rates recovered soon thereafter and 2009 and 2010 were relatively good years for dry bulk owners. The real crisis came in 2011 when the increased supply caused by the attractions of higher freight rates began to be delivered in larger amounts. Cape freight rates collapsed from US$300,000 per day to rates of US$3,000 per day, which did not even cover running costs. Similarly the value of new cape vessels on the water fell from around US$150 million each to around US$35 million. These extreme movements illustrate the operation of a perfectly competitive market governed by the laws of supply and demand. In 2004 it was calculated that a shortage of just 20 capesize vessels caused freight rates to rise from US$20,000 per day to more than US$50,000 per day. By the end of 2013 the rollercoaster ride was still continuing with freight rates

Table 7.1 Yearly TC averages

	BCITC	BPITC	BSITC	BHSITC
2003	$40,330.48	$20,063.52	N/A	$14,810.23
2004	$69,057.88	$35,736.19	N/A	$28,191.21
2005	$50,128.26	$24,700.71	N/A	$21,419.48
2006	$45,139.32	$23,778.47	$22,619.45	$19,425.15
2007	$116,049.44	$56,814.63	$47,449.01	$32,447.03
2008	$106,024.83	$49,014.23	$41,545.65	$29,281.62
2009	$42,656.26	$19,295.08	$17,338.03	$11,342.15
2010	$33,298.34	$25,041.29	$22,455.61	$16,427.39
2011	$15,639.46	$14,000.29	$14,400.50	$10,551.61
2012	$7,678.71	$7,684.16	$9,452.52	$7,626.15
2013	$14,580.46	$9,471.61	$10,275.11	$8,179.24
24/12/13	$38,998.75	$14,555.75	$15,195.13	$11,499.75

Source: Baltic Exchange.

recovering owing to an anticipated reduction in new supply from 2014 and increased expected demand. Cape freight rates were up to US$70,000 per day from some front haul routes.

In contrast to the Baltic Exchange's time charter averages (Table 7.1), which relate to the spot market, most long-term forecasters tend to use data relating to one-year time charters. One of the leading firms in this field is Marsoft who have provided the following data showing both mean and median time charter rates and prices. If we consider the cape market then there is a considerable difference between mean and median rates. For example the twenty-year mean rate is US$33,000 per day, while the median rate is US$20,000 per day. At the end of 2013 capes were fixing one-year time charters at approximately US$20,000 per day which is the exact median rate over the previous twenty years. Similarly, the median price of a ten-year vessel is US$23 million compared with an average price of US$34 million. These differences are not solely of academic interest. There have been several high-profile orders during late 2013 and in early 2014 for new cape vessels coupled with capital market transactions. It appears that some investment bankers and analysts may have used average figures to show that the purchase of these vessels at this time was at a low point from a historical and cyclical perspective. In fact, if the median earnings and prices had been used then this could show that such investments were not being undertaken in a low market. Indeed in some sectors purchases of vessels were at prices above the median level.

In the late months of 2013 the shipping markets have shown clear signs of improvement. However, we are also seeing a significant amount of new ordering, some of it based on the belief that prices are at a very attractive level from a historical point of view. The median analysis paints a different picture. There is one further factor that is an additional complication – namely, the technological changes in vessel designs which lead to substantial fuel savings, at least during the times the vessel is loaded and at sea. We have yet to see the full impact of this but it could prove to be significant and provide a comparative advantage to new "eco vessels". The degree of advantage will clearly depend on future bunker prices but the higher the prices the greater the advantage.

These results are based on Marsoft's quarterly data spanning the time periods noted above. At the time of writing, the observations for the fourth quarter of 2013 were preliminary; however, we do not believe that the conclusions are sensitive to revisions. The data are in nominal terms – no adjustment has been made for inflation. Tanker values have been normalized to a double-hull basis. The mean of the data is the simple average of each dataset. The median is the middle point of the data – half of the observations are smaller than the median and half are larger.

2 THE CAPITAL MARKET STRUCTURE FOR SHIPPING

Since the arrival of the Eurodollar markets around 1966, the market for financing vessels has been dominated by commercial bank lending. Around 80 per cent

Table 7.2 The capital structure of shipping – Marsoft data

Dry Bulk	1993–2013		1990–2013 (ex. 03/08)	
One-Year TC rate (USD/day):	Mean	Median	Mean	Median
Cape Size, 170,000 dwt	33,000	20,000	19,000	18,000
Panamax Size, 74,000 dwt	19,000	13,000	12,000	11,000
Handymax Size, 51,000 dwt	17,000	12,000	11,000	11,000
Handy Size, 27,000 dwt	12,000	8,000	8,000	7,000
Second-hand 10 year (USD mm):				
Cape Size, 170,000 dwt	34	23	24	21
Panamax Size, 74,000 dwt	22	16	16	15
Handymax Size, 51,000 dwt	20	16	15	14
Handy Size, 27,000 dwt	14	12	11	10
Tanker	**1993–2013**		**1990–2013 (ex. 03/08)**	
One-year TC Rate (USD/day):	Mean	Median	Mean	Median
VLCC	36,000	31,000	29,000	27,000
Suezmax	27,000	23,000	22,000	21,000
Aframax	21,000	19,000	18,000	17,000
LR1	19,000	17,000	16,000	17,000
MR2	17,000	15,000	14,000	14,000
Secondhand 10 year (USD mm):				
VLCC	55	42	41	38
Suezmax	41	35	32	30
Aframax	32	28	25	26
LR1	27	24	23	22
MR2	23	20	19	19
Container	**1993–2013**		**1990–2013 (ex. 03/08)**	
One-year TC rate (USD/day):	Mean	Median	Mean	Median
650 TEU, Geared	7,000	6,000	7,000	6,000
1000 TEU, Geared	9,000	9,000	8,000	9,000
1700 TEU, Geared	14,000	14,000	13,000	13,000
2000 TEU, Gearless	15,000	15,000	14,000	14,000

(continued)

Table 7.2 Continued

Container	1993–2013		1990–2013 (ex. 03/08)	
One-year TC rate (USD/day):	Mean	Median	Mean	Median
2500 TEU, Geared	18,000	19,000	17,000	18,000
4300 TEU, Gearless	25,000	26,000	23,000	26,000
Secondhand 10-year (USD mm):				
650 TEU, Geared	9	9	9	8
1000 TEU, Geared	12	12	11	11
1700 TEU, Geared	18	17	16	16
2000 TEU, Gearless	20	18	18	18
2500 TEU, Geared	26	25	23	24
4300 TEU, Gearless	36	35	33	34

Source: Marsoft Inc.

of the current debts of owners – or around $300 billion – has been provided by commercial banks. At their peak two German banks, namely HSH and Commerzbank, were providing close to $50 billion each to the shipping industry, and, in particular, the German Kommanditgesellschaft (KG) market. Similarly at its peak in 2008 RBS was providing $13 billion to the Greek market, making it by far the largest single lender to the Greek market at that time.

The so-called banking crisis, which affected most of the major banks involved in shipping, has left a void in funding caused by the reduction in commercial bank lending which has yet to be filled. Among the funding alternatives provided to date have been bonds, private equity houses, owners' equity and, perhaps most importantly, Export Credit Agencies (ECAs), especially in China, Korea and Japan, whose governments aim to support their shipyards. However, if ship prices have halved on average during this period then there has also been a fall in the amount of fresh ship finance required. In fact, this has fallen by more than half, as risk-averse banks have reduced loan to value rates from around 80 per cent to 60 per cent.

It was anticipated that the capital markets might have started to play a bigger role, but up to the end of 2013 there is limited sign of this as the amounts provided by both the capital markets and Private Equity firms are unlikely to have exceeded US$9 billion p.a. The reason for this is that during the boom years investment bankers brought too many shipping companies to market without sufficient analysis of the likely performance when the inevitable downturn arrived. This led to massive losses for many investors and a collapse of many share prices and indeed several Chapter 11 filings or major restructurings for once mighty companies including OSG (not a recent IPO company), Genmar, Excel, Torm and others.

Investors also lost money in the 1990s from the collapse of the so-called junk bonds. These instruments again were promoted by investment banks with structures that made no sense. Old vessels had bullet repayment profiles after ten years when the vessels would have been scrapped.

The conclusion of the above two events is that shipping and the capital markets are not an optimal mix in the long run, although there have been a few success stories such as Diana and Safe Bulkers. It does appear that public shipping companies have little comparative advantage over private companies in most cases. The reason for this is the cyclical nature of the industry. In 2008, analysts were encouraging owning companies to adopt a growth strategy and to invest at exactly the wrong time in the cycle. The boards of some of these companies may have been too willing to please market expectations rather than to follow their shipping instincts which would have led to the sale of vessels. Similarly, the prices of vessels were raised by the willingness and need of these companies to invest the monies raised in tonnage, especially tonnage on the water, in order to preserve earnings.

Historically, shipping has not proved particularly attractive to the capital markets principally because of its cyclical nature and also its perceived lack of transparency. Therefore, with very few exceptions, shipping companies have traded at a significant discount to net asset value (NAV) and this shows little sign of changing although, in 2008, some shipping companies were able to float at good premiums to NAV.

Table 7.3 provided by *Marine Money* Magazine, lists nearly 150 public shipping companies which they track. One can clearly discern the change in the fortunes of these companies. At the peak in 2007 these companies had a combined value of over US$350 billion, which exceeded the value of the aviation industry. However, by 2012 the value of these companies only just exceeded US$100 billion, or less than a third of the earlier figure, which is a slightly bigger fall than would have been expected based solely on the fall in the value of shipping asset prices. At the peak of the market some shipping companies achieved a premium over NAV of nearly 50 per cent. By 2012 many companies were trading at just 50 per cent of NAV.

However, 2013 witnessed the beginning of a rally in the price of shipping stocks. The Platou Shipping Index increased by 28 per cent in 2013 with dry bulk stocks rising by 80 per cent.

It seems, therefore, that the public shipping companies face even greater volatility than private ones. However, our view is that the Initial Public Offering (IPO) trend is not over for shipping and in the better shipping markets expected from 2014, we would anticipate that many more companies will be listed. Indeed, the ability of Scorpio to raise large amounts of money in the public markets is a testimony to this aided by the need of many investors in Wall Street to find profitable investment opportunities. The general feeling amongst many in Wall Street in 2013 was that vessel prices were historically low – and so were freight rates. Therefore, in a cyclical business the downward risk is low and the upward opportunity is high. This may or may not turn out to be the case, but the ability of many of these investors to realise their gain, say in 2015 when markets may be better, is likely to be limited. Time will tell.

Table 7.3 Marine Money list of public shopping companies

	2005	2006	2007	2008	2009	2010	2011	2012
AP Moller - Maersk Group	50886.04	41299.82	46847.85	21614.98	30996.27	42605.04	29087	32676.86
COSCO Holdings	na	na	59545.31	7859.824	12580.84	10456.97	7570.619	7118.642
Nippon Yusen Kaisha	7413.26	7522.07	9263.495	8229.218	5522.041	6777.514	7057.968	5738.165
Mitsui OSK Lines	7747.461	8159.9	14576.21	14845.01	6997.322	8572.99	7517.732	5650.07
China Shipping Container Lines	na	na	6893.044	2003.115	4798.561	5902.212	4485.69	4504.15
Orient Overseas International Limited	2122.674	3914.718	4602.952	1359.756	2929.331	6080.449	3655.155	4046.06
Kirby Corporation	1354.907	1956.912	2665.024	1568.74	1875.143	2359.318	3669.922	3502.355
Golar LNG Ltd.	868.67	839.1936	1494.781	454.4478	866.3243	1017.828	3566.668	2960.79
Teekay LNG Partners L.P.	na	1165.017	1005.307	666.7228	1385.413	2093.629	2151.406	2632.51
Neptune Orient Lines	2150.321	1985.757	4050.683	1157.836	2918.26	4005.416	2465.939	2275.725
Teekay Corp.	2847.902	3176.888	3872.198	1424.487	1687.228	2382.091	1837.153	2237.37
Teekay Offshore Partners LP	na	na	492.94	329.4525	752.115	1532.91	1878.758	2083.401
Wilh. Wilhelmsen ASA	na	na	na	na	na	na	1054.8	1832.992
Kawasaki Kisen Kaisha, Ltd.	161.7432	3509.612	5780.017	6198.342	2112.751	2992.543	3050.788	1808.587
Seacor Holdings Inc.	1690.139	3147.199	2981.962	2158.794	1724.165	2163.326	1862.2	1667.62
Golar LNG Partners LP	na	na	na	na	na	na	na	1558.17
Ship Finance International Limited	1236.066	1728.374	2015.709	803.8102	1078.474	1702.878	739.0742	1417.375
U-Ming Marine Transport	na	na	na	1041.925	1729.98	1705.298	1281.126	1326.829
Matson, Inc.	na	na	na	na	na	na	na	1251.162
D/S Norden A/S	1228.191	1942.485	4792.923	1449.981	1696.91	1645.273	966.2798	1173.739
Grindrod Limited	927.5007	993.8575	1540.153	746.3562	1090.955	1190.179	1026.608	1164.14
Pacific Basin Shipping Limited	595.4842	992.422	2553.593	793.5195	1399.068	1290.267	758.244	1084.987

(continued)

Table 7.3 Continued

	2005	2006	2007	2008	2009	2010	2011	2012
Costamare Inc	na	na	na	na	na	na	853.848	1041.216
Seaspan Corporation	na	1098.709	1409.179	593.852	624.5075	852.012	947.0664	981.036
Stolt-Nielsen SA	2170.72	2023.453	1883.051	572.2451	883.5983	1196.066	1151.131	961.4357
STX Pan Ocean Co., Ltd.	na	1150.281	4963.162	1529.731	1992.052	2002.121	1079.53	837.1685
Navios Maritime Partners	na	na	na	151.7464	471.9933	960.8495	818.2174	738.1508
DryShips Inc.	na	639.1749	2839.109	688.35	1631.497	2029.379	849.52	662.016
DFDS A/S	557.0036	959.165	1238.366	596.6801	551.7518	1199.313	919.6554	661.8929
CMB	1153.427	1493.84	2999.924	873.7402	1039.583	1152.334	759.3948	641.8428
Diana Shipping Inc.	na	838.7205	2339.838	957.7911	1179.121	985.1592	609.908	600.279
Exmar NV	653.6759	1081.037	1034.144	357.815	487.8168	459.8621	443.6268	593.3026
Capital Product Partners	na	na	na	193.5726	228.0682	367.356	425.2381	558.8394
Hoegh LNG Holdings Ltd.	na	na	na	na	na	na	na	545.9872
Algoma Central Corporation	297.9334	421.6005	538.8758	161.9662	291.1454	346.0657	388.5163	537.9666
Precious Shipping	391.9343	731.363	995.259	326.3397	587.7095	565.1455	543.4434	476.3516
Finnlines Plc	na	919.4724	901.0498	365.3647	463.056	536.0658	467.4715	473.5425
Nordic American Tanker Shipping Ltd	479.195	919.1131	983.7795	1160.089	1266.12	1220.338	567.127	463.05
Scorpio Tankers	na	na	na	na	na	na	187.5315	453.8313
Malaysian Bulk Carriers	444.5032	na	na	689.2713	940.5854	815.807	na	419.479
Odfjell ASA	1761.324	1530.62	1411.431	541.0159	733.871	733.1778	474.117	349.1101
Thoresen Thai	541.109	463.1536	1005.169	433.9778	729.5431	663.0284	329.0048	348.6537
Navios Maritime Holdings Inc.	na	333.4126	1303.547	317.5452	610.2877	536.2368	365.6037	347.9862
Danaos Corporation	na	na	1440.85	368.7039	243.3242	406.2014	367.026	301.4

Euronav	1523.229	1562.017	1840.837	706.7035	1134.868	943.4097	242.4793	295.5882
Golden Ocean Group	na	na	1662.39	177.8702	831.8883	641.323	288.8215	295.4186
Safe Bulkers Inc.	na	na	na	na	477.5251	583.6968	424.6311	257.5776
Frontline LTD	2837.364	2381.807	3591.6	2305.405	2127.108	1975.308	334.0194	253.8236
Teekay Tankers	na	na	na	317.5	272.96	641.5566	217.8176	242.44
Ultrapetrol Bahamas Limited	na	na	568.8654	94.16561	142.5287	192.5142	89.4298	231.66
Tsakos Energy Navigation (TEN)	703.2277	873.8901	1409.325	690.1327	552.2569	460.8	220.8838	211.65
Diana Containerships Inc.	na	na	na	na	na	na	na	194.4276
Regional Container Lines PCL	440.223	389.4684	651.1867	1174.4354	199.3817	376.2904	176.5119	188.5468
StealthGas Inc.	na	168.192	302.6167	106.6859	139.2144	na	79.323	163.358
Global Ship Lease, Inc.	na	na	na	na	77.6675	269.973	99.3147	161.406
Genco Shipping & Trading Ltd.	na	712.6097	1586.168	469.2932	712.624	517.68	245.4556	147.8618
d'Amico International Shipping	na	na	na	268.8545	235.3431	211.1062	86.33013	146.7557
Knightsbridge Tankers Limited	414.675	404.244	412.965	250.515	226.746	544.0561	333.9581	128.2575
Mercator Lines Singapore	na	na	na	129.9473	110.771	253.5274	228.2355	115.9351
Rickmers Maritime	na	na	na	114.9609	113.0725	119.2317	98.07559	114.4215
Jinhui Shipping & Transportation	209.2074	424.5238	898.7304	86.01442	370.5401	283.2607	126.1079	100.7454
Navios Maritime Acquisition Corp.	na	na	na	na	na	na	108.9988	97.6532
International Shipholding	na	na	156.5698	181.9707	188.3774	142.24	1074675	95.9136
IM Skaugen ASA	55.73247	191.2334	244.1847	140.2481	188.5393	172.8263	135.334	86.7409
Box Ships Inc.	na	na	na	na	na	na	na	85.813
Concordia Maritime	258.1742	381.5791	198.786	90.86144	113.0015	137.0073	89.81817	70.39803
Samudera Shipping Line Ltd	129.484	115.4002	148.5516	54.38948	80.41277	88.16557	53.97089	70.16216
Norwegian Car Carriers ASA	na	na	na	na	na	na	na	69.02471

(continued)

Table 7.3 Continued

	2005	2006	2007	2008	2009	2010	2011	2012
Baltic Trading Ltd	na	na	na	na	na	na	107.825	68.54
First Ship Lease	na	na	na	171.5191	255.6395	198.334	141.4432	64.15461
Goldenport Holdings	na	na	583.5328	106.817	122.3893	169.0511	96.05248	55.05286
Courage Marine Group	na	124.0734	288.1018	95.76829	143.1766	139.558	69.44594	49.02066
Euroseas	na	na	375.2364	131.4725	120.6196	na	73.2495	41.2412
DHT Holdings, Inc.	na	507.1521	367.5672	217.3841	179.1277	227.478	47.693	37.2912
Star Bulk	na	na	na	148.9506	172.3133	169.3047	71.5204	33.372
Eagle Bulk Shipping Inc.	na	622.506	1240.602	318.6781	307.5237	311.5488	59.22	24.96
Paragon Shipping	na	na	na	128.9055	232.91	191.6341	38.976	24.64
Globus Maritime	na	na	na	28.27956	43.58412	66.0474	33.431	17.2549
Eitzen Chemical	na	na	712.9614	212.9817	238.9871	237.225	30.25647	15.12797
Seanergy Maritime Holdings Corp.	na	na	na	na	na	100.9424	15.8112	12.4384
SinOceanic Shipping ASA	na	na	na	na	na	na	na	8.324071
Nordic Tankers	na	na	na	53.68626	35.84383	52.12851	15.00644	3.59883
Freeseas	na	na	124.458	29.42769	42.45376	24.31	2.7864	0.126
MISC	3936.846	8658.22	9768.249	9279.643	8756.819	na	10850.45	na
China Shipping Development	na	na	na	3801.683	5623.314	5164.83	3179.158	na
Alexander & Baldwin	2386.56	1888.884	2190.384	1027.46	1403.43	1653.239	1702.194	na
Yang Ming Marine Transport	1489.34	1317.584	1772.939	805.2906	965.7535	2265.721	1135.449	na
Overseas Shipholding Group	1991.988	2296.477	3036	1717.667	1180.365	1077.122	332.3813	na
Excel Maritime Carriers Ltd	224.7734	288.2553	804.9253	325.431	492.2764	479.2819	128.9195	na
D/S Torm	306.8667	420.3752	9370.0495	775.6842	715.2737	558.0321	47.46582	na

THE MARKET STRUCTURE OF SHIPPING AND SHIP FINANCE 263

Hellenic Carriers	na	na	na	11	59.21514	59.5811	29.43272	na
NewLead Holdings Ltd	na	260.5747	187.33	9.57	71.46	na	3.588	na
Wilh. Wilhelmsen Holding Group	1827.657	1823.124	1907.285	675.9258	1042.935	1484.198	na	na
NewCo3	na	na	na	na	na	na	na	na
NewCo4	na	na	na	na	na	na	na	na
NewCo5	na	na	na	na	na	na	na	na
Aegean Marine Petroleum Network	na	na	1630.078	691.0847	1181.887	497.6153	na	na
Chemoil Energy Limited	na	na	445.2063	152.8867	504.1196	439.4874	na	na
General Maritime Corp	1409.002	1152.085	783.1091	624.78	407.1535	291.1675	na	na
Pacific Shipping Trust	na	na	205.5176	85.51375	159.2325	215.2588	na	na
Horizon Lines, Inc.	na	853.8008	587.1973	104.7279	168.7209	134.3775	na	na
U-SEA Bulk Shipping A/S	na	na	na	na	189.8612	124.5327	na	na
K-SEA Transportation Partners LP	286.741	318.928	356.8543	435.7573	309.7203	94.1196	na	na
TBS International Limited	na	242.3078	917.5803	299.887	219.9414	90.422	na	na
Camillo Eitzen & Co ASA	na	425.4163	588.0895	76.18572	86.68258	78.64555	na	na
OceanFreight Inc.	na	na	na	55.07568	147.3585	76.6084	na	na
Trailer Bridge Inc	108.9248	101.3682	140.0766	43.56384	57.0816	33.1752	na	na
OSG America	na	na	na	147.6197	na	na	na	na
Maritrans Inc.	312.0351	na	na	na	na	na	na	na
DOF ASA	na	na	933.829	na	na	na	na	na
Deep Sea Supply	na	na	590.7738	na	na	na	na	na
Havila Shipping	na	na	345.6737	na	na	na	na	na
Dockwise	na	na	na	130.0838	639.9244	na	na	na
Gulf Navigation	na	na	na	301.8922	265.8946	na	na	na

(continued)

Table 7.3 Continued

	2005	2006	2007	2008	2009	2010	2011	2012
Omega Navigation	na	na	239.37	97.60751	na	na	na	na
Star Cruises Ltd	1418.421	1800.315	na	na	na	na	na	na
Carnival Corporation	43973.43	41837.46	na	na	na	na	na	na
Royal Caribbean Cruises Ltd	9478.887	9206.595	na	na	na	na	na	na
Hanjin Shipping	1462.299	na	3367.582	na	na	na	na	na
OMI Corporation	1294.168	1323.633	na	na	na	na	na	na
B+H Ocean Carriers	141.3209	103.4154	na	na	na	na	na	na
MC Shipping, Inc.	114.5405	85.95232	na	na	na	na	na	na
Bourbon	na	2745.163	3552.178	na	na	na	na	na
Quintana Maritime Limited	na	550.7863	1298.715	na	na	na	na	na
Trico Marine Services, Inc.	na	567.601	555.7813	na	na	na	na	na
Brostrom	na	7172.842	505.4359	na	na	na	na	na
Global Oceanic Carriers	na	25.68165	45.16533	na	na	na	na	na
BW Gas	na	1687.216	1279.422	189.1517	na	na	na	na
Tidewater, Inc.	2191.558	2916.592	3153.133	na	na	na	na	na
Hornbeck Offshore Services Inc	887.8377	912.5277	1157.912	na	na	na	na	na
Gulfmark Offshore, Inc.	603.4483	848.4588	1075.375	na	na	na	na	na
Farstad Shipping ASA	560.5192	842.044	1046.145	na	na	na	na	na
Solstad Offshore ASA	537.5886	829.7022	925.8771	na	na	na	na	na
Top Ships Inc.	345.3919	150.7949	141.1083	na	na	na	na	na
Arlington Tankers Ltd.	337.125	362.235	343.015	na	na	na	na	na
U.S. Shipping Partners L.P.	303.324	212.4938	239.3219	na	na	na	na	na

Premuda SPA	283.3166	291.0639	332.5369	na	na	na	na
American Commercial Lines Inc.	na	4053.955	1015.796	309.9495	156.5709	na	na
James Fisher & Sons, plc	318.9612	586.7256	649.2936	255.2368	353.1407	na	na
Noble Group	1845.841	1721.08	3586.804	2360.604	8890.706	na	na
Sinotrans Shipping Ltd	1658.49	1531.504	1971.936	963.2327	1853.254	na	na
PT Berlian Laju Tanker	414.1365	769.9435	1053.002	215.5579	356.6503	na	na
Star Reefers ASA	184.3191	235.7889	380.875	188.26	190.0026	na	na
Belships ASA	26.18093	47.51968	112.6074	17.45495	37.23608	na	na
Crude Carriers	na	na	na	na	na	na	na

Source: "Analysis of Public Shipping Company performance over the last ten years" *Marine Money* Magazine with permission of the author and the publisher.

3 BANKS, SHIPPING CYCLES AND THE SUPPLY AND DEMAND EQUATION

As we have seen at the start of this chapter, banks have been the most significant single factor in enabling the supply of tonnage to expand, financing around 80 per cent of the funds borrowed by the shipowners. Given the more limited supply of these funds since the banking crisis, new orders have reduced, notwithstanding the obvious price benefits. Although ECAs have met some of the shortfall, the perceived difficulty of borrowing has kept ordering to more reasonable levels and the increase in supply remains moderate. Indeed it is likely 2014 will start to see the fleet balance improve as the growth in supply in that year, less scrapping, is forecast to be just 2 per cent less than the growth in demand. New orders promoted by the better bulk market in Q4 2013 will not affect supply until late 2015 at the earliest.

Banks can also affect the supply by foreclosure activity or forbearance activity. Although foreclosures do not, in most cases, eliminate tonnage, they do cause vessels to become inactive for substantial periods. Similarly, forbearance has allowed many insolvent shipping companies to continue to trade.

As already noted, German banks have been the largest providers of loan capital to the shipping industry. The two largest German lenders have announced major reductions in their shipping activities and the attitude of regulators in this will also be critical. Will the regulators ask for more capital for shipping loans? Will they alter the LGD (loss given default ratios) from present industry averages of around 20 per cent to a higher figure? All these factors are negative for ship finance and will reduce the profitability of banking portfolios. A consequence of this may be less capital allocated to ship finance activities.

This is a suboptimal situation as banks should, at times like these, be expanding their portfolios at least from a risk–reward balance. Vessel prices are under their 15-year average and Loan to Value ratios (LTVs) are now closer to 60 per cent financing rather than the 80 per cent common in 2007. In addition, owners are willing to borrow short on long profiles, which reduces term liquidity premiums and funding costs for banks. Margins are at an historic high at around four times pre-2008 levels. However, many banks are unwilling or incapable of taking advantage of this attractive state of affairs and will no doubt return to the market at a less optimal time. Indeed, some may exit this activity as has been the case for nearly all of the US banks.

Our view is that over the next three years banks are unlikely to have a significant impact on the supply and demand equation for shipping, either through foreclosure activity or through a major expansion for financing newbuildings. However, it is possible that banks will increase their lending portfolios after a period of market recovery which will allow them to recover most of their bad debts. In the meantime by late 2013 we saw banks sell parts of their loan books to various hedge funds; in particular, Lloyds Bank, which had a portfolio of nearly US$20 billion at its peak in 2008, has now exited the industry and large parts of its portfolio have been sold to funds such as Oaktree Capital.

4 AN EVALUATION OF SHIP FINANCE CONSTRAINTS

Ship finance in the 'modern era' can be traced back to eighteenth-century England. Shipping was a leading activity although the word shipowner did not appear in the English language until around 1790 in the Newcastle registry. Before that we had merchants with shipping interests. These merchants sometimes owned large fleets and in 1771 Captain Williams, together with other shipping merchants, founded Williams Deacons Bank which, through a series of mergers and acquisitions, now comprises the shipping activities of The Royal Bank of Scotland.

In this book we have covered a variety of topics, including owners' chartering strategy and, by implication, ownership strategy. Some of the same principles can apply to ship finance. An owner's main concerns depending on the point he finds himself in the cycle are:

a. the availability of ship finance;
b. the price of ship finance;
c. the other terms attached; and
d. the consistency of the provision of ship finance.

Such decisions are just as critical as the buying and selling decisions as clearly, in most cases, ship finance is a dependency and, in terms of the ultimate result on the owner's balance sheets, possibly one of the two most important decisions.

The banking crisis, which has affected most banks since 2008, clearly impacted the provision of ship finance in many banks and has also led to owners having adopted mistaken strategies in many cases. This was not something that could have easily been forecast.

For example, by 2012 several banks had announced an exit or reduction in shipping activities, including several famous and large providers, especially in Germany, where Hamburg, up to 2008, was the largest single centre provider of ship finance in the world.

During the boom years of the China factor[1] 2003–08, the profitability of shipping reached levels unimaginable at the turn of the century. Front haul cape rates reached $300,000 per day in June 2008 but within five years were down to $3,000 per day in some cases. Such volatility, together with other factors, following the Lehman Brothers collapse caused all banks to review a number of long-term lending businesses, especially in the area of shipping.

The question now being openly asked is: will there be a future for ship finance in most banks? We have not hesitated in saying yes, but there will be changes to the structure of the business. In other words, we will have a significant shift in the terms of trade.

The poor market from mid-2010 to mid-2013 has brought about a change in most banks' risk appetite for shipping. This is very paradoxical as most risk officers happily concede that 2012 is a far better time to be considering ship finance loans than 2007.

In 2007 most banks were happy to lend 80 per cent at margins under 100 basis points (bp) for periods of up to ten years on 18-year profiles. Clearly this was risky even at the time as the value of some assets such as capes were three times their 15-year average. Banks and owners thought they might be protected with long-term charters but when the freight market collapsed, the charterers renegotiated. In most cases, a charter is a one-way option in favour of charterers. If charterers make money they keep the vessel. If they lose, they redeliver.

By 2012 most vessel asset prices had halved in value since 2007 and most banks have reduced loan to value (LTVs) to around two-thirds finance. Therefore the risk metrics have changed greatly as asset prices are in many cases at or lower than twenty-year averages and, therefore, the loan amount allows considerable security margin. Some cases are more extreme. For example, in 2007 a cape bulker on the water was worth $150m. In late 2012 it was worth $40m with the twenty-year average newbuilding price at around $60m. At what price should an owner buy and at what level should a bank finance?

The pricing of loans has also changed considerably over the years. When the eurodollar market opened approximately forty years ago the typical ship loan was priced at 250 basis points (bp). As we noted above, by 2007 margins had fallen to well under 100 bp for most strong ship owners although the fall was not linear.

With the start of the banking crisis in 2008 following the collapse of Lehman Brothers, the typical margin is now in the range of 300 to 450 bp and the typical loan period has halved from ten years to five, although with an unchanged profile. Whilst this has greatly improved banks' profitability, the increase in profitability has not been pro rata as banks have also faced massive increases in funding costs, as can be seen by their bond and Credit Default Swap (CDS) rates. This is especially true for long-term money as regulators focus more on leverage ratios and matched funding.

In analogy with Darwin's selection process only the strongest, most skilled banks will survive in ship finance, together with only the best owners, apart from state owned entities in the Far East that have easy access to capital.

The implication of this is significant in the short term but it again, in an economic sense, points to an adjustment towards equilibrium as expected in general equilibrium theory. From mid-2010 to mid-2013 the market had considerable excess supply. Simple economic theory dictates that supply needed to be restricted during this time period. Shipowners and shipyards did not abide by this principle as they hoped a non-zero sum game would apply whereby others would restrict the increase in supply while they took advantage of the perceived opportunities of low prices. The reality is different, however, and in practice the best way to restrict supply is to restrict the access of finance for shipping. This has a very significant effect on secondhand prices, as well as newbuilding prices, although the effects on the latter are mitigated by the availability of export credit agency (ECA) funding, especially in China and Korea. Indeed that very availability of ECA financing is driving owners to order newbuildings, which in turn increases supply and delays the recovery of freight rates.

In theory, if ship prices halve we would only need half the amount of ship finance in the world. But also if only half the ship finance is available ultimately prices would also halve until the restriction in supply at some point raised freight rates and earnings to such a point that the supply of finance ultimately increases.

Therefore, by withdrawing liquidity, banks are driving prices down, which has serious consequences for their balance sheets.

Let us consider optimal strategies for both owners and banks in the new capital-constrained world of the second decade of the twenty-first century, a period also characterised by low freight rates.

Why have so many banks announced their withdrawal from ship finance? Clearly there is no one single reason but the list would include one or more of the following:

a. long-term matched funding requirement
b. capital-intensive business
c. low historical margins
d. losses caused by the low freight market
e. EU and government policy which required reductions in balance sheets following state aid. This has greatly affected the German banks, who have been the largest providers of ship finance
f. Shipping is now considered a lower priority than other domestic business
g. Need to balance shipping exposure in relation to total assets. This affected the Scandinavian banks which had high shipping percentages and also at least one German Landesbank.

The problem for banks that wish to exit is the lack of buyers at any sensible prices. With ship finance totalling $300 billion, of which $80 billion is to the Greek market, it is clearly not possible to find alternative providers of finance, even in the medium term. However, the publicly stated wish of some banks to exit has been of great benefit to those banks that have chosen to stay. It has allowed them, in some cases, to increase their share of client wallet, thus improving funding gaps and Term Liquidity Premiums (TLP) rates, and at the same time rapidly improve the terms of trade. However, given the large amount of historical good loans on the books of most traditional shipping banks the return to profitability takes some time, although it is widely expected that by the end of this decade most leading shipping banks should be achieving risk-adjusted return on equity (RAROE) of between 15 per cent p.a. to 20 per cent p.a.

The surviving shipping banks will pay close attention to their capital and funding position, greatly assisted by the shortening of loan periods. With the inevitable improvement in freight markets the credit grades of shipping companies will improve thus reducing the necessary amount of capital under Basel III and leading to a significant increase in returns on capital. At this point it will be interesting to see if banks will return to shipping. Such a development is likely and when it does happen it may well be the case that we will see the same patterns repeated as the banks that return will not have the experienced shipping finance staff to avoid previous mistakes. Competition will increase, the terms of trade will worsen and another crisis will arrive when the freight markets dip and prices follow.

This has implications for owners' strategies. What should the prudent owner do to minimise risk? What is the optimal strategy? The following may seem obvious but it has clearly not been followed by the majority of owners in the past who have simply assumed a dollar from one bank is the same as a dollar from another, and often aim merely to minimise borrowing costs.

5 OPTIMAL STRATEGIES

1. Have a number of banks; say one bank for every five vessels. Try not to borrow more than $250m from each bank.
2. Give ancillary business only to banks lending to you. In the long run this will reduce your borrowing costs and increase availability of loans.
3. Try to avoid syndicates. They are wholly dysfunctional when it comes to problem solving and often the borrower becomes caught between vendettas between different banks and/or actions driven by internal risk appetites.
4. Always remember relationships are give and take. Try not to take advantage in "borrowers' markets" as this will be remembered in "lenders' markets". In 2006/07 and early 2008 borrowers often "pressurised" their banks into reducing their interest margins threatening to refinance. From late 2008 these actions were remembered.
5. As far as possible stick to "shipping banks". Avoid banks that place more emphasis on "investment banking". Try to establish long term relationships both with the institution and its staff. Although by and large this has always been true, the takeover of Deutsche Schiffsbank by Commerzbank and their unexpected withdrawal is an exception to this rule. Check the number of years of service staff have in ship finance and the years they have been with that institution.
6. Opt for banks based in countries where shipping and ship finance is important. They are more likely to stay the course and be under less pressure to favour domestic over global business.
7. Forge closer lending relationships with domestic banks but still remembering (1) above. Domestic Banks usually show preference to their domestic clients as evidenced by the treatment of German clients by German banks in 2009 onwards.
8. Irrespective of the health of shipping departments they will depend on the overall health of the institution to provide capital and funding. This must be monitored. A successful shipping department will still suffer if the parent bank is weak.
9. Look at the other clients of your bank. An owner knows the good and bad in the industry. Avoid banks that have a significant number of bad owners as clients, although we suspect all banks have at least one or two. A shipping bank with bad owners will be the first to fail in a low market causing problems for their good owners too. As the Greeks say "show me your friend and I will tell you who you are". In banking it is "show me your clients and I will tell you what kind of a bank you are".

If shipping within banks falls out of fashion one of the biggest future threats is that young bank staff will not be attracted to a career in ship finance. It will not seem an area with great potential and promotion prospects. This is potentially a very serious problem as experienced staff are the main factor in avoiding losses because of their industry knowledge. We very much doubt that any of the modern tools of Credit and Portfolio management have been reliable indicators of potential losses. Therefore a bank without the right staff will happily increase its portfolio in good times unaware of the full risks and more easily implode when the bad times inevitably return. No bank should consider ship finance unless it can find sufficient staff with at least 20 years' ship finance experience. This will be one of the main challenges to come with no easy solutions.

The first decade of the twenty-first century was the best ever for shipping. The second decade could be one of the worst, similar to the 1980s. We see the truth in the saying "every feast brings a famine". However, shipping will not die. 90 per cent of the world's trade is carried on ships and therefore shipping and shipping finance will always exist. What will change is the price of ship finance, its terms and the make up of lenders. However, memories tend to be short and no doubt the same mistakes of the past will be repeated in the next boom.

6 CONSPIRACY THEORIES IN SHIPPING

During the 1960s and 1970s a thesis developed that it was greatly beneficial to Japan Inc. for freight rates to be kept low as they benefitted both exports from Japan and imports into Japan. Therefore, the theory assumed that there was a great incentive for Japan to encourage its shipbuilders to overbuild and thus keep supply high relative to demand and freight rates low. This theory was shown not to have been correct, however, as Japan was unable to prevent periods of high freight rates and the growth of large Japanese shipping companies such as NYK and MOL would not have taken place if the conspiracy theory was correct.

By 2013 the theory had switched to China having a great incentive to keep freight rates low – it was argued that China is far more powerful than Japan even at its peak and therefore better able to keep freight rates low. However, as in the case of Japan, this argument is also incorrect based on the following calculations but far more marginal than in the original Japanese case.

If we take a weighted average of the dry bulk market, we have an average time charter rate of approx. US$10,000 per day in 2012, compared with a 20-year long-term average of US$20,000 per day. Assuming 10,000 global dry bulk vessels and China's share at 25 per cent with an 80 per cent utilisation gives a total saving to China of US$20 billion per annum. However, against this must be set the costs of keeping the supply high. This would require capacity growth annually of around 10 per cent or 70 million dwt. Assuming average vessel size of 100,000 dwt requires 700 vessels at a cost of, say, US$40 million per unit, leading to a gross cost of approx. US$28 billion compared with the perceived saving of US$20 billion. This makes it unlikely that China would risk the capital costs for no expected net gain.

The conclusion of the above is that whilst we are unlikely in the next decade to see freight rates as high as during the 2003–08 period, we are also unlikely to see a continued depression in shipping. Indeed, the most likely trend is a return to the traditional cycle of shorter cycles of high freight rates followed by longer periods of low but not catastrophic markets. As the book goes to print this is exactly what is happening.

8 THE FINANCIALISATION OF SHIPPING MARKETS

EXECUTIVE SUMMARY

In the 1950s and 1960s, business cycles in the US and other industrialised countries were demand-led. The stabilisation of the economy around potential output was, on the whole, successful through demand management (fiscal and/or monetary policy). In the following twenty years business cycles turned from demand- to supply-led, following abrupt changes in the price of oil. Three of these shocks were adverse or negative and one, in the mid-1980s, was fortuitous or positive. The shocks in the early and late 1970s were permanent, while the shock in the early 1990s was transient. Adverse supply shocks cause the rise of inflation and unemployment (stagflation), increase the amplitude of the cycles relative to demand-led ones and create a short-term trade-off for policymakers. From the early 1990s onwards cycles turned into asset-led ones, driven mainly by liquidity. The amplitude of asset-led cycles is even larger than that of supply-led cycles. The upswing of the cycle is usually long and pronounced, creating prosperity and euphoria, but the downswing leads to deep and protracted recession. Therefore, asset-led cycles exhibit the largest amplitude of all cycles.

The liquidity that has financed a series of bubbles in the last twenty years was created gradually. Financial deregulation and liberalisation laid the foundations for financial engineering, while central banks have pumped more liquidity into the financial system every time a bubble has burst, thereby perpetuating the bubble era. The expanding liquidity has resulted in the financialisation of shipping markets. In the first phase of financialisation the liquidity affected commodity prices, such as oil, iron ore and coal. In the second phase it is affecting vessel prices, effectively turning ships into commodities. Commodities attracted investors because of prolonged periods of backwardation or contango, which give rise to the possibility of profit with small risk. This was true at the beginning, but as more and more investors were attracted in commodities profits fell and risk soared. The advent of investors in the commodity markets increased price volatility and distorted the price mechanism. Prices convey a signal of market conditions. Unless higher commodity prices are caused by supply shocks, which can easily be detected because they are covered in the media, they imply improving market conditions. The advent of investors in commodity markets pushed prices higher than justified by economic fundamentals in the upswing of the cycle and lower in the downswing. The distortion of the price mechanism exacerbated the amplitude of the super shipping cycle. The accession of China into the World Trade Organisation in 2003, with its gradual rise to pre-eminence in world trade, signalled the advent of investors to commodity markets and caused a structural change in both the oil tanker and the dry markets. The structural change means that the interaction of demand and supply is insufficient to explain fluctuations in freight rates. Investor risk aversion is now an additional variable required to account for freight rate volatility. As a result of this change, freight rates are now behaving as asset prices, akin to equities, rather than prices that equilibrate demand and supply in a transport industry.

In the first phase of financialisation the influence of liquidity on shipping occurred indirectly through commodities. In the second phase of financialisation, outside investors are attracted to shipping itself, particularly the dry market. The consequence is that vessel prices are likely to rise more rapidly than is justified by freight rates, as investors are interested more in capital gains than profitability arising from the operation of the fleet.

This chapter is organised as follows. Section 1 analyses the evolution of business cycles from demand- to supply-led and finally to asset-led. Section 2 explains hedging and speculation in commodities, and the concepts of contango and backwardation. Sections 3 to 6 investigate the structural changes and the financialisation of the oil tanker and dry markets, while the final section summarises and concludes.

1 ASSET-LED BUSINESS CYCLES

In the period from the end of the Second World War to the first oil shock in 1973–74 business cycles in the US and most major industrialised countries were driven primarily by demand. In demand-led cycles the task of stabilising the economy around the potential output path is relatively easy, as fiscal and monetary policy have a direct bearing on demand. Easy policies when demand is below potential and tight policies when demand is above can, at least in principle, return the economy to its potential path. In the US, in the 1950s and 1960s fiscal policy rather than monetary policy was used rather successfully to stabilise the economy.

But in the following twenty years the two oil shocks of 1973–74 and 1978–80 turned the business cycles of the US and other major countries into supply-led. Stabilisation policies have a much harder task in supply-led cycles, as they operate on demand when the shocks hit the supply of the economy. Oil shocks cause stagflation, the simultaneous rise of unemployment and inflation, thereby creating a trade-off for policymakers. In the short run, policymakers can fight either the recessionary impact of the oil shock (unemployment) or the inflationary consequences. The optimal policy response for a country depends on the reaction of other countries. If the majority opt to fight unemployment, then the optimal response is to fight inflation. By following such a course of action, the country gains competitiveness in the medium term and takes advantage of the stimulus in world trade generated by the other countries. This was, for example, the reaction of Germany and Japan to the first oil shock, and it proved successful because the Anglo-Saxon countries pursued easy policies to fight unemployment, thereby boosting the overall levels of world trade. When all countries opt to fight inflation,

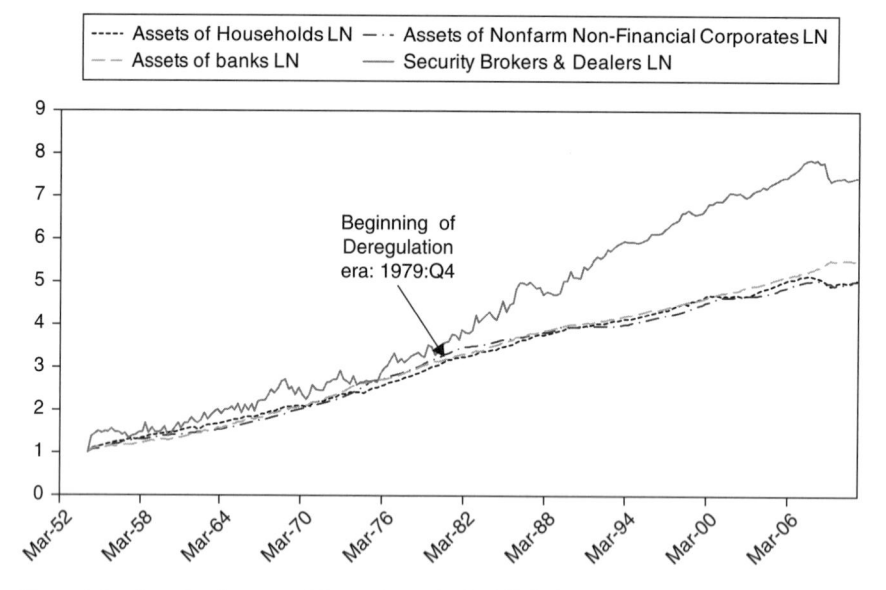

Figure 8.1 Growth of assets of four sectors in the US (in logs 1954 = 1)

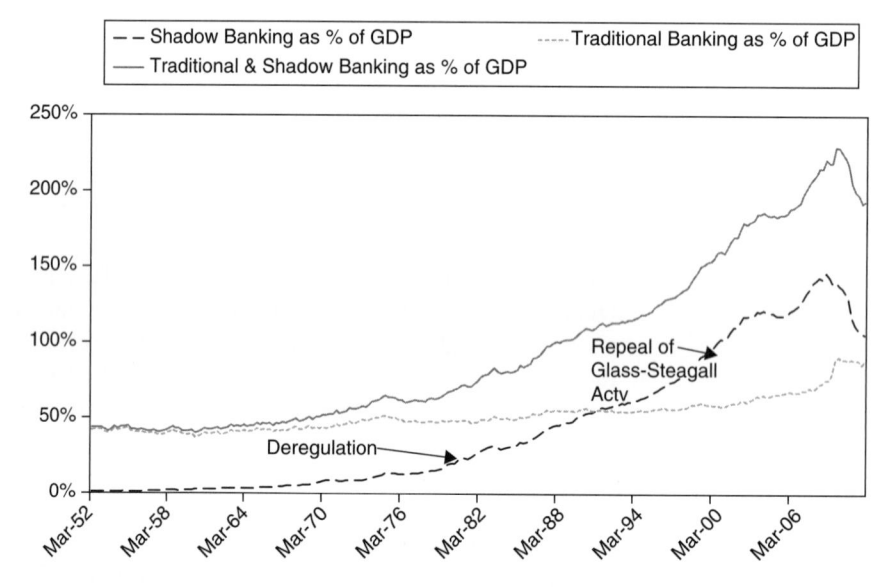

Figure 8.2 Liabilities of US shadow and traditional banking

as happened in the second oil shock, the optimal response is to assign monetary policy to inflation and fiscal policy to unemployment, as the US did. By doing so, a country protects itself from other countries exporting their inflation to the rest of the world, while simultaneously it fights unemployment by stimulating domestic demand.[1] However, even such a policy mix has long-term costs, as in the medium term it leads to a twin deficit in the budget and the current account of the balance of payments. A twin deficit would require the reversal of policies in the long run, which nonetheless may not be very costly if the rest of the world has recovered.

From the 1990s onwards, though, business cycles are no longer demand- or supply-led, they have been asset-led driven by liquidity. Asset-led business cycles, like the last two cycles in the US, or those experienced by Japan in the 1990s or the US in the 1930s, produce a larger variability in output than in inflation. In the upswing of the cycle output growth surpasses historical norms, giving the impression that the level of potential output growth has increased, thus creating a general feeling of euphoria and prosperity, as occurred in the US in the second half of the 1990s. In the downswing, however, the recession is deeper than normal, and, even more importantly, it lasts for a long time with many false dawns, as in the case of Japan. The bubble is usually pricked by rising interest rates, as the central bank attempts to control a relatively small increase in inflation. In fact, the more leveraged the economy, the smaller the interest rate hike necessary to prick a bubble. As asset prices fall, the past accumulation of debt becomes unsustainable and households and businesses engage in a debt reduction process by retrenching. This depresses demand, putting a new downward pressure on asset prices and thereby creating a vicious circle.

The current excessive liquidity in the world economy has been created by Japan in the 1980s and the development of shadow banking in the US and other countries since this period. The liquidity created in Japan was reflected in the money supply and credit statistics, in measures such as (M2+CDs). Throughout the 1980s this wide definition of the money supply grew at double-digit rates. The liquidity did not cause high inflation in the prices of goods and services, but rather high inflation in asset prices, in particular equities and residential and commercial real estate. Inflation in goods and services remained low, at less than 1 per cent, for most of the ballooning phase of the stock (electronics) and property bubbles, but ultimately crept up. When the Bank of Japan hiked rates to combat CPI inflation the electronics and property bubbles imploded, as invariably occurs. The liquidity was not drained from the system, however; instead it expanded in the

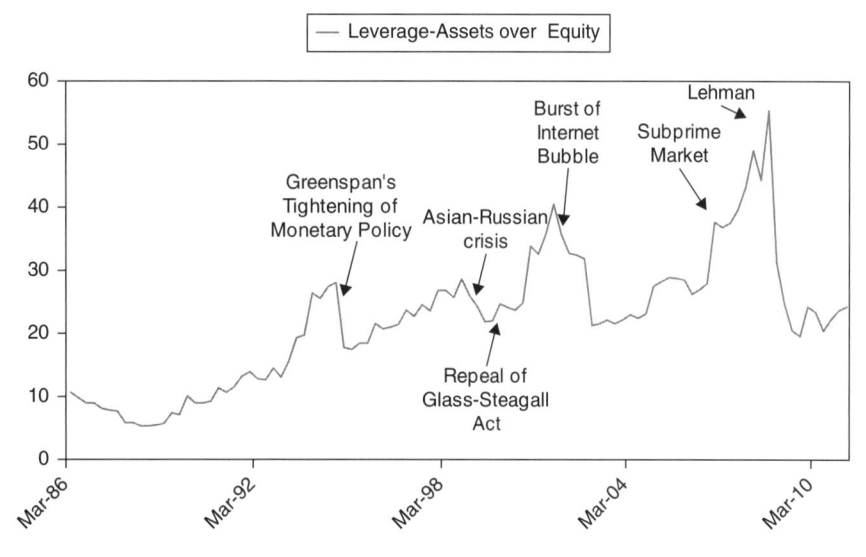

Figure 8.3 Asset leverage of US investment banks

late 1990s, in the first half of 2000s and in the 2010s. Following the example of the US after the financial crisis of 2007–08, the Bank of Japan stepped up the printing of money by adopting an explicit inflation target of 2 per cent.

This liquidity was channelled to Southeast Asia, creating similar bubbles in stocks and property that burst when the yen appreciated, resulting in the Asian-Russian crisis of 1997–98. In the 2000s, this liquidity expanded carry trade in risky assets and shipping benefitted (principally indirectly) from its relationship through commodities. The liquidity created in Japan could be monitored as it was reflected in the wide definition of the money supply and credit statistics. Consequently, policymakers in Japan were aware of it. On the other hand, the creation of liquidity through shadow banking in the US and other major countries before the financial crisis of 2007–08 was largely undetected, as it was not reflected in money supply and credit statistics. The liquidity that has financed non-traditional bubbles has been created in three stages: financial deregulation and liberalisation, financial innovation and monetary policy errors (see Arestis and Karakitsos, 2013). Financial deregulation and liberalisation laid the foundations for financial engineering, while central banks have pumped more liquidity into the financial system every time a bubble has burst, thereby perpetuating the bubble era.

Shadow or parallel banking consists of the security brokers and dealers (mainly the old investment banks) and the subsidiaries established by commercial banks, such as Structured Investment Vehicles (SIV) and Conduits, to move their loan portfolio off their balance sheets. The SIVs raised money from the London money market through credit lines that were made available by the mother banks. The SIVs used these funds to buy the loan portfolio (all forms of loans – mortgages, student loans, auto loans, and so on) of the mother banks, securitised it and sold it on to the money market mutual funds, hedge funds and other banks. Investment banks have also contributed to the securitisation process. Figure 8.1 shows the growth of assets in four US sectors: commercial banks; security brokers and dealers; households and the corporate sector. Until the deregulation of the financial system the assets of all four sectors grew at approximately the same pace, but thereafter the assets of security brokers and dealers grew much more rapidly, reflecting the increasing importance of securitisation. Figure 8.2 shows as a percentage of GDP the liabilities of the US commercial banks, which are regulated and have access to the discount window of the Fed, those of shadow banks, and the total. Since the beginning of 1990 the US shadow banking sector has overtaken the traditional commercial banks, and since the repeal of the Glass–Steagall Act in 1999 the liabilities of shadow banks skyrocketed.

As can be seen from Figures 8.1 and 8.2, the expansion of liquidity is a smooth function and therefore it is hard to imagine how this liquidity can be associated with the booms and busts of asset prices. The point is simply that it is not the liquidity, but its usage that exhibits large fluctuations. The trigger point for the ballooning and the burst of a bubble is fluctuations in the degree of risk aversion for risky assets. When risk appetite increases, more use is made of the existing liquidity and more is created. When risk appetite gives way to risk aversion less use of the existing liquidity is made and deleveraging (i.e. shrinking of balance sheets) occurs.

Figure 8.3 shows the degree of asset leverage of US investment banks, defined as the ratio of bank assets over capital. The fluctuations in asset leverage are triggered by corresponding changes in the degree of risk aversion. When risk appetite increases, more liquidity is used to expand assets and the degree of asset leverage increases, thus funding bubbles. When risk aversion rises, less liquidity is used and the degree of asset leverage is reduced, causing the burst of a bubble. By 1994 the degree of asset leverage of investment banks reached 30 times their capital. The tightening of monetary policy by the Fed under Greenspan in 1994, when the Fed funds rate soared in a series of steps from 3 per cent to 6 per cent, caused rising risk aversion and the degree of asset leverage fell to less than 20 times capital (see Figure 8.3). As the tightening of monetary policy succeeded in engineering a soft rather than a hard landing for the US economy risk appetite resumed and the degree of asset leverage increased gradually, nudging 30 times capital by the time of the Asian-Russian crisis in 1997–98. Risk aversion rose temporarily, but Greenspan lowered the Fed funds rate by 75 bps in three steps to 5.5 per cent, thereby halting the process of asset deleveraging. The repeal of the Glass–Steagall Act in 1999 triggered a new surge for risk appetite and asset leverage hit a new high of 40 times capital by the time of the burst of the internet bubble in 2000. Rising risk aversion in the course of the ensuing recession led to a drop of asset leverage to 20 times capital. But Greenspan lowered the Fed funds rate to 1 per cent to deflect the deflationary impact of the burst of the internet bubble and when the economy recovered he removed the accommodation bias very gradually. This fuelled the housing bubble, which ballooned under the subprime market and the use of financial engineering through Collateralised Debt Obligations (CDO). This led to a new high at 55 times capital in asset leverage by the time of the Lehman Brothers collapse in 2008. Rising risk aversion since the credit crisis erupted in mid-2007 has lowered asset leverage to 20 times capital. The liquidity in the financial system has been drained only partially, as the Fed initiated three quantitative easing (QE) programmes of expanding its balance sheet (through the printing of money). The liquidity in the system poses a threat to financial stability as another stock market bubble in the US is in the making.

A well-functioning banking system should create liabilities (that is, liquidity) of something less than the size of GDP. The combined liquidity in the US economy alone had hit 230 per cent of GDP by the end of 2008 (see Figure 8.2). This huge liquidity has financed a series of bubbles: internet, housing, commodities and shipping. But since the burst of the US housing bubble some deleveraging has taken place, in spite of the printing of money by the Fed.

Leading to the financial crisis, the main beneficiaries of this excessive liquidity were banks and other financial institutions. They channelled this liquidity to stocks, Emerging Countries and unregulated markets – shipping, commodities and, in particular, oil. The last phase of the commodities and shipping bubbles occurred in the first half of 2008 after the burst of the housing bubble and the downfall of equities. The excess liquidity stopped funding houses and stocks and was channelled to three new areas: oil, other commodities and shipping. Whereas the direct impact on shipping is small, the indirect impact through the oil and commodities markets is large.

2 HEDGING AND SPECULATION

In this section we explain what attracted investors into the commodities markets, using oil as an example. We analyse the oil trading strategies that are associated with backwardation and contango. These strategies have played a dramatic role in distorting the price mechanism in the oil tanker market, resulting in a misallocation of resources. Similar strategies in other commodity markets, such as iron ore and coal, distorted the price mechanism in the dry bulk market. As a result of these distortions there is a premium or discount in freight rates in the both dry and the wet markets.

For the sake of clarity, it is assumed that market participants are risk neutral and there are no transaction costs from either entering or liquidating a position. The assumption of risk neutrality implies that investors do not care about risks as long as on average they come out with a profit.

Investors in the oil market can be classified into two opposite camps. First there are those who believe that there is an ample supply of oil today, but that there will be a shortage in the future because oil reserves would soon run out or the demand for oil would increase faster than before because of a ferocious appetite for oil consumption by rapidly expanding countries like China. Investors in this group expect the price of oil to rise in the future. In the opposite camp are investors who believe that there is a shortage of oil today because of China, but that there will be a plentiful supply in the future as cheaper sources of energy are developed. Investors in this camp expect a lower oil price in the future. These investors can make money by developing two alternative trading strategies. The first involves an arbitrage in the spot market, while the second involves an arbitrage between the futures and the spot markets. Two separate equilibrium conditions are derived from these trading strategies.

An investor who expects the spot price of oil to rise six months from now will borrow money and buy oil today in order to sell it six months later. If this strategy is to be profitable, the difference between the price of oil six months from now and that of today should cover at least the cost of money. However, in practice there are two more considerations. First, physical possession of oil for the six-month period requires storage facilities and therefore the cost of storage should be added to the interest rate cost. Second, if the investor is an oil company that wants to increase its market share, then one has to take into account the profit that the company would make by selling the oil six months from now at a lower price than its competitors. This profit is called the 'convenience yield'. The convenience yield also includes other profits, such as foregone earnings for failing to meet an unexpected increase in demand in the future. If we define as 'cost of carry' the interest cost plus the storage cost less the convenience yield, then in order for the first trading strategy to be profitable, the difference between the expected spot price and the spot price today should exceed the cost of carry.

An investor who expects the difference between the expected spot price and the spot price today to be less than the cost of carry will wait to buy oil six months from now. The first type of investor, by buying now and selling later, would push up today's spot price and push down the spot price six months from now. The second type of investor, by deferring the purchase of oil, would push down the spot price now and push up the spot price six months from now. In equilibrium, the expected spot price should equal the spot price today plus the cost of carry.

The second alternative trading strategy is executed through the futures market. An investor who expects the price of oil to rise can buy a futures contract today with delivery of the physical in six months. For the trade to be profitable the current futures price should be higher than the investor's expectation of the spot price six months from now, because, in principle, he can buy the oil in the spot market six months later to fulfil the obligation of the futures contract. An investor who expects the price of oil to fall can short sell a futures contract now. For the trade to be profitable the current futures price should be lower than the investor's expectation of the spot price six months from now. The first type of investor would push up both the futures price and also the six-month spot price. The second type of investor would push down both the futures price and the six-month spot price. In equilibrium, the futures price of a six-month contract should equal the expected spot price six months from now.

The two trading strategies do not seem to yield the same result. The second strategy appears to be superior because in reality a futures contract is settled with cash without physical delivery and therefore does not seem to involve storage costs or the convenience yield. The cost of carry for a futures contract seems to involve only the cost of money. But this is not true because if an investor in the futures market does not take into account that the true cost of carry also involves storage costs and the convenience yield, they can easily be toppled by an arbitrageur who takes the opposite position in the physical market (trading strategy I).

The equivalence of the two trading strategies can be verified succinctly in mathematical terms. Let S_t denote the spot price of oil today; $E_t(S_{t+6})$ the expectation of the spot price six months from now as with information at time t; F_t the current futures price of oil with delivery six months from now; CC the cost of carry; r the interest rate; SC storage costs; and CY the convenience yield. With this notation the first trading strategy involves the following equilibrium condition:

$$E_t(S_{t+6}) = S_t + CC \tag{8.1}$$

$$CC = r + SC - CY \tag{8.2}$$

The second trading strategy involves the following equilibrium condition:

$$F_t = E_t(S_{t+6}) \tag{8.3}$$

The two equilibrium conditions imply a third one, which mathematically is obtained by substituting (8.1) into (8.3).

$$F_t = S_t + CC \tag{8.4}$$

The last equilibrium condition implies a third trading strategy, which is the combination of the first two trading strategies: Buy the oil today in the spot market taking physical delivery now; and at the same time sell a futures contract. The obligation of the futures contract to sell the physical six months from now would be fulfilled by selling the oil that was bought in the spot market and stored for six months.

With these trading strategies as a background it is now easy to define the concepts of contango and backwardation. When the current price of a six-month futures contract exceeds the spot price there is contango. This will happen only if the cost of carry is positive, which, in turn, requires that the sum of the cost of money and the storage cost exceeds the convenience yield. Formally, there is contango

$$\text{if } F_t > S_t, \text{ then } CC > 0, \text{ which implies that } r + SC > CY \qquad (8.5)$$

Backwardation is a situation where the opposite is true. Formally, there is backwardation

$$\text{if } F_t < S_t, \text{ then } CC < 0, \text{ which implies that } r + SC < CY \qquad (8.6)$$

Over a short period of time the storage cost and the convenience yield do not change drastically, as they reflect both the characteristics of the commodity and the strategies of participants in the industry. According to the US Energy Information Agency (EIA) storing oil would cost a company about $1.50 and $4.00 per barrel per year depending on whether one owns or rents the storage facility. This is a trivial amount of money. For gasoline in the US (or petrol in the UK), the corresponding costs would be $2 and $6 per barrel per year, or $0.01 per gallon per month. Therefore, the cost of carry is very sensitive to the interest rate, but not to storage costs and the convenience yield. So, with low interest rates it is likely that a market that is predominantly in contango will turn into backwardation. With high interest rates it is likely that a market that is predominantly in backwardation will turn into contango.

The cost of carry may be an important determinant of contango or backwardation as far as the decisions of speculators are concerned. But for the normal market participants (namely, the consumers and producers of a commodity) hedging is the decisive factor. Consumers are prepared to pay a premium over the spot price to reduce the uncertainty associated with future costs. This gives rise to contango. Producers are also prepared to pay a premium over the spot price, which means accepting a lower price than the spot, to reduce the uncertainty of future income. This gives rise to backwardation. Thus, in contango markets consumer hedging dominates producer hedging and vice versa in backwardation markets. From this perspective, a commodity market would be in contango or backwardation depending on whether consumers are more risk averse than producers. If, on balance, consumers are more risk averse than producers, then the market will be in contango. By contrast, if producers are more risk averse than consumers, then the market will be in backwardation.[2] Which side would dominate depends on the price elasticity of demand. If the commodity is price inelastic, consumers would buy it irrespective of whether it is cheap or expensive, provided the expenditure on the commodity is a small percentage of the total. Hence, consumers have only a small incentive to hedge. As the proportion of this expenditure in the total increases, so does the incentive to hedge. However, in the case of the commodity producers, this may be their only source of income. Hence, producer hedging would dominate consumer hedging and the market would be in backwardation. For example, in the case of many agricultural products, such as oranges, the normal state is backwardation. For steel producers, who are consumers of iron ore, the price elasticity of iron ore is very inelastic, since they cannot substitute iron ore in the production of steel. Moreover, the expenditure on iron ore is a large proportion of the total cost. Therefore, the hedging incentive of steel mills is high. For the big oil companies, the incentive to hedge is low because they can easily pass on the extra cost to consumers. By contrast, for the oil-producing countries the incentive to hedge is high. Hence, for the oil market the normal condition is backwardation. The conclusion that emerges from this analysis is that depending on the commodity the normal state can be either backwardation or contango.

However, a prolonged period of either backwardation or contango in a commodity market would attract speculators, for whom the cost of carry is the decisive factor. The commodities markets were highly inefficient before 2003, that is, there were deviations from the equilibrium condition (8.4). This implies profit opportunities with small risk (that is, a high risk–reward (Sharpe) ratio) – an ideal situation for investors.

The entrance of speculators into the commodities market is supposed to make it more efficient by eliminating unexploited profit opportunities. In terms of equation (8.5) speculators are supposed to enforce the equality of equation (8.4). It was on the grounds of this argument that regulators allowed investors to enter the commodities market. But the outcome was the reverse of what was intended: huge volatility in the prices of commodities and the misallocation of resources in the oil tanker and dry bulk markets.

3 THE FINANCIALISATION OF THE OIL TANKER MARKET

The excessive liquidity present in the world economy in the 2000s affected the oil tanker market through three main channels: the price of oil, the dollar and 'contango or oil-storage trade'. In economic analysis the price mechanism provides a faster and more reliable indicator of the prospects of a market than volumes of trade. In the oil tanker market this role is played by the price of oil. In normal conditions when the price of oil rises it suggests that demand for seaborne trade improves, and vice versa. The exception to this rule arises when the price of oil rises because of geopolitical risks. Concerns about the disruptions of supply as a result of of wars, sanctions such as those imposed on Iran, and oil embargoes such as those which occurred in the 1970s can send the price of oil skyrocketing without that implying any improvements in the market conditions for seaborne trade. In practice, it is easy to distinguish between demand and supply shocks because the latter tend to attract media coverage. On the other hand, the impact of the channelling of the excessive liquidity to oil and other commodities from 2003 onwards is difficult to discern. This liquidity has distorted the oil price mechanism by making demand grow faster than was justified by economic fundamentals in the ballooning phase of the commodities bubble. Similarly, when liquidity was finally withdrawn in 2011, demand was growing at a smaller rate than was justified by economic fundamentals.

It is easy to see how the value of the dollar affects the oil tanker market. A fall in the dollar is associated with rising risk appetite for risky assets, including oil and other commodities. Such carry trade would push the price of oil up without implying improved conditions for seaborne oil trade. The influence of contango or oil storage trade is more complicated and requires some further analysis. In the rest of this section this matter is considered in more detail.

Speculators have an incentive to be long or short in oil depending on whether there is a premium or discount in the spread of the futures over the spot price of oil. When the futures price exceeds the spot price (that is, there is a premium) investors prefer to take a short position, whereas when there is a discount they will prefer to take a long position. The premium is usually called contango, whereas the discount is backwardation.[3] Because the futures price must converge to the expected future spot price at the expiry of the futures contract, in contango markets futures prices are falling over time, whereas in backwardation markets futures prices are rising. In contango markets investors with a long position are penalised by the premium, but those with a short position are rewarded by that amount. Hence, contango markets create a bias for short positions and are therefore associated with the prospect of falling markets in the future. An uptrend may persist in the short term, but the larger the premium (that is, the bigger the contango), the more likely the market would fall in the long term. By the same token, in backwardation investors with a short position are penalised by the discount, whereas those with a long position are rewarded by that amount. Hence, backwardation implies a rising market in the long term, although the downtrend may persist in the short term.

Some people believe that backwardation is abnormal, and that when it occurs it suggests insufficient supply in the corresponding (physical) spot market. However, many commodities markets are frequently in a state of backwardation, especially when the seasonal aspect is taken into consideration, as in the case of perishable and/or soft commodities.[4] But Keynes (1930, chapter 29) argued

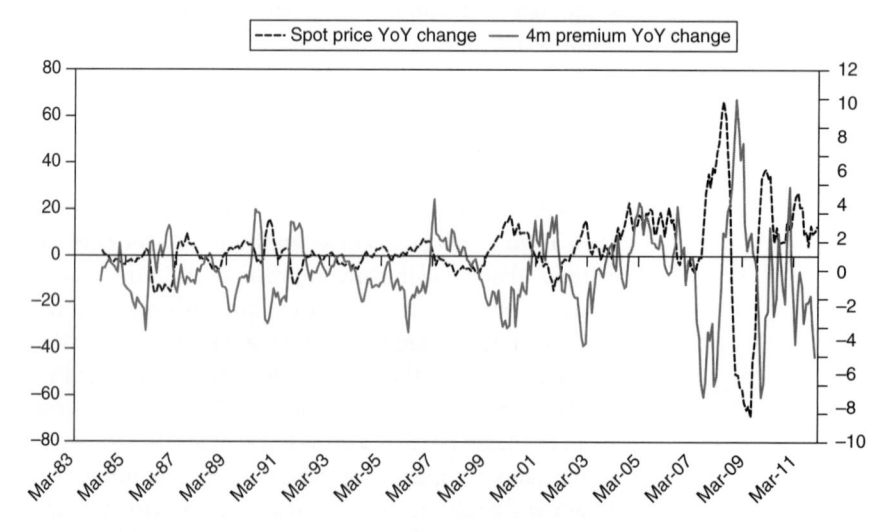

Figure 8.4 Oil premium and spot oil price (YoY change)

that in commodity markets backwardation is not an abnormal market situation because producers of commodities are more prone than consumers to hedge their price risk.[5]

Figure 8.4 confirms the coexistence of contango and backwardation in the oil market and the trading opportunities that arise from these patterns. The graph plots yearly changes (not percentages) in the dollar price of oil (LHS) along with yearly changes in the premium of the four-month futures spread over the spot price of oil (RHS). The graph highlights the strong negative correlation of these two variables; when the premium increases, the spot oil price falls and vice versa.

In spite of Keynes' verdict, investors in financial markets believe that when a market is backwardated, it indicates a perception of a current *shortage* in the underlying commodity, but expectations of ample supply in the future. Similarly, when a market is in contango, it indicates a perception of a current supply *surplus* in the commodity, but expectations of future shortage. The persistency of these patterns gave rise to 'oil-storage trade' or 'contango trade'. In contango trade oil companies and financial institutions buy oil for immediate delivery, store it in storage tanks, and sell contracts for future delivery at a higher price. Unless expectations change, when delivery dates approach, they close out existing contracts and sell new ones for future delivery of the same oil. Oil-storage trade is only successful if the oil market is in contango, as it is cheap to buy now when there is ample supply and sell it later on when a perceived shortage arises. Oil-storage trade grew impressively from 2005 onwards. This is exemplified by the number of Exchange Traded Funds (ETF), which between 2005 and 2010 rose from two to 95 with total assets rising from $4 billion to nearly $100 billion.

In recent years the advent of investors into oil and other commodities increased price volatility and made them more unpredictable. The market attracts all kind of investors: long term, short term, contrarian and noise traders or herd investors. Long-term investors take the view that oil prices will rise over the long term whereas short-term investors are pouncing on pricing anomalies. Herd investors are simply following trends or fads. The oil-storage trade gives rise to excessive volatility in the oil market.[6] By buying now in the spot market and simultaneously selling short the futures, the spot price rises, while the futures price falls. When contango is replaced by backwardation and vice versa these trends are exacerbated. At the peak of oil-storage trade in mid-2008, when contango had reached new highs of nearly $4, oil companies and speculators alike had a financial incentive to save oil in storage tanks for sale later on. The fall of the US economy off the cliff in mid-2008 triggered a sudden sale of all stored oil, as there was a shift in expectations from a bull to a bear market. The unwinding of contango trading led to an unprecedented fall in the price of oil from $145 to $30 in the second half of 2008.

In the initial phase of the oil-storage trade speculators hired oil tankers to store oil before selling it onto the market. It is speculated that a little less than 10 per cent of oil tankers were used to store oil rather than transport it. Oil storage shifted from oil tankers to onland tank farms, as it was cheaper to store oil in the latter than the former. The US EIA estimated that holding crude oil would cost a company between $1.50 and $4.00 per barrel per year depending on whether it owns or rents storage. For gasoline, the costs would be beween $2 and $6 per barrel per year, or $0.01 per gallon per month.

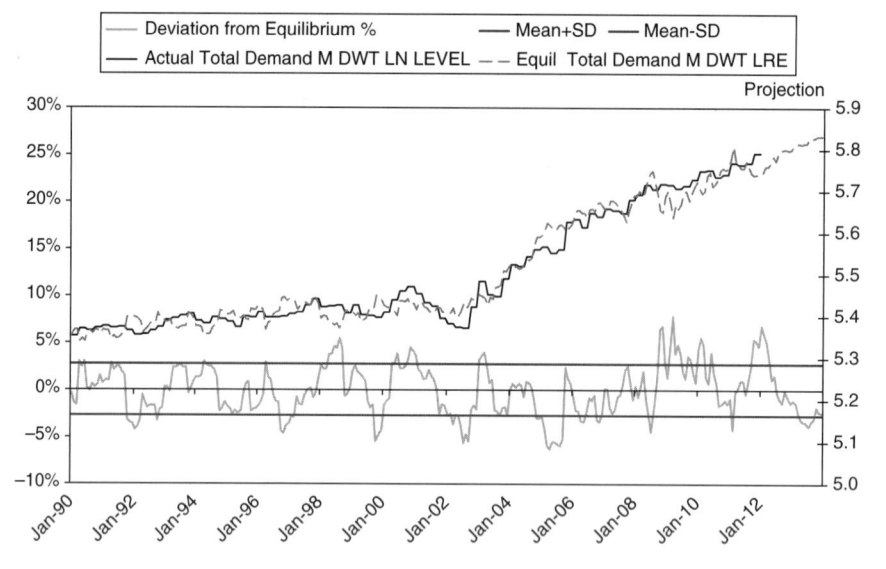

Figure 8.5 Demand – long-run equilibrium

Oil-storage trading is profitable if the future price exceeds the spot price by the cost of carry. In equilibrium, the contango should not exceed the cost of carry, because producers and consumers can compare the futures contract price against the spot price plus the cost of carry, and choose the better one. Arbitrageurs can sell one and buy the other for a theoretically risk-free profit. But in the real world the herd syndrome of investors can make contango exceed the cost of carry for around 12 months.

4 A STRUCTURAL CHANGE IN THE OIL TANKER MARKET

In traditional analysis the demand for seaborne oil trade depends on world oil consumption and the distance covered between production and consumption centres. So, the two variables should be moving in tandem. However, as Figure 8.6 illustrates the demand for oil moves more or less in a straight line, meaning that it has been expanding by a fixed rate, as the graph is plotted in natural logarithms. Yet the demand for seaborne oil trade is a kinked curve, suggesting a structural change (see Figure 8.6). Until the end of 2002 the demand for seaborne trade was expanding at a moderate pace; but from 2003 onwards the uptrend is much steeper. The popular explanation is that this structural change is due to China. China's growth rate accelerated from 2003 onwards and therefore so did its demand for raw materials and oil. In fact, the world demand for oil shows a slightly faster pace since 2003. But this is not sufficient to explain the very pronounced increase in the demand for seaborne oil trade (see the trends on each line in Figure 8.6). The underlying reason is that at the end of 2002 China's share of global oil consumption was only 7 per cent. By the end of 2011 this share had increased by 3.8 per cent to 10.8 per cent. This is a small change in world oil consumption to account for the very large increase in seaborne oil trade. A more profound reason for this structural change in the demand for seaborne oil trade is the excessive liquidity in the world economy.

5 SOLVING THE PUZZLE OF THE STRUCTURAL CHANGE

Now we are ready to combine all the pieces that account for the structural change in the oil tanker market. The excessive liquidity has affected the oil tanker market through the dollar, the oil price and the contango or oil-storage trade. Accordingly, the demand for seaborne oil trade depends upon the following economic fundamentals: global oil consumption, the value of the dollar, the price of oil and the oil-storage trade captured by the premium/discount of the four-month futures spread over the spot price of oil. These economic fundamentals can explain 96 per cent of the variation in the seaborne oil trade in the last 22 years, but, more importantly, can account for the kinked demand curve. In other words, when the influence of excessive liquidity, through the dollar, the price of oil and the oil-storage trade, is added to world oil consumption there is no structural change (see Figure 8.5). Table 8.1 gives some basic statistics of the oil tanker market, while Table 8.2 provides a quantitative assessment of each factor to the change in demand for seaborne oil trade.

Between 2003 and 2011 the change in demand for seaborne oil trade increased by 30 per cent (see Tables 8.1 and 8.2). In this period world oil consumption increased by 11 per cent. The traditional approach to the oil tanker market attributes to China the structural change in the demand for seaborne oil trade. However, the new framework for analysing the oil tanker market attributes only 28 per cent of this change to world oil consumption. Therefore, even if the entire increase in world oil consumption is attributed to China, it accounts for less than one-third of the increase in demand for seaborne oil trade. Two-thirds of the change in the demand for seaborne oil trade is accounted for by excessive

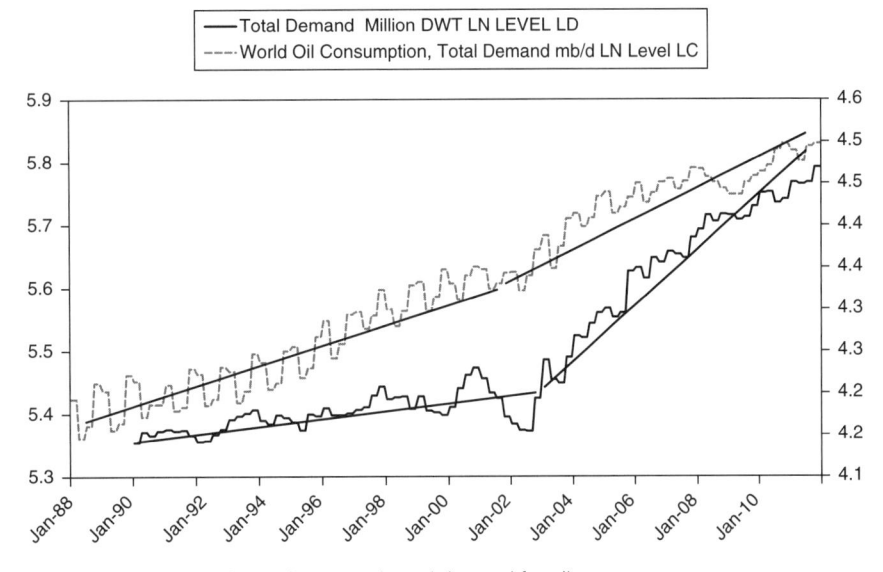

Figure 8.6 Demand for seaborne trade and demand for oil

Table 8.1 The oil tanker market in the last eight years

	Percent change between 2003 and 2011	Average rate of growth over the period	Standard deviation	Max over the period	Min over the period
Demand for Seaborne trade	30%	4.3%	2.3%	8.1%	0.9%
Fleet expansion M dwt	40%	4.9%	1.2%	7.0%	3.3%
Average Earnings $pd	−75%	30,423	12,103	44,591	13,907
Secondhand prices $ per dwt	8%	583	157	794	405
Newbuilding prices $ per dwt	28%	564	108	739	379

Table 8.2 Decomposition of change in demand for seaborne trade between 2003 and 2011

	Percent change between 2003 and 2011	Contribution to Demand growth	Percent Contribution to Demand growth
World oil consumption	11%	8.6%	28%
US dollar	−23%	8.1%	27%
Oil price	112%	8.2%	27%
Oil storage trade	3.6%	4.3%	14%
Other factors		1.1%	4%
Total demand change		30.3%	100%

liquidity. Of the three factors that account for the impact of liquidity, the dollar accounts for 27 per cent, the price of oil for another 27 per cent and the oil storage trade for 14 per cent (see Table 8.2). The implication of this new framework is that the oil tanker market has changed from a fundamental transport industry to a risky asset, akin to equities and commodities. This has the further implication that the prospects of the oil tanker market depend on the appetite for risky assets in addition to economic fundamentals in the shipping market, such as world oil consumption. Failure to appreciate these fundamental changes can lead to erroneous conclusions about the prospects for the oil tanker market.

6 THE FINANCIALISATION OF THE DRY BULK MARKET

In the long run, the demand for shipping services in the dry cargo market has kept pace with the growth in the net fleet stock. Since 1990 demand and supply in the dry market have more than doubled (see Figure 8.7). Two distinct trends can be identified. The first is from 1990 to 2003, when demand and supply grew at a modest pace. But from 2003 onwards a steeper trend can clearly be discerned with both demand and supply increasing at a much faster pace. In spite of the long-run relationship between demand and supply, in the short run there have been systematic periods of excess demand or excess supply, which, according to the mainstream model, along with the price of bunkers account exclusively for the variability of freight rates. Figure 8.8 shows the fleet capacity utilisation (CU) and the Baltic Dry Index (BDI), a surrogate measure of freight rates. The most remarkable feature of this relationship is the increased sensitivity of freight rates to the fleet CU through time. From 1990 to 2003 the responsiveness of BDI to CU was low. Nonetheless, the general pattern that freight rates increase when there is excess demand can be verified. But from then onwards the sensitivity has nearly doubled and this coincides with the improved fortunes in the dry market. This feature can be verified by comparing the relationship between BDI and CU in two subperiods: 1990 to 2003, and 2003 to 2010. The coefficient of CU in the second period is almost double that in the first; in the period 1990–2003 the coefficient is 11.3, whereas in the post-2003 period it is 20.9. These coefficients measure the elasticity of BDI with respect to CU. Thus, in the pre-2003 period a one per cent increase in CU leads in the long run to an 11.3 per cent increase in BDI, but in the post-2003 it leads to an increase of 21 per cent. The second observation is that the volatility of BDI has nearly trebled in the post-2003 relative to the pre-2003 period: the standard deviation soared from 23 per cent to 63 per cent in the second period. Finally, these equations show that the importance of CU in explaining the volatility of BDI has increased markedly, from 67 per cent prior to 2003 to 80 per cent afterwards. What accounts for this sharp increase in the sensitivity of freight rates to the fleet capacity utilisation?

Two factors account for the increased responsiveness of freight rates to market conditions; first, the rise of China to pre-eminence in world trade; and second,

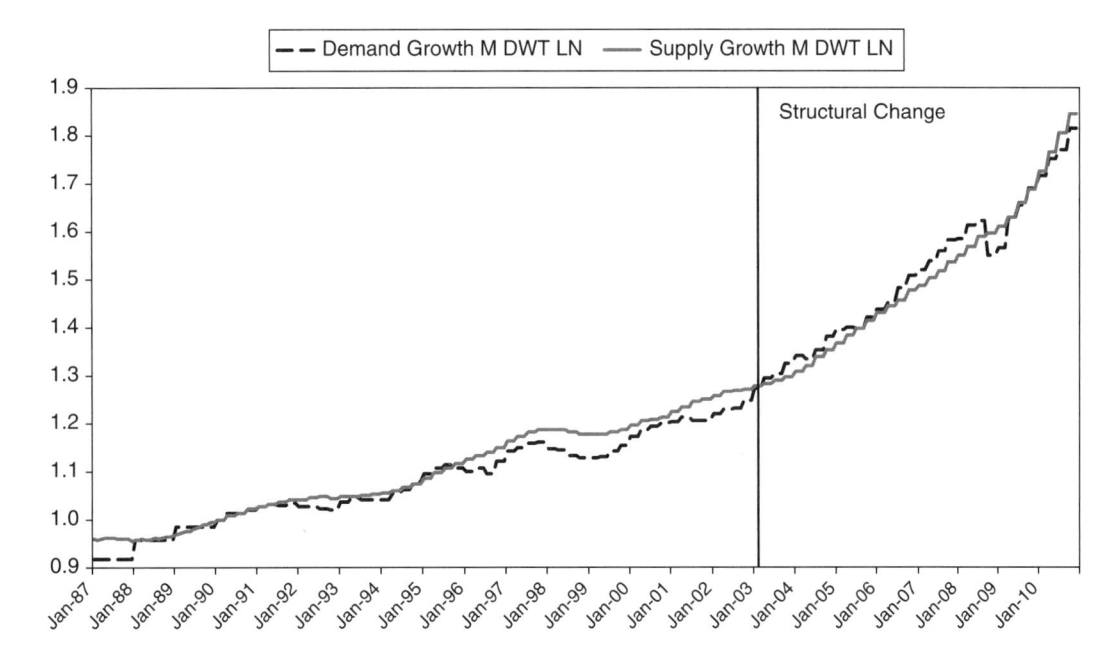

Figure 8.7 Demand, supply growth in dry market

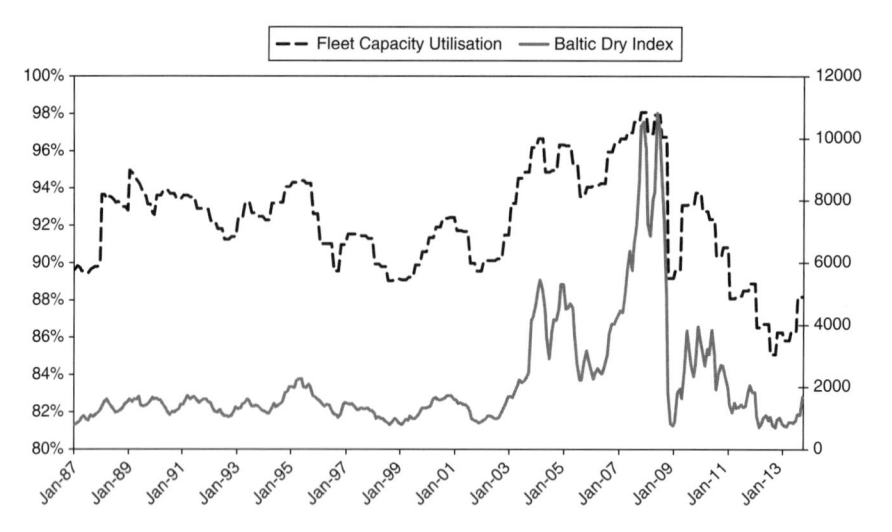

Figure 8.8 Fleet capacity utilisation and BDI

the excessive liquidity in the world economy that has financed a series of bubbles in the last ten years, such as the internet, housing, commodities and shipping. The increasing role of China in the dry market has been made possible by a shift of production from the western world to low-wage countries such as China and India. The shift has been gradual, but it has occurred mainly in waves. In every recession more companies shift production to China in order to survive the stifle competition at home. These relocation trends are captured in the statistics of Foreign Direct Investment (FDI). The influx of FDI has been prompted by changes in China's policies that welcomed the relocation of foreign companies as a means of acquiring 'know-how'. This shift of policy coincides with China's efforts to promote growth through exports. This export-led strategy was successful in the first decade of the new millennium.

If China is responsible for the improved fortunes of the dry market in the period of 2003–08, then there should be a structural change not just in freight rates but also in the demand for dry. Figure 8.9 verifies that this is indeed the case. Demand grew at a trend of around 2 per cent to 2003, but at 7 per cent thereafter. Thus the graph provides support of a kinked demand curve, evidence of a structural change. The popular explanation for these structural changes in the dry market is that they are due to the advent of China to pre-eminence in world trade. But the data do not support this hypothesis because the structural changes in China do not coincide with the structural changes in the dry market in 2003. Thus, Chinese steel production incurred two structural changes in 2001 and 2006. As Figure 8.10 shows, steel production grew at around 8 per cent in the period 1987–2001, 17 per cent in the period 2001–06 and at 12 per cent thereafter. Similarly, there were also two structural changes in cargo handled in major ports in China, a direct measure of the importance of China in world trade. These structural changes occurred in 1998 and 2007 (see Figure 8.11). In the period 1985–98 the trend in cargo handled at major ports in China grew at the comparatively moderate pace of 8 per cent. This rate more than doubled (17 per cent) in the period 1998–2007, but abated to 12 per cent thereafter (see Figure 8.11). The three-year delay in the change of trend in steel production following the change in trend in cargo handled at major ports in China is reasonable. China managed first to attract foreign direct investment, which helped its export sector and hence the economy, and then went on to develop its domestic demand, which required more steel. Therefore, one needs to look elsewhere for an explanation of the structural changes in the dry market.

Although the role of China is widely recognised in the shipping industry there is a second factor, which is largely overlooked and less understood. China is important to the extent that it provided a selling story for the surge in commodity prices in 2003–08 and 2009–11. The surge in commodity prices was triggered by China, but it was accentuated by banks channelling liquidity into commodities.

In the last ten years or so the nature of the dry market has changed from a fundamental transport industry to a risky asset market, akin to equities, commodities, corporate bonds and currencies. From this perspective, freight rates should be viewed as asset prices or derivatives rather than as prices that equilibrate the demand for and supply of shipping services (see Chapter 2 for a theoretical justification of such an approach). The main characteristic of an asset market is that it is a discounting mechanism of the future implications of current events. Asset prices are shaped by expectations of economic fundamentals. In this framework the fundamentals of demand and supply can no longer account fully for freight rates. There is a premium or discount in freight rates that depends on liquidity and the degree of aversion for risky assets. In good times there is a premium as liquidity is channelled into commodities, as occurred in the boom years of 2003–08 and again in 2009–11. But in bad times, as in the period 2011–13, freight rates are at a

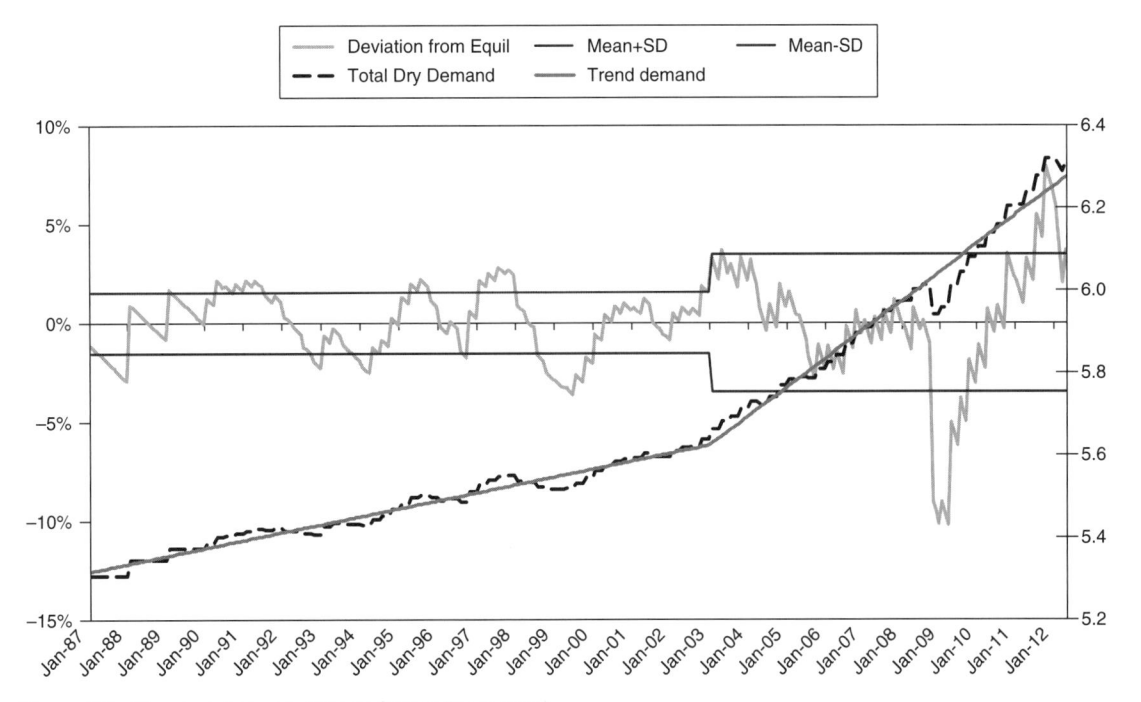

Figure 8.9 The deviations of demand from a linear trend

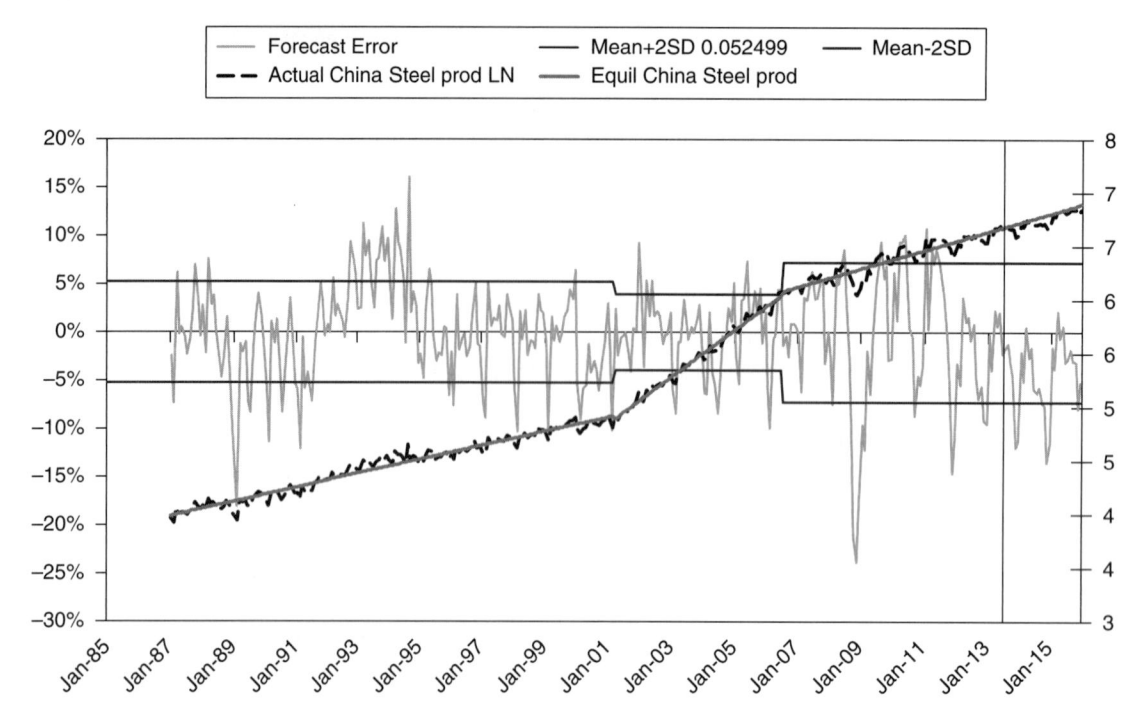

Figure 8.10 Deviations of China steel production from trend (8%, 17%, 12%)

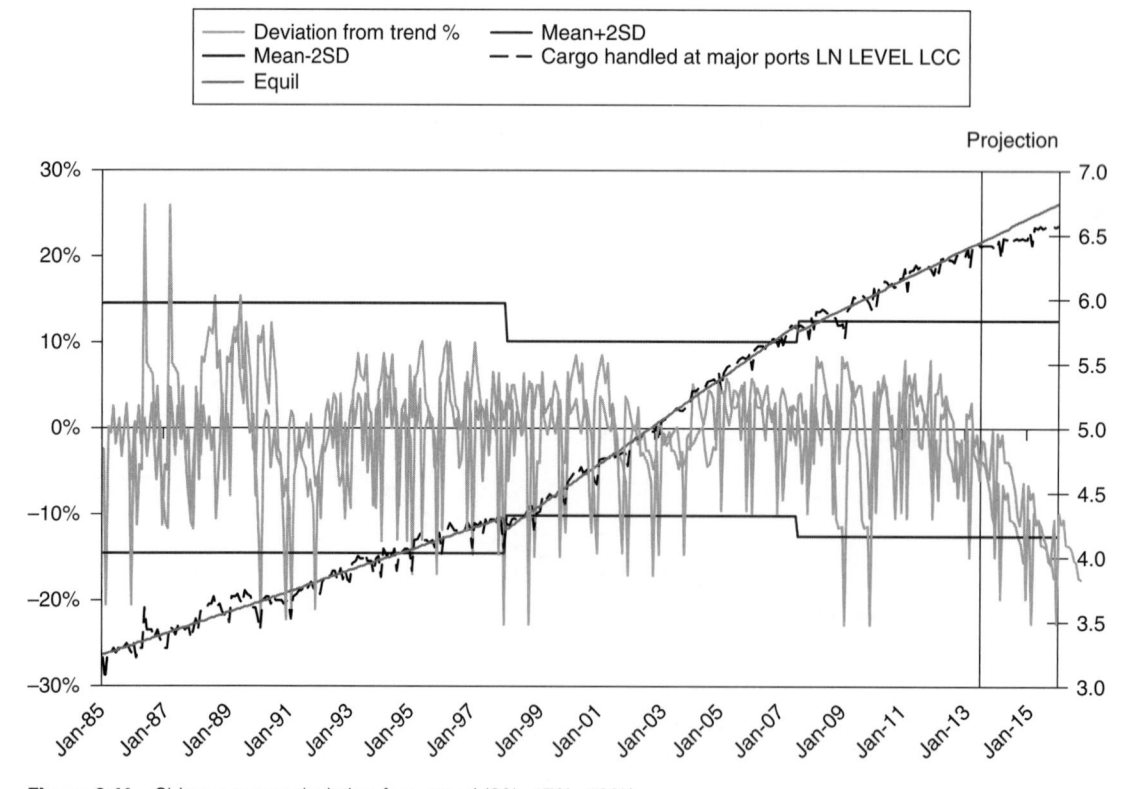

Figure 8.11 Chinese cargo: deviation from trend (8%, 17%, 12%)

discount to economic fundamentals, as liquidity is withdrawn from commodities and other risky assets. This change has occurred gradually as business cycles are no longer demand- or supply-led, but rather asset-led, driven by liquidity. This has made shipping cycles more coincident with business cycles and subject to the same forces that drive business cycles.

This theory implies that both charterers and owners form expectations of future freight rates on the basis of current information. As new information comes in (for example, that China would not reflate its economy) both agents infer the future implications for demand and supply for shipping services and compute the present value of the gains or losses and hence the equilibrium freight rates that would result from the new information (see Chapter 2 for more details). This is exactly the same pricing principle as in the case of all other risky assets, such as equities and commodities. The current asset price reflects the future implications of a change in current economic fundamentals and discounted back to today. Thus, freight rates and equity prices are determined by the same principles. Both are risky assets. A risky asset is different from a risky business. Shipping has always been a risky business. Freight rates have also become risky assets (derivatives). There is no reason to assume that in 2003 there was a sudden change in the way that both agents formed such expectations. Quite the contrary, the same principles have applied all along. So, what happened?

In 2003 investment banks began to invest in commodities by setting up their own proprietary trading desks. They were motivated to do so by the observation that contango and backwardation were permanent states of affairs in many commodity markets. This observation meant that banks can make profits by taking short positions in contango markets and long positions in backwardated markets without assuming any excessive risk. Retail investors were also lured to this lucrative game through ETFs. These expanded dramatically durng this period: In 2005 there were only a few ETFs trading in commodities with assets of less than $3 billion; by 2009 these assets had expanded to more than $100 billion.

The entrance of investment banks and retail investors into the commodities market distorted the signals of demand and supply to charterers and owners.

Box 8.1 Freight rates as asset prices

An owner was reporting in Feb 2013: "one vessel is delayed by ten days in US West coast Colombia River grain loading; another ship is delayed by thirty days in Paranagua-Brazil grain loading and before that was delayed about twenty six days discharging cement in W. Africa and another in Iraq discharging grains by 14 days, yet freight rates are depressed". How is this possible?

The nature of the dry market has changed from a fundamental transport industry to an asset market. Freight rates are depressed not because there is no cargo, but because sentiment is bad as a result of the failure of policymakers to address the problem of inadequate growth in the world economy.

The higher prices associated with the scenario of an increase in demand (as opposed to an increase in supply) based on the widespread observation that China was growing much faster than before led to an overoptimistic assessment of the demand for dry. Giant commodity producing companies, such as H. B. Billiton, Rio Tinto and Vale of Brazil, expanded capacity to accommodate the seemingly huge increase in demand, only to curtail it later on. After the burst of the commodity bubble, commodity prices started falling faster than justified by economic fundamentals as liquidity was withdrawn from commodities and invested back into US stocks. This again distorts the signals of demand and supply to owners and charterers, but on this occasion by creating gloomy expectations. Astute owners (see Box 1) have observed that there is sufficient cargo yet freight rates are depressed.

7 CONCLUSIONS

In 2003 there was a structural change in both the oil tanker market and the dry market. This coincides with the financialisation of the commodities market triggered by the advent of institutional investors. The price of oil, the dollar and contango or oil storage trade account for the impact of liquidity on the oil tanker market. Liquidity, along with oil consumption, account for the structural change in the oil tanker market.

The trend in the demand for dry increased from 2 per cent to 7 per cent. The supply of shipping services also increased, albeit with a lag, ultimately leading to an excess supply. Freight rates became more responsive to fleet capacity utilisation (or the demand–supply balance). The volatility of freight rates nearly trebled; and fluctuations in fleet capacity utilisation account for a larger proportion of the variance of freight rates. Two factors account for this structural change: the rise of China to pre-eminence in world trade; and the channelling of liquidity into commodities. The role of China bears greater responsibility for the increase in demand, whereas liquidity is more responsible for the structural change in freight rates.

These structural changes in the wet and dry markets have changed the nature of shipping from a fundamental transport industry to an asset market, akin to equities, bonds, currencies and commodities. Freight rates should now be viewed as asset prices or derivatives rather than as prices that equilibrate the demand and supply of shipping services. This means that freight rates can be higher or lower than justified by economic fundamentals, namely from the level consistent with demand and supply. There is a premium over the fundamental price when risk appetite increases; and there is a discount when there is rising risk aversion.

9 THE INTERACTION OF BUSINESS AND SHIPPING CYCLES IN PRACTICE

EXECUTIVE SUMMARY

Chapter 6 shows that shipping cycles are caused by business cycles. Actual and expected macroeconomic developments shape expectations of shipping market conditions, which affect in turn the outcome of the bargaining between owners and charterers over current freight rates. Given a shipyard delivery lag of two years, the supply of shipping services is largely predetermined by past expectations of current demand for shipping services. As a result, large swings in expectations of demand cause a disproportionate increase in the supply of fleet in the boom years. When the euphoria dissipates, the installed fleet leads to lower fleet capacity utilisation, thereby creating a shipping cycle. Volatile expectations can be rational, as a result of cyclical developments in macroeconomic variables, or irrational, what Keynes called 'animal spirits', a situation where economic fundamentals remain unchanged and yet expectations swing from optimism to pessimism or vice versa.

As is shown in Chapter 5, expectations of macroeconomic variables can be approximated by expectations of real interest rates, as these affect aggregate demand in the economy (real GDP growth) and, consequently, inflation. Central banks, by changing nominal interest rates, shape the expected path of future real interest rates with a view of steering growth and inflation to their target levels. This process affects the expectations of owners and charterers about market conditions in shipping, and therefore the shipping cycle. This is the underlying theory of the interaction of business and shipping cycles. In this chapter we examine whether this theory is consistent with empirical evidence. In doing so we analyse the business cycles of the US, Japan, Germany and China and assess their impact in explaining shipping cycles.

The conclusion is that in the 1980s and the 1990s the business cycles of Japan, and to a lesser extent of Germany, shaped shipping cycles. But throughout the 1980s, the business cycles of Japan were primarily determined by developments in world trade because of its dependence on exports. World trade in turn was affected by the US because of its huge impact on the world economy. In the 1990s, the business cycles of Japan were also affected by domestic developments and, in particular, by the ramifications of the burst of the property and the equity bubbles in 1989. But the US continued to affect the world economy and therefore Japan's exports. In the 2000s, China supplanted Japan in its dependence on world trade – an export-led economy. The US has shaped the last two business cycles of the world economy and consequently of China in the twenty-first century. Therefore, although China now determines the long-term growth of the demand for shipping services and, in particular, the dry market, the US continues to trigger fluctuations in this demand and therefore accounts for the shipping cycles.

The theory of freight rates, developed in Chapter 2, shows that freight rates should be viewed as asset prices and the financialisation of shipping, analysed in Chapter 8, confirms that this has become the case in the twenty-first century. Section 1 offers evidence of freight rates as a leading indicator of business cycles, which confirms their role as asset prices. The rest of this chapter is organised as follows. Section 2 examines the stylised facts of shipping cycles, which are explained gradually in the rest of this chapter. Section 3 analyses the official classification of business cycles, using the US as an example because of its might in affecting world cycles, and puts forward an alternative approach that enables the distinction to be drawn between 'signal' and 'noise' in identifying trends and discerning the reversal of trends. This approach helps to compare the actual and optimal conduct of US monetary policy in business cycles and shows how correct expectations of interest rates can be formulated in shipping (section 4). Section 5 examines why US cycles differ and how this information helps to explain the stylised facts of shipping cycles. Sections 6 and 7 analyse the business cycles of Japan and Germany, the latter as a leading indicator of the rest of Europe. It shows that business cycles are mainly the result of how central banks respond to unexpected shocks in the economy and to the maturity of business cycles. It is this kind of information that helps participants in shipping form more elaborate expectations of interest rates without the need of a particular model. Section 8 uses the business cycle analysis of the US, Japan and Germany to explain the stylised facts on shipping cycles until the 2000s. The cycles since then are explained by China, which has supplanted Japan in pre-eminence in world trade. Although China is now the factory of the world economy, thereby explaining the long-term growth rate of demand for shipping services, the trigger for the fluctuations in demand has been the US because of its significance in shaping world business cycles. It will take time for China to reform its economy from export-led to domestic-led, a reform that has been endorsed at the Third Plenum of the Party in November 2013.

The irrational volatility of expectations of demand for shipping services – the 'animal spirits' – is due to a herd syndrome that induces a swing in expectations between optimism and pessimism governed by greed and fear. This swing is related to uncertainty about macroeconomic developments. Therefore, a full explanation of the stylised facts of shipping cycles requires an extension of the business cycle analysis to conditions of uncertainty and the role of the cost and availability of credit (ship finance). The theory of the fleet capacity expansion under uncertainty, analysed in Chapter 3, provides the basis for this extension. Section 9 explains how uncertainty about demand can lead to overcapacity, as owners may decide to wait until the recovery is sustainable before investing. Uncertainty in demand also helps to explain why secondhand prices usually lead newbuilding prices in the course of the cycle. If the recovery of demand is largely unanticipated or the rebound is perceived as transient rather than permanent, owners would buy ships in the secondhand market rather than new ones. When the recovery is perceived as permanent newbuilding prices catch up with secondhand ones.

The availability of credit makes shipping cycles even more pronounced than would otherwise be the case. Banks and other credit providers are highly pro-cyclical; the loan portfolio increases in the upswing of the cycle and decreases in the downswing. This is due to the myopic attitude of credit institutions in granting credit according to the collateral value of the loan, which is highly pro-cyclical. Therefore, ship finance increases the amplitude of shipping cycles; a topic analysed in section 10. The credit market for shipping, as in other industries, does not clear through prices (interest rates), leading to credit rationing (an excess demand for credit) in the downswing of the shipping cycles. This makes the availability of credit an additional channel to interest rates in shipping cycles. The reasons for credit rationing – a rational choice by the commercial banks or imposed by the central bank – are also analysed in section 10. The final financial factor that explains the large amplitude of shipping cycles is the reliance of investment on corporate profits (internal finance). Corporate profitability is highly cyclical, thereby making shipping cycles more pronounced – another topic that is discussed in section 10. Finally, section 11 summarises and concludes.

1 FREIGHT RATES AS A LEADING INDICATOR OF BUSINESS CYCLES

The interpretation of freight rates as asset prices in Chapter 2 implies that they can act as a leading indicator of world economic activity. In the 'Great Recession' of 2008–09 central banks, anxious for evidence that their measures started to have an impact on economic activity, looked at freight rates and took comfort from the recovery of the BDI. The BDI bottomed in December 2008 before rebounding sharply in the first half of 2009, thereby signalling that there was an upturn in world trade (see Figure 9.1). The recovery of the BDI preceded even stock prices by three months. The S&P 500 bottomed in early March and then rose sharply, signalling an imminent recovery of the economy (see Figure 9.1). In fact, the US and other economies entered the recovery phase in mid-2009. Thus the BDI acted as a six-month leading indicator of the trough of the business cycle, whereas the S&P 500 as a three-month leading indicator.

> **Box 9.1 Freight rates as a leading indicator of business cycles**
>
> An owner in late 2012 observed: if shipping is leading business cycles in the same manner as stock prices do and sometimes even faster, what should be my guidance to shipping developments?
>
> The answer is economic policy. Forming expectations of the likely conduct of monetary and fiscal policy would enable the owner to predict, to a reasonable degree, the business cycle and then through a structural model of shipping, like the K-model, the impact on the demand-supply balance in the dry market and the other key variables, freight rates and vessel prices.

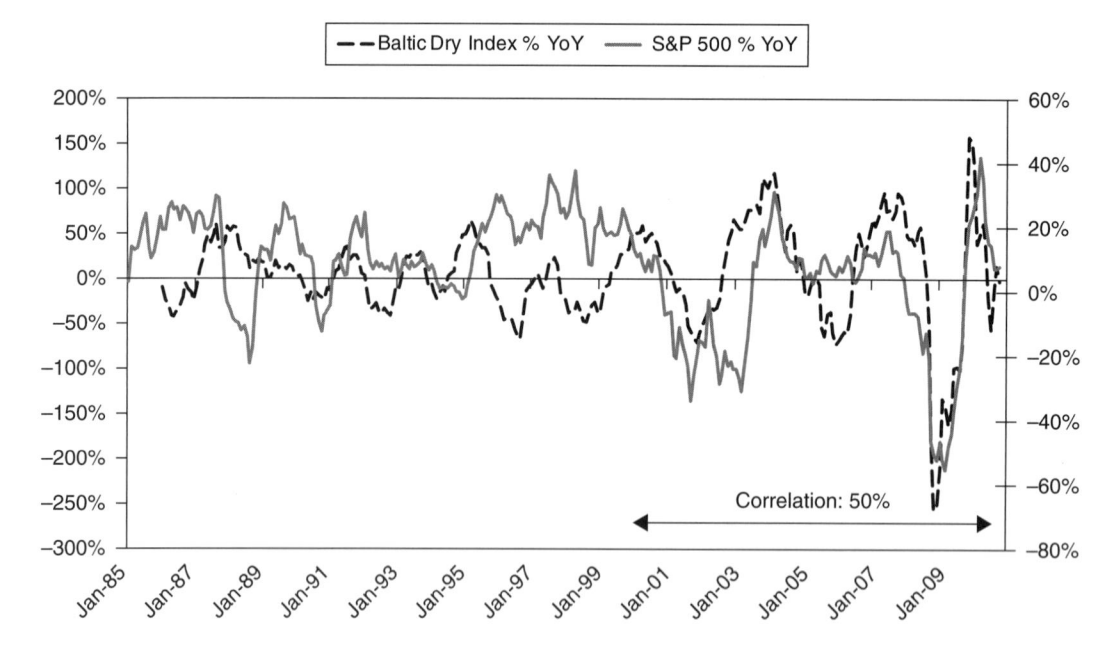

Figure 9.1 BDI and S&P 500

Figure 9.1 shows the increasing correlation of the BDI with the S&P 500 index over the past decade. The graph confirms that the shipping market acts increasingly like a leading indicator of economic activity. The association has been particularly pertinent over the past ten years or so, with the correlation rising to 50 per cent. The relationship is not a causal one. Movements in the S&P 500 do not cause changes in freight rates, and vice versa. Instead, the increasingly close association of the two variables, as of late, is the result of common determinants, which have their roots in the financialisation of the shipping markets. The common factor that drives equity prices and freight rates is liquidity, as explained in the previous chapter.

2 SHIPPING CYCLES: THE STYLISED FACTS

Table 9.1 shows the various cycles of freight rates in the dry market. As the BDI began in 1985, we have used for the entire period the one-year time charter rate. Each cycle is measured from peak-to-peak and is divided into three phases: recession, recovery and expansion.[1] A cycle may contain a double-dip recession, namely a partial recovery that does not reach the previous peak followed by another dip. The recovery is measured from the deeper of the two troughs to the peak of the previous cycle. The expansion phase is measured from the end of the recovery to the new peak. Employing these criteria there have been five identifiable cycles since 1980. Cycles have varied in length from 50 to 106 months with an average of 84 months (or seven years); the longest cycle was from May 1995 to March 2004, but it contained a double-dip recession from November 2000 to November 2001. The average recession has lasted 36 months, but they have varied in length from 23 to 45 months. The deepest recession was in 2008. The BDI fell a staggering

Table 9.1 Shipping cycles

Shipping Cycle (Peak to Peak)	Length of Shipping Cycle (months)	Recession Cycle (peak to end of recovery)	Recession Cycle Length (Months from Peak to Recovery)	Recession	Length of the Recession (months between Peak and Trough)	Depth of the Recession (Peak to Trough % fall)	Recovery	Recovery Length (Time between trough and previous peak level)	Expansion (Recovery to Peak)	Expansion Length (months)	Expansion Strength
Apr 80– Apr 88	96	Apr 80– Apr 88	96	Apr 80– Dec 82	32	−65%	Dec 82– Apr 88	64	0	0	0
Apr 88– May 95	85	Apr 88– Dec 94	80	Apr 88– Jan 91	33	−37%	Jan 91– Dec 94	47	Dec 94– May 95	5	12%
May 95– Mar 04	106	May 95– Oct 03	101	May 95– Feb 99	45	−51%	Feb 99– Oct 03	56	Oct 03– Mar 04	5	102%
Mar 04– May 08	50	Mar 04– Mar 07	36	Mar 04– Feb 06	23	−46%	Feb 06– Mar 07	13	Mar 07– May 08	14	94%
May 08–				May 08– Feb 12	45	−94%					
Average	84		78		36	−59%		45		6	52%

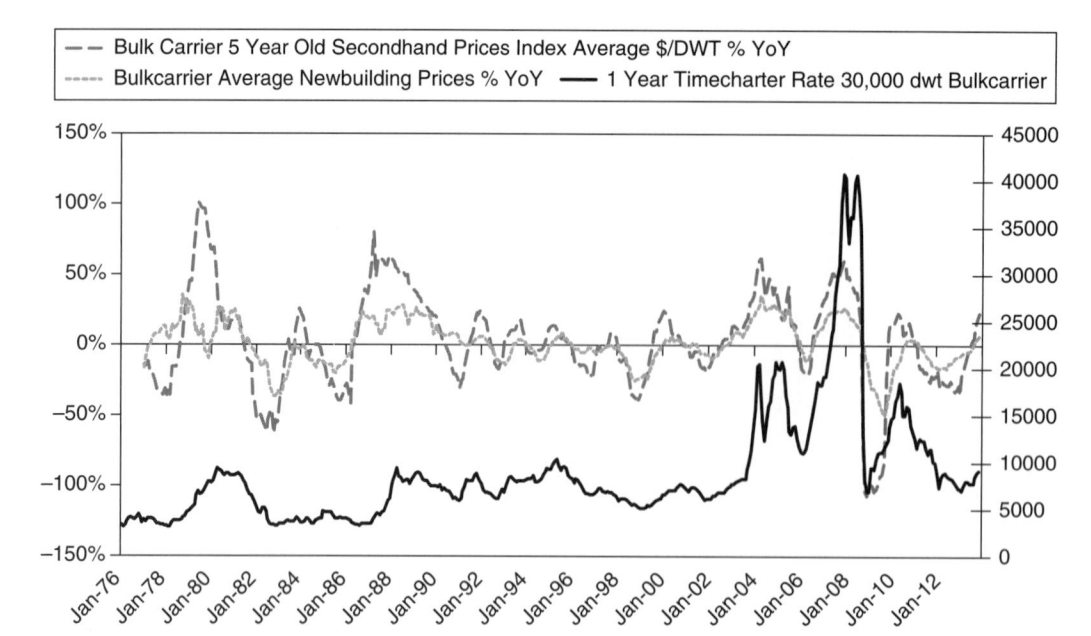

Figure 9.2 TC rates, SH prices and NB prices

94 per cent compared with an average fall of 59 per cent. Over the same period the one-year time charter rate fell by 83 per cent. The shallowest recession was between 1988 and 1991 when freight rates fell 37 per cent. The average recovery has lasted 45 months, but they have varied in length from 13 to 64 months. The longest recovery was in the 1980–88 cycle and this also contained a double-dip recession. The average expansion phase lasted for six months and freight rates advanced 52 per cent. But again there is a large variation. The expansion phase has ranged from zero to 14 months with the smallest profit being zero and the largest 102 per cent.

In summary, each shipping cycle is different in the sense that it has unique characteristics. But all cycles share some common features. All cycles are deep compared to those in the stock market. Although the average recession lasts for three years, the recovery is slow. It takes double the time of the recession to return to the previous peak level. Hence, the recession–recovery cycle is nearly four years. The following sections will attempt to explain these stylised facts by exploring the interaction of business and shipping cycles and the impact of uncertainty and finance on shipping cycles.

A shipping cycle is characterised by the movement of freight rates, which determine cash flows. The interrelationship of freight rates, secondhand and newbuilding prices is also a matter of investigation. Although freight rates are most of the time leading prices, secondhand prices are either a leading indicator or a coincident indicator of newbuilding prices (see Figure 9.2). The explanation of these stylised facts is the subject matter of this chapter. We begin with the interrelationship of shipping and business cycles, using the US because of its influence on world cycles.

3 US BUSINESS CYCLES

In the textbook treatment the business cycle is divided into three phases: recession, recovery and expansion. The recession is defined as the period in which the level of Gross Domestic Product (GDP) is contracting, therefore moving from peak to trough. The depth of the recession is measured by the percentage fall in the level of GDP between the peak and the trough. The recovery is defined as the period between the trough of GDP and the time needed to return to the previous peak level. This would mark a complete cycle as we have returned to where we started from – recession and recovery make one full cycle. The expansion phase is the period that follows the recovery phase until the next peak of GDP.

The rate of growth of GDP relative to the previous quarter (q-o-q) captures these turning points in the level of GDP (as it is an approximation of the first time derivative). Therefore, the rate of growth of GDP helps to discern the different phases of the business cycle. For example, the peak of the business cycle would be identified with one quarter lag as the new level of GDP would be lower and its rate of growth (q-o-q) negative.[2] As there may be temporary drops in the level of GDP conventional wisdom has defined a recession as a period of at least two consecutive quarters of negative growth (q-o-q). However, this is not necessary as the drop in the first quarter might outweigh the recovery in the second quarter, and might then be followed by a further drop in the third quarter. This is indeed what happened in the 2001 and 1960 US recessions. The change in the rate of growth of GDP between two consecutive quarters (which is an approximation of the second time derivative) would help to measure the intensity of the recession. If the change in the (q-o-q) rate of growth of GDP is negative, the recession is deepening; if it is positive the recession is easing. Obviously, before the trough of the recession is hit the second time derivative must become positive. In layman's terms the recession must ease with the economy showing signs of stabilisation, before the bottom is hit. Table 9.2 offers the characteristics of the US business cycles in the post-Second World War period.

The textbook treatment, however, is a crude approach compared with that used by the National Bureau of Economic Research (NBER), which is the official body for determining the chronology of the US business cycles. According to the NBER, 'a recession is a significant decline in economic activity spread across the economy, lasting for more than a few months, normally visible in production, employment, real income, and other indicators' (NBER 2008, p. 1). Because a recession is a broad contraction of the economy, not being confined to one sector, the NBER emphasizes economy-wide measures of economic activity and does not use a fixed rule. Instead, it draws its conclusion by considering a number of variables, such as employment, industrial production, the real disposable income of households less transfer payments, real consumer expenditure and wholesale-retail trade sales. The NBER believes that domestic production and employment are the primary conceptual measures of economic activity. In determining the chronology of business cycles with quarterly data the NBER uses the two most reliable comprehensive estimates of aggregate domestic production. These are the quarterly estimate of real GDP and the quarterly estimate of real Gross Domestic Income (GDI), both produced by the Bureau of Economic Analysis. Conceptually,

Table 9.2 The US business cycles

	Cycle Length (Quarters from Peak to Peak)	Characteristics	Recession	Depth of the Recession (Peak to Trough % fall)	Length of the Recession (quarters between Peak and Trough)	Recovery	Recovery Length (Time between trough and previous peak level)
1948–IV–1950–I	5	Double dip recession	1948–IV–1949–II	−1.7%	2	1949–II–1950–I	3
1953–II–1954–IV	6	V-type recovery	1953–II–1954–I	−2.7%	3	1954–I–1954–IV	3
1957–III–1958–IV	5	V-type recovery	1957–III–1958–I	−3.7%	2	1958–I–1958–IV	3
1960–I–1961–II	5	V-type recovery	1960–I–1960–IV	−1.6%	3	1960–IV–1961–II	2
1969–III–1971–I	6	Double dip recession	1969–III–1970–IV	−0.6%	5	1970–IV–1971–I	1
1973–IV–1975–IV	8	False Recovery 74–II	1973–IV–1975–I	−3.4%	5	1975–I–1975–IV	3
1980–I–1981–I	4	First Leg of Recession	1980–I–1980–III	−2.2%	2	1980–III–1981–I	2
1981–1–1983–II	9	Second Leg of Recession	1981–I–1982–III	−2.4%	6	1982–III–1983–II	3
1990–I–1992–I	7	Anaemic Recovery	1990–II–1991–I	−1.5%	3	1991–I–1992–I	4
2001–I–2001–IV	4	Anaemic Recovery	2000–IV–2001–III	−0.6%	3	2001–III–2001–IV	1
2007–IV–2009 II		Anaemic Recovery	2007–IV–2009–II	−4.6%	7	2009–II–?	
Average	5.9			−2.3%	3.4		2.5

the two are the same, because sales of goods and services generate income for producers and workers equal to the value of the sales. However, because the measurement on the product and income sides proceeds somewhat independently, the two actual measures differ by a statistical discrepancy.

In determining the beginning of the 2008 recession using quarterly data the NBER noticed that the income side gave an unequivocal signal of the beginning of the recession in 2007Q4, but the production side was ambiguous as the level of GDP in 2008Q2 was higher than in 2007Q4. Nonetheless, using monthly data the NBER noted that all series under consideration gave signals that the recession started anytime between November 2007 and June 2008. The NBER concluded that 'a peak in economic activity occurred in the U.S. economy in December 2007. The peak marks the end of the expansion that began in November 2001 and the beginning of a recession. The expansion lasted 73 months; the previous expansion of the 1990s lasted 120 months' (NBER 2008, p. 1). In reaching that conclusion the NBER noted that employment (the payroll measure, which is based on a large survey of employers) peaked in December 2007 and fell in every subsequent month of 2008. Moreover, the real disposable income of households less transfer payments peaked in December 2007 and although it oscillated until June 2008 it remained below the December 2007 peak. The deduction of transfer payments from the disposable income of households places the data closer to the ideal real gross domestic income. The calculation of real disposable income less transfer payments was made from the nominal magnitudes through the GDP-deflator. The latter is available only in quarterly data and the NBER interpolated the series to arrive at monthly inflation adjusted data.

Despite the significance of the (q-o-q) growth rate of GDP it is a volatile measure of economic activity as it oscillates around the rate of growth on the year earlier period (y-o-y) (see Figure 9.3). Hence, the (q-o-q) measure exaggerates current trends and in that way, most of the time, it is misleading as these trends are soon reversed. The (y-o-y) measure, on the other hand, underlines current trends, but it is bound to miss turning points. A cursory look at Figure 9.3 confirms the advantages and drawbacks of the (y-o-y) rate over the (q-o-q) measure. The former is a reliable indicator of current trends; but it is a lagging indicator especially of the trough and, on some occasions, of the peak of the business cycles. The (q-o-q) measure, on the other hand, is a leading indicator of turning points, but it is an unreliable indicator of current trends. For instance, the peak of the latest cycle is the third quarter of 2007. This is confirmed by both the (q-o-q) and the (y-o-y) measures (see Figure 9.3). The (q-o-q) rate shows clearly that the economy decelerated very rapidly and fell into recession in the first quarter of 2008, whereas according to the (y-o-y) measure the recession did not begin until the third quarter of 2008. Despite the early warning property of the (q-o-q) rate, its unreliability is emphasised by suggesting that the economy rebounded in the second quarter of 2008 and then fell again into recession in the third quarter, when in fact the (y-o-y) rate became negative.

The oscillations of the (q-o-q) growth rate around the (y-o-y) rate imply that the former mean reverts on the latter. The mean-reverting property (which in layman's terms means that what goes up must come down) helps to distinguish between 'signal' (that is, systematic factors, whose impact is long-lasting) and

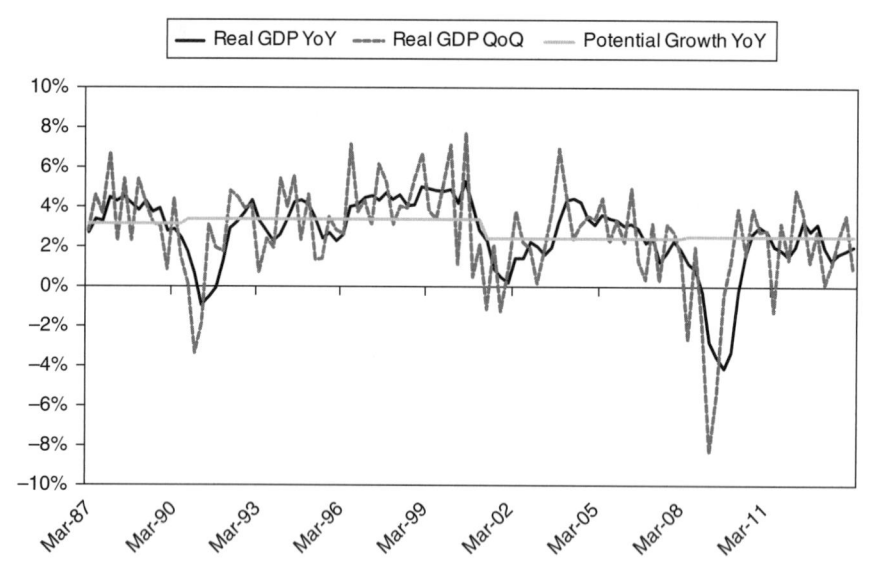

Figure 9.3 US real GDP in the last business cycle

'noise' (that is, factors which are purely random and whose impact dissipates very quickly). When the current trend is solidly in place the (q-o-q) measure is simply noise and, therefore, it should be ignored; but when there is suspicion that the trend might change the (q-o-q) measure provides a signal. Therefore, both measures are important in identifying trends and discerning reversal of trends when they are also used in conjunction with other information.

Moreover, the (y-o-y) measure bears a close relationship to inflation and therefore it provides an insight into the future conduct of monetary policy. Thus, in what follows we concentrate on the (y-o-y) rate of growth of GDP as we are interested in embedded trends, which have implications for inflation and monetary policy and, therefore, for shipping. We are conscious that the (y-o-y) rate would miss turning points and, in particular, the bottom of the cycle by approximately one quarter. By concentrating on the rate of growth of GDP than its level the cycle does not simply consist of recession and recovery; it also converts the expansion phase into a cycle. Hence, the previous division of the business cycle into three phases, namely recession, recovery and expansion, does not adequately describe the entire cyclical nature of the economy. In what follows an alternative division of the cycle is employed, which is more informative than the traditional textbook treatment and has implications for the formation of forward expectations in shipping.

Figure 9.4 divides the business cycle into four or five phases according to the peak and the trough of the rate of growth of GDP and the peak and trough of inflation. Point A represents the trend rate of growth of GDP, which is otherwise called the rate of growth of potential output, as it shows the maximum rate of growth with steady inflation. The economy can obviously grow at a faster pace than potential, but at the cost of higher inflation. When growth exceeds potential

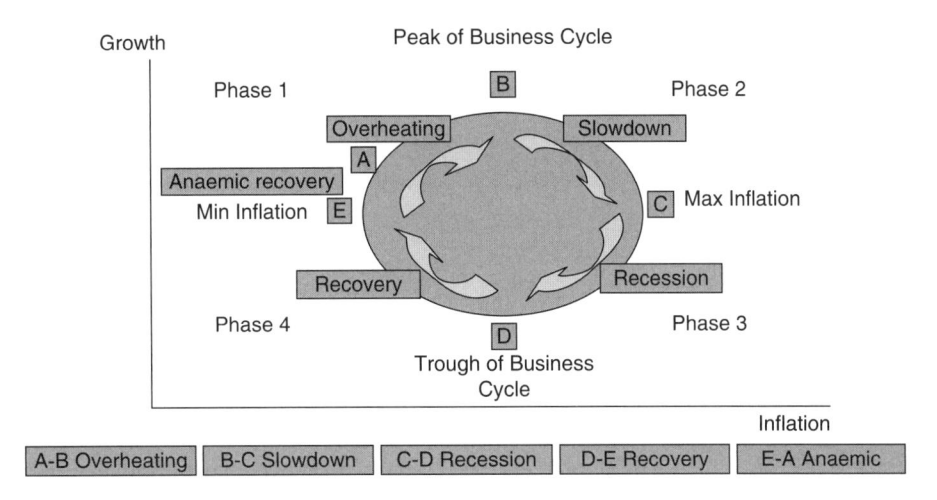

Figure 9.4 Business cycle

the economy is overheated and when growth falls short of potential the economy operates with spare capacity. Inflation rises with a lag when the economy becomes overheated, and falls, again with a lag, when there is spare capacity. Thus, point A shows the rate of inflation that corresponds to the rate of growth of potential output. Accordingly, point A represents a bliss point as this is the maximum rate of growth without an acceleration of inflation. Hence, point A is the target point of monetary policy. Central banks aim to minimise fluctuations around point A by moving interest rates up or down. The more successful central banks are, the higher the probability that as a result of their actions they would prolong the business cycle. Point B is the peak of the cycle, while point D is the trough. Point C is the maximum inflation in the cycle and point E the lowest inflation. The path of the economy in the business cycle is represented by a clockwise movement. As the economy moves from A to B it becomes overheated and inflation creeps up with a lag of approximately one year. Phase (I) is therefore called the overheating phase. In this phase growth rises and inflation also increases, thus giving rise to positive correlation between inflation and growth.

As the economy moves from B to C it cools down (growth slows), but inflation continues to increase. In this phase (II), which is called the slowdown phase, the correlation between inflation and growth is negative; growth decelerates but inflation goes up. As the economy moves from C to D there is recession. Inflation falls and growth becomes increasingly negative. Hence, the correlation between inflation and growth is again positive in phase (III). As the economy moves from D to E, the recovery commences, but inflation continues to fall. Thus, in the recovery phase (phase IV) the correlation between inflation and growth is again negative. As the economy moves from E to A, it returns back to normality. However, phase (V) might or might not be observed in the real world depending on the buoyancy of the recovery. From the 1950s through to the 1980s the recovery was very buoyant, meaning that growth exceeded potential very quickly. In all these cycles phase (V) was not observed. However, from the 1990s onwards the recovery has been anaemic, meaning that growth did not exceed potential for some time. In these cycles phase (V) has been observed. Accordingly, phase V has been dubbed anaemic recovery. The inability of the economy to grow faster than potential implies a jobless recovery with unemployment continuing to climb. In an anaemic recovery business investment is also likely to be subdued, thereby increasing the likelihood that the recovery will falter (see Arestis and Karakitsos 2010). In the anaemic recovery phase the correlation of inflation and growth is again positive, as both are on the increase. This switch of correlation from positive to negative in the various phases of the business cycle does not indicate that the economy is suffering from stagflation. Stagflation is a symptom of a supply shock, usually a huge and permanent increase in the price of oil, as in the 1970s and the 1980s, and lasts for more than a single phase of the business cycle. The changing pattern of correlation in the various phases is an inherent property of the cycle. This is the underlying rationale.

In phase (I) the economy is overheated and bottlenecks emerge in both the output and input markets. Demand is rising faster than potential output and firms respond by increasing both prices and output. In order to increase production and meet the extra demand, firms have to pay higher wages to employ the existing labour force more intensely through overtime (increasing the number of working hours) or to attract the required additional skilled labour force, which, however, is becoming increasingly scarce as the overall level of demand in the economy is rising. Moreover, as demand is rising fast and firms find it difficult to meet the extra demand, they become less worried about losing market share and are increasingly inclined to raise their profit margins above normal. This is adding to the cost pressures and inflation accelerates. Hence, in phase (I) as growth increases inflation is rising, although with a lag of around one year, thereby resulting in a positive correlation.

The negative correlation between inflation and growth in the slowdown phase is due to the persistence of inflation or the price-inertia – a characteristic of most industrialised countries. This persistence is due to two main factors. The first is uncertainty. Given the cost of hiring and firing, firms do not immediately lay off workers in response to a fall in demand, as they are not sure that this is a permanent situation. Hence, in the short run demand is falling while employment lags behind. This results in a reduction in productivity and, therefore, an increase in unit labour cost, which is pushing up inflation. However, in the medium term employment is adjusting to the perceived permanent fall in demand. The second reason for the inflation-persistence is due to wages, which continue to rise in the slowdown phase. As inflation increases, backward-looking expectations boost expected inflation. The more backward- as opposed to forward-looking expected inflation is, the higher expected inflation induces workers to demand higher wages to protect their real value. Wage inflation does not moderate for most of the slowdown phase and this exacerbates the increase in unit labour cost. Firms respond to the increased labour cost by passing it onto the consumers via raising prices, thereby setting in motion a wage–price spiral. In most situations the wage–price spiral is not explosive (that is, it does not lead to hyperinflation), as central bank action contains expected inflation. As unemployment is rising and profits fall the effects of the wage–price spiral taper off.

The positive correlation of inflation and growth in the recession is due to the dynamics of inflation and growth. Inflation usually peaks when the economy is in the neighbourhood of the recession. In the slowdown phase the deceleration of the economy gathers pace. In this process an increasing number of firms are convinced that the reduction in demand is permanent and, accordingly, fire workers to adjust the labour force to a falling level of demand, thereby containing the rise in unit labour cost through increases in labour productivity. Moreover, although the cost of hiring and firing deters firms from instantly adjusting the labour force to demand conditions in the goods market, it is only sensible that as the process of falling demand gathers pace, they are less deterred in adjusting employment. Therefore, unemployment rises throughout the slowdown phase, albeit initially at a small pace. During the recession unemployment soars and this forces workers to moderate their wage demands for fear of losing their job and as outside opportunities deteriorate. Hence, the rate of change of unit labour cost is declining as both wage inflation is reduced and productivity rises through the adjustment of employment. It is through this process that inflation peaks at C. Moreover, this process gathers pace throughout the recession. In parallel, as the economy falls in recession firms find it increasingly difficult to pass on to the consumers the increased unit labour cost because their profits are squeezed. Hence, an increasing number of firms absorb through their profit margins the higher unit labour cost as they strive for survival in trying to maintain their market share. This second force also causes inflation to peak at C. Throughout the recession these two forces reinforce each other. Hence, both wage inflation and price inflation fall. In phase (III), therefore, the correlation between inflation and growth is positive – both are falling.

The correlation between growth and inflation is once again negative in the recovery phase – growth increases while inflation subsides. The recovery phase is symmetric to the slowdown phase but with the picture reversed. The reasoning for the negative correlation is, therefore, similar to that in phase (II). As the economy recovers, firms are hesitant in increasing employment because they are unsure if the recovery is sustainable. This uncertainty leads firms to increase the working hours or employ temporary staff. Accordingly, productivity rises and unit labour cost falls, thereby reducing inflation. Moreover, profitability is low and the costs of hiring deter firms from increasing employment. However, as the recovery firms and confidence builds up there is an increase in employment. Thus, as the recovery matures the reduction in unit labour cost moderates and the gains on the inflation front are reduced. This will ultimately lead to inflation bottoming out at point E.

Finally, the correlation between inflation and growth is positive in phase (V). In this phase the economy is returning back to normality with both growth and inflation rising. We can thus summarize the relationship (correlation) between inflation and growth in the course of the business cycle as being either positive or negative. In phases (I), (III) and (V) the correlation is positive, while in phases (II) and (IV) it is negative.

4 US MONETARY POLICY IN THE COURSE OF THE BUSINESS CYCLE

The business cycle is one of the two major systematic factors that induce a cyclical pattern in shipping. The other is economic policy. Although both fiscal policy and monetary policy have a major impact on the business cycle, over the past twenty years or so the former has been subordinated to the latter in the everyday management of the economy. Nonetheless, at crucial points in time when the economy is hit by recession fiscal policy has been used aggressively, at least in the US. This has happened in almost every recession in the post-Second World War period. The fiscal stimulus has been 1–2 per cent of GDP and in the last few cycles it has been applied faster than in the past. In recent times fiscal policy has been eased the moment data confirm that the economy has fallen into recession. This has soothed the recession pains exactly when it hurts the most and as a result it has made recent recessions shallower than would otherwise have been the case. But in spite of the use of fiscal policy at an early stage of the downturn, it is widely considered to be a burden in demand management because of the difficulty in cutting public spending or increasing taxation, although the required tightening is only a fraction of the stimulus.[3] For example, if subsidies to the personal sector, such as increases in unemployment benefits, are provided in a cyclical downturn to alleviate the impact of the recession on poor families, their reversal later on would be resisted, thereby creating a political cost to the government that can be exploited by the opposition. Moreover, any change in the stance of fiscal policy is time consuming, as it takes time for lawmakers to pass the necessary legislation. By contrast, changes in monetary policy can be very timely and as central banks are now independent there is no political cost to the governing party from increases in interest rates. Hence, monetary policy has become widely accepted as the pre-emptive instrument of daily demand management. In Chapter 5 we have spent a great deal of effort in explaining how monetary policy is formulated in theory. In this section we re-examine this issue in practice.

A central bank formulates monetary policy with the view of achieving, in the medium term, its statutory targets of inflation and/or growth or employment. The priority of inflation relative to that in relation to growth varies according to the remit of each central bank. A tough, anti-inflation or simply hawkish central bank has a very high relative inflation priority. In some extreme cases, such as the European Central Bank (ECB), the priority on growth may even be zero. A balanced central bank, such as the Fed, assigns equal priority to inflation and growth. Finally, a dove or 'wet' central bank, such as the Fed in the 1970s up until Paul Volcker, or the Bank of England until Thatcher, assigns a higher priority to growth than to inflation. However, the characterisation of a central bank as balanced, tough or dove (wet) is only pertinent at point A in terms of Figure 9.4, that is, when monetary policy is neutral, as every central bank would alter the relative inflation priority in the course of the business cycle. Thus, only at point A in Figure 9.4 is one comparing like with like.

Nonetheless, in the course of the business cycle the speed at which and the degree to which a central bank moves still both depend on this characterisation. As the economy moves from A to B in terms of Figure 9.4 the economy becomes overheated and this is a harbinger for future higher inflation. A balanced central bank would move quickly and decisively to eliminate any overheating before it has a chance of rekindling inflation. Sometimes a balanced central bank may even act pre-emptively in anticipation that the economy would become overheated. Such action might be guided by some leading indicators, such as the stock market, the supply of money, bank lending, orders and surveys of business or consumer confidence. Such pre-emptive tightening may be necessary as it takes about a year for the economy to respond to higher interest rates and two years for inflation. A balanced central bank may be too quick in its pre-emptive action or it may overreact by hiking rates too much. On the other hand, it might act too little, too late, thereby causing positive inflation surprises. Nonetheless, policy errors for a balanced central bank are likely to be symmetric, largely obeying the normal distribution. By contrast, for a tough or wet central bank any policy errors are likely to be skewed. For a tough central policy errors are likely to be skewed towards fewer positive inflation surprises, but at the expense of more spare capacity and, accordingly, of higher unemployment. For example, the Bundesbank in the 1990s was not prepared to tolerate any overheating and the economy operated with spare capacity throughout most of this period. As a result, inflation was kept under strict control but unemployment remained elevated throughout the 1990s. On the other hand, a wet central bank is more likely to wait longer before it acts or hike interest rates by less than would be optimal. For this reason, with a wet central bank overheating will lead to higher inflation more frequently than with a balanced one. Accordingly, for a wet central bank any policy errors are likely to be skewed towards more frequent inflation surprises, but lower unemployment at least in the short to medium run, albeit not necessarily in the long run. The experience of the US and the UK in the 1970s is a prime example of inflation accelerating to double-digit figures for the sake of combating unemployment. However, the benefits to unemployment were short term and unemployment shot up in the long run. Thus, it is fair to say that a balanced central bank would be nearer to the optimum than either a tough or a wet one. This bias towards inflation or growth (and unemployment) is likely to distort the clear picture that interest rates peak at C and bottom at E. With a tough central bank interest rates are likely to peak after point C and bottom before point E. With a wet central bank interest rates are likely to peak before point C and bottom after point E. Only with a balanced central bank interest rates are likely to peak at C and bottom at E.

Although the natural course of the economy in terms of Figure 9.4 is a clockwise movement, shocks can distort this movement, forcing the economy to move anti-clockwise (or, more precisely, to loop around) or enter abruptly another phase. The most commonly observed shocks in the real world are a sudden large change in the price of oil and changes in economic policy. In practice, the attempt by policymakers to steer the economy on a target (desired) path, for example to bring the economy back to point A, can be considered as a shock which forces the economy to loop around (move anti-clockwise). Whereas exogenous shocks are unpredictable, the behaviour of the policymakers in the course of the cycle is systematic and, therefore, predictable. The central bank stabilisation task is to prolong the business cycle and minimise the amplitude of the business cycle. This can only be achieved by prolonging phase (I) of the business cycle. The more proactive a central bank is in eliminating any overheating before inflation rekindles, the more likely that it will be successful in its stabilisation task. A successful pre-emptive tightening occurring just before the overheating emerges will not choke it off immediately as it takes time for the policy change to have an impact on the economy. Thus, overheating would emerge but, if policy is properly timed and its dosage is nearly optimal, then the overheating would last for a year. In the second year the economy would cool down with some spare capacity generated. By the end of the second year the economy is likely to approach again potential growth. The cooling-down phase is called 'soft landing' as inflation remains tamed in both the upswing and the downswing of this mini-cycle. This looping around potential output growth adds two more years to the length of the business cycle. Two conditions have to be met for a successful soft landing. Monetary policy has to pre-emptive and the degree of tightening should be nearly optimal. Second, the rate of growth of unit labour cost has to be falling or low by the standards of the cycle. If unit labour cost is high and rising, then the chance of a soft landing is minimal. Thus, the perfect soft landing in the US economy occurred in 1994–95, when Greenspan tightened monetary policy throughout 1994, hiking rates from 3 per cent to 6 per cent. The tightening was pre-emptive as the overheating emerged in 1994. But in 1995 the economy cooled down with growth less than potential. Inflation remained muted, as unit labour cost was extremely low throughout this period, thereby making the soft landing possible. Monetary policy was subsequently eased and returned to a neutral stance, while the economy returned to potential output at the end of 1995. By contrast, the tightening during 2006–07 under Bernanke had very little chance of success, as unit labour cost was on the increase. These principles further help in forming forward-looking expectations in the shipping market and explain the shipping cycles.

5 WHY DO US BUSINESS CYCLES DIFFER?

Table 9.2 provides a summary of the US business cycles in the post-Second World War era; Table 9.3 lists the length of each phase in these business cycles, while Figure 9.5 is an empirical version of Figure 9.4; it identifies the various phases of the business cycles since 1960. There is a difference, however, between the two aforementioned tables. Table 9.2 follows the official definition of measuring cycles by looking at the level of GDP. In this approach the cycle is defined as the time it takes the economy to move from peak to peak. The minimum level of GDP defines the trough of the recession, which is defined as the period between the peak and the bottom. The recovery is defined as the period of time between trough and the time it gets to return to the previous peak. The expansion phase

Table 9.3 The US business cycles

Cycle trough-to-trough	Phase IV years	Phase I years	Phase II years	Phase III years
1961–70	<2	5½	2	1
1970–75	<2	<1	2	<1
1975–82	<2	2½	2	<3
1982–91	4½	1	3	<1
1991–01	7	2	<1	1
2001–09	3	3	1	1

is defined as the time the economy spends from the recovery point to the next peak. Table 9.3, on the other hand, applies these concepts to the rate of growth (y-o-y) rather than the level of GDP. This approach is more informative as it takes into account that the economy exhibits an uneven pattern of growth due to random shocks and equally, if not more importantly, to the fluctuations of growth around potential because of the fine-tuning of the economy by the policymakers. However, the drawback of this approach is that the peak of the cycle is not easily discernible as it does not coincide with the maximum rate of growth, but with the last peak before the recession.

A cursory look at Table 9.3 and Figure 9.5 shows that each cycle is different than the others; and the length of each phase varies from cycle to cycle. The shortest cycle was less than five years, in the first half of the 1970s, and the longest was in the 1990s, more than ten years, exceeding the 1960s cycle by a whisker. The length of a cycle will depend on both its nature and the shocks that hit the economy. It also depends on which phase of the cycle the economy happened to be in when it was hit by a shock. We can distinguish three types of business cycles – demand-, supply- and asset-led. The five cycles after the war until the early 1970s and the early 1990s cycle were demand-led. The two cycles in the early 1970s and early 1980s were supply-led. The last two cycles – in the early and the late 2000s – are asset-led.

The default cycle is demand-led, which means that in the absence of shocks the cycle is demand-led. Usually, these cycles last as long as demand expands until inflation gets out of control. The policymakers then react by tightening fiscal and/or monetary policy to rein in inflation, thereby causing a recession. The average length of the first five post-war cycles was 16 months; the average recession lasted for nine months and the depth of the recession was –2.1 per cent of GDP; the recovery was mainly buoyant (V-shaped) and quick (an average of seven months).

Supply-led business cycles are caused by supply shocks, such as the major oil shocks in the early and late 1970s. Once these shocks occur, the default demand-led cycle is turned into a supply-led one. The average length of the two supply-led cycles in the early 1970s and early 1980s was 31 months, which is double that of the average demand cycle. The average recession was nearly 20 months, which is again more than double the average demand cycle. The average depth of the recession was –2.7 per cent, a little deeper than the average demand cycle (–2.1 per cent). The recovery, however, was again buoyant (V-shaped) and quick, albeit a bit longer than the average demand-cycle (nine months compared with seven). It should be noted that there are other supply shocks, such as a permanent improvement in productivity caused by the widespread use of computers in manufacturing and services (1990s) or new legislation that supported the introduction and enhanced the adoption of 'flexible labour markets', shorthand for increasing the ability of firms to fire easily and the weakening of collective wage negotiations (1980–90s). However, the effect of these shocks is gradual and its impact can be observed between cycles than within the same cycle. Whereas asset-led business cycles were common before the twentieth century there are only two episodes before the 2000s, these being Japan in the 1990s and the US in the 1930s. The early 2000s US downturn and that which occurred in the late 2000s are due to the burst of a bubble; the internet bubble in the former incident and the property

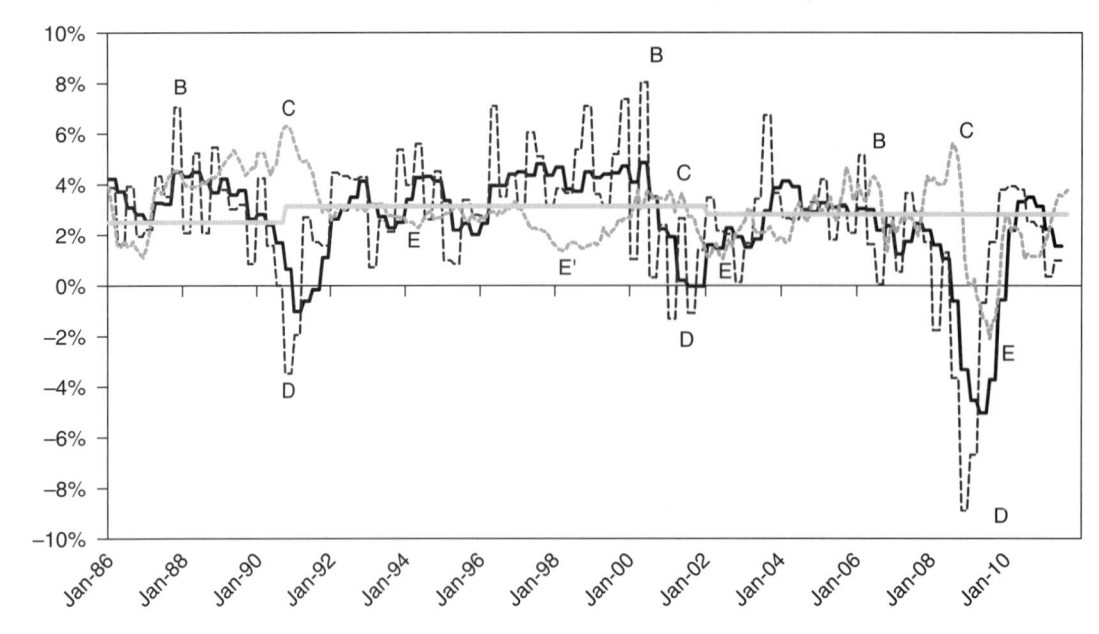

Figure 9.5 US business cycles

bubble in the recent one. The characteristic of asset-led cycles is that the expansion phase is very long and inflation is subdued, but the recession is long and deep.

In the absence of shocks and hence for the default demand cycle, the wild card that decides the length of a cycle is the length of the overheating stage (phase I). Table 9.3 shows that in recent decades the length of phase (I) has varied from less than a year to nearly six years; but this is mainly due to shocks. In the absence of shocks, the length of phase (I) depends on the number of soft landings the central bank can engineer in the course of a business cycle. In this context a soft landing is defined as a pre-emptive tightening of monetary policy (and/or fiscal policy) when the economy is overheated, which manages to cool the economy down before inflation had a chance to accelerate and get out of control. Assuming that the central bank is competent, its ability to create soft landings depends on the nature of the cycle. If the cycle is demand-led then they are relatively easy to engineer. The reason is that if the boom is caused by increased demand, say a reduction in the personal sector saving ratio or a surge in capital spending because of increased confidence in the corporate sector, then monetary policy has a good chance of cooling the economy down because it has a direct bearing on the level of demand. After all, monetary policy is an instrument of demand management. Hence, the likelihood of success in such circumstances is rather high. It involves estimating the degree of overheating a year ahead and raising interest rates sufficiently to choke it off. In a supply-led cycle the chances of a soft landing are negligible because the shock causes stagflation (stagnation and inflation). In such a case the tightening of the central bank aggravates the deceleration of the economy, usually turning it into a recession.

The clearest example of a soft landing is the US experience between 1994 and 1996. The Fed tightened monetary policy throughout 1994 by lifting interest rates from 3 per cent to 6 per cent on the accurate projection that the economy would become overheated that year. The tightening was pre-emptive, as inflation fell in the first half of 1994 and increased in the second half remaining, on average, stable over the entire period. The higher interest rates succeeded in cooling down the economy (achieving growth less than potential) in 1995 (see Figure 9.5). However, a central bank cannot go on engineering soft landings forever, even if it is extremely competent. The reason is that the overheating has a ratchet effect on inflation. Every time the economy becomes overheated the underlying level of inflation is creeping up. When the economy cools down, inflation subsides but it bottoms at a slightly higher level than the previous occasion. Hence, after a number of soft landings, in reality just one, the central bank has to tighten severely as the last overheating unleashes a wage–price spiral. In this case the tightening leads to a hard landing (that is, to a recession). The business cycles in the 1950s and 1960s, up until the first oil shock in 1973–74 (OPEC-I), were demand-led for all industrialised countries and soft landings were, at least conceptually, relatively easy. Hence, the cycles tended to be rather long. For example, phase (I) in the 1961–70 cycle was almost six years long. Demand management (either through monetary, but mainly through fiscal policy at the time) until OPEC-I was relatively successful.

The length of phase (I) depends not only on the ability of policymakers to engineer soft landings but also on random shocks. For example, German reunification occurred when the economy was in phase (I), which meant that it was prolonged to four and a half years. If the shock that hits the economy is a supply one and the economy happened to be in the overheating or slowdown phase, then this phase would be cut short, with the economy quickly entering into a recession. For example, in the US the shortest phase (I) occurred in the cycle of 1970–75 (see Table 9.3). In this cycle inflation bottomed in mid-1972, point E, and growth peaked in the first quarter of 1973, point B, making phase (I) only nine months long. The slowdown in the economy was turned into a recession in the aftermath of the quadrupling of the price of oil (OPEC-I) following the Arab–Israeli War of 1973. This turned the cycle into a supply-led one. The increase in the price of oil exacerbated the recession, making it longer and deeper, while at the same time the inflation rate at its peak, point C, doubled relative to the previous cycle.

The second example of a supply shock toppling demand management occurred in the cycle that lasted for something more than seven years from the beginning of 1975 to the second half of 1982. Phase (IV) was less than two years, up to the end of 1976. Phase (I) was extended until the end of 1978 thanks to the locomotive strategies adopted by the US, Germany and Japan. In this scheme one of these countries would act as the locomotive in reflating the world economy and then pass on that role to the other after a while so that the costs of reflation, mainly in the form of current account deficits, would be shared by all of them. The second oil shock of 1979 (OPEC-II) in which the price doubled forced the economy into another recession in 1980 and exacerbated the upward trend of inflation. Point C was reached at the end of 1979, meaning that phase (II) again lasted less than two years. The sharp change in the conduct of monetary policy caused the second leg of the 1980–82 recession (see below) prolonging phase (III) to two and a half years.

The length of the deceleration stage (phase II) depends on the flexibility of the labour market and the extent of overheating in phase (I). In general, the greater the flexibility of the labour market and the smaller the extent of the previous overheating the shorter phase (II) would be. The underlying rationale is simple. The length of phase (II) depends on the ferocity of the wage–price spiral. This captures the struggle between workers and firms over income distribution. Inflation is a mechanism for redistributing income. Those with weak industrial muscle or without any strong political voice, such as pensioners, see their real income eroded by higher prices. For those employees whose jobs are relatively secure the primary concern is the preservation of the real wage rate. For the unemployed the concern is employment. For a number of different reasons those who care about their real wage rate outweigh those who care about employment and unemployment increases throughout the slowdown and recession phases (see the Appendix of Chapter 5 for more details).

If the workers fight for preserving their real wage rate when the economy decelerates and unemployment is rising, then they intensify the wage–price spiral and prolong the slowdown phase (II). This would be the case if demand conditions in the labour market play a relatively small role, while deviations from the target real wage are more important (what is called 'real wage resistance'). Legislation in the labour market, minimum wages, employment protection legislation and trade union power and practices weaken the importance of demand conditions in the labour market and enhance real wage resistance. Moreover, unemployment benefits and social welfare influence the target real wage rate. If large unemployment benefits are paid, then the supply of labour is reduced (the incentive to participate in the labour force declines) and demand conditions in the labour market play a smaller role, thereby prolonging the slowdown phase (II). The high natural rate of unemployment in Europe in the 1980s is usually regarded as having being caused, to a large extent, by the relatively high rates of unemployment benefits. On the other hand, the structural changes in the UK labour market in the Thatcher years provide evidence that the impact of the wage–price spiral can be reduced, thus shortening the length of phase (II). In the US downturn in the 1990s firms started laying off workers much earlier than the early 1980s downturn, thereby allowing greater role for demand conditions in the labour market.

The effects of the wage–price spiral depend not only on workers but also on firms. If firms find it difficult to pass on the labour cost increases to the consumers, then the wage–price spiral is reduced, thereby shortening phase (II). The cost of firing and hiring is important in this respect as also is the degree of competition. If the market structure of the economy is competitive rather than monopolistic, then firms will find it difficult to pass on the labour cost to the consumers, as they will be struggling to maintain market share. Hence, cost increases will tend to be absorbed in profit margins, thereby shortening phase (II).

In the absence of shocks the length of phase (II) tends to be less variable than other phases as it reflects structural characteristics of the economy, for example, the degree of competition, and the labour market, in particular. For most industrialised countries it is between one and two years (see Table 9.3). In the US, the average length of phase (II) is two years. The exception to this rule was the 1980s business cycle in which phase (II) lasted for three years – from the fourth quarter of 1987 (the peak of the cycle coincided with the stock market crash of 1987) to the end of 1990. At the time the Fed was trying to engineer a second soft landing, the first one being in 1984–86. The Fed had tightened for a year till February 1989 with the Fed funds rate raised from 6.5 per cent to 9.75 per cent. After this time the Fed began to lower rates to avoid the slowdown turning into a recession. In all probabilities the Fed would have succeeded in engineering a second soft landing as the economy had decelerated to the rate of growth of potential output by mid-1990. However, the surge in the price of oil as a result of the Iraqi invasion in Kuwait in August 1990 and the resultant surge in the price of oil, albeit short-lived, exacerbated the inflationary pressures and turned the soft landing into a hard one (the recession of 1990–91).

Phase (III) is the shortest in the cycle, being usually one year in length. In the absence of shocks it does not vary a great deal as it comes after the breaking of the wage–price spiral. The only notable exception to this rule was the 1975–82 cycle where phase (III) lasted for two and a half years. But this was due to a substantial shift in the objective function of the Fed. The then chairman, Paul Volcker, put an end to the commitment to full employment in order to eradicate the inflationary pressures which soared to an all-time high following the second oil shock in 1979–80. The two successive oil shocks in the 1970s, along with the commitment of the policymakers in the Anglo-Saxon world to full employment, which was a legacy of the Great Depression, led to a persistent surge in inflation. As the stylised facts suggest, inflation peaked at much higher levels in every successive cycle despite its cyclicality. In other words, the two oil shocks and the commitment to full employment fuelled inflation expectations in the 1970s and led to explosive inflation. It was in this climate that Paul Volcker decided to break the inflation psychology by creating another recession the moment the last one had finished. The Fed tightened for a second time after inflation had peaked at the beginning of 1980, causing the second leg of the 1980–82 recession. Had the Fed not tightened, phase (III) would have been less than a year as the economy emerged from the first leg of the double-dip recession.

Phase (IV), the recovery phase, is long and variable. In the last thirty years it has varied from just over a year to seven years. In the absence of shocks the length of phase (IV) is usually two years long. Its length depends on the depth and the length of the preceding recession and the buoyancy of the recovery. The rationale is very simple as phase (IV) is symmetric to phase (II) with the picture reversed. The deeper and the longer is the preceding recession, the greater are the productivity gains and hence the larger the reduction in unit labour cost and, therefore, the fall in inflation. Moreover, the more anaemic the recovery, the more hesitant firms are in increasing permanent employment. Therefore, the more anaemic the recovery, the larger the productivity gains and hence the larger the decline in unit labour cost and therefore the fall in inflation. If the recovery is anaemic, as it was in the early 1990s, and hence firms are doubtful about the sustainability of the recovery, they will respond by increasing the working hours of the existing labour force or alternatively they will hire more temporary staff and will delay the hiring of permanent staff.

If an adverse supply shock hits the economy, such as a drastic increase in oil prices, while it is in phase (IV) the inflationary consequences would be rather subdued while the recessionary effects will be large. The reason is that unemployment is high and profitability is low. Therefore, firms are willing to absorb the higher cost of energy in their profit margins and workers are prepared to accept a real wage cut. For example, the increase in the price of oil in 1980 when the US economy was emerging from the first leg of the recession caused inflation to increase only slightly and for just a short period – less than two quarters. However, the economy fell into a double-dip recession, although this was caused, in addition, by the Fed tightening.

But supply shocks need not be adverse. There are instances in which they are fortuitous. For example, the economy emerged from the 1981–82 recession in a very buoyant way, thus becoming almost immediately overheated (see Figure 9.5). The then Fed Chairman, Paul Volcker, tightened monetary policy not with a view of engineering a soft-landing, as Greenspan did in the mid-1990s, but because he wanted to eradicate the extremely high inflationary expectations, which were, to a large extent, responsible for the double-digit inflation of the mid-1970s and early 1980s. The Fed tightened monetary policy aggressively from early 1983 till late in the summer of 1984, with the Fed funds rate rising from 8.5 per cent to 11.5 per cent on evidence that the economy was becoming overheated. Such tightening would have been sufficient to throw the economy back into recession. Indeed, the deceleration of the economy was dramatic and continued until the beginning of 1987 (see Figure 9.5). However, the economy managed a soft rather than a hard landing because of a fortuitous supply shock – the collapse of the price of oil in 1984–85. The OPEC cartel was in conflict mainly with other non-OPEC oil producers, but there was also conflict within the cartel. Saudi Arabia increased production to drive non-compliant oil producer countries or cartel cheaters out of the market. The price of oil fell to less than $10 per barrel. This decreased inflation and boosted the incomes of households and the profits of companies, thereby ameliorating the impact of the slowdown and enabling the Fed to lower interest rates to less than 6 per cent by late 1986. The economy not only avoided a recession in 1985–86, but recovered to exceed potential (overheating) in 1987. Had it not been for the fortuitous supply shock, the economy would have been in phase (I), as inflation had previously bottomed in mid-1983. However, the collapse of the price of oil prolonged phase (IV) from one year to four and a half years.

Fortuitous shocks can also come from the demand side. For example, in the 1991–2001 cycle inflation had bottomed in mid-1994 and the US economy was in phase (I). However, the Asian-Russian crisis of 1997–98 triggered a collapse in the prices of many commodities and industrial supplies. This lowered US inflation from 3.3 per cent at the beginning of 1997 to 1.4 per cent by the spring of 1998, producing the lowest inflation in the 1990s. The Fed lowered interest rate pre-emptively from 5.5 per cent to 4.75 per cent to bolster confidence and avert a worldwide recession. The combination of falling inflation and lower interest rates maintained growth intact. However, the economy was already overheated and remained so, thus paving the way for a revival of inflation and presaging the end of the business cycle in early 2000. Had it not been for the Asian-Russian crisis, phase (IV) would have been three years long, but the shock prolonged it to seven years.

Phase (V) may not be observed if the recovery is buoyant with the economy immediately becoming overheated. If, however, the recovery is anaemic with growth not exceeding potential for a long time, then phase (V) is detectable. Phase (V) was not observed in all recoveries in the post-war period until the recession in the early 1990s. However, since that period all of the recoveries have been anaemic. This new feature of the recent recoveries may be the result of globalisation and the adoption of flexible labour markets. Companies take advantage of globalisation and shift production to low wage-countries when the home country is in recession or in recovery, thus causing the recovery at home to become anaemic. Similarly, flexible labour markets have made companies more cautious to hire during a recovery until they are convinced about the sustainability of the upturn. There is evidence that this hesitancy has turned recoveries anaemic.

6 THE BUSINESS CYCLES OF JAPAN

Over the course of the last thirty years Japan has experienced seven business cycles. For shipping, industrial production is more important than GDP and, therefore, this variable is used to measure economic activity and divide business cycles.

Table 9.4 breaks down business cycles in accordance with the standard classification of slowdown, recession, recovery and expansion. This is the methodology that is used to divide the various cycles in Table 9.1 for the US economy. It enables a comparison of the shipping cycles with Japan's business cycles. At a glance, it can be verified that Japan played a key role in accounting for the cycles in the dry bulk market in the 1980s and the 1990s. Table 9.5 divides the cycles of Japan into the four phases of overheating, slowdown, recession and recovery and offers details about the depth of the recession and the inflation cycles. The conduct of monetary policy in the various phases of the business cycle can serve as guidance on how expectations should be formed in shipping.

The first cycle is from 1980 to 1984. Industrial production peaked at nearly 9 per cent in early 1980 as the Bank of Japan hiked interest rates to 9 per cent

Table 9.4 Japan business cycles

Cycle	Slowdown	Recession	Recovery
Apr 80–Jul 84	Apr 80–Feb 81	Mar 81–Mar 83	Apr 83–Jul 84
Jul 84–Jun 88	Jul 84–May 86	Jun 86–Jan 87	Feb 87–Jun 88
Jun 88–Jun 97	Jun 88–Oct 91	Nov 91–Jul 94	Aug 94–Jun 97
Jun 97–Sep 00	Jun 97–Jan 98	Feb 98–Jul 99	Aug 99–Sep 00
Sep 00–Dec 06	Sep 00–May 01	Jun 01–Sep 02	Oct 02–Dec 06
Dec 06–Jun 10	Dec 06–Aug 08	Sep 08–Jan 10	Feb 10–Jun 10
Jun 10–Jun 12	Jun 10–Mar 11	Apr 11–Feb 12	Mar 12–Jun 12

to combat ramping inflation following the second oil shock in the late 1970s. Exports, the driving force of Japan, also fell precipitously in response to the world recession. As a result, the economy fell into the first leg of recession, in line with the US. Inflation fell rapidly and the Bank of Japan eased monetary policy. The economy rebounded for a while, but as the Fed tightened aggressively and the US led the world into a double-dip recession, Japan was also dragged down through a slump in exports. The economy rebounded with the rest of the world economy in early 1983, but the recovery only lasted for just over a year. The US was again responsible for the slowdown of Japan from mid-1983 onwards, as the Fed under Volcker hiked interest rates from early 1983 to the autumn of 1984 to fight rekindling inflation. This slowed the world economy with Japan's exports once again being the casualty. Although the Fed loosened monetary policy from 1984 until the end of 1986, the US economy continued to lose steam, hurting Japanese exports. Industrial production in Japan came to a standstill in mid-1986 and there was then a mild recession until the spring of 1987. The slowdown in the US was affected not just by the slowdown and the recession in Japan, but also the sharp appreciation of the yen with the effective exchange rate (a trade-weighted basket of currencies) appreciating from 75 to 120, following the Plaza Accord of 1985. This led to a loss in competitiveness, which hurt Japanese exports. The US economy rebounded in the course of 1987, helping Japan to recover through a surge in exports. Industrial production growth accelerated to double-digit figures until the second half of 1988. But then the US economy started to slow, as the Fed, under its new chairman, Alan Greenspan, tightened monetary policy dragging down Japan once again. The dollar appreciated across the board and the yen fell following the Fed tightening. Japan's effective exchange rate fell from 120 to 90, and this moderated the weakness of exports. Industrial production growth fell from 11 per cent in mid-1988 to 2.5 per cent in mid-1990.

The conclusion is that although Japan shaped the shipping cycles throughout the 1980s, Japan's cycles were triggered in turn by the US business cycles. Japan continued to play a key role in shipping cycles in the 1990s, but its business cycles were affected less by the US and more by domestic developments following the burst of the electronics bubble. The Bank of Japan started tightening monetary policy in mid-1989 as inflations increased modestly from 1 per cent to 4 per cent. The discount rate was lifted from 2.5 per cent to 6.0 per cent, a level that was maintained until mid-1991. By then the equity and property bubbles had been pricked by the high interest rates and the US was already in recession, following the surge in the price of oil because of the Iraq War. In less than six months from the peak of economic activity in the spring of 1991 Japan fell into a deep and protracted recession, with industrial production falling 7 per cent.

In the first half of the 1990s the yen nearly doubled in value. The effective exchange rate appreciated from around 90 to more than 170. This sharp appreciation of the yen was triggered by the sale of foreign assets and the repatriation of profits into Japan to cover the losses from the burst of the stock and property bubbles. The yen appreciation accentuated the recession and caused negative inflation (deflation). Once more the US affected Japan in 1995. The Fed tightened monetary policy from February 1994 till the end of the year as the US economy showed signs of overheating. This pre-emptive tightening cooled the US economy

Table 9.5 Japan cycle phases

	Cycle turning points		Growth	Inflation	Cycle	Phase
	Apr–80	B	9%		1	Overheating
Sep–80		C		8.7%		Slowdown
	Mar–83	D	–1.9%			Recession
Sep–83		E		0.9%		Recovery
	Jul–84	B	10.1%		2	Overheating
Aug–85		C		3.0%		Slowdown
	Aug–86	D	–0.7%			Recession
Jan–87		E		–1.0%		Recovery
	Jun–88	B	11%		3	Overheating
Jan–91		C		4.0%		Slowdown
	Dec–92	D	–6.9%			Recession
Oct–95		E		–0.7%		Recovery
	Jun–97	B	6.2%		4	Overheating
Oct–97		C		2.6%		Slowdown
	Oct–98	D	–8.6%			Recession
Nov–99		E		–1.1%		Recovery
	Sep–00	B	6%		5	Overheating
Jan–01		C		–0.3%		Slowdown
	Jan–02	D	–11.7%			Recession
Feb–02		E		–1.6%		Recovery
	Dec–06	B	5.3%		6	Overheating
Jul–08		C		2.3%		Slowdown
	Jun–09	D	–29.6%			Recession
Oct–09		E		–2.5%		Recovery
	Jun–10	B	24.2%		7	Overheating
Mar–12		C		0.5%		Slowdown
	Aug–11	D	–5.7%			Recession

down in 1995 and this dampened Japanese exports for a while. But as the US economy gathered steam in 1996 and beyond Japan rebounded until mid-1997. But then the Asian-Russian crisis of 1997–98 dragged Japan into an even deeper recession than in the first half of the 1990s. Industrial production fell more than 8

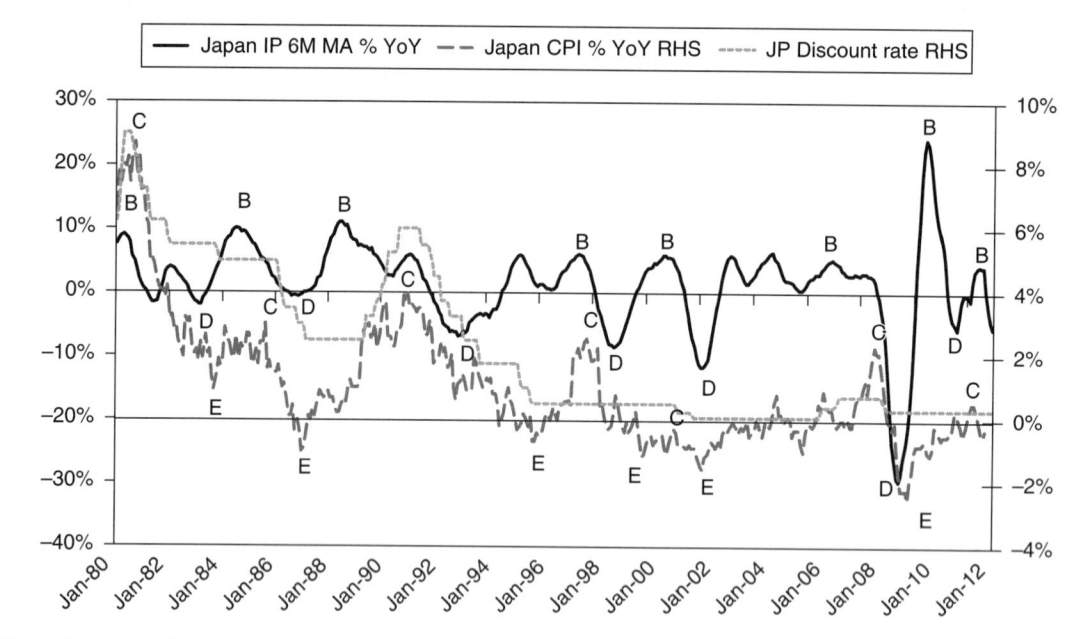

Figure 9.6 Japan business cycles

per cent and the recession lasted for one and a half years until mid-1999. Deflation became a permanent state of affairs in Japan as the yen appreciated sharply. The effective exchange rate soared from 105 in August 1998 to 160 in September 2000.

Japan rebounded once more as the US economy went into overdrive with growth exceeding 5 per cent in the last few years of the old millennium despite the yen appreciation. The extremely high growth in the US was attributed at the time to large structural productivity gains associated with information technology, but this was wrong. The boom was triggered by the internet bubble, which gave the impression of a structural improvement in productivity, whereas *ex post* it proved to be purely cyclical. As the internet bubble burst and the US fell into a mild recession, Japan entered a third recession in just ten years. This was deeper than the previous two recessions, with industrial production plunging more than 12 per cent. As the US economy led the world economy into recovery from 2003 onwards, Japan recovered and deflation was beaten for a while.

Easy monetary policy in the US and the rest of the world fuelled bubbles in commercial and residential real estate. This helped Japan to recover and experience a relatively long period of relative prosperity of four years, something that had not happened since the burst of the stock and housing bubbles at the end of 1989. Nonetheless, growth was modest, ranging from 2 per cent to 5 per cent. The US economy peaked at the beginning of 2004, as the fiscal stimulus faded and the Fed began to tighten monetary policy from mid-2004 onwards. Japan peaked at the end of 2006, but remained in positive growth territory until the spring of 2008. As the US economy fell off the cliff in the second half of 2008, so did Japan. Industrial production fell by a staggering 30 per cent in the Great Recession In the new millennium China became an increasingly important trading partner of Japan. Hence, the recession of 2008 in Japan was triggered by both the US and China. But it is fair to say that the weakness of China was caused by the US.

Therefore, Japan's business cycles were determined largely by the US. In the 1980s Japanese business cycles were shaped almost entirely by the US, as Japan relied on exports for growth. In the 1990s, the influence of the US diminished, as Japan faced the ramifications of the burst of the stock and housing bubbles. But factors from abroad continued to play a role, including the US slowdown in the mid-1990s and the Asian-Russian crisis of 1997–98. The US helped Japan to recover from the recession triggered by the Asian-Russian crisis, but was also responsible for the cycles in the new millennium, either directly or indirectly, through its influence on China.

7 THE BUSINESS CYCLES OF GERMANY

Statistics for the euro area and for (unified) Germany commence in 1991 and, therefore, prior to this date West Germany is used as a proxy for the euro area. In the statistics reported below Germany is used after 1991 and West-Germany prior to this date. Whereas GDP is used in business cycles, in the current analysis industrial production is employed, as it is more relevant for shipping cycles.

Since the early 1960s Germany has had ten full business cycles. It is now in its eleventh cycle (see Figure 9.7 and Table 9.6). The first cycle was of average length, lasting for five years (from mid-1964 to mid-1969). The cycle is V-shaped: a sharp fall in economic activity followed by a buoyant recovery. At the peak in mid-1969 industrial production was nearly double the growth of the previous peak. The next cycle was slightly shorter (lasting four instead of five years). The slowdown was sharp, but the economy did not experience a proper recession. Industrial

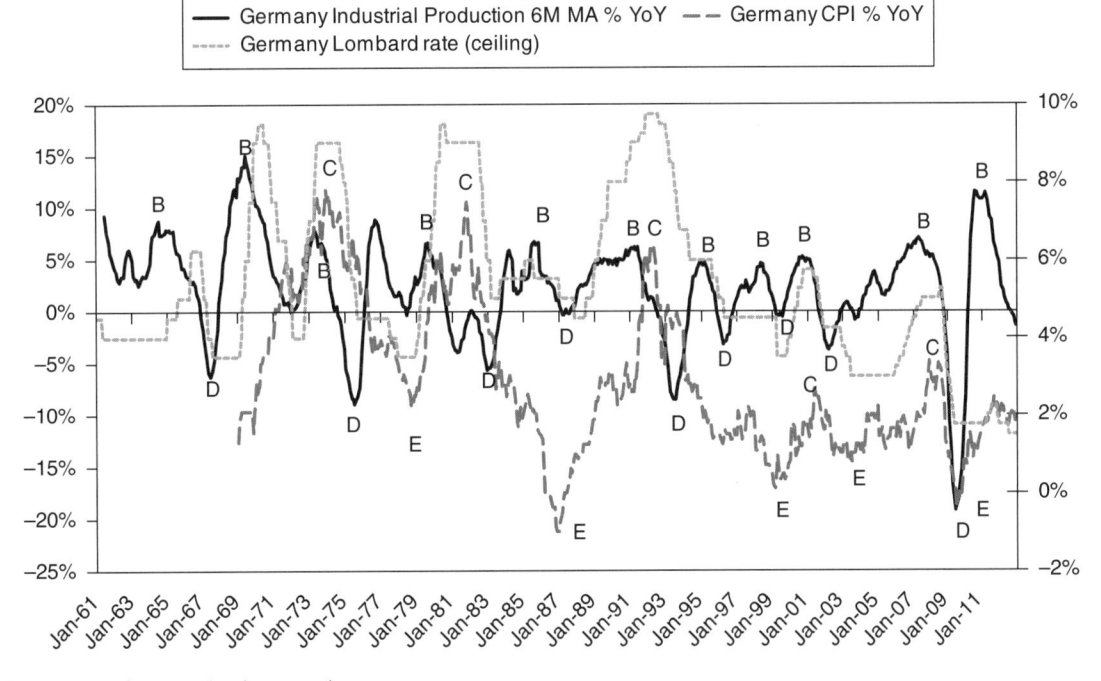

Figure 9.7 Germany business cycles

Table 9.6 Germany business cycles

	Slowdown	Recession	Recovery
Aug 79 – Aug 85	Aug 79 – Aug 80	Sep 80 – Aug 83	Sep 83 – Aug 85
Aug 85 – Jun 91	Aug 85 – Feb 87	Mar 87 – Jul 87	Aug 87 – Jun 91
Jun 91 – Dec 94	Jun 91 – Jun 92	Jul 92 – Apr 94	May 94 – Dec 94
Dec 94 – May 98	Dec 94 – Oct 95	Nov 95 – Sep 96	Oct 96 – May 98
May 98 – Sep 00	May 98 – Feb 99	Mar 99 – Jul 99	Aug 99 – Sep 00
Sep 00 – Apr 07	Sep 00 – Aug 01	Sep 01 – Nov 03	Dec 03 – Apr 07
Apr 07 – Mar 11	Apr 07 – Sep 08	Oct 08 – Mar 10	Apr 10 – Mar 11

production came to a standstill at the beginning of 1972 and then rebounded, retaining the V-shape. The Bundesbank reversed the easing of monetary policy from September 1972, a year earlier than the first oil shock. The Lombard rate was hiked rapidly from 4 per cent, reaching 9 per cent in August 1973. As a result of the tight monetary policy economic activity peaked in the spring of 1973 and slowdown followed. The first oil shock in the fourth quarter of 1973 struck the the economy when it was in the slowdown phase. Inflation heightened temporarily, hitting nearly 8 per cent at the end of 1973.

The economy was already in the slowdown phase at the time of the first oil shock and was heading for a recession anyway as a result of the tight monetary policy imposed by the authorities. The economy had already peaked six months earlier than inflation and as a result of the quadrupling of the price of oil it plunged into a deep recession with industrial production growth registering -9 per cent in mid-1975. Thus, the length of the slowdown phase was seven months. Actual inflation peaked when the economy entered the deep recession. The economy reached the trough of the recession in mid-1975, lasting for one year. The Bundesbank lowered interest rates until the beginning of 1978, as inflation continued to abate. In September 1978 it reached a low of just over 2 per cent. This is a prime example that monetary policy in Germany was formulated in accordance with developments in actual inflation.

The recovery from the recession in the first half of the 1970s was again V-shaped, lasting for a little longer than a year. But the recovery faltered and industrial production came to another standstill in mid-1978. Thus, the recovery phase (phase IV) lasted for just over three years. The economy rebounded once more with the help of the locomotive strategies of the US, Germany and Japan (see above) with growth in industrial production peaking in August 1979. Following the second oil shock, the economy fell for the first time into a double-dip recession that lasted for exactly three years (until September 1983). This was the longest, but not the deepest recession of all cycles. Inflation did not peak before the economy was deep into recession in October 1981. Phase II (from the peak of the economy to the peak of inflation) was the longest of all cycles, lasting for more than two years. The second leg of the double-dip recession was caused partly by the Bundesbank which, in turn, was trying to protect the value of the mark in response to the Fed's very tight monetary policy.

Germany recovered from the recession of 1980–83, but only for a short period. In late 1985 growth in industrial production peaked and the slowdown succumbed to recession in 1987. The main reason was the sharp appreciation of the Deutsche Mark following the Plaza Accord in 1985 in which the major central banks agreed to coordinate their intervention in the foreign exchange markets in selling the dollar. Weak growth and prolonged recessions for the majoroty of the 1980s kept the unemployment rate extremely high, not just in Germany but throughout Europe. Many analysts attributed the stubbornly high level of unemployment to the welfare state and inflexible labour markets, a framework which came to be known as 'Euro-sclerosis'. However, the truth is different. Throughout the 1980s there was no fiscal stimulus to domestic demand, while the Bundesbank kept interest rates high to stem the weakness of the Deutsche Mark against the dollar and curb imported inflation from the US. When in 1983–84 the Fed under Volcker tightened monetary policy to prevent the resurgence of inflation the Bundesbank followed suit. The Lombard rate was cut to 5 per cent in the spring of 1983, a very high level considering that inflation at the time was less than 3 per cent. When the Fed hiked rates, the Lombard rate was lifted to 6 per cent, thus choking off economic activity. As a result of these actions industrial production came to a standstill at the beginning of 1987. Whereas the Fed reversed the rate hikes when the price of oil started a free fall, the Bundesbank kept rates high. The result was that inflation in Germany became negative, showing that monetary policy was unduly tight. When fiscal policy was turned easy for the first time with the aim of integrating East Germany, following the collapse of the Berlin Wall, growth in the economy accelerated for the first time to above potential levels and unemployment fell rapidly. This is the best evidence against the claim that the high level of unemployment in the 1980s is due to Euro-sclerosis. It is simply due to deficient demand because of extremely tight monetary and fiscal policy.

Throughout the era of easy fiscal policy the Bundesbank continued to hike interest rates. The economy peaked in the first half of 1991 and finally succumbed to recession a year later. The recovery was short lived, however, as Europe had embarked on the task of convergence in monetary magnitudes with the aim of preparing for monetary union. The reining in of inflation was the number one priority throughout Europe. Germany fell again into recession in the first half of 1996. A modest and short-lived recovery was followed by yet another recession in industrial production following the Asian-Russian crisis of 1997–98. The recovery was once more anaemic and short lived, lasting until the early 2000s. This time the recession is due to the US, which dragged the rest of the world down with the burst of the internet bubble.

The last cycle was relatively long compared to the 1980s and the 1990s, a little less than seven years. Germany and Europe benefitted from the low interest rates offset by the European Central Bank (ECB). The interest rates may not have been low for German standards, but they certainly were for the rest of Europe. Debt soared in the periphery, financing housing bubbles in Spain and Ireland and even state bubbles, as in Greece. The credit boom in the rest of Europe was an indirect

8 AN EXPLANATION OF THE STYLISED FACTS OF SHIPPING CYCLES

The above information can help us to explain the various shipping cycles in recent decades. Table 9.1 summarises the shipping cycles and Figure 9.8 provides a visual overview. Using the same notation as in business cycle analysis we denote the peak of the shipping cycle by B and the trough by D. We thus identify four cycles from 1980 to 2008, while the fifth one is under progress.

Although we do not have data prior to 1976, we suspect that freight rates peaked sometime in the world recession of 1973–75 and remained low until the beginning of 1978. The boom, however, lasted for only two years, as the world economy plunged into recession in 1980–82. As a result of the world recession, the dry bulk market fell also into recession from April 1980 to December 1982. In the recession freight rates slumped 65 per cent and they did not regain the previous peak until April 1988, namely for eight whole years. Moreover, the cycle ended in April 1988 with no expansion phase. This is the worst period ever experienced by the dry bulk market. For six whole years freight rates hovered around the bottom. The reason for this horrific period is mainly due to Japan and, to a lesser extent, to Europe. As China is now the most important country for the dry market, in those days it was Japan. Fluctuations in Japan's industrial production account for 44 per cent of the variance of BDI in the 1980s and the 1990s, whereas fluctuations in Germany's industrial production account for just 11 per cent. The reason that

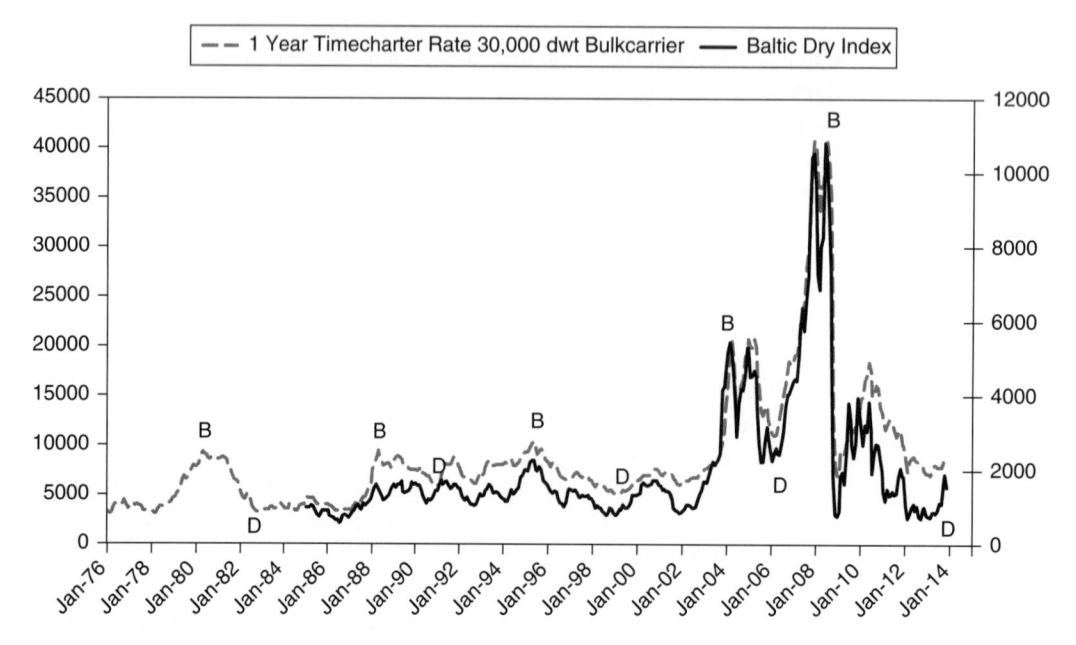

Figure 9.8 Shipping cycles

Japan played a more important role than Germany in that time period is that steel production in Japan was (and still is) more than twice the level of that in Germany.

Japan did not recover throughout the majority of the 1980s. It slowed down from mid-1984 and fell into recession in the second half of 1986. Germany also peaked in late 1985 and slowed dramatically, with industrial production coming to a standstill in 1987. The reason why both countries suffered at the same time is because their currencies appreciated sharply following the Plaza Accord of 1985. Japan and Germany rebounded in 1987 and with them the dry bulk market after the Louvre Accord in 1987, which showed that the dollar had fallen enough.

The second recession in the dry bulk market occurred in the period April 1988 to January 1991. The market recovered for just six months and then again fell into a double-dip recession. Freight rates did not recoup the previous peak until the end of 1994. Once again, this is caused by the slowdown of Japan from June 1988 to October 1991 and the ensuing recession from November 1991 to mid-1994, as a result of the burst of the bubble in equities and real estate. On top of that, the world economy fell into recession following the Iraqi invasion of Kuwait in mid-1990, thereby further contributing to the doldrums of the dry market.

The third recession in the dry market took place in the period May 1995 to February 1999. Germany peaked in December 1994 and slowed down dramatically, falling into recession from November 1995 to September 1996. Japan also peaked in mid-1997 and fell into recession to mid-1999 following the Asian-Russian crisis of 1997–98. The rebound in freight rates was short-lived as the world economy fell into recession in 2001–02 following the burst of the internet bubble. Japan remained in recession until late 2002, while Europe did not begin to recover until the end of 2003.

In the new millennium, China has emerged to pre-eminence in world trade and supplanted Japan as the most important country that drives the demand for dry. The boom in freight rates in the last two cycles is due not only to China, but also to the advent of investors in commodities, as explained in Chapter 8. However, China is also an export-led economy – as Japan was from the 1970s onwards – and as a result the US sets the tone for its business cycles. Although China aims to change the economy from export-led to domestic-led, this might take a number of years. Until then, the business cycles of China would be shaped by the US.

9 UNCERTAINTY-LED SHIPPING CYCLES

So far, we have concentrated on the role of business cycles in explaining shipping cycles. But in explaining shipping cycles uncertainty plays just as important a role as business cycles. In this section we turn our attention to uncertainty-led shipping cycles.

The theory of fleet capacity expansion under uncertainty, developed in Chapter 3, offers an insight into the causes of shipping cycles. If investment is irreversible, and if decisions may be postponed, then there may be a gain by waiting in an uncertain environment. The decision not to invest is equivalent to the purchase of an option. By not investing, the firm foregoes an expected profit stream, but this enables it to make more profitable choices later on. The most likely reason for waiting is uncertainty about demand conditions. Thus, if owners are faced with demand uncertainty they may prefer to wait rather than invest until conditions improve. Consider, for example, a firm that contemplates capacity expansion in a recession. The decision makers may be uncertain as to the depth and the length of the recession and, therefore, they may decide that it is better to wait before investing. Similarly, consider a firm with a similar decision choice in the recovery stage of the business cycle. If the decision makers feel that the recovery is not sustainable they would again prefer to wait.

Keynes emphasized the uncertainty of expectations as a primary source of instability. Expectations may be erratic, fluctuating between optimism and pessimism. The term 'animal spirits' describes a situation where economic fundamentals remain unchanged and yet expectations swing from optimism to pessimism. A herd syndrome may induce optimism or pessimism to all firms in the economy.[4] Thus, when economic conditions are bad – like a recession or at the early phase of a recovery – there is general pessimism and many firms may be deterred from investing. This would be particularly true if decision makers are risk averse, which evidence seems to suggest that this is, indeed, the case. It is, therefore, natural that most firms will defer investment until economic conditions are good. This will usually be the case when the economy is on a sustainable path to recovery. Given the shipyards' gestation lags, investment expenditure will come too late in the cycle. Hence, overcapacity is the most likely outcome. In other words, firms will defer investment until economic conditions have improved. But by delaying investment, the capacity would be in place too late, that is, when the downturn has started, thereby resulting in overcapacity. This overcapacity would act as a drag to prices in the recovery, thereby generating shipping cycles.

Uncertainty can also explain the complex interrelationship of freight rates, newbuilding and secondhand prices in the course of the cycle. Newbuilding prices are most of the time a coincident or lagging indicator of secondhand prices (see Figure 9.2). This is not surprising, as changes in the near-term outlook for the market alter the demand for and the supply of existing vessels, but not those in the new vessel market. Conditions in the new vessel market will be altered only when the change in the outlook for shipping is permanent (long-term) rather than transitory (short-term). A short-term profit opportunity might be realised by the rapid purchase of a secondhand vessel that would only cost a fraction of the new one and then selling it once the improved environment comes to an end. To place an order for a new vessel, on the other hand, an owner must be convinced that the improved market conditions are permanent rather than transitory. Hence, most of the time the secondhand market is acting as a leading indicator of a reversal of short-term market trends. The two markets of new and secondhand vessels will give the same signal (that is, they are coincident indicators) if the perceived changes in the market outlook are permanent.

Secondhand prices might act as a leading, coincident or even lagging indicator of freight rates (see Figure 9.2). It all depends on whether owners' expectations adjust faster than those of the shippers. Although both owners and shippers are forming forward-looking expectations, the former are likely to be more proactive than the latter. Shippers have fewer degrees of freedom than owners, since to take advantage of perceived changes in freight rates they need to coordinate on the demand for and the supply of the goods to be transported. This implies that owners are better placed than shippers to act proactively. Sometimes, owners perceive that market conditions will change for the better or worse before they actually do so and they are willing to act upon it by buying or selling a vessel in the secondhand market. Once actual conditions change, freight rates will adjust. In this case secondhand prices will adjust first and freight rates will follow, thereby making the former a leading indicator of the latter. In other times, owners might be more hesitant and their actions might not be proactive. Instead, they might wait for market conditions to improve before they act. On this occasion secondhand prices are a lagging indicator of freight rates.

The most important factor that determines whether owners' expectations adjust faster or slower than actual market conditions is uncertainty. When uncertainty is high, owners are more likely to wait for actual conditions to change before they act – they are reactive to market conditions. When uncertainty is low owners are more likely to be proactive. The level of uncertainty, in turn, depends, first, on the gap between expected and actual fleet capacity utilisation and second on market volatility in freight rates, newbuilding and secondhand vessel prices. The higher the positive gap between expected and actual fleet capacity utilisation, the higher the probability that owners will be proactive, thereby making secondhand prices a leading indicator of freight rates.

The higher the volatility in these asset prices, the higher the probability that owners will be reactive rather than proactive, thereby making secondhand prices a lagging indicator of freight rates. As we saw in Chapter 8, in asset bubbles volatility is low in the euphoria years when the bubble balloons. This causes a disproportionate expansion of the fleet size. However, in the aftermath of the burst of a bubble volatility soars, thereby making owners reactive to market conditions.

10 FINANCE-LED SHIPPING CYCLES

In the theoretical model of optimal fleet expansion, developed in Chapter 3, owners can borrow as much as they like at the prevailing rate of interest. But this is simply unrealistic. Firms, especially small to medium-sized ones, cannot borrow as much as they would like at the prevailing rate of interest. Put it differently, there is an excess demand for credit or there is credit rationing. The reason for credit rationing is that the lender (that is, a bank) regards the capital project of the borrower (the owner) as carrying a high risk of default. Rather than raising the interest rate to clear the market banks prefer to turn this demand for credit down.

But why should banks turn down this demand for credit? If they relied on the price mechanism the natural reaction would have been to raise the interest rate they charge on the basis that this project carries more risk. However, raising the interest rate would create problems of 'moral hazard'. This means that the owners that would not accept the terms of the loan would be the honest or risk-averse ones on the basis that at that level of the interest rate the investment would be unprofitable. On the other hand, the reckless or risk-seeking owners would simply borrow at the higher interest rate because they do not expect to pay back if the project turns out badly. Hence, raising the interest rate to clear the market invites risk seekers and turns down risk-averse firms, which is clearly the opposite of what a bank should do. Thus credit rationing – that is, the application of non-price criteria – is a rational method for the banks to clear the credit market.

But why does credit rationing intensify the pro-cyclical pattern of investment? The answer is simple. The degree of credit rationing is not independent of the business cycle. Banks tend to be less cautious in their lending in a boom. In the upswing of the cycle borrowers find it easy to serve their debt and their demand for new credit increases as there is widespread optimism. Moreover, bank profits rise and the value of the collateral upon which loans are made available is also increased. Hence, the balance sheet of the banks looks healthy in the upswing of the cycle. This induces banks to relax their strict criteria for lending and the amount of credit rationing is reduced. Conversely, in the downswing banks are more cautious in their lending. Although the demand for new credit declines as economic conditions deteriorate, borrowers find it difficult to service their debt and the number of ship foreclosures and bankruptcies rises. This reduces bank profits. In addition, the value of the collateral falls and, therefore, the balance sheet of the banks worsen, as the loan is removed from the asset side of the balance sheet, while reserves are built on the liability side. Banks become more cautious about new lending and, therefore, the amount of credit rationing is increased. This phenomenon has been termed a 'credit crunch' and applies particularly to small or medium-sized firms. In the 1990s downturn there was evidence of a credit crunch in all major leveraged economies (in particular, the US, the UK, Japan, the Nordic countries and Canada). The credit crunch was more severe in all those countries, which experienced asset deflation.

A second type of credit rationing occurs when the central bank imposes credit limits on commercial banks. These may take the form of a limit on the growth rate of loans per period, say 5 per cent per annum, or a limit on the proportion of loans to deposits. It should be noted that such actions would have an abrupt and severe effect on the economy, as the US in the early 1980s. The Fed used such controls to cool the economy down and beat inflation, which had risen to double-digit figures. The effect of such limits on the availability of bank credit had a severe dent on economic activity – as, for example, in the 1980 recession. But the opposite is also true. A central bank can relax credit limits to kick-start the economy. The Barber boom in the UK in the early 1970s is a classic example of this. But as in the case of credit controls their relaxation will have sizeable and quick results in the economy. In both cases of credit rationing – a rational choice by the commercial banks or imposed by the central bank – the availability of credit becomes an important channel of monetary policy, in addition to interest rates (see Jaffee and Stiglitz, 1990, for a survey). The disadvantage of such credit controls, which operate directly on the availability of credit, is that their effect is too large and too quick.

The final factor which explains the pro-cyclical movement of investment and its large volatility is corporate profitability. Empirical evidence suggests that outside funding for investment from banks, the bond market and the equity markets is rather small. Firms rely instead on retained earnings (that is, their own savings) which exceed 50 per cent of earnings for all firms and are even higher for small to medium-sized firms. Empirical evidence for the US (see, for example, Gertler and Hubbard, 1989, Fazzari et al., 1988) suggests that retained earnings account for more than two-thirds of all funds used to finance investment. In the UK, internal funds account for 60 per cent of the financing needs of industrial and commercial companies. The implication is that corporate earnings are at least as important as the cost of capital in determining investment decisions. Corporate earnings vary pro-cyclically – rising in a boom and falling in a recession and they are in addition volatile. Given the strong link of earnings with investment it follows that the latter will vary pro-cyclically with a lot of volatility.

11 SUMMARY AND CONCLUSIONS

Shipping cycles are caused by business cycles. World recessions cause recessions in the dry market. In world recessions the US is a leading indicator of shipping cycles. But the business cycles of a particular country, like Japan in the 1980s and the 1990s, can also lead to recessions in the dry market. In the 1980s and the 1990s, primarily the business cycles of Japan, and to a lesser extent those of Germany, were the main drivers of shipping cycles. In the new millennium the pattern has not changed, because the role of Japan is now played by China. Although the direct cause of shipping cycles may be a particular country, Japan or China, they reflect the indirect impact of the US.

Forward-looking expectations in the dry market can be formed with the help of the conduct of monetary policy in the major economic regions of the US, China, Europe and Japan. Central banks adopt tight monetary policy in the overheating and slowdown phases, thus serving as leading indicators of the peak of the dry market. Central banks switch into easy monetary policy when expected or actual inflation is about to turn around and this heralds the recovery of the world economy. Shipping though might lag behind the recovery of the world economy, if the previous boom has been largely unanticipated and has led to an overexpansion of fleet capacity.

10 INVESTMENT STRATEGY

EXECUTIVE SUMMARY

The investment strategy implications of this book are that an owner should have the optimum fleet and be in the spot market during a bull shipping market; and should keep the fleet at a minimum to cover fixed expenses and be in the period market during a bear shipping market. The strategy switches at the turning points of the shipping cycle. Chapters 6 and 9 show that a necessary condition for owners buying ships and moving from the period to the spot market is when the world economy starts to rebound, while a sufficient condition is when the fleet capacity utilisation has bottomed. A practical way to decide whether these conditions are met, is the distribution of earnings (see below).

The conditions for selling ships and moving from the spot to the period market are more complicated and the distribution of earnings is of no help. This book shows that these decisions depend on the length of the shipping cycle. Shipping cycles are caused by business cycles and, therefore, an owner should sell ships and move from the spot to the period market when the business cycle comes to an end. This occurs when the cycle has matured to such an extent that the economy has become overheated and inflation deviates significantly from the central bank target or asset price inflation has become a bubble that has burst.

Whereas the business cycles of countries like Japan in the 1980s and the 1990s and China in the twenty-first century have shaped the trend in the demand for shipping services, the US has always been the trigger for the end of the world business cycle and, therefore, of shipping cycles. Therefore, developments in the world economy and central bank action, particularly of the Fed, are the key to deciding when to take profits by selling ships or moving from the spot to the period market.

The identification of these conditions can be eased with the use of structural models, which are based on the theoretical models of business and shipping cycles developed in all previous chapters.

This chapter is organised as follows. The next section analyses the distribution of earnings and shows how it can be used as a buying signal of ships and moving to the spot market. But the distribution of earnings is of no use as to when to sell ships or move from the spot to the period market. We then offer an example of how structural models can ease these decisions. In section 3 we present a case study to highlight these principles.

1 THE MAJOR DECISIONS IN SHIPPING

There are two main decisions to be made in the business of shipping: when to buy and sell ships and when to be in the spot or period market. Profits can be made out of these two decisions, which sometimes can be lucrative. Such profits have lured many investors in the shipping industry, but only a few have survived the ups and downs. It requires great skills, deep understanding of what is involved and luck to be a successful entrepreneur in shipping. This point can be verified by looking at the statistics of average earnings in the dry market over the period 1985–2012 (Figure 10.1) and in the oil tanker market in the period 1973–2012 (Figure 10.2).

Sometimes in this time period a shipowner made as little as $5,000 per day (p/d), but on occasions he made as much as $65,000. Although the average for the period was nearly $15,000, the median[1] was nearly $10,000. In fact, the median suggests that half of the time he made less than $10,000, but the other half of the time he made much more. Around 30 per cent of the time he made between $7,500 and $10,000. The median describes more accurately what we commonly consider to be the norm; whereas the mean is affected by extreme values. As in this period the mean is higher than the median it is affected more by extremely high than extremely low values.

The distribution of earnings in this period has zero probability of arising out of a normal distribution (in which the mean and the median coincide, the skewness coefficient is zero, and the kurtosis coefficient is 3). Earnings are very volatile; the standard deviation is almost $12,000. The skewness coefficient measures the symmetry of the distribution of earnings around the mean; while the kurtosis coefficient measures the existence of fat or thin tails (namely, the probability that extreme values occur more or less frequently than 5 per cent of the time). The kurtosis coefficient of earnings is more than 3, implying the existence of fat

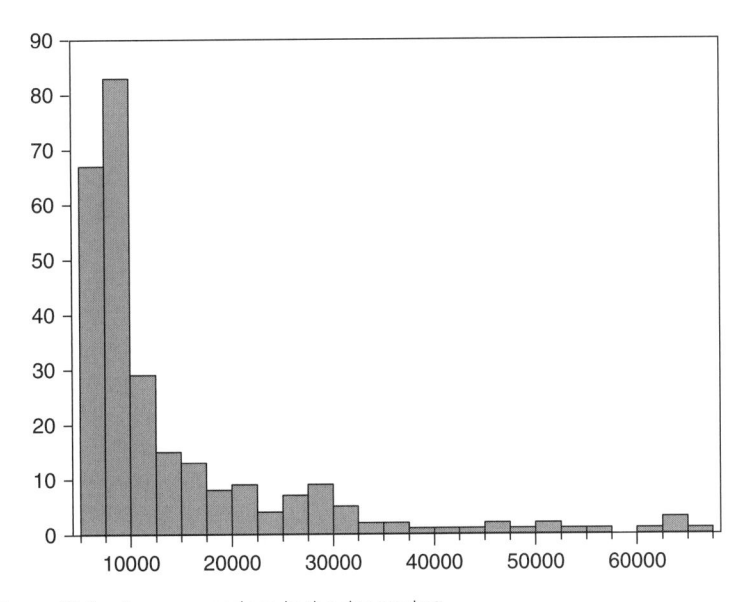

Figure 10.1 Average earnings in the dry market

Figure 10.2 Average earnings in the oil tanker market

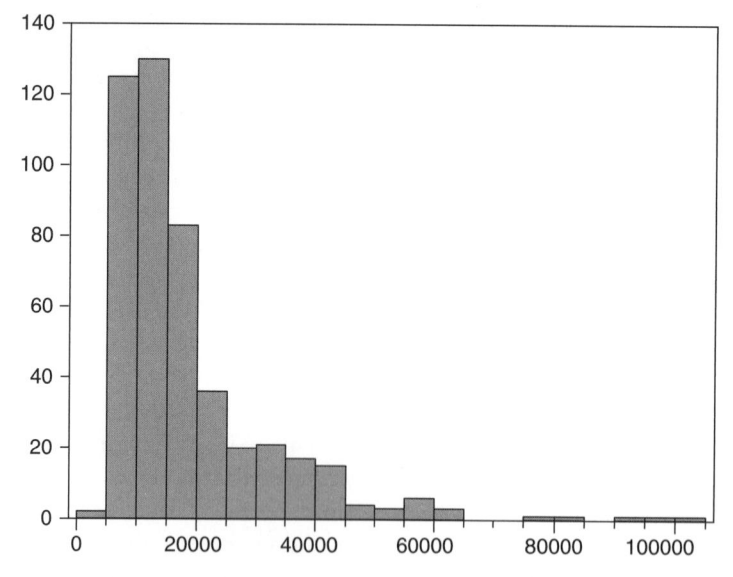

Series: EL	
Sample 1973M01 2012M02	
Observations 470	
Mean	18353.16
Median	13986.50
Maximum	103960.0
Minimum	4817.000
Std. Dev.	13890.79
Skewness	2.458532
Kurtosis	11.41512
Jarque-Bera	1860.256
Probability	0.000000

tails. Therefore, the distribution of earnings is asymmetric with fat tails and high volatility.

A similar picture emerges from the distribution of earnings in the oil tanker market. The mean of average earnings is more than $18,000, compared with a median of $14,000, confirming that the distribution is skewed towards low values with high earnings being the outliers. The skewness coefficient of the two distributions of earnings is nearly the same (2.4 vs 2.5). But the kurtosis coefficient is higher for the oil tanker market than the dry (11.4 vs 8.6), suggesting more extreme outliers for the oil tanker market than for the dry. The standard deviation of the oil tanker market is slightly higher than the dry.

The distribution of earnings suggests that successful owners are those that can afford to wait for a long time until they grasp a good opportunity to make exceedingly high earnings. Such opportunities do occur invariably, but owners have to be patient. Keeping costs low to remain profitable in poor economic conditions of low earnings is the key to success. In many cases, the cost of servicing the debt may be more important than operating costs. This implies that the degree of leverage and the volatility of interest rates play a crucial role in the success of the business. In demand- and supply-led business cycles the peak of inflation signals a period of falling and low interest rates. In asset-led business cycles, however, the cost of money may remain high despite policy rates being cut by the central bank. Therefore, in asset-led cycles an owner should wait longer before buying ships or moving to the spot market than in either demand- or supply-led cycles.

The distribution of earnings also suggests that an experienced shipowner would know when to buy as in the bottom of every cycle earnings usually revisit $5,000. Unfortunately, even an experienced owner would not know when to sell because the upper end of earnings would always be a surprise. This is the paradox of earnings. The low end of the distribution is known with near certainty, but the

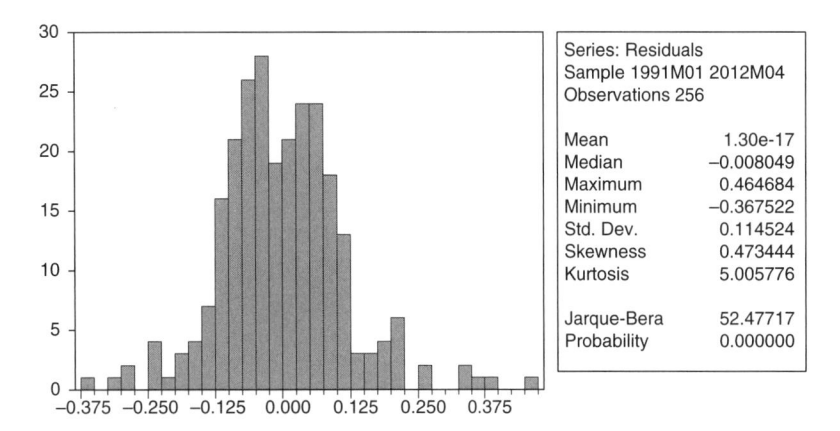

Figure 10.3 The errors of a model of average earnings in the oil tanker market

Series: Residuals	
Sample 1991M01 2012M04	
Observations 256	
Mean	1.30e-17
Median	−0.008049
Maximum	0.464684
Minimum	−0.367522
Std. Dev.	0.114524
Skewness	0.473444
Kurtosis	5.005776
Jarque-Bera	52.47717
Probability	0.000000

upper end is difficult to guess. Hence, experienced shipowners know when to purchase a ship, but not when to take profits.

This is a direct consequence of the fact that the distribution of earnings is not normal, which implies that there is no easy way of judging if even higher earnings can be achieved. Misjudging the high end of earnings can trigger a premature sale of the vessel or quitting the spot market for the sake of the period market. Thus there is no guarantee that an owner would reap the full benefits of his long and patient waiting, even if he is experienced. The history of earnings that comes with experience offers little guidance of future earnings outliers.

Structural models that integrate the macro models of the major economies with shipping can be of help in improving the main shipping decisions. To illustrate, assume a structural model in which freight rates (or earnings) in the oil tanker market are a stable function of the fleet capacity utilisation and the factors that determine the degree of risk aversion of investors. Such a model has a margin of error (one standard deviation) of 14 per cent. This is large in absolute terms, suggesting the difficulty in forecasting accurately earnings. Nonetheless, the margin of error is small relative to the volatility of earnings, which is treble that of equities. The standard deviation of earnings is 53 per cent compared with 15 per cent volatility for equities.

Structural models, such as the one considered here, can help to ameliorate the problems that arise from the paradox of earnings. Structural models assess whether the observed level of earnings is justified and therefore offer a yardstick against which to judge whether earnings have an even higher potential or whether is time to take profits by selling ships. They do that by identifying the set of economic variables that determine earnings. Such a model computes the equilibrium or fair value of earnings, conditional upon current economic fundamentals. By comparing actual earnings with the equilibrium or fair value one obtains an objective measure of whether current earnings are high or low. The deviation of actual earnings from the equilibrium or fair value offers a yardstick against which to measure outliers. If actual earnings are more than two standard deviations higher than those justified by economic fundamentals, then this is indeed an outlier and the market is likely to correct. If actual earnings are lower than those justified by economic fundamentals, then the market is likely to have a further upside. These structural models are constructed in such a way so that their errors (that is, the deviation of actual earnings from those justified by economic fundamentals) are much closer to the normal distribution than the historical distribution of earnings.

Figure 10.3 provides the distribution of errors of the aforementioned structural model. The distribution is not normal, but it is a big improvement compared with the historical distribution of earnings. The model can be of use to owners as it complements their experience. The mean is zero and it is equal to the median, as in a normal distribution. The distribution of errors is much more symmetric than the historical distribution. The skewness coefficient is 0.5 compared with 2.4 for the historical distribution of earnings. The distribution of errors has less fat tails than the historical distribution of earnings. The kurtosis coefficient is 5 compared with 8.6 for the historical distribution of earnings. Thus, the distribution of errors offers a reasonable yardstick against which to judge outliers. It is certainly a big improvement on judging from experience whether to take profits.

Structural models can go one step further. By projecting economic fundamentals into the future one can see whether the equilibrium value of earnings is likely to go up or down.

2 CASE STUDY: WHEN TO INVEST IN THE DRY MARKET[2]

A private equity fund, called HOPE-FUND, becomes interested in investing in the dry market, attracted by the returns of 2003–08. HOPE-FUND selects a maritime company, called SAIL, to manage its fleet and provide investment advice. SAIL, using the methodology developed in this book, submits to HOPE-FUND its investment advice in the third quarter of 2012 on when to buy ships and whether they should be secondhand (SH) or newbuilding (NB) ones. In evaluating SAIL's investment advice HOPE-FUND builds a model based on best practice in the Street.

There is a stark contrast in the dry market investment strategy between SAIL and HOPE-FUND. According to SAIL, the fleet should reach 70 per cent of the target fleet capacity in 2013 made up of SH vessels, while according to HOPE-FUND the investment should be delayed until 2015. A third-party adviser is employed to analyse the underlying assumptions and methodology employed by each group so that a consensus strategy can be reached. This is the report of the third-party adviser.

The diverse views on investment strategy stem from different assessments regarding the prospects of the demand for dry, freight rates, NB and SH prices. In the HOPE-FUND model, the demand for dry is assumed to grow at nearly half the pace of the SAIL model. This affects the projections of freight rates and SH prices, giving rise to a gloomy outlook, as both variables depend on fleet capacity utilisation. In the HOPE-FUND model, the fleet capacity utilisation rises until 2015 by one-third of the SAIL model (3.5 per cent compared with 9 per cent). In the HOPE-FUND model, the fleet capacity utilisation reaches the level of the SAIL model two years later, at the end of 2016. In the HOPE-FUND model,

NB prices are assumed to grow at the paltry rate of 2.2 per cent throughout the projection period 2013–18 compared to the 15 per cent rise in just the first two years, that is forecast by SAIL. The poor prospects of NB prices, along with the slow recovery of the fleet capacity utilisation, are responsible for HOPE-FUND conclusions that SH prices do not show any gains until 2015. SH prices increase at the modest pace of around 10 per cent in the years 2015–16 and around 15 per cent in 2017–18. This is in stark contrast with the SAIL view according to which SH prices shoot up by more than 50 per cent in 2013 and 2014. Finally, freight rates stay stagnant in the HOPE-FUND model until 2015, whereas they rise sharply in the SAIL model.

The HOPE-FUND gloomy outlook for the dry market relative to SAIL is due to methodological issues centring mainly on two areas of modelling the dry market: demand for dry and NB prices. The reliability of a model to predict the future depends on how well it explains the past. In this respect the SAIL model is far superior to the HOPE-FUND model. The margin of error of the HOPE-FUND model of dry demand is 8 per cent compared with 1 per cent in the SAIL model. The corresponding margin of error for NB prices is at best 15 per cent for the HOPE-FUND model compared with 3.1 per cent for the SAIL model. In the worst case the HOPE-FUND model of NB prices is simply based on arbitrary assumptions about the profit margin of the shipyard industry.

Although the SAIL model is subject to fewer errors in forecasting the demand for dry, the pessimistic HOPE-FUND prospects might still materialise. The SAIL model, though, offers a realistic assessment of risks by composing a main and risk scenario and assigning a probability to each one of them based on the likelihood that the assumed policy assumptions materialise. The main scenario assumes a small fiscal drag in the US coupled with austerity measures in Europe and a delay of a fiscal stimulus in China because the new leadership takes over in March 2013 with no serious action expected until the second half of 2013. These factors cause slower growth in the world economy in 2013 and hence in the demand for dry until the second half of 2013. Strong growth in the world economy is then predicted to return in 2014 and beyond, as policy is redirected towards growth and/or the impact of tight policies fades away.

The risks to the main scenario stem from the US being unable to withstand the fiscal contraction, political instability in Europe resulting in deeper recession, the lack or delay of a policy stimulus in China, currency war among major central banks in weakening their currencies and resurgence in the price of oil because of tensions with Iran over its nuclear policy. In the risk scenario, the US fiscal contraction is double that assumed in the main scenario; the stimulus in China is half that of the main scenario, the recession in the euro zone is twice as deep as in the main scenario and the yen returns to ¥80 because of currency war. SAIL assigns 70 per cent probability to the main scenario.

The next section provides a non-technical assessment of methodological issues. Appendix 1 provides a detailed and rather technical treatment of the methodological issues of the dry demand model and offers ways through which the HOPE-FUND model can be improved. Appendix 2 deals with the methodological issues of NB prices and again suggests ways to improve the HOPE-FUND model.

3 A NON-TECHNICAL ASSESSMENT OF METHODOLOGICAL ISSUES

In the HOPE-FUND model, the demand for dry is based on a rule of thumb that relates demand to world GDP, adjusted for port congestion and China's coastal trade. The major component of demand is a fixed multiple of world GDP, while the adjustments for port congestion and China's coastal trade play a minor role. Appendix 1 shows that this model offers a poor explanation the past, which is a prerequisite for reliable projections. The HOPE-FUND model has a margin of error of 8 per cent in explaining annual data over the period 2001–11. The adjustments of port congestion and China's coastal trade do not help the model in explaining the past because of the low volatility and the small contribution of these factors to total demand. A log-linear model based on world GDP has a slightly lower margin of error (of 7.3 per cent) than the original HOPE-FUND model. But both models are far from the SAIL model of demand, which has a margin of error of just 1 per cent for an even longer period 1989–2012 using quarterly data, which are more volatile and therefore more difficult to explain than annual data.

The distinguishing factor between the two models is that in the HOPE-FUND model demand is a fixed multiple of world GDP, whereas the multiplier in the SAIL model is variable. The value of the multiplier depends on whether each major economy is overheated or operates with spare capacity, the policy mix in each major economy, the availability of credit and the cost of money; the level of stocks of raw materials and finished goods; and business and consumer confidence. A fiscal stimulus would boost growth more if: there is spare capacity in the economy; stocks are low; monetary policy is accommodative; the cost of money is low; credit is available; and confidence is low. In this case, the multiplier in the first year of the stimulus may be as high as 3 (instead of 1.43, as assumed in the HOPE-FUND), fading in the second and third years to less than 1. In the SAIL model, instead of arbitrarily assuming the value of the multiplier for each year, the factors that affect its value are allowed to enter explicitly in the demand model.

The HOPE-FUND model of NB prices is based on another rule of thumb, namely that prices are a mark-up on costs. Three types of costs are considered: the price of hot rolled steel sheet, which is assumed to account for nearly 30 per cent of total cost; China's labour cost adjusted for productivity, which is assumed to account for more than 13 per cent of total cost; and the cost of vessel equipment (machines and the like), which accounts for 57 per cent of total cost. The profit margin is equal to zero for the years 2013–16, 5 per cent for 2017 and 10 per cent for 2018. The HOPE-FUND's gloomy projections of NB prices adversely affect the outlook for SH prices, because they are modelled as a fraction of NB prices. This fraction, in turn, depends on fleet capacity utilisation, which, because of the gloomy outlook for demand, does not recover making any investment in shipping unattractive.

The track record of the HOPE-FUND model can be tested by applying the above mentioned rule of thumb on the assumption that the profit margin is constant through time. Appendix 2 shows that this model gives rise to an average margin of error of 15 per cent relative to 3.1 per cent in the SAIL model.

Appendix 2 shows that the HOPE-FUND model can be improved by considering a log-linear model of NB prices just on the price of hot rolled steel. Of the three cost components, namely equipment, labour and hot rolled steel, only the latter matters. Variations in equipment and labour costs do not contribute to the explanation of the large variability of NB prices. The volatility of NB prices is 28 per cent, whereas the volatility of the cost of equipment is low, because PPI inflation in the US, Europe and Japan, where most of this equipment is manufactured are stable relative to NB prices. China's labour cost is affected by the tightness of the labour market, which depends in turn on the rate of growth of the economy, which is stable relative to NB prices. Accordingly, from the three cost components only the price of hot rolled steel contributes directly to the variability of NB prices. This model has a margin of error of 14.7 per cent, which is by a whisker smaller than the original HOPE-FUND model and is consistent with the stylised facts of a highly pro-cyclical profit margin, rising in the upswing and falling in the downswing.

By contrast, the SAIL model bypasses the aforementioned problems by making the profit margin a function of fleet capacity utilisation, the demand–supply balance in the shipyard industry (shipyard capacity), as well as SH prices, as shipyards raise profit margins when SH prices are increasing. This model has a margin of error of 3.1 per cent, which gives credence to its projections.

This analysis casts doubts on the conclusion of HOPE-FUND that NB would increase by only 2.2 per cent in 2014. The gloomy projections are the result of arbitrary assumptions on the profit margin. On the other hand, the SAIL model prediction that prices increase by 15 per cent seems more plausible.

APPENDIX 1: THE DEMAND FOR DRY

According to HOPE-FUND, the demand for dry would grow at 5.5 per cent until 2015 and 6 per cent for the three-year period of 2015–17. According to SAIL, demand would grow at nearly double this pace in 2013 and 2014 (see Figure 10.A1 for the two projected paths). SAIL does not provide a forecast for a period longer than two years, because it is conditional on economic policy and central banks have a two-year horizon. The sharp contrast of the first two-year prospects of demand is largely responsible for the different investment strategies of the two groups. As both groups model the demand for dry rather than assume its pace, the differences arise from methodological issues.

In the HOPE-FUND model, the demand for dry is a constant multiple (1.43 times) of the rate of growth of world GDP, adjusted for port congestion and China's coastal trade. The last two items play a minor role in arriving at the total demand figure. With world growth assumed to be sluggish in 2013 by most economists, including SAIL for the first half of 2013, the HOPE-FUND model's gloomy outlook for the demand for dry seems sensible. Yet there is a critical modelling assumption in the HOPE-FUND methodology, namely that the demand for dry in each year is a fixed multiple of world GDP, which is not shared by SAIL. According to SAIL, the multiplier is variable rather than fixed. A fiscal stimulus would boost growth more if: there is spare capacity in the economy; stocks are

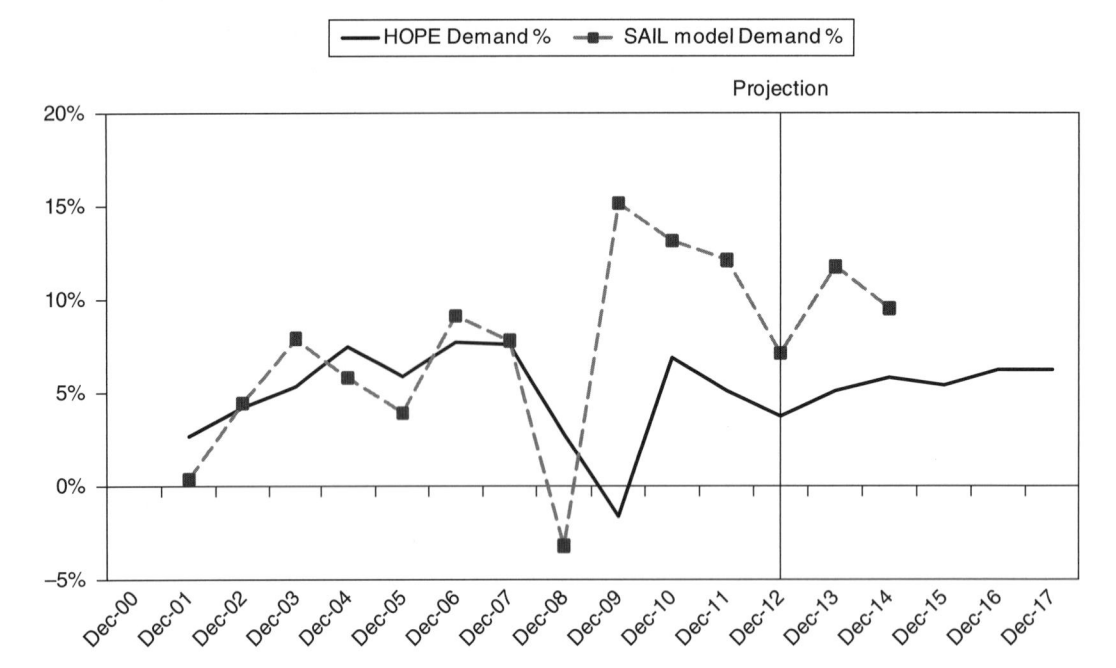

Figure 10.A1 HOPE and SAIL demand

low; monetary policy is accommodative; the cost of money is low; credit is available; and confidence is low. In this case, the multiplier in the first year of the stimulus may be high as 3, fading in the second and third years to less than 1. In the SAIL model, instead of assuming the value of the multiplier for each year, the factors that affect the value of the multiplier are allowed to enter explicitly in the demand model.

Figure 10.A1 plots the trajectories of the demand growth implied by each model both in the historical and the projected period. As can be seen, the two models do not differ substantially in the historical period 2001 to 2007, when the volatility of world GDP and world trade was low. But the models differ significantly from 2008 onwards, when there was a heightened volatility in world economic activity and world trade. The volatility of demand In the SAIL model is much higher than in HOPE-FUND. This higher volatility is a testament of the variable multiplier in the SAIL methodology as opposed to the fixed multiplier in HOPE-FUND. Thus, the SAIL model suggests that demand plunged in the 2008 recession to −3.2 per cent (y-o-y) and soared to more than 15 per cent in 2009, following the fiscal and monetary stimuli adopted simultaneously in all major economies, the low level of stocks, the huge spare capacity and the low confidence. By contrast, according to HOPE-FUND the demand fell only to −1.6 per cent and recovered to just 6.9 per cent. This is a gap of 8.5 per cent in the HOPE-FUND model compared to 18.5 per cent in the SAIL model.

The validity of each model in projecting the rate of growth of demand can be tested empirically by posing the question of which model can explain better the historical data. The trajectories of the HOPE-FUND demand model are plotted in Figure 10.A2, along with the actual historical errors and the margin

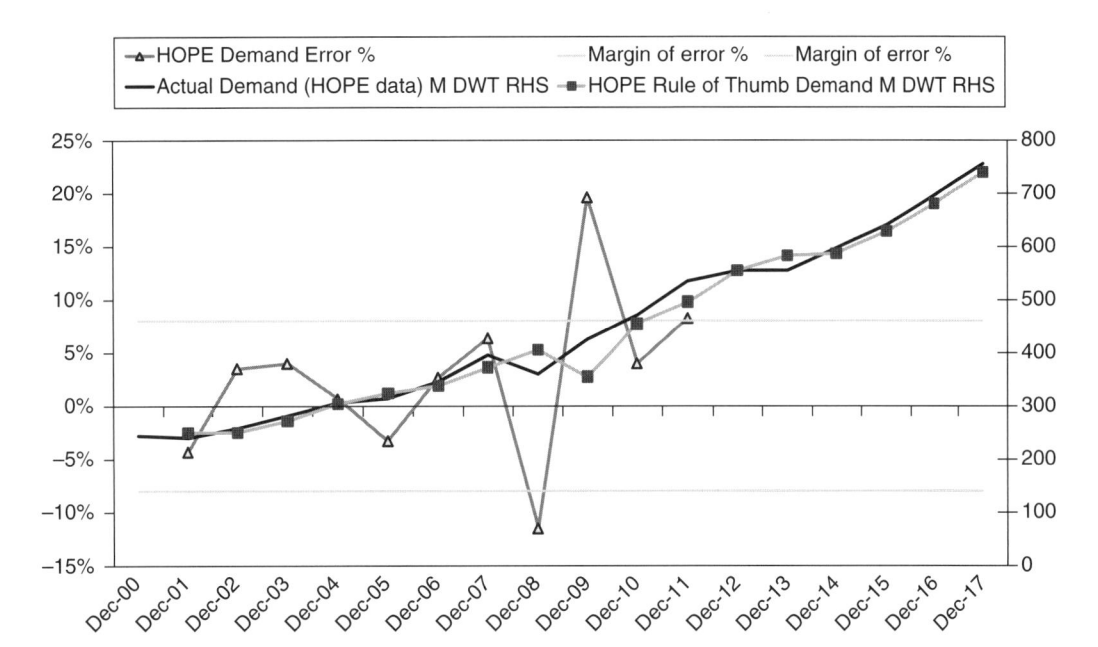

Figure 10.A2 HOPE rule of thumb demand

of error based on one standard deviation. The demand trajectories are obtained by applying the rule of thumb of a fixed multiplier (1.43 times) of world GDP and applying a bias favourable to HOPE-FUND. HOPE-FUND does not provide historical data on port congestion and China's coastal trade, but in the projection period, it assumes that the sum of these factors contributes between 0.4 per cent and 0.8 per cent to the demand figures. In these simulations we assume that these two factors contribute 0.5 per cent to the growth of demand in a way that favours HOPE-FUND. Thus, when HOPE-FUND underpredicts (that is, when the actual demand figure exceeds the HOPE-FUND figure) we add the 0.5 per cent adjustment; and when HOPE-FUND overpredicts (that is, when the actual figure falls short of the HOPE-FUND figure) we subtract the 0.5 per cent adjustment. With these favourable assumptions to HOPE-FUND the margin of error[3] in the HOPE-FUND model is 8 per cent compared with the volatility of demand of 28 per cent in the historical period 2001–11.

The HOPE-FUND model does not benefit by disaggregating the impact of world GDP, port congestion and China's coastal trade on the demand for dry. A log-linear model of the demand for dry on world GDP level would give a slightly better model than the one currently used by HOPE-FUND. Figure 10.A3 plots a similar set of statistics for this alternative HOPE-FUND model. The margin of error of this revised model is slightly smaller than the original HOPE-FUND (7.3 per cent compared with 8 per cent). Using the HOPE-FUND assumptions of world GDP, this log-linear model would give a more improved outlook for demand than the original model. Demand would grow 7.5 per cent in 2014 and 2015 and 8.6 per cent in 2016 and 2017 compared with 5.8 per cent in the original model. Yet even this improved model implies a nearly 10 per cent error in predicting demand in 2011 and more than 7 per cent for 2012. These errors

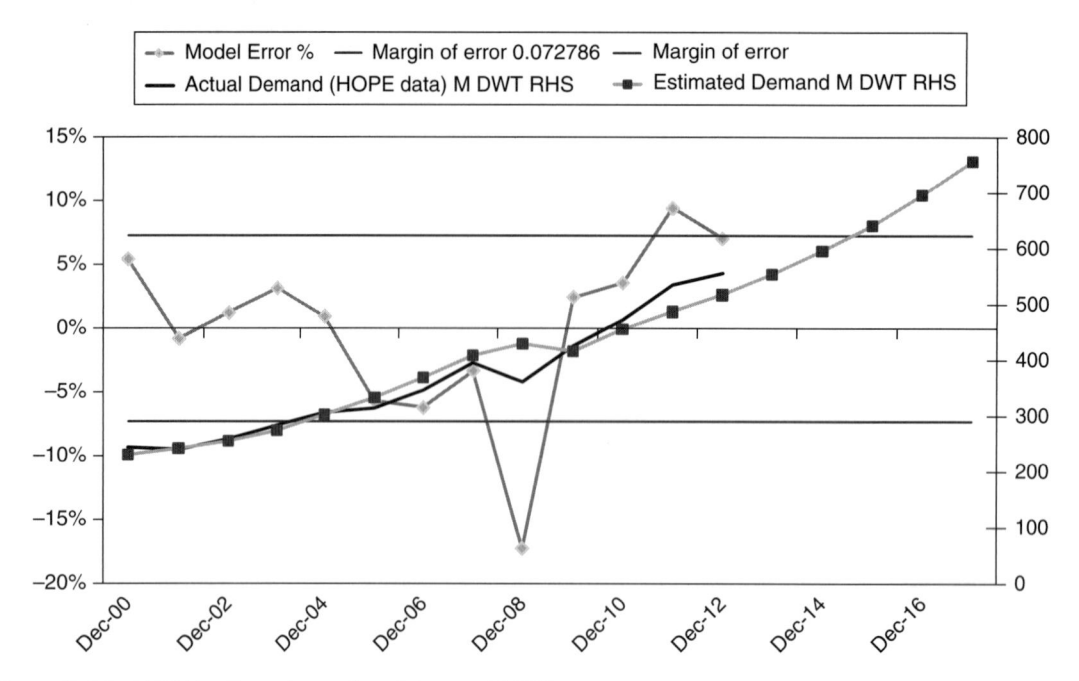

Figure 10.A3 HOPE log-linear demand model on world GDP

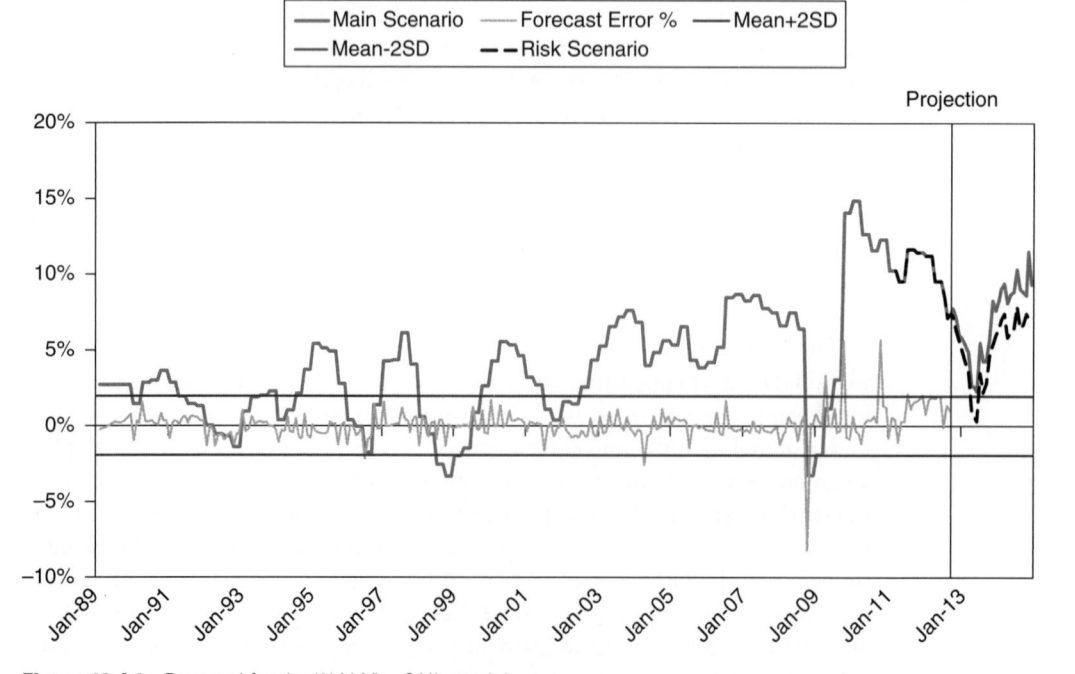

Figure 10.A4 Demand for dry (% YoY) – SAIL model

suggest that HOPE-FUND is seriously underpredicting demand and, therefore, biasing its investment strategy towards waiting, instead of taking action now.

The SAIL model, on the other hand, with its emphasis on variable multipliers has a margin of error of just 1 per cent compared with 8 per cent and 7.3 per cent in the original and revised HOPE-FUND models. Figure 10.A4 plots a similar battery of model statistics for the SAIL model. The graph plots two standard deviations of the margin of error in the SAIL model, which demands accuracy at 95 per cent probability instead of one standard deviation in the HOPE-FUND model, which demands accuracy of only 66 per cent. The model properties of the SAIL model are plotted over a much longer period of time (1989–2012) and using quarterly data to show that the model has been valid under varied conditions of world GDP and world trade with the wider volatility implicit in quarterly data than annual data. Figure 10.A4 shows that actual demand has exceeded the 2 per cent error in only five quarters during the entire period from 1989 to 2012. The errors in the SAIL model are evenly distributed within two standard deviations (see Figure 10.A4).

The conclusion is that the demand projections of the SAIL model are more reliable than those of the HOPE-FUND.

APPENDIX 2: NB PRICES

In the HOPE-FUND model NB prices are a mark-up on costs. Three types of costs are considered in the model: the price of hot rolled steel, which is assumed to account for nearly 30 per cent of total cost; China's labour cost adjusted for productivity, which accounts for more than 13 per cent of total cos; and the cost of vessel equipment (machines and the like), which accounts for 57 per cent of total cost. Unfortunately, HOPE-FUND does not provide historical data for the various cost components, making it very difficult to validate the historical track record of the model. Nonetheless, there are data for the rate of growth of wages in China. In line with HOPE-FUND, it is assumed that productivity has been growing at 7 per cent per annum to derive data on China's unit labour cost. Equipment costs are related to PPI-inflation in the US, Europe and Japan, where most of this equipment is manufactured. We take Germany's PPI data as representative of equipment costs. Data for the price of hot rolled steel are available from Reuters. Using these data, with the HOPE-FUND assumptions about the contribution of each cost component to the total and assuming a fixed profit margin, we compare actual NB prices with the HOPE-FUND rule of thumb.

Figure 10.A5 plots the rate of growth of NB prices according to the HOPE-FUND rule of thumb that prices are a fixed mark-up on total costs, along with the actual rate of growth of NB prices and the forecast error for the period 2000–12. The HOPE-FUND model has a margin of error of 15 per cent on the assumption that the profit margin is constant.

More doubts emerge about the validity of HOPE-FUND model because the volatility of NB prices is very high (28 per cent), whereas the volatility of the cost of equipment is low, associated with PPI-inflation in the US, Europe and Japan, where most of this equipment is manufactured. China's labour cost is affected by the tightness of the labour market, which depends in turn on the rate of growth of the economy. Hence, it is unlikely that variations in equipment and labour costs

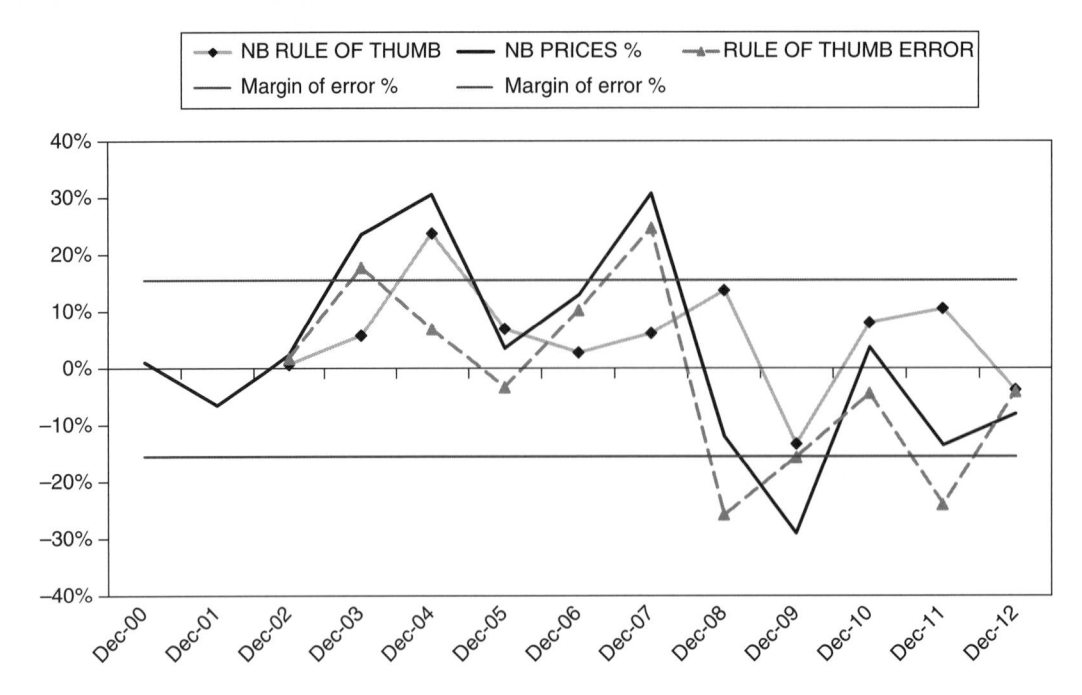

Figure 10.A5 HOPE NB rule of thumb

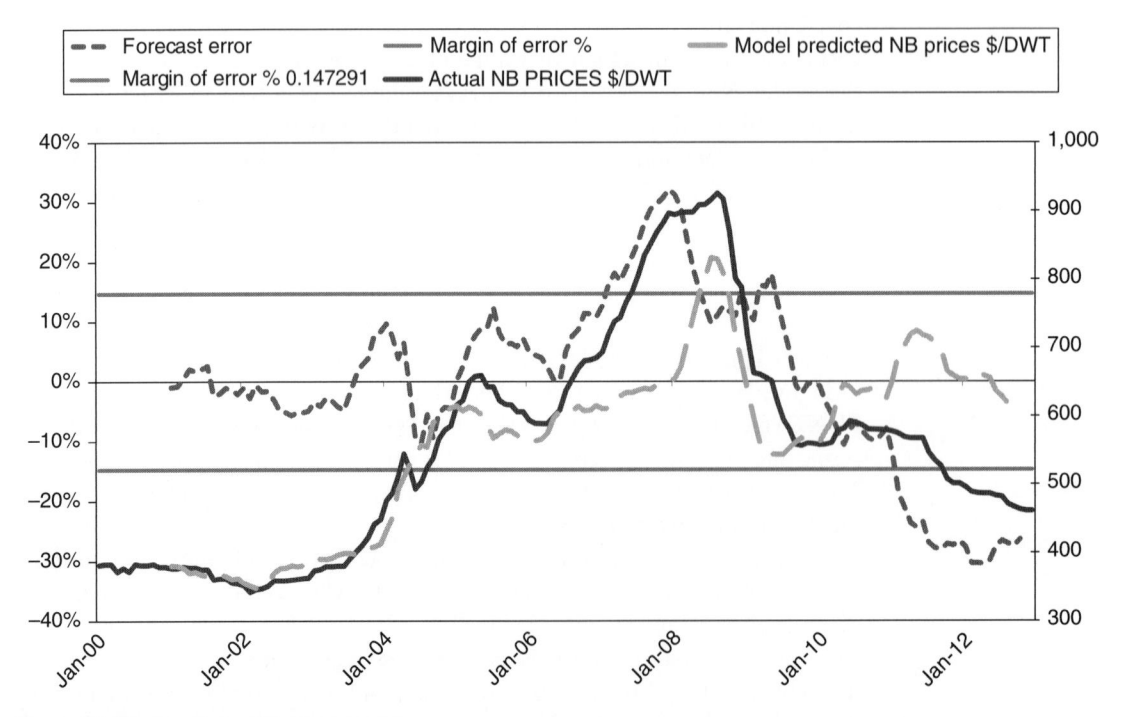

Figure 10.A6 Log-linear NB prices model

can account for the large volatility of NB prices. Accordingly, from the three cost components only the price of hot rolled steel can directly contribute to the variability of NB prices. To test this hypothesis a log-linear model is built of NB prices on the price of hot rolled steel (see Figure 10.A6).

This log-linear model has a slightly smaller margin of error than the original HOPE-FUND model (14.7 per cent relative to 15 per cent), implying approximately the same average profit margin This casts doubt on the usefulness of the HOPE-FUND model in forecasting NB prices, as the projections depend on arbitrary assumptions about the profit margin of the shipyard industry. The HOPE-FUND model assumes that the profit margin is equal to zero for the years 2013–16, 5 per cent for 2017 and 10 per cent for 2018.

By contrast, the SAIL model bypasses the aforementioned problems by making the profit margin a function of the combined demand–supply balance in the freight market (fleet capacity utilisation) and the demand–supply balance in the shipyard industry (shipyard capacity), as well as SH prices, as shipyards can raise profit margins when SH prices are increasing. This model has a margin of error of 3.1 per cent, which gives credence to its projections (see Figure 10.A7). The errors in the SAIL model are evenly distributed within two standard deviations over the longer period of 1989–2012, using quarterly data. It is worth noting that the margin of error of a quarterly model is higher than that of an annual model.

This analysis casts doubts on the conclusion of HOPE-FUND that NB would increase by only 2.2 per cent in 2014. The SAIL model instead predicts 15 per cent increase, which enables a sharper increase in SH prices. The overall conclusion is that the projection of NB prices should be based on the SAIL model.

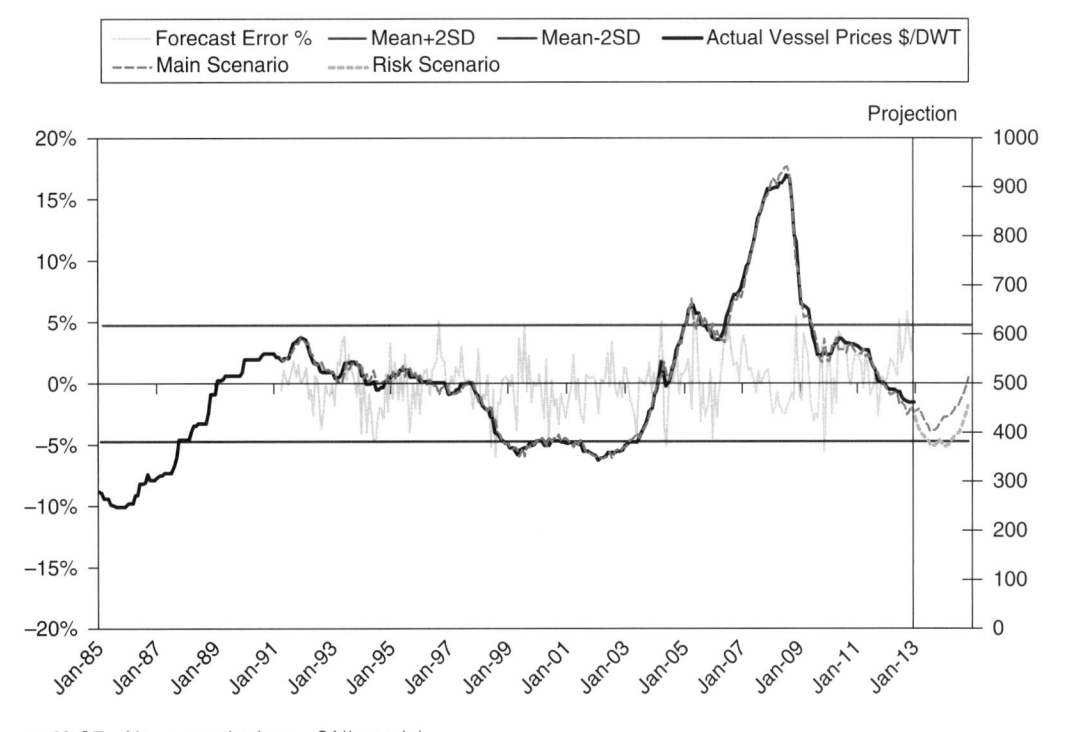

Figure 10.A7 New vessel prices – SAIL model

NOTES

2 THE THEORETICAL FOUNDATIONS OF THE FREIGHT MARKET

1. Stopford (2009) recognises the invalidity of the homogeneity assumption, but does not take the logical implication of rejecting the perfectly competitive model of freight rates.
2. In a zero-sum game of two players the profit of one player is the loss of the other.
3. This is not the sum of the demand functions that each owner confronts in the market, but the sum of the demand functions of all charterers. The former is a perfectly elastic demand curve, whereas the latter is negatively sloped, resembling, but not matching, the shape of the aggregate market demand curve.
4. The revenue function of an individual owner is $R = P Q_i$, where P is a constant. Hence, the average revenue curve $AR = R/Q_i = P$ and marginal revenue $dR / dQ_i = P$.
5. The second-order condition for maximum profits requires that the marginal cost is rising at the maximum profit. This implies that the MC curve cuts the AVC curve from below.
6. Cardinal utility coincides with the layman's conception that the monetary scale can be used as a basis for evaluations and decisions. This is illustrated by the attempt of businessmen to assign monetary values to intangible assets, such as 'goodwill'. The nineteenth-century economists W. Stanley Jevons, Leon Walras and Alfred Marshall considered utility measurable, just as the weight of objects is measurable. The decision maker was assumed to possess a cardinal measure of utility, that is, s/he was assumed to be capable of assigning to every outcome or combination of outcomes a number representing the amount or degree of utility associated with it. The numbers representing amounts of utility could be manipulated in the same fashion as weights. Accordingly, the differences between utility numbers could be compared, and the comparison could lead to statements such as 'A is preferred to B twice as much as C is preferred to D'.
7. The postulate of rationality is equivalent to the following statements: (1) for all possible pairs of alternatives A and B the (decision maker) knows whether s/he prefers A to B or B to A, or whether s/he is indifferent between them; (2) only one of the three possibilities is true for any pair of alternatives; (3) if the (decision maker) prefers A to B and B to C, s/he will prefer A to C.
8. This is true for $A > 0$, $B > 0$. This is the archetype of all bargaining games (Nash, 1950; Harsanyi, 1956).
9. This is, strictly speaking, true only in *cardinal* analysis. As has been shown by Shapley (1969), no resolution of the two-person bargaining problem can be made on the basis of *ordinal* utility alone. In terms of Figure 2.5 this means that an exact solution of the bargaining problem may be impossible from just the *ordinal* information

depicted in the figure. The reason for this sweeping statement is that the origin can be moved by a linear transformation, thereby changing the value of the bargaining solution.
10. It is assumed that $g(X_1, X_2)$ is continuous, has continuous first- and second-order partial derivatives and that the first-order partial derivatives are strictly positive, whereas the second-order partial derivatives are negative. In a *cardinal* utility analysis the first partial derivatives are defined as the marginal utilities of the payoffs X_1 and X_2. The marginal utility of a payoff is frequently referred to as the increase in utility resulting from a unit increase in its payoff. In *cardinal* analysis such derivatives have numerical interpretation. However, the first partial derivatives of an *ordinal* utility function cannot be given a cardinal (numerical) interpretation. Moreover, it is assumed that $g(X_1, X_2)$ is a regular strictly quasi-concave function. The latter condition implies that all points on a line segment connecting any two points on an iso-utility curve lie on a higher iso-utility curve. Thus, all points on the line AB lie on a higher iso-utility curve than U^0 (see Figure 2.6).
11. The rate of payoff substitution is frequently referred to in economics as the *marginal rate of substitution* (Hicks, 1946).
12. The iso-utility curves in Figure 2.6 are rectangular hyperbolas. Such curves become asymptotic (that is, they are parallel) to each axis at low values of X_1 and X_2. Thus lines D and F are asymptotic to the iso-utility curve U^0. Similarly, lines E and F are asymptotic to U^1. The asymptotic lines define the limits beyond which it is impossible to squeeze the other player and yet achieve agreement.
13. For a solution of the bargaining problem the utility function need not be cardinally measured. An ordinal utility function will suffice.
14. For a given level of utility, U^0, the equation of the iso-utility curve that corresponds to (2.3) is

$$X_2 = \left(\frac{U^0}{A}\right)^{1/\beta} X_1^{\frac{-(1-\beta)}{\beta}} \qquad (2.4)$$

It is apparent from equation (2.4) that the bargaining power of the owner is an increasing function of $1/\beta$. For $X_1 = 1$, the second term on the right-hand side of equation (4) will always give 1, while the first term would be raised to a higher power, the lower β is (provided $U^0 > A$). Hence for $X_1 = 1$, X_2 will be larger (and therefore it will lie on a higher iso-utility curve), the larger is $(1/\beta)$.
15. Although the two iso-utility curves in Figure 2.7 do not intersect, in general there is no guarantee that this would be the case.
16. For simplicity it is assumed that expectations of future freight rates are homogeneous. This means that the owner and the charterer form exactly the same expectations about future freight rates. This would be the case if both players shared the same information and processed it in the same manner. This will be true of the information about all economic and shipping variables that have become available up to time t; but it is not necessarily true of the processing process, as this implies that both players share the same macro and dry cargo models and imputed the same objective function for all policymakers. In the more realistic case in which the owner and the charterer process the common information differently, the two expectations of freight rates would not coincide. This would complicate the bargaining process, but the assumption of homogeneous expectations can be defended by invoking the postulate of rationality. If both players are 'rational', they would learn from their own mistakes and incorporate such mistakes in the new round of forming expectations.

This will enable convergence of the process, so that in time the two players will converge to the same expectations. Such convergence process is guaranteed by the adaptive expectations scheme

$$E_t Y(t+1) - E_{t-1} Y(t) = \gamma [Y(t) - E_{t-1} Y(t)] \quad 0 < \gamma < 1 \qquad (2.5)$$

The left-hand side of equation (2.5) shows the revision of expectations about Y, as new information arrives between time $t-1$ and t. The term in the parenthesis on the right-hand side represents the error or the mistake of the previous expectation. Hence, expectations are revised in proportion of the previous mistake. If in the last period a player under-predicted the actual Y, the expectation in this period will be revised up by the proportion γ of the previous mistake. If there was an over-prediction, the new forecast would be revised down. As γ is between zero and 1, the revision of expectations is only a fraction of the previous mistake. The higher γ is, the bigger the correction and hence the faster the convergence of expectations. Equation (2.5) implies that the expectation of Y is formed as an average of past observations of Y with exponentially declining weights. This interpretation follows from the solution of equation (2.5), which is a first order difference equation.

$$E_t Y(t+1) = (1-\gamma) \sum_{j=0}^{\infty} \gamma^j \cdot Y_j \qquad (2.6)$$

The exponential decay means that the recent past carries more weight than the distant one.

17. In the world recession of 2008–09 freight rates rebounded first among all other leading indicators, such as the S&P 500, acting as a leading indicator of world economic activity (see Chapter 9 for details).
18. See Stopford (2009, Table 6.2) for a detailed definition of these costs.
19. In the bond market the 'return' is defined as the holding period yield, the spot yield and the yield to maturity. The term structure of interest rates is, therefore, expressed in these three alternative definitions of the 'return'. The holding period yield is the return over a particular holding period, which may cover one or more periods. The (one period – that is, one month, one quarter or one year) holding period yield is defined as the return over the period resulting from capital gains and coupons (interest payments). If P^n is the price of a coupon bond maturing in n periods and C is the coupon paid in this period, then the *ex post* one-period holding period yield, H^n, is defined as

$$H^n_{t+1} = \frac{P^{n-1}_{t+1} - P^n_t}{P^n_t} + \frac{C_t}{P^n_t} \qquad (2.19)$$

In this definition the holding period yield is the return accruing from t to $t+1$ and is payable at the beginning of period $t+1$, but after the lapse of one period the maturity of the bond has been reduced by one period. The first term on the right-hand side represents the capital gains (or losses) from holding the bond for one period, while the second term represents the return from the payment of the coupons).

The spot yield, r, for an n-period pure discount bond or zero coupon bond (such as Treasury bills) is defined from the equation

$$P^n_t = F / (1+r)^n \qquad (2.20a)$$

where P^n is the current price of the bond and F is the face value or redemption price on maturity, usually $100. Zero coupon bonds are issued below par and the return (spot yield) is the implicit interest rate that increases the price P^n to par, F, in n-periods. Spot yields can also be defined for coupon bonds from

$$P_t^n = \frac{C}{1+r_1} + \frac{C}{(1+r_2)^2} + \cdots + \frac{C+F}{(1+r_n)^n} \qquad (2.20b)$$

Each term in (2.20b) is regarded as a spot yield of a zero coupon bond paying C in periods 1 to $n-1$ and $C+F$ in period n.

The yield to maturity, R, is the constant discount rate that ensures in equilibrium the equality of the current price with the discounted present value of the future coupon payments, $\underline{C}$, and the redemption price, F

$$P_t^n = \sum_{i=1}^{n}\left(\frac{C}{(1+R)^i} + \frac{F}{(1+R)^n}\right) \qquad (2.21)$$

Thus, the difference between spot yields and the yield to maturity is that the former vary with time, whereas the latter is constant.

It makes no difference whether the term structure of interest rates is expressed as the holding period yield, spot yields or the yield to maturity (see Cuthbertson, 1996).
20. See equation (5.17) and how it is derived from (5.25) and (5.26) in Chapter 5.

3 THE SHIPYARD, SCRAP AND SECONDHAND MARKETS

1. The elasticity of substitution is defined as the percentage change of K divided by the percentage change in S. This is equal to the product of the negative of the derivative times the average speed of each vessel, namely as $-dK/dS \times S/K$. The derivative is measured by the tangent on each point of an isoquant, such as the tangents at points A and B in Figure 3.1.
2. Formally, the optimum is found by maximising the supply of shipping services (output) subject to the constraint of a given cost TC^1:

$$V = F(K,S) + \lambda(TC^1 - PB \cdot d \cdot f(S) - r \cdot K - b) \qquad (3.9)$$

V is the value of the function to be maximised and $\lambda \neq 0$ is an undetermined Lagrange multiplier. To obtain the maximum the first order derivatives with respect to K, S and λ are set equal to zero:

$$\frac{\partial V}{\partial S} = F_S(K,S) - \lambda \cdot PB \cdot d \cdot f'(S) = 0 \qquad (3.10a)$$

$$\frac{\partial V}{\partial K} = F_K(K,S) - \lambda \cdot r = 0 \qquad (3.10b)$$

$$\frac{\partial V}{\partial \lambda} = TC^1 - PB \cdot d \cdot f(S) - r \cdot K - b = 0 \qquad (3.10c)$$

Moving the price terms on the right of the first two equations and diving the first by the second the optimality condition is

$$\frac{F_S(K,S)}{F_K(K,S)} = \frac{PB \cdot d \cdot f'(S)}{r} \quad \text{or} \quad \frac{F_S}{PB \cdot d \cdot f'(S)} = \frac{F_K}{r} = \lambda \quad (3.10d)$$

First-order conditions state that the ratio of the marginal products of K and S must be equated with the ratio of their prices. The Lagrange multiplier is equal to the derivative of supply of shipping services (output) to cost with prices constant and quantities variable.

3. It is worth noting that Jiang and Lauridsen (2012) in their empirical work find that freight rates are quantitatively the most important factor affecting newbuilding prices. This is true, according to our model, only when $\sigma > 1$.
4. A 'squeeze' between demand and supply occurs in asset (financial) markets, when investors close 'short' positions in evidence that demand increases instead of decreasing as they had assumed. But by doing so, investors add to demand; hence the 'squeeze'.
5. The long-run demand curve for vessels is described by equation (3.33) and the long-run supply curve by equation (3.32). See Appendix 1 for more details.
6. The supply curve of shipyards coincides with the portion of the marginal cost that exceeds average total cost. The supply curve is upward sloping, namely shipyards would produce more ships at higher vessel prices. The implicit assumption of the supply curve is that shipyard capacity is fixed. The marginal cost is increasing because with fixed capacity shipyards would have to pay more for the variable factors of production, mainly labour, to increase production.
7. This relationship can be described by the following equation, where SCU denotes the shipyard capacity utilisation and CU the fleet capacity utilisation:

$$SCU = \lambda \cdot CU \quad \lambda > 0 \quad (3.19)$$

8. This elasticity is computed from an empirical model of the demand-supply balance in the shipyard industry on the dry fleet capacity utilisation rate. Using this model with the latest data available in September 2012 on the dry fleet capacity utilisation at 85 per cent and the shipyard demand-supply balance at -28 million dwt, the elasticity of the shipyard capacity with respect to the fleet capacity utilisation is 6.4 million dwt.
9. This basic economic principle can be stated mathematically as follows.

$$\dot{P} = a \cdot SCU \quad a > 0 \quad (3.20)$$

The dot over the variable P denotes the rate of change of prices (that is, the first-order time derivative). The coefficient a measures the speed at which prices adjust to a given excess demand (supply) balance in the shipyard industry.
10. The dynamic adjustment model can be stated as follows:

$$\ddot{P} = c_1 \cdot \dot{SCU} + c_2 \cdot \dot{Z} + c_3 \cdot u_{t-1} + c_4 \cdot \ddot{P}_{t-i} \quad (3.21)$$

The vector Z denotes the explanatory variables of the system of equations (3.12) or (3.13) in Appendix 1, equation (3.8) and equation (3.9); u the residuals of this system, which measure the degree of disequilibrium of NB prices from these economic fundamentals; all other symbols are the same as before. Two dots over a variable indicate the second time derivative. The coefficients c_1, c_2 and c_4 in equation (3.21) should be positive, whereas the coefficients c_3 should be negative.
11. For a derivation of the supply curve for one factor of production see equation (3.34) in Chapter 6. The generalisation to three factors of production is straightforward.

12. Such an approach has been adopted by Merikas et al (2008). The study is noteworthy because it investigates empirically the determinants of the ratio of secondhand to newbuilding prices.
13. When $\sigma = 1$ (or $\rho = 0$) the CES production function is reduced to a Cobb–Douglas production function with constant returns to scale, that is, $Q = A\,K^a\,S^{1-a}$. In other words, only when a Cobb–Douglas function is assumed the elasticity of output and relative prices are equally important.
14. These variables are called the state-variables of the system, as for each value of the state variables one obtains a short-run equilibrium. As the state-variables are adjusting through time, another short-run equilibrium is obtained. The system would converge into a steady state (long run equilibrium), when all state-variables have stopped adjusting. As the time path of the state-variables is described by a set of differential equations, the steady state of the system is obtained at the point where all derivatives in the differential equations are equal to zero.
15. Technically, it is said that the actual fleet lies on the notional supply curve of shipyards.
16. A saddlepoint is a point that is simultaneously a minimum for one curve and a maximum for another curve, like the unique point on a horse saddle from where the name is derived.

4 THE EFFICIENCY OF SHIPPING MARKETS

1. If X_i is a discrete random variable with probability of occurrence p_i then the expected value of X, denoted by $E(X)$, is defined as

$$E(X) = \sum_{i=1}^{\infty} p_i X_i$$

If X is a continuous random variable assuming values in the range $(-\infty < X < \infty)$ with a continuous probability distribution $f(X)$ (e.g. normal distribution) then

$$E(X) = \int_{-\infty}^{\infty} X f(X) dX$$

The conditional expectation based on the information set I_t is defined as

$$E(X|I_t) = \int_{-\infty}^{+\infty} X f(X) dX$$

2. In statistics, the bias (or bias function) of an estimator (or a forecast) is the difference between this estimator's expected value and the true value of the parameter being estimated (or forecast). An estimator (or a forecast) with zero bias is called unbiased estimator (or forecast).
3. The transformation is also necessary from a statistical point of view. When the term structure of freight rates is expressed in levels (as in equation (2.23) in Chapter 2), estimation of (4.12) is not possible because all variables are non-stationary. A non-stationary variable is broadly speaking one that includes a trend (a time trend or a stochastic trend). Estimation of models with non-stationary variables gives rise to spurious correlation and invalidates statistical inferences base on t and F-statistics. The transformation of the expectations theory of the term structure of freight rates into a spread and changes of spot rates makes ordinary least squares estimation valid.

Both variables, the spread and changes in spot rates, are stationary. The Statistical Appendix explains all these concepts.
4. These models are analysed in the Statistical Appendix at the end of this chapter.
5. It is worth noting that the error term u in (4.13) is now a moving average process of order $\tau-2$, $MA(\tau-2)$, as the rational expectations forecast errors ε are combined with the errors of equation (4.12). Although the estimation does not require the use of instrumental variables, the Generalised Method of Moments is recommended to correct for possible heteroscedasticity. Kavussanos and Alizadeh (2002a) use this method in their empirical work.
6. To see these differences, assume that π is an AR(1) process:

$$\pi_t = \rho \pi_{t-1} + \varepsilon_t \quad \varepsilon_t \approx N(0, \sigma^2)$$

In this process the conditional mean is $\rho \pi_{t-1}$, whereas the unconditional mean is zero. The conditional variance of π is σ^2, whereas the unconditional variance is $\sigma^2/(1-\rho^2)$, which is a constant.
7. A statistical distribution is usually described by the first four moments: the mean (first moment); the variance or standard deviation (second moment); the skewness index (third moment); and the kurtosis index (fourth moment). In standard statistical analysis the mean of a statistical time series is modelled as an AR process. The innovation of the Engle ARCH/GARCH approach is to model the variance or standard deviation as an AR process.
8. To see this assume that all the roots of the characteristic equation $1 - \beta(L) = 0$ lie outside the unit circle. In this case equation (4.57) can be written as a distributed lag of past squared residuals η:

$$E_t(\sigma^2_{t+1}) = w\left[1 - \sum_{i=1}^{q} \beta_i\right]^{-1} + \sum_{i=1}^{\infty} \delta_i \eta^2_{t-i}$$

9. There are some exceptions to this consensus (see Adland et al., 2004; Tvedt, 2003). The last author argues that freight rates are stationary even in levels if they are converted from US dollars to yen.
10. An attempt to test the appropriateness of expectations generating mechanism is carried out in Berg-Andreassen (1997). He tests through cointegration analysis based on Augmented Dickey–Fuller tests and Johansen's likelihood ratio tests for (a) the Zannetos hypothesis; (b) the lagged Zannetos hypothesis; (c) the Koyck-lag hypothesis; (d) the rational expectations hypothesis; and (e) the Conventional Wisdom Hypothesis. He defines the latter as a hypothesis that time charter rates are a function only of *changes* in spot rates and finds that this scheme outperforms the rest.
11. The first-order Taylor series expansion of a function $f(x)$ evaluated at $x = a$ is given by:

$f(a) + f'(a)(x - a)$, where the first-order derivative denoted by f' is also evaluated at $x = a$.
12. See the Statistical Appendix for an explanation of non-stationarity. If a series is integrated of order one, denoted by $I(1)$, it becomes stationary by taking the first difference.
13. The difference between two $I(1)$ non-stationary variables is an $I(0)$ variable (that is, stationary). See the Statistical Appendix.
14. This cointegrating vector is postulated in Kavussanos and Alizadeh (2002b). But there are other possible cointegrating vectors, which are based on an explicit

structural model of the newbuilding and secondhand markets. See below for justification and references.

15. The variance definition is the sum of squared deviations from the mean, but as the mean is zero, the variance is simplified to (A.3b).
16. The covariance of ε_t and ε_{t-j} (for any j different than zero) is defined as

$$\text{cov}(\varepsilon_t, \varepsilon_{t-j}) = E(\varepsilon_t - E\varepsilon_t)(\varepsilon_{t-j} - E\varepsilon_{t-j}) = E(\varepsilon_t \cdot \varepsilon_{t-j}).$$

The simplification in the last term comes from the assumption that the mean is zero for all t. The covariance can be computed from the correlation coefficient ρ, which is defined as the covariance divided by the standard deviation of ε_t and ε_{t-j}.

$$\rho = \text{cov}(\varepsilon_t, \varepsilon_{t-j}) / \{[\text{var}(\varepsilon_t)^{-1/2}] \cdot [\text{var}(\varepsilon_{t-j})^{-1/2}]\} \quad \text{or}$$

$$\text{cov}(\varepsilon_t, \varepsilon_{t-j}) = \rho \cdot \text{var}(\varepsilon_t) = \rho \cdot \sigma^2$$

As the standard deviation of ε_t is equal to that of ε_{t-j} the denominator is simply the variance of ε_t. Hence, the covariance would only be zero, if the correlation coefficient is zero.

17. A mathematically oriented reader can easily see that the coefficient β should be within the unit circle (that is, ± 1), as equation (A.1) is a first-order difference equation. See the next footnote for more details.
18. The requirement of the condition (absolute) $\beta < 1$ is also apparent from the solution of the first-order difference equation (A.1) derived through back substitution of (A.1)

$$y_t = a[1 + \beta + \beta^2 + \cdots] + \beta^n y_{t-n} + \sum_{i=0}^{n-1} \beta^i \varepsilon_{t-i} = \frac{a}{1-\beta} + \sum_{i=0}^{n-1} \beta^i \varepsilon_{t-i} \quad (A.7)$$

In deriving this relation we have used that the second term on the right-hand side tends to zero as n tends to infinity and that the first term in the parenthesis on the right-hand side is the sum of a geometric series with declining weights at the rate β. The sum is only equal to $1/(1-\beta)$, if $\beta < 1$.

Using (A.7) the variance of y is

$$\text{var}(y_t) = E(y_t - \mu)^2 = E\left(\sum_{i=0}^{n-1} \beta^i \varepsilon_{t-i}\right) = \sigma^2[1 + \beta^2 + \beta^4 + \cdots] = \frac{\sigma^2}{1-\beta^2} \quad (A.6b)$$

In deriving (A.6b) we replaced $a/(1-\beta)$ in (A.7) with μ and made use of the hypothesis that the term in the parenthesis $[1 + \beta^2 + \beta^4 + \cdots]$ is the sum of a geometric series with declining weights at the rate β^2. The sum is only equal to $1/(1-\beta^2)$, if $\beta < 1$.

19. The concept of stationarity is equivalent to stability. A time series variable y is stable if for any exogenous shock it returns to its equilibrium value.
20. In a spurious regression two variables are seemingly related through a third variable, such a time trend. The correlation is high, but this is due to the trend. When the trend is removed the two variables are not correlated.
21. The Akaike Information Criterion (AIC) measures the trade-off between the 'goodness of fit' and the information loss from the complexity of the model. The AIC is defined as:

$$\text{AIC} = 2p - 2\ln(L) \quad (A.23)$$

where p is the number of coefficients estimated in the statistical model and L is the maximised value of the likelihood function. Hence, the AIC is a measure of the relative quality of the statistical model. The AIC is used in the literature.
22. Until the advent of the error correction methodology, the Box–Jenkins statistical approach advocated using rates of change, as these are stationary variables. Economists had favoured the use of levels of variables as economic theory was expressed in levels. Thus, the ECM methodology reconciled the two alternative approaches by making use of both levels and rates of change.
23. The OLS estimator converges at $T^{1/2}$, whereas the superconsistency implies convergence at T.
24. These concepts apply to any stationary stochastic process. Accordingly, if a variable needs to be differenced d-times to become stationary the formulae A36–A40 apply to it as well. Such a series is said to follow an ARIMA(p, d, q) process, where p denotes the order of the autoregressive process, AR(p), q the order of the moving average process MA(q) and d the degree of integration, $I(d)$. The ARIMA model includes the other two classes MA, AR and ARMA. For example, an ARMA process is and ARIMA with $d = 0$.
25. This sort of VAR model has been advocated by Sims (1980) for estimating dynamic models without imposing economic restrictions, that is, as a purely statistical model.
26. The reason Δz appears as Δz_{t-k+1} in (A.46) is because $z_{t-k} = z_{t-1} - \Delta z_{t-1} - \cdots - \Delta z_{t-k+1}$

5 BUSINESS CYCLES

1. In statistics, a function like (5.1) is called an autoregressive function (see the Statistical Appendix, Chapter 4). A variable, Y, has an autoregressive structure if it is a function of past values of Y. Thus, if $f(B)$ is nth order polynomial function in the backward operator (defined as $LY_t = Y_{t-1}$ and, in general, $L(n)Y_t = Y_{t-n}$), then $f(L)Y = 0$, has an autoregressive structure of nth order.
2. A stationary series has a finite mean and variance (they do not depend on time). In a non-stationary series the mean or the variance is a function of time.
3. The equation is an ARMA(2, 12) restricted to lags 8 and 12 only and estimated with an exact maximum likelihood method. The t-statistics for the right-hand side variables starting with the constant respectively are: 0.24; 12.63; −3.98; −2.24; and −3.00. $\bar{R}^2 = 0.87$; S.E. of Regression $= 0.75\%$; $DW = 2.07$.
4. The equation is an ARMA(12, 12) estimated with an exact maximum likelihood method. The t-statistics for the right-hand side variables starting with the constant respectively are: 4.39; 8.49; −2.05; −2.01; −1.87; 3.84; and −1.66. $\bar{R}^2 = 0.9$; S.E of Regression $= 0.6\%$; $DW = 1.85$.
5. Campbell and Mankiw (1987) estimate a similar equation, but with the quarter-on-growth rate rather than the year-on-year.
6. In the General Theory Keynes attributed inertia to wages, but not to prices. But he later changed his mind, as he became aware of Dunlop's (1938) empirical findings that real wages are, if anything, pro-cyclical rather than counter-cyclical. Originally, the emphasis was on *nominal* wage-price rigidity, but later on the emphasis shifted to *real* rigidity.
7. Lucas (1976) argued that traditional macroeconometric models take as given observed correlations of variables. But these correlations would have been different if another policy had been pursued. Thus, it is naïve to try to predict the effects of a change in economic policy entirely on the basis of relationships observed in historical data. These correlations can be expected to change when new policies are introduced, invalidating predictions based on past observations.

8. Kydland and Prescott (1982) is the seminal paper of RBC models. This approach to business cycles is surveyed in Cooley (1995).
9. Appendix 1 provides a summary of these models so that the reader can get a feeling of this voluminous, but extremely important literature that provides the basis for policy intervention in business cycles.
10. However, the view that 'Macro is good' has been modified recently in view of the 'Great Recession' (see, for example, Blanchard et al., 2010). It is indeed the case that the 'great recession' has forced the profession to seriously begin to reexamine the theoretical and policy propositions of the NCM. Blanchard (2011a) argues that the crisis 'forces us to do a wholesale reexamination of those principles' (p. 1).
11. For a step-by-step derivation of the two equations see, Bernanke, Gertler and Gilchrist (1998), Gali (2008) and Woodford (2003).
12. The ½ in front of the objective function is set for convenience, as it disappears when the first order partial derivatives with respect to the output gap and inflation are obtained.
13. For an assessment of the benefits of commitment see Clarida, Gali and Gertler (1999) and Gali (2008). Such benefits may accrue under rational expectations, where agents take into account the likely response of the central bank to shocks. The case of no commitment is not as restrictive as it sounds, as central banks in the real world make an effort to affect private sector expectations favourably by announcing how they would respond to unexpected shocks. For example, the Fed as from late 2012 has started announcing the feedback law upon which it operates. Such announcements in effect make financial markets do the work of the central bank, as asset prices adjust to the level consistent with the policy objective of the central bank.
14. See Clarida, Gali and Gertler (1999) for a step-by-step derivation of the interest rate feedback law.
15. See, for example Clarida, Gali and Gertler (1999).
16. As potential output is fixed, we use output (or income) Y instead of the output gap, x.
17. Arestis and Karakitsos (2004) predicted that the internet-bubble would be transformed into a housing bubble that would burst when the yield curve becomes inverted. Both predictions were accurate.
18. Arestis and Karakitsos (2013) discuss at length the new regulatory framework in the US, the UK, Europe and Basel III and argue that it would be insufficient to prevent the new equity bubble that is in the making. When the bubble bursts the world economy would fall into another severe recession.
19. The other major arm of demand management policies is monetary policy.
20. Supply-side economics aims to influence the rate of growth of potential output through research and development, labour policies that affect the labour participation rate, such as taxes on employment and capital and policies that affect structural unemployment, such as matching the skills of the unemployed with those of vacant jobs.
21. There is a school of economic thought that rejects this notion. In 'real business cycle analysis' the economy does not have a potential rate of growth, given by the productive capacity of the economy, but a constantly changing rate of growth. This is a minority view and major institutions like central banks and the CBO stick to our view of business cycles.
22. The multiplier is the first partial derivative of GDP with respect to the particular instrument of fiscal policy. It gives the total impact of the particular instrument of fiscal policy on GDP spread over a number of periods.
23. The sum of a series, where each term is smaller than the one before by a particular factor, is equal to the first term divided by 1 less the rate of increase of the series (e.g. in this case 0.8).
24. See equation (2.10) in Karakitsos (1992).
25. See equation (2.17) in Karakitsos (1992).

26. The K-model is an integrated econometric model of the macroeconomy, financial markets and shipping for the US, Europe, Japan and China (see Arestis and Karakitsos, 2004, 2010 and 2013 for details).
27. This view is also shared by all major central banks (for example the Fed, the ECB and the BoE) and major International bodies, such as the IMF, the OECD and the CBO.
28. A steady or non-accelerating inflation rate is obtained when the second time-derivative of the price level (e.g. the Consumer Price Index) is zero. This is consistent with any rate of inflation, because inflation is simply the first order time derivative of prices.
29. In mathematics, a spline is a sufficiently smooth polynomial function that is piecewise defined (a different function is defined in each piece) and possesses the property of smoothness at the points where the polynomial pieces connect (called the *knots*). In a spline a different curve is fitted in each piece (defined as an interval between any two knots) of the overall curve, such as a linear function between the first two knots and a parabola between the third and the fourth knots. In spite of the connection of different functions, the overall function is not discontinuous, as there is a smooth connection around each knot.
30. These two factors are derived from a production function for the overall economy, which for a given technology describes the way labour (L) and capital (K) are combined to produce a composite good (the goods and services produced in the economy as measured by GDP). Technological progress (A) is a positive function of time, as it improves through time, thus increasing output for a given capital and labour stock. Whereas (average) labour productivity is measured as the output per man (or output per hour worked), (average) total-factor productivity measures the productivity of all factors: labour, capital and technology. A Cobb–Douglas or a CES production function makes all these determinants explicit. Thus assuming a Cobb–Douglas production with constant returns to scal.

$$Y = A(t)K^a L^{1-a} \Rightarrow \frac{Y}{L} A(t)\left[\frac{K}{L}\right]^a \tag{5.53}$$

Average labour productivity (Y/L in the LHS of the equation) is equal to the product of the technological change and the capital-labour ratio (i.e. the amount of capital available to each worker). In economics, it is marginal productivity that matters, not average productivity. Mathematically, (marginal) labour productivity is the partial derivative of output with respect to labour (that is, the marginal improvement in output due to a marginal increase in labour) with all other factors remaining unchanged, while total factor productivity is the total derivative, where all factors that affect output are allowed to change. Thus, (marginal) total factor productivity is the improvement to output due to a change in all factors, not just labour. Total-factor productivity measures the improvement to output stemming from an increase in labour, augmented and improved capital and technology.

31. Equation (5.54) is derived from (5.53) by taking the natural logs of both sides and then computing the total time differential, which yields rates of growth for all determinants:

$$\ln Y = \ln A(t) + a \cdot \ln K + (1-a) \cdot \ln L$$
$$\Rightarrow \frac{1}{Y}\frac{dY}{dt} = \frac{1}{A(t)}\frac{dA(t)}{dt} + a\frac{1}{K}\frac{dK}{dt}$$
$$+ (1-a)\frac{1}{L}\frac{dL}{dt} \Rightarrow \bar{y} = q + a \cdot k + (1-a) \cdot l$$

32. The approximation is the result of two common assumptions in growth models and growth accounting, namely that the production function is homogeneous of degree one (which means that a given percentage increase in the factor inputs yields the same percentage increase in output) and that firms are cost minimisers under perfect competition (see Denison, 1985 and Gordon, 1999).

6 THE THEORY OF SHIPPING CYCLES

1. The solution of (6.9) for $\theta = 2$ is obtained from the characteristic equation

$$x^2 - x - \lambda \cdot \rho = 0 \tag{6.10}$$

The two roots, denoted by x_1 and x_2 are given by

$$x_1, x_2 = \frac{1}{2}[1 \pm (1 + 4\lambda\rho)^{1/2}] \tag{6.11}$$

If, $1 + 4\lambda\rho < 0$ or $-\lambda\rho > (1/4)$ (6.12)

then the dynamic equation (6.9) has two conjugate complex roots and exhibits cyclical *fluctuations around the fleet equilibrium value, K^e. The dynamic evolution of the fleet, around K^e*, is given by the equation

$$K_t = K^e + \lambda\rho^{(1/2)\cdot t}[A_1 \cos\varphi \cdot t + A_2 \cdot \sin\varphi \cdot t] \tag{6.13}$$

The coefficients A_1 and A_2 are arbitrary constants (expressible in terms of initial values K_0 and K_1). The solution of (6.9) can be written in an alternative way.

$$K_t = K^e + A\lambda\rho^{(1/2)\cdot t} \cdot \cos(\varphi \cdot t - \varepsilon) \quad A = +\left(A_1^2 + A_2^2\right)^{(1/2)} \tan\varepsilon \frac{A_2}{A_1} \tag{6.14}$$

2. According to the Efficient Market Hypothesis the returns of all assets, by properly incorporating a risk premium where necessary, are equalised. Returns are adjusted instantly to clear unanticipated movements in demand and supply (see Chapter 3).
3. There are some notable exceptions, such as Haralambides et al. (2005) and Jin (1993).
4. The total revenue of the fleet is equal to $F \times K \times S$ and the fuel bill for the fleet is equal to $PB \times K \times S^a$. By dividing both sides of the profit equation by the fleet, equation (6.25) is obtained, where the profit per ship is defined.
5. This can easily be verified by considering the conditions under which . It is obvious from equation (6.51) that this would be true, if the following inequality is satisfied.

$$\dot{cu} > 0 \Rightarrow \dot{q} - \mu \cdot ps + \mu_1 \cdot pt + \mu_2 \cdot psc > \dot{q} - \mu \cdot ps^* + \mu_1 \cdot pt + \mu_2 \cdot psc$$

$$\text{or} \quad ps < ps^* \tag{6.55}$$

Notice that the right-hand side of (6.55) is zero because when SH prices are in their long-run equilibrium then $cu = 0$, in accordance to equation (6.52).

6. A permanent increase in the price of steel or the scrap price has similar effects to the increase in demand. Hence, Figure 6.8 can serve in addition to explaining the dynamic adjustment of the system to these exogenous shocks as well.

7. The derivation for a single factor of production is given by equations (6.32)–(6.34) in this chapter.
8. Later on in this section we relax the assumption that scrapping is proportional to the net fleet and allow scrapping to become an economic decision. In this case the net fleet is the result of the interdependence of all four markets, including scrapping.
9. In deriving equation (6.87) it is assumed that both the demand and the supply functions for new ships are log-linear. Assume instead that the supply is linear of the form: $K^s = c_1 \, PN$. Equating the demand with the supply of new vessels and solving for PN we have

$$PN = \frac{A_1}{c_1} \cdot Q \cdot \left[\frac{FR}{UC}\right]^\sigma$$

Taking the total differential of log-linear version of this equation we have an alternative form of (6.87):

$$\Delta pn = \Delta q + \sigma \, (\Delta fr - \Delta uc) \qquad (6.87a)$$

Therefore, when we combine a log-linear demand with a linear supply the output elasticity of the newbuilding price is one. For this reason in the simulations we assume that the output elasticity of the newbuilding price is one.

7 THE MARKET STRUCTURE OF SHIPPING AND SHIP FINANCE

1. In 2002 China entered the WTO and the five years following this witnessed an unprecedented increase in manufacturing and infrastructure projects which led to a massive rise in freight rates in all three major sectors of shipping namely dry bulk owing to the imports of iron ore, tankers owing to the increased energy demands and container vessels to export the finished products from the new low-cost Chinese factories.
2. We have seen with many banks that they have long term funding issues as such funds are more expensive. In turn, they have to charge borrowers more. Most banks prefer to lend for a maximum of five years although the profile of the loan can remain much longer thus resulting in a high balloon after 5 years. To the extent that clients place deposits with their lending bank this can reduce the TLP charge and allow the ship finance areas of the bank to charge the borrower a lower interest margin.

8 THE FINANCIALISATION OF SHIPPING MARKETS

1. With tight monetary policy (high or rising interest rates), a country attracts capital inflows that cause its currency to appreciate. This reduces imported inflation and consequently mitigates the inflationary impact of higher oil prices. The higher exchange rate makes its exports dearer, which increases inflation in other countries because of rising imported inflation. In this sense by pursuing a tight monetary policy a country exports its inflation to the rest of the world.
2. Viewed from this angle is what prompted Keynes (in the *Treatise of Money*, chapter 29) to argue that backwardation arises when producers are more prone to hedge their price risk than consumers.
3. The contango term originated in mid-nineteenth-century England and is believed to be a corruption of "continuation", "continue" or "contingent". In the past, in the

London Stock Exchange contango was a fee paid by a buyer to a seller when the buyer wished to defer settlement of the trade they had agreed. The charge was based on the interest foregone by the seller not being paid. Similarly, backwardation was a fee paid by a seller wishing to defer delivering stock they had sold. This fee was paid either to the buyer, or to a third party who lent stock to the seller.
4. A "soft commodity" is a commodity such as coffee, cocoa, sugar, corn, wheat, soybean and fruit. This term generally refers to commodities that are grown, rather than mined.
5. The academic dispute on the subject has been fierce (see Bodie and Rosansky, 1980).
6. The standard deviation of the price of oil from 1983 to 2000 was only $5.7, but from 2000 onwards the standard deviation increased by more than fivefold to $27.7.

9 THE INTERACTION OF BUSINESS AND SHIPPING CYCLES IN PRACTICE

1. The methodology is the same as that used in business (or economic) cycles and the reader should consult section 4 for an analysis and detailed explanations. If a contraction lasts for less than six months, it is regarded as a transient drop rather than a recession.
2. In the expansion phase the first time derivative is positive. At the peak it becomes zero and in the recession negative. At the bottom it becomes again zero and in the recovery positive. The second time derivative, which is approximated by the change in the (q-o-q) growth rate, would be negative in the initial stage of the recession when the recession deepens. It will become zero when the economy stabilises and then positive as the recession eases. As the period of a positive second derivative precedes the bottom of the recession it serves as a leading indicator of the trough of the cycle.
3. See the section on fiscal policy in Chapter 5 for more details.
4. Scarsi (2007) finds evidence of a herd syndrome amongst owners. In particular, owners make mistakes when they ignore market trends, following their personal intuition or imitate their competitors.

10 INVESTMENT STRATEGY

1. The median is the value of average earnings that divides the distribution of earnings into two equal halves.
2. The case study is based on real life data. We have used pseudonyms to protect the identities of the parties involved.
3. The term margin of error is, technically speaking, the root mean square error. This is the standard deviation of the sum of the square of errors (namely actual minus model predicted value) over the historical period.

BIBLIOGRAPHY

Abel, A.B. (1983) "Optimal Investment under Uncertainty", *American Economic Review*, 73, 228–33.

Abel, A.B. (1984) "The Effects of Uncertainty on Investment and the Expected Long run Capital Stock", *Journal of Economic Dynamics and Control*, 7, 39–53.

Adland, R. and Cullinane, K. (2005) "A Time Varying Risk Premium in the Term Structure of Bulk Shipping Freight rates", *Journal of Transport Economics and Policy*, 39(2), 191–208.

Adland, R. and Strandenes, S. (2006) "Market Efficiency in the Bulk Freight Market Revisited", *Maritime Policy & Management*, 33(2), 107–17.

Adland, R., Koekebakker, S. and Sodal, S. (2004) "Non-stationarity of Freight Rates Revisited". Paper presented at the Proceedings for the Annual International Association of Maritime Economists, Izmir, Turkey.

Akerlof, G. (1984) "Gift Exchange and Efficiency Wages: Four Views", *American Economic Review*, 74, 79–83.

Akerlof, G. A. and Miyazaki, H. (1980) "The Implicit Contract Theory of Unemployment Meets the Wage Bill Argument", *Review of Economic Studies*, 47(2), 321–38.

Akerlof, G.A. and Yellen, J. (1985) "A Near Rational Model of Business Cycle with Wage and Price Inertia", *Quarterly Journal of Economics*, 100, Supplement, 823–38.

Akerlof, G. and Yellen, J. (1987) "The Fair Wage-Effort Hypothesis and Unemployment", mimeo, University of California, Berkeley.

Alchian, A. (1969), "Information Costs, Pricing and Resource Unemployment," *Economic Inquiry*, 7, 109–28.

Alizadeh, A.H. and Nomikos, N.K. (2011) "Dynamics of the Term Structure and Volatility of Shipping Freight Rates", *Journal of Transport Economics and Policy*, 45(1), 105–28.

Arestis, P. and Karakitsos, E. (2004), *The Post-Bubble US Economy: Implications for Financial Markets*, London and New York Palgrave Macmillan.

Arestis, P. and Karakitsos, E. (2007) "Unemployment and the Natural Interest Rate in a Neo-Wicksellian Model", in P. Arestis and J. McCombie (eds), *Unemployment: Past and Present*, London and New York: Palgrave Macmillan.

Arestis, P. and Karakitsos, E. (2010) *The Post 'Great Recession' US Economy: Implications for Financial Markets and the Economy*, Basingstoke: Palgrave Macmillan.

Arestis, P. and Karakitsos, E. (2013), *Financial Stability in the Aftermath of the Great Recession*, Basingstoke: Palgrave Macmillan.

Arnott, R. and Stiglitz, J. (1982) "Labour Turnover, Wage Structures and Moral Hazard", *Working Papers* 496, Queen's University, Department of Economics.

Artis, M.J. and Karakitsos, E. (1983) *Memorandum of Evidence on International Monetary Arrangements*, Fourth Report from the Treasury and Civil Service Committee on International Monetary Arrangements, HC 21-III (London: HMSO), pp. 142–206.

Azariadis, C. (1975) "Implicit Contracts and Underemployment Equilibria", *Journal of Political Economy*,

Azariadis, C. (1983), "Employment with Asymmetric Information", *Quarterly Journal of Economics* (Supplement), 157–72.

Azariadis, C. and Stiglitz, J. (1983), "Implicit Contracts and Fixed Price Equilibria", *Quarterly Journal of Economics* (Supplement), 2–22.

Baily, M., "Wages and employment under uncertain demand", *Review of Economic Studies*, 41, 37–50.

Ball, L. (1997), "Efficient Rules for Monetary Policy", *NBER Working Paper*, 5952.

Ball, L. and Romer, D. (1989a) "Are Prices Too Sticky?", *Quarterly Journal of Economics*, 104, 3, 507–24.

Ball, L. and Romer, D. (1989b) "The Equilibrium and Optimal Timing of Price Changes", *Review of Economic Studies*, 56, 2, 179–98.

Barro, R.J. (1972), "A Theory of Monopolistic Price Adjustment", *Review of Economic Studies*, 39, 1, 17–26.

Barro, R.J. (1972), "Long term Contracting, Sticky Prices and Monetary Policy", *Journal of Monetary Economics*, 3, 3, 305–16.

Beenstock, M. (1985). "A theory of Ship Prices", *Maritime Policy and Management*, 12, 31–40.

Beenstock, M., and Vergottis, A. (1989a) "An Econometric Model of the World Market for Dry Cargo Freight and Shipping", *Applied Economics*, 21, 339–56.

Beenstock, M., and Vergottis, A. (1989b) "An econometric model of the world tanker market", *Journal of Transport Economics and Policy*, 23, 263–80.

Beenstock, M. and Vergottis, A. (1993) *Econometric Modelling of World Shipping*, London: Chapman & Hall.

Berg-Andreassen, J.A. (1997) "The Relationship Between Period and Spot Rates in international Maritime Markets", *Maritime Policy and Management*, 24, 335–50.

Blanchard, O.J. (2009) "The State of Macro", *Annual Review of Economics*, 1, 209–28.

Bernanke, B., Gentler, M. and Gilchrist, S. (1999) "The Financial Accelerator in a Quantitative Business Cycle Framework", *American Economic Review*, 91, 2, 253–7.

Blanchard, O.J. (1986) "The Wage–Price Spiral", *Quarterly Journal of Economics*, 101, 3, 543–65.

Blanchard, O.J. (2011a) "Rewriting the Macroeconomists' Playbook in the Wake of the Crisis", *IMFdirect, The International Monetary Fund's Global Economic Forum*, 4 March (Washington, DC: IMF).

Blanchard, O., Dell'Ariccia, G. and Mauro, P. (2010) "Rethinking Macroeconomic Policy", *IMF Staff Position Note*, SPN/10/03, Washington, DC: International Monetary Fund.

Blanchard, O.J. and Fischer, S. (1990), *Lectures in Macroeconomics*, Cambridge, MA: MIT Press.

Blanchard, O.J. and Kiyotaki, N. (1987) "Monopolistic Competition and the Effects of Aggregate Demand", *American Economic Review*, 77, 4, 647–66.

Blanchard, O.J. and Summers, L. (1986), "Hysteresis and the European Unemployment Problem", *NBER Macroeconomics Annual*, 15–77.

Bodie, Z. and Rosansky, V. (1980) "Risk and Return in Commodity Futures", *Financial Analyst's Journal*, 36, 3–14.

Bollerslev, T. (1986) "Generalised Autoregressive Conditional Heteroscedasticity", *Journal of Econometrics*, 31, 307–27.

Bresnahan, T. F., (1989) "Empirical Studies of Industries with Market Power", in R. Schmalensee and R. Willig (ed.), *Handbook of Industrial Organization*, edition 1, volume 2 (Amsterdam: Elsevier), 1011–57.

Cabellero, R.J. (1991) "Competition and the Non-Robustness of the Investment-Uncertainty Relationship", *American Economic Review*, 81, 279–88.

Calvo, G. (1979) "Quasi Walrasian Theories of Unemployment", *American Economic Review*, 69, 102–7.

Calvo, G. (1983) "Staggered Prices in a Utility Maximising Framework", *Journal of Monetary Economics*, 12, 3, 383–98.

Campbell, H. (2007) "Analysis of Public Shipping Company Performance over the Last Ten Years", *Marine Money*.

Campbell, J.Y. and Mankiw, G.N. (1987) "Are Output Fluctuations Transitory?", *Quarterly Journal of Economics*, 102, 4, 857–80.

Campbell, J.Y. and Shiller, R.J. (1987) "Cointegration and Test of Present Value Models", *Journal of Political Economy*, 95, 1062–88.

Campbell, J.Y. and Shiller, R.J. (1988) "Stock Prices, Earnings and Expected Dividends", *Journal of Finance*, 43, 3, 661–76.

Campbell, J.Y. and Shiller, R. J. (1991) "Yield Spreads and Interest rate Movements: A Bird's Eye View", *Review of Economic Studies*, 58, 495–514.

Caplin, A.S. and Spulber, D.F. (1987) "Menu Costs and the Neutrality of Money", *Quarterly Journal of Economics*, 102, 4, 703–25.

Carlin, W. and Soskice, D. (2005) "The 3-Equation New Keynesian Model – A Graphical Exposition", *Contributions to Macroeconomics*, 5, 1, 1–36.

Carlin, W. and Soskice, D. (2006) *Macroeconomics: Imperfections, Institutions and Policies*, Oxford: Oxford University Press.

Carlton, D.W. (1986) "The Rigidity of Prices," *American Economic Review*, 76, 4, 637–58.

Cecchetti, S.G. (1986) "The Frequency of Price Adjustment: A Study of Newsstand Prices of Magazines", *Journal of Econometrics*, 31, 3, 255–74.

Clarida, R., Galí, J. and Gertler, M. (1998) "Monetary Policy Rules in Practice: Some International Evidence", *European Economic Review*, 42, 2, 1033–68.

Clarida, R., Galí, J. and Gertler, M. (1999) "The Science of Monetary Policy: A New Keynesian Perspective", *Journal of Economic Literature*, 37, 4, 1661–707.

Clarida, R., Galí, J. and Gertler, M. (2000) "Monetary Policy Rules and Macroeconomic Stability: Evidence and Some Theory", *Quarterly Journal of Economics*, 115, 1, 147–80.

Congressional Budget Office (2001) "CBO's Method for Estimating Potential Output", Washington DC.

Congressional Budget Office (2012) *The Economic Impact of the President's 2013 Budget*, Washington DC.

Cooley, T.F. (1995), *Frontiers of Business Cycle Research*, Princeton, NJ: Princeton University Press.

Council of Economic Advisers (1997) *Economic Report of the President*, Washington DC.

Craine, R. (1989) "Risky Business: The Allocation of Capital", *Journal of Monetary Economics*, 23, 201–18.

Cuthbertson, K. (1996) *Quantitative Financial Economics*, New York: John Wiley & Sons.

De Long, J.B. (1997) "America's Peacetime Inflation: The 1970s", in C. Romer and D. Romer (eds), *Reducing Inflation: Motivation and Strategy*, Chicago: University of Chicago Press.

Denison, E.F. (1985) *Trends in American Economic Growth, 1929–82*, Washington, DC: Brookings Institution.

Dertouzas, M., Lester, R.K. and Solow, R. (1990) *Made in America*, New York: Harper Paperback Edition.

Dixit, A.K. (1992) "Investment and Hysteresis" *Journal of Economic Perspectives*, 6, 107-132.

Dixit, A.K. and Pindyck, R.S. (1994) *Investment under Uncertainty*, Princeton, NJ: Princeton University Press.

Driver, C. and Moreton, D. (1991) *Investment, Expectations and Uncertainty*, Oxford: Basil Blackwell.

Dunlop, J.T. (1938) "The Movement of Real and Money Wage Rates", *Economic Journal*, 48, 413–34.

Eckbo, B.E. (1977) "Risikopreferanser blant noen Scandinavishe tankrederier for og under krisen I tankmarkedet", *Institute of Shipping Research, Norwegian School of Economics and Business Administration*, Bergen, Norway.

Engle, R. F. (1982) "Autoregressive Conditional Heteroskedasticity with Estimates of the Variance of UK Inflation", *Econometrica*, 50, 987–1007.

Engle, R.F. and Granger, C.W. (1987) "Cointegration and Error-correction: Representation, Estimation, and Testing", *Econometrica*, 55, 251–276.

Engle, R.F., Lilien, D.M. and Robins, R.P. (1987) "Estimating Time-Varying Risk Premia in the Term Structure: the ARCH-M Model", *Econometrica*, 55, 2, 391–407.

Espinosa, M. and Rhee, C.Y. (1987) "Efficient Wage Bargaining as a Repeated Game", mimeo, Harvard University.

Fama, E.F. (1970) "Efficient Capital Markets: A Review of Theory and Empirical Work", *Journal of Finance*, 25, 2, 383–423.

Fama, E.F. (1991) "Efficient Capital Markets: II", *The Journal of Finance* 46, 5, 1575–617.

Fazzari, S.M., Hubbard, R.G. and Petersen, B.C. (1988) "Financing Constraints and Business Investment", *Brookings Papers on Economic Activity*, 1, 141–206.

Fischer, S. (1977a) "Long term Contracts, Rational Expectations and the Optimal Money Supply", *Journal of Political Economy*, 85, 1, 191–206.

Fischer, S. (1977b) "Wage Indexation and Macroeconomic Stability", in.: Karl Bruner and Allan Meltzer (eds), *Stabilization of the Domestic and International Economy*, Carnegie-Rochester Conference Series on Public Policy, vol. 5, Amsterdam: North-Holland, pp. 107–47.

Gali, J. (2008) *Monetary Policy, Inflation and the Business Cycle*, Princeton, NJ: Princeton University Press.

Galí, J. and Gertler, M. (2007) "Macroeconomic Modelling for Monetary Policy Evaluation", *Journal of Economic Perspectives*, 21 (4), 25–45.

Gertler, M. and Hubbard, R.G. (1989) *Financial Factors in Business Fluctuations*, NBER Working Paper No. 2758.

Glen, D.R. (1997) "The Market for Secondhand Ships: Further Results on Efficiency Using Co-integration Analysis", *Maritime Policy and Management*, 24, 245–60.

Glen, D.R. (2006) "The Modelling of Dry Bulk and Tanker Markets: A Survey", *Maritime Policy & Management*, 33, 5, 431–45.

Glen, D.R. and Martin, B.T (1998) "Conditional Modelling of Tanker Market Risk Using Route Specific Freight Rate", *Maritime Policy and Management*, 25, 117–28.

Glen, D.R. and Martin, B.T. (2005) "A Survey of the Modelling of Dry Bulk and Tanker Markets", *Shipping Economics Research in Transportation Economics*, Volume 12, 19–64.

Glen, D.R., Owen, M. and Van der Meer, R. (1981) "Spot and Time-charter Rates for Tankers 1970-1977", *Journal of Transport Economics and Policy*, 15, 45–8.

Goodfriend, M. and King, R.G. (1997) "The New Neoclassical Synthesis and the Role of Monetary Policy', in Bernanke, B.S. and Rotemberg, J.J. (eds.), *NBER Macroeconomics Annual: 1997*, Cambridge, MA: MIT Press.

Gordon, D.F. (1974) "A Neoclassical Theory of Keynesian Unemployment", *Economic Inquiry* 12, 431–49.

Gordon, R.J. (1983) "Price Inertia and Policy Ineffectiveness in the United States, 1890–1980", NBER Working Papers 0744.

Gordon, R.J., (1982) "Price Inertia and Policy Ineffectiveness in the United States, 1890–1980", *Journal of Political Economy*, 90, 6, 1087–117.

Gordon, R.J. (1990), "What is New Keynesian Economics", *Journal of Economic Theory*, 28, 3, 1115–71.

Gordon, R.J., (1999), "U.S. Economic Growth Since 1870: One Big Wave?", *American Economic Review*, 89, 2, 123–8.

Gottfries, N. and Horn, H. (1987) "Wage Formation and the Persistence of Unemployment", *Economic Journal*, 97, 877–86.

Granger, C.W.J. (1966) "The Typical Spectral Space of an Economic Variable", *Econometrica*, 34, 1, 150–61.

Greene, W.H. (1997) *Econometric Analysis*, 3rd edition, Prentice Hall.

Grossman, S.J. and Hart, O. (1983) "Implicit Contracts under Symmetric Information", *Quarterly Journal of Economics* (Supplement), 123–56.

Grossman, S.J. and Stiglitz J.E. (1980) "The Impossibility of Informational Efficient Markets", *American Economic Review*, 66, 246–53.

Haavelmo, T. (1960) *A Study in the Theory of Investment*, Chicago: University of Chicago Press.

Hale, C. and Vanags, A. (1989) "Spot and Period Rates in the Dry bulk Market: Some Tests for the Period 1980–1986", *Journal of Transport Economics & Policy*, 23, 3, 281–91.

Hale, C. and Vanags, A. (1992) "The Market for Secondhand Ships: Some Results on Efficiency Using Cointegration", *Maritime Policy & Management*, 19, 131–40.

Hamilton, J.D. (1994) *Time Series Analysis*, Princeton, NJ: Princeton University Press.

Hansen, L.P. (1982) "Large Sample Properties of Generalised Method of Moments Estimators", *Econometrica*, 50, 1029–52.

Haralambides, H.E., Tsolakis, S.D. and Cridland, C. (2005) "Econometric Modelling of Newbuilding and Secondhand Ship Prices", *Research in Transport Economics*, 12, 65–105.

Harsanyi, J.C. (1956) "Approaches to the Bargaining Problem before and after the Theory of Games: A Critical Discussion of Zeuthen's, Hicks and Nash's Theories", *Econometrica*, 24, 144–157.

Hartman, R. (1972) "The Effects of Price and Cost Uncertainty on Investment", *Journal of Economic Theory*, 5, 258–66.

Hashimoto, M. (1981), "Firm-Specific Investment as a Shared Investment", *American Economic Review*, 475–82.

Hawdon, D. (1978), "Tanker Freight Rates in the Short and Long Run", *Applied Economics*, 10, 203–17.

Henry, S.G.B., Karakitsos, E. and Savage, D. (1982) "On the Derivation of the Efficient Phillips Curve", *Manchester School*, L, 2, 151–77.

Hicks, J.R. (1946), *Value and Capital*, 2nd edition, Oxford: Clarendon Press.

Jaffee, D. and Stiglitz, J. (1990) "Credit Rationing", *Handbook of Monetary Economics*, 2, 837–88.

Jiang, L. and Lauridsen, J.T. (2012) "Price Formation of Dry Bulk Carriers in the Chinese Shipping Industry", *Maritime Policy & Management*, 39, 3, 339–51.

Jing, L., Marlow, P.B. and Hui, W. (2008) "An Analysis of Freight rate Volatility in Dry Bulk Shipping Markets", *Maritime Policy & Management*, 35, 3, 237–51.

Johansen, S. (1996) *Likelihood Based Inference in Co-integrated VAR Models*, Oxford: Oxford University Press.

Juselius, K. (1995) "Do Purchasing Power Parity and Uncovered Interest rate Parity Hold in the Long Run? An Example of Likelihood Inference in a Multi-variate Time-series Model", *Journal of Econometrics*, 69, 211–40.

Karakitsos, E. (1992) *Macrosystems: The Dynamics of Economic Policy*, Oxford: Blackwell.

Karakitsos, E. (2008) "The 'New Consensus Macroeconomics' in the Light of the Current Crisis", *Ekonomia*, 11, 2, 89–111.

Karakitsos, E. (2009) "Bubbles Lead to Long-term Instability", in Giuseppe Fontana, John McCombie and Malcolm Sawyer (eds.), *Macroeconomics, Finance and Money: Essays in Honour of Philip Arestis*, Basingstoke: Palgrave Macmillan.

Karakitsos, E. and Rustem, B. (1984) "Optimally Derived Fixed Rules and Indicators", *Journal of Economic Dynamics and Control*, 8, 1, 33–64.

Karakitsos, E. and Rustem, B. (1985) "Optimal Fixed Rules and Simple Feedback Laws in the Design of Economic Policy", *Automatica*, 21, no. 2, 169–80.

Katz, L. (1986) "Efficiency Wage Theories: A Partial Evaluation", *NBER Macroeconomics Annual*, 235–276.

Kavussanos, M. (1996a) "Comparison of Volatility in the Dry-cargo Ship-sector", *Journal of Transport Economics and Policy*, 30, 67–82.

Kavussanos, M. (1996b) "Price Risk Modelling of Different Size Vessels in the Tanker Iindustry", *Logistics and Transportation Review*, 32, 161–76.

Kavussanos, M. (1996c) "Highly Disaggregate Models of Seaborne Trade: An Empirical Model for Bilateral Dry-cargo Trade Flows", *Maritime Policy and Management*, 23, 27–43.

Kavussanos, M. (1997) "The Dynamics of Time-varying Volatilities in Different Size Second-hand Ship Prices of the Dry-cargo sector", *Applied Economics*, 29, 433–43.

Kavoussanos, M.G. and Alizadeh, A.H. (2002a) "The Expectations Theory of the Term Structure and Risk Premiums in the Dry Bulk Shipping Freight Markets", *Journal of Transport Economics and Policy*, 36, 2, 267–304.

Kavoussanos, M.G. and Alizadeh, A.H. (2002b) "Efficient Pricing of Ships in the Dry Bulk Sector of the Shipping Industry", *Maritime Policy and Management*, 29, 3, 303–30.

Keynes, J.M. (1936) *The General Theory of Employment, Interest and Money*, London: Macmillan.

Keynes, J.M. (1930) *A Treatise on Money*, London: Macmillan.

Kiyotaki, N. (1988) "Multiple Expectational Equilibria Under Monopolistic Competition", *Quarterly Journal of Economics*, 103, 4, 695–713.

Koopmans, T.C. (1939) "Tanker Freight Rates and Tankship Building", *Haarlem, Holland*.

Kydland and Prescott (1982) "Time to Build and Aggregate Fluctuations", *Econometrica*, 50, 1345–70.

Layard, R. and Nickell, S. (1986) "Unemployment in Britain", *Economica*, 53, 121–69.

Leibenstein, H. (1957) *Economic Backwardness and Economic Growth*, New York: Wiley.

Leontief, W. (1946) "The Pure Theory of the Guaranteed Annual Wage Contract", *Journal of Political Economy*, 54, 1, 69–76.

Lindbeck, A. (1987) "Union Activity, Unemployment Persistence and Wage Employment Ratchets", *European Economic Review*, 31, 1/2, 157–67.

Lindbeck, A. and Snower, D. (1986) "Wage Setting, Unemployment and Insider–Outsider Relations", *American Economic Review*, 76, 235–9.

Lindbeck, A. and Snower, D. (1988) *The Insider–Outsider Theory of Employment and Unemployment*, Cambridge, MA: MIT Press.

Lorange, P. and Norman, V.D. (1973) "Risk Preferences in Scandinavian Shipping", *Applied Economics*, 5, 49–59.

Lucas, R.E. (1976) "Econometric Policy Evaluation: A Critique", *Carnegie-Rochester Conference Series on Public Policy*, 1, 19–46.

Mankiw, N.G. (1985) "Small Menu Costs and Large Business Cycles: A Macroeconomic Model of Monopoly", *Quarterly Journal of Economics*, 100, 529–39.

McDonald, I. and Solow, R. (1981) "Wage Bargaining and Employment", *American Economic Review*, 71, 896–908.

McDonald, I. and Solow, R. (1985) "Wages and Employment in a Segmented Labor Market", *The Quarterly Journal of Economics*, 100, 4, 1115–41.

MacKinnon, J. (1991), "Critical Values for Co-integration Tests", in R.F. Engle and C.W.J. Granger (eds), *Long run Economic Relationships*, Oxford: Oxford University Press.

Malkiel, B. (1992), "Efficient Market Hypothesis", in P. Newman, M. Milgate and J. Eatwell (eds), *New Palgrave Dictionary of Money and Finance*, London: Macmillan.

Mandelbrot, B. (1963) "The Variation of Certain Speculative Prices", *The Journal of Business*, 36, 4, 394–419.

Mankiw, N.G. and Miron, J.A. (1986) "The Changing Behaviour of the Term Structure of Interest Rates", *Quarterly Journal of Economics*, 101, 2, 211–28.

McDonald, R. and Siegel, R. (1986) "The Value of Waiting to Invest", *Quarterly Journal of Economics*, 101, 707–28.

Merikas, A.G., Merika, A.A. and Koutroubousis, G. (2008) "Modelling the Investment Decision of the Entrepreneur in the Tanker Sector: Choosing between a Secondhand Vessel and a Newly Built One", *Maritime Policy and Management*, 35, 5, 433–47.

Meyer, L.H. (2001) "Does Money Matter?", *Federal Reserve Bank of St. Louis Review Quarterly Review*, 83, 5, 1–15.

Mortensen, D. and Pissarides, C. (1994) "Job Creation and Job Destruction in the Theory of Unemployment", *Review of Economic Studies*, 61, 3, 397–415.

Muth, J.F. (1961) "Rational Expectations and the Theory of Price Movements", *Econometrica*, 29 3, 315–335.

Nakamura, E. and Steinson, J. (2006), *Five Facts about Prices: A Reevaluation of Menu Costs Models*, Cambridge, MA: Harvard University Press.

Nickell, S.J. (1978) *The Investment Decisions of Firms*, Cambridge University Press.

NBER (2008) Determination of the December 2007 Peak In Economic Activity, *NBER*.

Pissarides, Christopher (2000) *Equilibrium Unemployment Theory*, 2nd edn, Cambridge, MA: MIT Press.

Nash, J.F. (1950), "The Bargaining Problem", *Econometrica*, 18, 155–62.

Okun, A,M, (1962) *Potential GNP, its measurement and significance*, Cowles Foundation, Yale University.

Okun, A.M. (1975) *Equality and Efficiency: The Big Trade-off*, Washington, DC: Brookings Institution Press.

Oswald, A. (1985) "The Economic Theory of Trade Unions: An Introductory Survey", *Scandinavian Journal of Economics*, 87, 2, 160–93.

National Bureau of Economic Research, Business Cycle Dating Committee (2008) *Determination of the December 2007 Peak in Economic Activity*.

Pencavel, J. (1985) "Wages and Employment under Trade Unionism: Microeconomic Models and Macroeconomic Applications", *Scandinavian Journal of Economics*, 87, 2, 197–225.

Phelps, E. S., et al. (1970) *Microeconomic Foundations of Employment and Inflation Theory*, New York: Norton.

Phelps, E.S. and Winter, S.G. (1970), "Optimal Price Policy under Atomistic Competition", in E.S. Phelps, et al., *Microeconomic Foundations of Employment and Inflation Theory*, New York: Norton.

Phelps E, and Taylor J.B (1977) "Stabilizing Powers of Monetary Policy Under Rational Expectations", *Journal of Political. Economy*, 85, 1, 163–90.

Phillips, P.C.B. (1987) "Time Series Regression with a Unit Root", *Econometrica*, 55, 277–301.

Phillips, P.C.B. and Perron, P. (1988) "Testing for a Unit Root in Time Series Regression", *Biometrica*, 75, 335–46.

Paraskevopoulos, D. Karakitsos, E, and Rustem, B. (1991) "Robust Capacity Planning Under Uncertainty", *Management Science*, 37(7), 787–800.

Pindyck, R.S. (1988) "Irreversible Investment, Capacity Choice and the Value of the Firm", *American Economic Review*, 79(4), 969–85.

Pindyck, R.S. (1991) "Irreversibility, Uncertainty and Investment", *Journal of Economic Literature*, 29, 1110–52.

Prescott, E. (1986) "Theory Ahead of Business Cycle Measurement", *Federal Reserve Board of Minneapolis Quarterly Review*, 10.

Roberts, H. (1967) Statistical versus clinical prediction of the stock market, Unpublished manuscript.

Rotemberg, J.J. (1982) "Monopolistic Competition and Aggregate Output", *Review of Economic Studies*, 49, 517–31.

Rotemberg, J.J. and Woodford, M. (1997) "An Optimization-Based Econometric framework for the Evaluation of Monetary Policy", *NBER Macroeconomics Annual 1997*, Cambridge, MA: National Bureau of Economic Research.

Rotemberg, J.J. and Woodford, M. (1998) "Interest Rate Rules in an Estimated Sticky Price Model", NBER Working Paper, 6618.

Salop, S. (1979), "A Model of the Natural Rate of Unemployment", *American Economic Review*, 69, 117–25.

Sargent, T.J. and Wallace, N. (1975) "Rational Expectations, the Optimal Monetary Instrument and the Optimal Money Supply Rule", *Journal of Political economy*, 83, 241–54.

Scarsi, R., (2007) "The Bulk Shipping Business: Market Cycles and Shipowners' Biases", *Maritime Policy & Management*, 34, 6, 577–90.

Shapiro, C. and Stiglitz, J. (1984) "Equilibrium Unemployment as a Discipline Device", *American Economic Review*, 74, 433–44.

Shapley, L.S. (1969) "Utility Comparison and the Theory of Games", *La Decision*, 251–63.

Sheshinki E. and Weiss Y. (1983) "Optimum Pricing Policy under Stochastic Inflation", *Review of Economic Studies*, 50, 3, 513–29.

Sims, C.A. (1980), "Macroeconomics and Reality", *Econometrica*, 48, 1–48.

Solow, R.M. (1956) "A Contribution to the Theory of Economic Growth". *Quarterly Journal of Economics*, 70, 1, 65–94.

Solow, R. (1979) "Another Possible Source of Wage Stickiness", *Jounral of Macroeconomics*, 1, 79–82.

Solow, Robert M, (1985) "Insiders and Outsiders in Wage Determination", *Scandinavian Journal of Economics*, 87, 2, 411–28.

Stigler, G.J. and Kindahl, J.K. (1970) "Comparison of Indexes for Individual Commodities", *The Behaviour of Industrial Prices*, Cambridge, MA: National Bureau of Economic Research, Inc., 56–70.

Stiglitz, J. (1976) "Prices and Queues as Screening Devices in Competitive Markets", *IMSSS Technical Report* 212, Stanford University.

Stiglitz, J. (1982) "Information and Capital Markets", *NBER Working Papers* 0678, National Bureau of Economic Research, Inc.

Stiglitz, J. (1986) "Theories of Wage Rigidities", in J.L. Butkiewicz et al. (eds), *Keynes' Economic Legacy: Contemporary Economic Theories*. New York: Praeger.

Stopford, M. (2009), *Maritime Economics*, 3rd ed, London: Routledge.

Strandenes, S.R. (1984) "Price Determination in the Time Charter and Secondhand Markets", *Discussion Paper* 0584, Norwegian School of Economics and Business Administration, Bergen, Norway.

Strandenes, S.R. (1986) "Norship: A Simultaneous Model of Market in Bulk Shipping", *Discussion Paper* 11, Norwegian School of Economics and Business Administration, Bergen, Norway.

Summers, L.H. (1987) "Investment Incentives and the Discounting of Depreciation Allowances", in M. Feldstein (ed.), *The Effects of Taxation on Capital Accumulation*, Chicago: University of Chicago Press.

Svensson, L.E.O. (1997) "Inflation Forecast Targeting: Implementing and Monitoring Inflation Targets', *European Economic Review*, 41, no. 6, 1111–46.

Svensson, L.E.O. (1999) "Inflation Targeting as Monetary Policy Rule", *Journal of Monetary Economics*, 43, 4, 607–54.

Svensson, L.E.O. (2003) "What is Wrong with Taylor Rules? Using Judgement in Monetary Policy through Targeting Rules", *Journal of Economic Literature*, XLI, 2, 426–77.

Taylor, J.B. (1980) "Aggregate Dynamics and Staggered Contracts", *Journal of Political Economy*, 88, 1, 1–24.

Taylor, J.B. (1993) "Discretion versus Policy Rules in Practice", *Carnegie-Rochester Conference Series on Public Policy*, December, 195–214.

Taylor, J.B. (1999) "Staggered Price and Wage Setting in Macroeconomics", in J.B. Taylor and M. Woodford (eds), *Handbook of Macroeconomics*, Amsterdam: Elsevier.

Tinbergen, J. (1931) "Ein Schiffbauzyclus?", *Weltwirtschaftliches Archiv*, 34, 152–64.

Tinbergen, J. (1934) "Scheepsruimte en Vrachten", *De Nederlandsche Conjunctuur*, March, 23–35.

Tsolakis, S. D., Cridland, C. and Haralambides, H. (2003) "Econometric Modelling of Secondhand Ship Prices", *Maritime Economics and Logistics*, 5, 347–77.

Tvedt, J. (2003) "A New Perspective on Price Dynamics of the Dry Bulk Market", *Maritime Policy and Management*, 30, 3, 221–30.

Veenstra, A.W. (1999a) "The Term-Structure of Ocean Freight Rates", *Maritime Policy and Management*, 26(3), 279–93.

Veenstra, A. (1999b) *Quantitative Analysis of Shipping Markets*, Delft: Delft University Press.

Vergottis, A. (1988) "An Econometric Model of World Shipping", Ph.D. Thesis, City University, London.

Weiss, A. (1980) "Job Queues, and Layoffs in Labour Markets with Flexible Wages", *Journal of Political Economy*, 88, 526–38.

Woodford, M. (1999) "Optimal Monetary Policy Inertia", NBER Working Paper Series, No. 7261, Cambridge, MA: National Bureau of Economic Research.

Woodford, M. (2003) *Interest and Prices*, Princeton, NJ: Princeton University Press: Princeton.

Woodford, M. (2009) "Convergence in Macroeconomics: Elements of the New Synthesis", *American Journal of Economics – Macro*, 1, 1, 267–79.

Wright, G. (1999) "Long run freight rate relationships and market integration in the wet bulk carrier shipping sector", *International Journal of Transport Economics*, 26, 3, 439–46.

Wright, G. (2003) "Rational Expectations in the Wet Bulk Shipping Market", *International Journal of*
Transport Economics, 29, 3, 309–18.

Wright, G. (2011) "Quantifying Time-varying Term risk premia in Shipping Markets", *Journal of Transport Economics and Policy*, 45, 2, 329–40.

Yellen, J. (1984) "Efficiency Wage Models of Unemployment", *American Economic Review*, 74, 2, 200–5.

Zannetos, Z.S. (1966) *The Theory of Oil Tankship Rates*, Cambridge, MA: MIT Press.

INDEX

Aframax tankers   13
   time charter rate   256
aggregate fleet   53
Akaike Information Criterion (AIC)   115, 138, 349–50
animal spirits   6, 153, 293, 294, 324
arbitrage *see* freight market efficiency
ARCH model *see* autoregressive conditional heteroscedastic (ARCH) model
ARIMA process *see* autoregressive integrated moving average (ARIMA) process
ARMA *see* autoregressive moving average
Asian–Russian crisis   277, 321
asset growth   274
asset leverage   275, 277
asset prices   58–61, 268
   as freight rates   291, 295
   rational bubbles   59
asset-led business cycles   273–7, 309–10
augmented Dickey–Fuller (ADF) test   138–9, 144, 348
autocorrelation function   137
autoregressive conditional heteroscedastic (ARCH) model   122–3, 348
autoregressive integrated moving average (ARIMA) process   145–7, 350
autoregressive models   107
   univariate time series   132–6

VAR model *see* vector autoregression (VAR) model
autoregressive moving average (ARMA)   123, 131, 145–7, 160, 350
average earnings   329–32
   dry bulk market   329
   oil tanker market   330, 331
average variable cost   16, 20

backwardation   280, 281, 282
Baltic Dry Index   287, 288, 295, 296
Baltic Exchange time charter averages   254–5
Baltic Indices   27
Bank of England   172, 306
banking crisis   257, 267
banks   266
   central *see* central banks
   choice of   270–1
   credit rationing   326
   foreclosures   266
   withdrawal of funding   268–9
   see also shipping financialisation
bargaining (contract) curve   22–3, 25, 27, 37, 38
bargaining power   25–7, 343
   and economic conditions   27–9, 38
bargaining utility function   23–5
Barro price-output model   207
Beenstock–Vergottis model   1, 2–4, 15, 16, 84, 86, 209–10, 216–32, 250
   dynamic adjustment   227–32
   steady-state properties   224–7
   summary   218–24
best-linear-unbiased estimators (BLUE)   133–4
Blanchard–Kiyotaki price-output model   205
block exogeneity test   115, 126, 129, 131
bond market   326, 344–5
   junk bonds   258
Box–Jenkins approach   148, 350
bubbles   154, 180, 272, 309–10, 321
   asset price   59
Bundesbank   307, 319–21
bunker costs   1, 3, 17–18, 28–30, 32–3, 36, 38–9, 44, 45, 47, 48, 67–8, 78, 81, 209–10, 212, 217, 226, 229, 234, 237
business cycles   5, 153–208, 350–3
   asset-led   273–7, 309–10
   demand-led   309
   economic interpretation   160–6
   fiscal policy   181–7
   freight rates as leading indicator   295–6
   generation of   159–60
   Germany   319–22
   integrated model   210–11, 232–41
   interaction with shipping cycles   241–9, 293–327, 355
   Japan   315–19
      phases   317

business cycles – *continued*
  monetary policy   175–9
  New Consensus
    Macroeconomics
    see New Consensus
    Macroeconomics (NCM)
  nominal rigidities   199–208
    efficiency wages   165,
      202–4
    implicit contracts   199–200
    nominal price and wage
      rigidity models   204–8
    real price rigidity models
      204
    unions vs insider-outsider
      models   200–2
  phase length   312–13
  phases   302–3
  potential output   191–8
    statistical estimates   192
    supply-side
      determinants   192–7
  shocks *see* shocks
  soft landings   311, 314
  statistical explanation
    155–60
  supply-led   309–10
  USA   299–305
    characteristics   308–15

Calvo model   170–1
capacity utilisation *see* fleet
    capacity utilisation
Capesize dry bulk carriers   13,
    14, 30, 108
capital asset pricing model
    (CAPM)   34–5, 99, 130
capital market structure   255–65
cardinal utility   22–3, 25, 342–3
central banks   46, 172–3, 247,
    293, 306–7, 352
  Bank of England   172, 306
  Bundesbank   307, 319–21
  European Central Bank   172,
    306, 321
  Federal Reserve System   306,
    313, 316, 321
chain rule of forecasting   107
charter rate *see* time charter rate

charterers   2, 7, 11–12, 16, 18–19,
    21–33, 35, 37–8, 69,
    253–4, 268
  expectations   221–2, 237–8,
    291, 293
China   32, 254, 268, 271, 273,
    293–4
  cargo   290
  dry bulk market   288–91
  foreign direct investment   288
  steel production   290
China Factor   267
Cobb–Douglas production
    function   193, 347, 352
cointegrating tests
  freight market efficiency
    117–19
  Johansen method   147–52
  ship prices   129–30
  single equations   139–45
cointegrating vectors   98–9,
    119–20, 131, 142, 348
collateralised debt obligations
    277
conduits   276
conspiracy theories   6, 253, 271
constant elasticity of substitution
    function   80–1
consumption   168–9
container market   14, 85
  return on capital   254
  secondhand market   257
  time charter averages   256–7
contango   280, 281, 282, 354–5
contract (bargaining) curve
    22–3, 25, 27, 37, 38
contracts   22–3
  implicit   199–200
contrarian owners   34, 50, 53,
    56, 61, 77, 89, 283
convenience yield   278–80
conventional wisdom
    hypothesis   348
correlogram   137
costs
  average variable   16, 20
  bunker *see* bunker costs
  marginal   20, 61, 79, 82, 102,
    171, 191, 205, 240, 342

  menu   205–6
  sunk   54, 82
  see also price
credit default swap (CDS)   268
credit rationing   295, 325–6

decision-making   329–32
demand shocks   153–4, 155–6,
    157, 178–9, 201, 206–7,
    249–50, 314–15
demand and supply   16–17, 29,
    41, 64, 66, 213, 222–3,
    266, 346
  dry bulk market   287, 289,
    335–9
  dynamic adjustment   232
  elasticity   213–14, 234–5
  excess demand   66
  excess supply   66
  flow demand   48
  long run   20
  new vessels   80–1, 234–5
  order cancellations   65
  output gap   245
  Say's Law   161, 162, 177
  shipping services   234–5, 238
  short-run   19
  stimulus   182–3
  stock demand   48
  world orderbook   64
  see also shipping services supply
demand-led business cycles   309
demolition   14–15, 69–70,
    224–5, 244
  elasticity   241–2
deterministic trends   133, 136,
    138, 144–6
Dickey–Fuller test   137–8, 144
dry bulk market   254
  average earnings   329
  components of   15
  demand and supply   287, 289,
    335–9
  financialisation   286–92
  fleet capacity utilisation   287,
    288
  hierarchical structure   14
  investment in   332–3
  recessions   323

return on capital   254
secondhand   256
shipping cycles   297
time charter averages   256
Dutch auctions   11, 21
dynamic adjustment   346
  Beenstock–Vergottis
    model   227–32
  demand   232
  fleet   51–2, 63, 89–91, 92, 214,
    215, 216
  fleet capacity utilisation
    77–8, 230
  investment   92
  newbuild prices   51–2, 63,
    89–91
  saddlepoint path   90–1
  secondhand price   230, 246
  see also newbuild (NB) prices;
    shipping cycles
dynamic stochastic general
  equilibrium (DSGE)
  models   164–5, 166

economic conditions   31, 57
  and bargaining power   27–9,
    38
  growth   159
  see also GDP; macroeconomics;
    monetary policy
efficiency
  freight market see freight market
    efficiency
  informational   101, 116, 125,
    131
  semi-strong   102
  shipping market see shipping
    market efficiency
efficiency wages   165, 202–4
Efficient Market Hypothesis
  (EMH)   59–61, 95, 97,
  100–3, 353
  freight market efficiency   106,
    125
elasticity
  of demand   213–14, 234–5
  of demolition   241–2
  of freight rates   249
  of investment   163

of substitution   167–8, 170,
  236, 244, 345–6
of supply   212, 214, 243
wage   191
Engle–Granger representation
  theorem   143
Engle–Granger two-stage
  procedure   131, 141, 144,
  147
error correction model (ECM)
  141–3
Euro-sclerosis   321
eurodollar market   255, 268
European Central Bank   172,
  306, 321
event studies   103
exchange traded funds   283
exogenous shocks   3, 13, 42, 68,
  175, 198, 237, 307, 349
exogenous variables   1, 3, 30, 31,
  38, 39, 40, 147, 148, 174,
  209, 210, 217, 225, 237
expectations   39, 81–4
  conditional   101
  freight market efficiency
    107–8
  freight rates   343–4
  generation of   40, 154
  homogeneous   102, 343
  long-term   30, 39
  medium-term   30
  pure expectations hypothesis
    34
  rational see rational
    expectations
  rule of iterated expectations
    101
  short-term   39
  subjective   102
  uniform   221–2
export credit agencies   257, 268

Federal Reserve System   306,
  313, 316, 321
Feynman, Richard   22
finance-led shipping cycles
  325–7
financial liberalisation   2, 272,
  276

financialisation see shipping
  financialisation
fiscal policy see monetary policy
fleet
  adjustment see dynamic
    adjustment
  aggregate   53
  capacity utilisation see fleet
    capacity utilisation
  net   68–70
  optimal   44–53, 76–80
  shadow price   79
  stock   47, 48, 78, 286
  target   46, 49, 56–8
fleet capacity utilisation   17, 32,
  38, 41, 72, 225, 240, 244,
  254
  case study   56–8
  dry bulk market   287, 288
  dynamic adjustment   77–8,
    230
  and freight rates   246, 248
  investment under uncertainty
    53–6
  long-run   47–53, 229, 231
  optimal   44–53, 76–80
  short-run   44–7
Fletcher, Gordon   163
flow demand   48
foreclosures   266, 326
freight forward aggreements
  (FFAs)   102
freight market efficiency
  105–26, 217–18
  cointegrating tests   117–19
  empirical evidence   125–6
  expectations-generating
    mechanisms   107–8
  time-varying risk premia   34,
    120–4
    ARCH model   122–3, 348
    GARCH model   123–4,
      348
    VAR models see vector
      autoregression (VAR)
      model
  weak-form   117–19
  see also Efficient Market
    Hypothesis

freight rates   2, 11–40, 49, 77, 212–13, 238, 294, 342–5
  as asset prices   291, 295
  bargaining power   25–9
  bargaining utility function   23–5
  contract/bargaining curve   22–3
  dynamic adjustment   215, 216
  elasticity   249
  equilibrium   78
  expectations   343–4
  and fleet capacity utilisation   246, 248
  game-theoretic model   21–32
  generation of expectations   40
  good/bad times   26
  as indicator of business cycles   295–6
  spot market   13, 32, 243–4
  term structure   33
  time charter rate *see* time charter rate
  traditional model   16–21, 84–7
Friedman, Milton   163–4
fuel consumption   44, 45, 219
full information maximum likelihood (FIML) method   149
futures market   74, 86, 102, 222, 227, 278–9

game-theoretic model of freight rates   21–2, 37, 234
  bargaining power   25–9
  bargaining utility function   23–5
  comparison with conventional model   29–32
  contract/bargaining curve   22–3
GARCH model *see* generalised autoregressive heteroscedastic (GARCH) model
Gaussian white noise   134

GDP   155, 157–8, 160
  business cycle generation   159–60
  forecasts   159
  growth   181–2, 299, 301–2
  stimulus   182–3
generalised autoregressive heteroscedastic (GARCH) model   123–4, 348
generalised method of moments (GMM)   112, 348
German *Kommanditgesellschaft* market   257
Germany   293
  Bundesbank   307, 319–21
  business cycles   319–22
Golden Age of Capitalism   163
goods market   161–3, 166, 167, 171, 193–4, 204, 207, 208, 305
Granger causality   115, 131
Great Depression   162, 163, 164, 313
Great Recession   180, 244, 295, 351
Greece
  financial crisis   95, 96, 98, 118, 120, 187
  ship owners   253–4
Greenspan, Alan   180, 277, 314
gross domestic income (GDI)   299, 301
gross domestic product *see* GDP

Handymax dry bulk carriers   13, 14, 30
  secondhand price   256
  time charter rate   120, 256
Handysize dry bulk carriers   13, 14, 30
hedging   273, 280–1
herd owners   34, 56, 61, 68, 294, 324
heteroscedasticity   112, 134
  ARCH model   122–3
  GARCH model   123–4
Hicks, John   163
homogeneous expectations   102, 343

homoscedasticity   112, 134
hurdle rate   55

implicit contracts   199–200
indexation puzzle   201
inflation   154, 161, 171, 175, 237
  and economic growth   304–5
  and interest rates   179, 245
  and output gap   173–4
  targeting   199
informational efficiency   101, 116, 125, 131
initial public offering (IPO)   258
innovations   102, 122, 125, 126
insider–outsider models   200–2
integrated model   210–11, 232–41
  properties   237–41
interest rates   162–3, 344
  and inflation   179, 245
  and investment   185
  natural   176, 178, 180
  nominal   5, 174, 176, 244, 247
  real   5, 32, 163, 178
  smoothing   176, 236
  term structure   33
investment   328–41, 355
  decision-making   329–32
  dry bulk market   332–3
  dynamic adjustment   92
  methodological issues   334–5
  under uncertainty   53–6
investment banks   258
  asset leverage   275, 277
iso-utility curves   24–7, 38, 343
isocost curves   46–7
isoquants   43–4

Japan   254, 271, 275, 293, 322–3
  business cycles   315–19
    phases   317
  excess liquidity   275–6
Johansen method of cointegration   147–52
joint hypothesis problem   97, 103
junk bonds   258

K-model   352
Kamsarmax dry bulk carriers   13

Keynes, John Maynard   6, 153, 293, 324
Keynesian economics   161–4, 166, 350
knots   352
Korea   268
Koyck-lag hypothesis   348

labour supply   169–70
labour unions   200–2
'leaning against the wind'   173
Lehman Brothers   254, 268, 277
Life-Cycle Hypothesis   163, 165, 167
liquidity   180, 269, 272, 275
   excess   275–6
liquidity preference hypothesis   34
liquidity trap   163, 186
loan to value (LTV)   266, 268
Lombard rate   320, 321
long-run equilibrium   87–92, 142, 176
   NCM nodel   177
   shipyard   88
   spot market   142
   time charter market   142
loss given default (LGD) ratio   266
Louvre Accord (1987)   323

macroeconomics   2–4, 161, 350
   Classical system   161–2
   Keynesian   161–4
   Neoclassical Synthesis   163–4
   New Consensus see New Consensus Macroeconomics
marginal costs   20, 61, 79, 82, 102, 171, 191, 205, 240, 342
marginal rate of substitution   343
marginal revenue   82
Marine Money, public shipping company list   259–65
market efficiency see freight market efficiency
market structure   253–71
Marsoft   254, 255, 256

martingale model   97, 100, 103, 104
mean function   123–4, 130
menu costs   205–6
Modigliani, Franco   163
monetary policy   162–3, 176, 187
   and business cycle generation   175–9
   intervention   162
   New Consensus Macroeconomics   172–5
   US President's budget (2013)   188–91
   see also economic conditions
money market   162, 163, 185, 186
moral hazard   203, 325
moving average autoregressive process see autoregressive moving average
moving average (MA) process   145–7
multicollinearity   143
multiplier effects   162, 181–7

NAIRU   193
National Bureau of Economic Research (NBER)   299, 301
natural interest rate   176, 178, 180
NCM see New Consensus Macroeconomics (NCM)
neo-Wicksellian model see New Consensus Macroeconomics (NCM)
Neoclassical Synthesis   163–4
net asset value   258
net fleet   68–70
New Consensus Macroeconomics (NCM)   153–5, 161, 166–75
   consumption   168–9
   labour supply   169–70
   mathematical expression   167
   monetary policy   172–5
   reformulation   179–81
New Keynesian economics   166
New Keynesian Phillips curve   170–1

newbuild (NB) prices   41–2, 49, 222, 239–40, 244, 298, 339–41
   dynamic adjustment see dynamic fleet adjustment
   efficiency   127
   and SH prices   70–6
   and shipyard capacity   61–8
nominal interest rate   5, 174, 176, 244, 247
nominal rigidities   199–208
   efficiency wages   165, 202–4
   implicit contracts   199–200
   nominal price and wage rigidity models   204–8
   real price rigidity models   204
   unions vs insider-outsider models   200–2
non-accelerating inflation rate of unemployment see NAIRU
non-stationarity   135, 138, 139, 348
normal returns   103

oil premium   282
oil prices   156, 278–80, 281
   dollar   283
   speculation   282
   spot   282, 283
oil shocks   273, 311, 313
oil storage trading   283–4
oil tanker market   13
   average earnings   330, 331
   demand for seaborne trade   285–6
   financialisation   281–4
   return on capital   254
   secondhand market   256
   structural changes   284–6
   time charter averages   256
   VLCCs   13
Okun's Law   193–4, 195, 198
optimal capacity utilisation   44–53, 76–80
   see also fleet capacity utilisation
optimal fleet
   long run   47–53, 76–80
   short run   44–7

optimal speed   219–20
    expectations   81–4
    long run   47–53, 76–80
    short run   44–7
order book   64
order cancellations   65
ordinal utility   22–3, 25, 342–3
ordinary least squares (OLS)   112, 140, 144
orthogonality   101, 116, 125, 131
output   154
    actual   157
    potential   157, 191–8
    statistical estimates   192
    supply-side determinants   192–7
output gap   158, 170, 171, 172, 176, 237
    and demand for shipping   245
    and inflation   173–4
overcapacity   6, 56, 62, 65, 67–8, 294, 324
owners *see* ship owners

Panamax dry bulk carriers   13, 14, 30, 108
    secondhand price   256
    time charter rate   256
parallel banking *see* shadow banking
payoff substitution, rate of   24
perfect foresight spread   108, 111–12, 116
period market *see* time charter market
Permanent Income Hypothesis   163, 167
perpetual time charter   85
Perron sequential testing procedure   139
persistency effect   155, 165, 247
Phillips curves   164, 166, 198
    long-run   178
    New Keynesian   170–1
    short-run   177, 178
Phillips Z-test   139, 144
Phillips–Perron test   137, 139, 144
Platou Shipping Index   258

Plaza Accord (1985)   316, 321
policy intervention   162
preferred habitat hypothesis   34, 105, 117
price
    asset *see* asset prices
    flexibility   161
    menu costs   205–6
    newbuild *see* newbuild (NB) prices
    rigidity
        nominal   204–8
        real   204
    scrap *see* scrap price
    secondhand *see* secondhand price
    shadow   78–80
    ship *see* ship prices
    stickiness   166
private information, tests for   103
productivity shocks   165
profit   19
    excess   99, 121, 122
    maximisation   28
public shipping company list   259–65
pure expectations hypothesis   34

quantitative easing   277
quantity theory of money   163

random walk model   104–5, 135–6
    with drift   135–6
    without drift   135–6, 140
rational bubbles   59
rational expectations   1, 83, 107–8, 131, 164, 250, 348, 351
rationality   342
real interest rate   5, 32, 163, 178
real price rigidity models   204
real wage rate   312
real wage resistance   312
required rate of return   55
return predictability, tests for   103
risk appetite   5, 35–6, 267, 270, 276–7, 282, 292

risk aversion   36, 54, 76, 99, 124, 200, 273, 276–7, 292, 331
risk premium   34, 35–6, 75–6
    negative   37
    time-varying   34, 120–4
risk-adjusted return on equity (RAROE)   269
rule of iterated expectations   101

Samuelson, Paul   163
saving   183
Say's Law   161, 162, 177
Schwartz Information Criterion   115
scrap price   12, 15, 68–70, 224, 226, 241–2
    demolition *see* demolition
secondhand price   13, 15, 70–6, 220–1, 228, 235–6, 242–3, 298, 324–5
    container ships   257
    dry bulk carriers   256
    dynamic adjustment   230, 246
    tankers   256
semi-strong efficiency   102
shadow banking   274, 276
shadow price of fleet   78–80
ship owners   18–19, 21–33, 253–4
    and central banks   46
    contrarian   34, 50, 53, 56, 61, 77, 89, 283
    expectations   221–2, 237–8, 291, 293
    herd   34, 56, 61, 68, 294, 324
    risk minimisation   269–70
    smart   67–8
ship prices   58–61, 126–32
    cointegrating tests   129–30
    empirical evidence   131–2
    market efficiency tests   128–9
    newbuild *see* newbuild (NB) prices
    secondhand   70–6
    time-varying risk premia   130–1
    weak-form efficiency   129–30
shipping banks   270

shipping cycles   6, 209–50, 266,
       296–8, 353–4
   Beenstock–Vergottis model   1,
       2–4, 15, 16, 84, 86,
       209–10, 216–32
   characteristics   322–3
   dry bulk market   297
   finance-led   325–7
   integrated model   210–11,
       232–41
   interaction with business cycles
       241–9, 293–327, 355
   Tinbergen–Koopmans
       model   2, 3, 5, 16, 29, 37,
       209, 211–16
   uncertainty-led   323–5
   see also business cycles
shipping financialisation
       272–92, 354–5
   asset-led business cycles
       273–7
   capital market structure
       255–65
   constraints   267–70
   dry bulk   286–92
   hedging and speculation
       278–81
   market structure   253–5
   oil tankers   281–6
   optimal strategies   270–1
shipping market efficiency
       95–152, 347–50
   Efficient Market Hypothesis
       59–61, 95, 97, 100–3
   freight markets   105–26
       cointegrating tests
           117–19
       Efficient Market Hypothesis
           106, 125
       empirical evidence   125–6
       expectations-generating
           mechanisms   107–8
       time-varying risk premia
           34, 120–4
       VAR models *see* vector
           autoregression (VAR)
           model
       weak-form efficiency
           117–19

Martingale model   104
random walk model   104–5
ship prices   126–32
   cointegrating tests   129–30
   empirical evidence   131–2
   market efficiency tests
       128–9
   time-varying risk
       premia   130–1
   weak-form efficiency
       129–30
   VAR models *see* vector
       autoregression (VAR)
       model
   variance inequality test   98
shipping market structure
       253–71, 354
shipping services supply
   long run   47–53
   order cancellations   65
   shipyard capacity   61–8
   short run   42–7
   world orderbook   64
shipyard capacity   61–8
   analysis of   62–3
   long-run equilibrium   88
   and new build price   64–7
shipyard market   12, 15, 41–92,
       222–3
shocks   155–6
   and business cycle generation
       175–9
   demand-side   153–4, 155–6,
       157, 178–9, 201, 206–7,
       249–50, 314–15
   exogenous   3, 13, 42, 68, 175,
       198, 237, 307, 349
   oil price   273, 311, 313
   persistent/permanent   153,
       155, 156–8
   private sector   156
   productivity   165
   supply-side   272, 311, 314
   transitory   153, 155, 156–7
short-term profit   3
smart owners   67–8
soft commodities   282, 355
soft landings   311, 314
Solow growth model   192–3

Solow residual   193
speculation   278–81
   oil prices   282
spline   192, 352
spot market   13, 32, 121, 125,
       237, 344–5
   freight rates   13, 32, 243–4
   long-run equilibrium   142
   time charter averages
       254–5
spurious correlation   139, 140–1,
       148, 347
stagflation   164, 272, 273, 310
Standard & Poor's Index   296
stationarity   135, 349
   tests for   136–9
steel production   290
stochastic trends   136, 140
stock
   demand   48
   fleet   47, 48, 78, 286
Strandenes model   84–5
strong-form efficiency   102
structured investment vehicles
       276
subjective expectations   102
Suezmax oil tankers   13
   term charter rate   256
sunk costs   54, 82
superconsistency   144
supply *see* demand and supply
supply shocks   272, 311, 314
supply-led business cycles
       309–10, 351
Supramax dry bulk carriers   13,
       14, 30
   charter price   120

tankers *see* oil tankers
target fleet   46, 49
   case study   56–8
term liquidity premium (TLP)
       269
term premium   33, 34
term structure   33
time charter averages   254–5
   container ships   256–7
   dry bulk   120, 256
   oil tankers   256

time charter rate   13, 16, 27, 32–7, 39, 60, 95–9, 105–6, 109, 114–21, 124–7, 131, 298
  long-run equilibrium   142–3
  perpetual time charter   85
time-varying risk premia   34, 120–4
  freight market efficiency   120–4
    ARCH process   122–3
    GARCH process   123–4
  ship prices   130–1
Tinbergen–Koopmans model   2, 3, 5, 16, 29, 37, 209, 211–16, 250
  dynamic fleet adjustment   214
trading strategies   278–81
traditional banking   274

uncertainty, investment under   53–6
uncertainty-led shipping cycles   323–5
unemployment   305
  efficiency wage hypothesis   165, 202–4
uniform expectations   221–2
unit root tests   136–9
univariate time series   132–6
USA   293–4
  asset growth   274
  asset leverage of investment banks   275, 277

business cycles   299–305
  characteristics   308–15
  Federal Reserve System   306, 313, 316, 321
  GDP growth   301–2
  Glass-Steagall Act   276
  monetary policy   305–8
  National Bureau of Economic Research   299, 301
  shadow banking   274
  traditional banking   274
utility
  cardinal   22–3, 25, 342–3
  ordinal   22–3, 25, 342–3

VAR *see* vector autoregression
variance bounds test   115, 117
variance inequality test   98, 115, 116
vector autoregression (VAR)
  model   97–8, 107, 108–10, 125–6, 145–7, 350
  block exogeneity test   115
  perfect foresight spread   108, 111–12, 116
  restricted   110, 113
  ship prices   128–9
  unrestricted   113
  volatility tests   115
  Wald test   114–15, 126, 129
very large crude carriers (VLCCs)   13

very large ore carriers (VLOCs)   61
vessels
  demand for   80–1
  sizes   13
Volcker, Paul   306, 313, 314, 316, 321

wage
  bargaining   200
  efficiency   165, 202–4
  elasticity   191
  flexibility   161
  real   312
  rigidity   200–2, 204–8
  stickiness   202, 208
wage–price spiral   312
Wald test   114–15, 126, 129
Walrasian equilibrium   203
weak-form efficiency   102
  cointegrating tests   117–19, 129–30
  freight markets   118–19
  ship prices   129–30
wet market   14
white noise   134, 138, 140, 143
Williams Deacons Bank   267
Wold's decomposition theorem   146–7
World Trade Organisation   254, 273

Zannetos hypothesis   348
zero-sum game   342

Printed and bound by CPI Group (UK) Ltd, Croydon, CR0 4YY